The $k \cdot p$ Method

Dr. Lok C. Lew Yan Voon
Wright State University
Physics Dept.
3640 Colonel Glenn Highway
Dayton OH 45435
USA
lok.lewyanvoon@wright.edu

Dr. Morten Willatzen
University of Southern Denmark
Mads Clausen Institute for
Product Innovation
Alsion 2
6400 Soenderborg
Denmark
willatzen@mci.sdu.dk

ISBN 978-3-642-10090-1 e-ISBN 978-3-540-92872-0
DOI 10.1007/978-3-540-92872-0
Springer Dordrecht Heidelberg London New York

Cover design: WMXDesign GmbH

Springer is part of Springer Science+Business Media (www.springer.com)

Über Halbleiter sollte man nicht arbeiten, das ist eine Schweinerei; wer weiss ob es überhaupt Halbleiter gibt.

–W. Pauli 1931

Foreword

I first heard of $k{\cdot}p$ in a course on semiconductor physics taught by my thesis adviser William Paul at Harvard in the fall of 1956. He presented the $k{\cdot}p$ Hamiltonian as a semiempirical theoretical tool which had become rather useful for the interpretation of the cyclotron resonance experiments, as reported by Dresselhaus, Kip and Kittel. This perturbation technique had already been succinctly discussed by Shockley in a now almost forgotten 1950 Physical Review publication. In 1958 Harvey Brooks, who had returned to Harvard as Dean of the Division of Engineering and Applied Physics in which I was enrolled, gave a lecture on the capabilities of the $k{\cdot}p$ technique to predict and fit non-parabolicities of band extrema in semiconductors. He had just visited the General Electric Labs in Schenectady and had discussed with Evan Kane the latter's recent work on the non-parabolicity of band extrema in semiconductors, in particular InSb. I was very impressed by Dean Brooks's talk as an application of quantum mechanics to current real world problems. During my thesis work I had performed a number of optical measurements which were asking for theoretical interpretation, among them the dependence of effective masses of semiconductors on temperature and carrier concentration. Although my theoretical ability was rather limited, with the help of Paul and Brooks I was able to realize the capabilities of the $k{\cdot}p$ method for interpreting my data in a simple way. The temperature effects could be split into three components: a contribution of the thermal expansion, which could be easily estimated from the pressure dependence of gaps (then a specialty of William Paul's lab), an effect of the nonparabolicity on the thermally excited carriers, also accessible to $k{\cdot}p$, and the direct effect of electron-phonon interaction. The latter contribution could not be rigorously introduced into the $k{\cdot}p$ formalism but some guesses where made, such as neglecting it completely. Up to date, the electron-phonon interaction has not been rigorously incorporated into the $k{\cdot}p$ Hamiltonian and often only the volume effect is taken into account. After finishing my thesis, I worked at the RCA laboratories (Zurich and Princeton), at Brown University and finally at the Max Planck Institute in Stuttgart. In these three organizations I made profuse use of $k{\cdot}p$. Particularly important in this context was the work on the full-zone $k{\cdot}p$, coauthored with Fred Pollak and performed shortly after we joined the Brown faculty in 1965. We were waiting for delivery of spectroscopic equipment to set up our new lab and thought that it would be a good idea to spend idle time trying to see how far into the Brillouin zone one could extend the $k{\cdot}p$ band structures: till then the use of $k{\cdot}p$ had been confined to the close neighborhood of band edges. Fred was very skilled at using the early computers available to us. We, of course, were aiming at working with as few basis states as possible, so we started with 9 (neglecting spin-orbit coupling). The bands did not look very good. We kept adding basis states till we found that rather reasonable bands were obtained with 15 $\mathbf{k} = 0$ states. The calculations were first performed for germanium and silicon, then they were generalized to III-V compounds and spin-orbit coupling was added. I kept the printed computer output for energies and wave functions versus **k** and used it till recently for many calculations. The resulting Physical Review publication of Fred and myself has been cited nearly 400 times. The last of my works which uses $k{\cdot}p$ techniques was published in the Physical Review in 2008 by Chantis, Cardona, Christensen, Smith, van Schilfgaarde, Kotani, Svane and Albers. It deals with the stress induced linear terms in **k** in the conduction band minimum of GaAs. About one-third of my publications use some aspects of the $k{\cdot}p$ theory.

The present monograph is devoted to a wide range of aspects of the $k{\cdot}p$ method as applied to diamond, zincblende and wurtzite-type semiconductors. Its authors have been very active in using this method in their research. Chapter 1 of the monograph contains an overview of the work and a listing of related literature. The rest of the book is divided into two parts. Part one discusses $k{\cdot}p$ as applied to bulk (i.e. three-dimensional) "homogeneous" tetrahedral semiconductors with diamond, zincblende and wurtzite structure. It contains six chapters. Chapter 2 introduces the $k{\cdot}p$ equation and discusses the perturbation theoretical treatment of the corresponding Hamiltonian as applied to the so-called one-band model. It mentions that this usually parabolic model can be generalized to describe band nonparabolicity, anisotropy and spin splittings. Chapter 3 describes the application of $k{\cdot}p$ to the description of the maxima (around $\mathbf{k} = 0$) of the valence bands of tetrahedral semiconductors, starting with the Dresselhaus, Kip and Kittel Hamiltonian. A problem the novice encounters is the plethora of notations for the relevant matrix elements of p and the corresponding parameters of the Hamiltonian. This chapter lists most of them and their relationships, except for the Luttinger parameters γ_i, κ, and q which are introduced in Chap. 5. It also discusses wurtzite-type materials and the various Hamiltonians which have been used. In Chap. 4 the complexity of the $k{\cdot}p$ Hamiltonian is increased. A four band and an eight band model are presented and Löwdin perturbation theory is used for reducing (through down-folding of states) the complexity of these Hamiltonians. The full-zone Cardona-Pollak 15 band Hamiltonian is discussed, and a recent "upgrading" [69] using 20 bands in order to include spin-orbit effects is mentioned. Similar Hamiltonians are also discussed for wurtzite.

In order to treat the effects of perturbations, such as external magnetic fields, strain or impurities, which is done in Part II, in Chap. 5 the $k{\cdot}p$ Hamiltonian is reformulated using the method of invariants, introduced by Luttinger and also by the Russian group of Pikus (because of the cold war, as well as language difficulties, it took a while for the Russian work to permeate to the West). A reformulation of this method by Cho is also presented. Chapter 6 discusses effects of spin, an "internal" perturbation intrinsic to each material. Chapter 7 treats the effect of uniform strains, external perturbations which can change the point group but not the translational symmetry of crystals.

Part II is devoted to problems in which the three-dimensional translational symmetry is broken, foremost among them point defects. The $k{\cdot}p$ method is particularly appropriate to discuss shallow impurities, leading to hydrogen-like gap states (Chap. 8). The $k{\cdot}p$ method has also been useful for handling deep levels with the Slater–Koster Hamiltonian (Serrano et al.), especially the effect of spin-orbit coupling on acceptor levels which is discussed here within the Baldereschi–Lipari model. Chapter 9 treats an external magnetic field which breaks translational symmetry along two directions, as opposed to an electric field (Chap. 10) which break the translational symmetry along one direction only, provided it is directed along one of the 3d basis vectors. Chapter 11 is devoted to excitons, electron hole bound states which can be treated in a way similar to impurity levels provided one can separate the translation invariant center-of-mass motion of the electron-hole pair from the internal relative motion. Chapters 12 and 13 give a detailed discussion of the applications of $k{\cdot}p$ to the elucidation of the electronic structure of heterostructures, in particular confinement effects. The $k{\cdot}p$ technique encounters some difficulties when dealing with heterostructures because of the problem of boundary conditions in the multiband case. The boundary condition problem, as extensively discussed by Burt and Foreman, is also treated here in considerable detail. The effects of external strains and magnetic fields are also considered (Chap. 13). In Chap. 12 the spherical and cylindrical representations used by Sercel and Vahala, particularly useful for the treatment of quantum dots and wires, are also treated extensively. Three appendices complete the monograph: (A) on perturbation theory, angular momentum theory and group theory, (B) on symmetry properties and their group theoretical analysis, and (C) summarizing the various Hamiltonians used and giving a table with their parameters for a few semiconductors. The monograph ends with a list of 450 literature references.

I have tried to ascertain how many articles are found in the literature bases bearing the $k{\cdot}p$ term in the title, the abstract or the keywords. This turned out to be a rather difficult endeavor. Like in the case of homonyms of authors, the term $k{\cdot}p$ is also found in articles which have nothing to do with the subject at hand, such as those dealing with pions and kaons and even, within condensed matter physics, those referring to dielectric susceptibilities at constant pressure κ_p. Sorting them out by hand in a cursory way, I found about 1500 articles dealing in some way with the $k{\cdot}p$ method. They have been cited about 15000 times. The present authors have done an excellent job reviewing and summarizing this work.

Stuttgart
November 2008

Manuel Cardona

Preface

This is a book detailing the theory of a band-structure method. The three most common empirical band-structure methods for semiconductors are the tight-binding, the pseudopotential, and the $k \cdot p$ method. They differ in the choice of basis functions used to represent Schrödinger's equation: atomic-like, plane-wave, and Bloch states, respectively. Each have advantages of their own. Our goal here is not to compare the various methods but to present a detailed exposition of the $k \cdot p$ method.

One always wonder how a book got started. In this particular case, one might say when the two authors were postdoctoral fellows in the Cardona *Abteilung* at the Max Planck Institut für Festkörperforschung in Stuttgart, Germany in 1994–1995. We started a collaboration that got us to use a variety of band-structure methods such as the $k \cdot p$, tight-binding and ab initio methods and has, to date, led to over 50 joint publications. The first idea for a book came about when one of us was visiting the other as a Balslev research scholar and, fittingly, the final stages of the writing were carried out when the roles were reversed, with Morten spending a sabbatical at Wright State University.

This book consists of two main parts. The first part concerns the application of the theory to bulk crystals. We will spend considerable space on deriving and explaining the bulk $k \cdot p$ Hamiltonians for such crystal structures. The second part concerns the application of the theory to "perturbed" and nonperiodic crystals. As we will see, this really consists of two types: whereby the perturbation is gradual such as with impurities and whereby it can be discontinuous such as for heterostructures.

The choice of topics to be presented and the order to do so was not easy. We thus decided that the primary focus will be on showing the applicability of the theory to describing the electronic structure of intrinsic semiconductors. In particular, we also wanted to compare and contrast the main Hamiltonians and $k \cdot p$ parameters to be found in the literature. This is done using the two main methods, perturbation theory and the theory of invariants. In the process, we have preserved some historical chronology by presenting first, for example, the work of Dresselhaus, Kip and Kittel prior to the more elegant and complete work of Luttinger and Kane. Partly biased by our own research and partly by the literature, a significant part of the explicit derivations and illustrations have been given for the diamond and zincblende semiconductors, and to a lesser extent for the wurtzite semiconductors. The impact of external strain and static electric and magnetic fields on the electronic structure are then considered since they lead to new $k \cdot p$ parameters such as the deformation potentials and g-factors. Finally, the problem of inhomogeneity is considered, starting with the slowly-varying impurity and exciton potential followed by the more difficult problem of sharp discontinuities in nanostructures. These topics are included because they lead to a direct modification of the electron spectrum. The discussion of impurities and magnetic field also allows us to introduce the third theoretical technique in $k \cdot p$ theory, the method of canonical transformation. Finally, the book concludes with a couple of appendices that have background formalism and one appendix that summarizes some of the main results presented in the main text for easy reference. In part because of lack of space and because there exists other excellent presentations, we have decided to leave out applications of the theory, e.g., to optical and transport properties.

The text is sprinkled with graphs and data tables in order to illustrate the formal theory and is, in no way, intended to be complete. It was also decided that, for a book of this nature, it is unwise to try to include the most "accurate" material parameters. Therefore, most of the above were chosen from seminal papers. We have attempted to include many of the key literature and some of the more recent work in order to demonstrate the breadth and vitality of the theory. As much as is possible, we have tried to present a uniform notation and consistent mathematical definitions. In a few cases, though, we have decided to stick to the original notations and definitions in the cited literature.

The intended audience is very broad. We do expect the book to be more appropriate for graduate students and researchers with at least an introductory solid state physics course and a year of quantum mechanics. Thus, it is assumed that the reader is already familiar with the concept of electronic band structures and of time-independent perturbation theory. Overall, a knowledge of group representation theory will no doubt help, though one can probably get the essence of most arguments and derivations without such knowledge, except for the method of invariants which relies heavily on group theory.

In closing, this work has benefitted from interactions with many people. First and foremost are all of our research collaborators, particularly Prof. Dr. Manuel Cardona who has always been an inspiration. Indeed, he was kind enough to read a draft version of the manuscript and provide extensive insight and historical perspectives as well as corrections! As usual, any remaining errors are ours. We cannot thank our family enough for putting up with all these long hours not just working on this book but also throughout our professional careers. Last but not least, this book came out of our research endeavors funded over the years by the Air Force Office of Scientific Research (LCLYV), Balslev Fond (LCLYV), National Science Foundation (LCLYV), the Danish Natural Science Research Council (MW), and the BHJ Foundation (MW).

Dayton, OH
Sonderborg
November 2008

Lok C. Lew Yan Voon
Morten Willatzen

Contents

Acronyms

BF	Burt–Foreman
CC	Chuang–Chang
DKK	Dresselhaus–Kip–Kittel
DM	diamond
FBZ	first Brillouin zone
GJ	Gutsche–Jahne
KDWS	Koster–Dimmock–Wheeler–Statz
LK	Luttinger–Kohn
MU	Mireles–Ulloa
QW	quantum well
RSP	Rashba–Sheka–Pikus
SJKLS	Sirenko–Jeon–Kim–Littlejohn–Stroscio
SV	Sercel–Vahala
WZ	wurtzite
ZB	zincblende

Chapter 1
Introduction

1.1 What Is $k \cdot p$ Theory?

Bir and Pikus [1] made the interesting observation, since the physics of semiconductors is (for the most part) governed by the carriers in the extrema of the various energy bands, that: first, only the neighborhoods of the band extrema are important and, second, the qualitative physics should be governed by the shape of these energy surfaces – a property that should be readily obtainable from symmetry arguments. The first observation has led to the common view of $k \cdot p$ theory as a perturbative theory as explified by the seminal work of Dresselhaus et al. [2] and Kane [3], while the second one has manifested itself through the power of symmetry analysis such as the method of invariants introduced by Luttinger [4].

The fact that one can go beyond the neighborhood of band extrema (by not using perturbation theory) was already demonstrated by Cardona and Pollak in 1966 [5], when they obtained realistic band structures for Si and Ge using a full-zone $k \cdot p$ theory. Thus, $k \cdot p$ theory is an empirical band-structure method with a basis of band states. This definition can be extended to nonperiodic systems such as impurities [6] and nanostructures [7] by replacing the Bloch phase factor by an envelope function.

1.2 Electronic Properties of Semiconductors

The subtitle of the book is the electronic properties of semiconductors. Specifically, we will show how the $k \cdot p$ method can be used to obtain the band structure of four types of semiconductors: those with a diamond, a zincblende, and a wurtzite structure, and of their nanostructures (Fig. 1.1). For those not familiar with their band structures, we hereby present a very brief description.

Diamond and zincblende are closely related and they will, therefore, be discussed together. They both have cubic symmetry with two atoms per primitive unit cell (see Appendix B for details). They are mostly the group-IV elements such as Si, the III-V compounds such as GaAs, and a few II-VI compounds such as CdTe. Most of them are direct-gap semiconductors, though some are indirect (e.g., Si and GaP)

L.C. Lew Yan Voon, M. Willatzen, *The* k · p *Method*,
DOI 10.1007/978-3-540-92872-0_1, © Springer-Verlag Berlin Heidelberg 2009

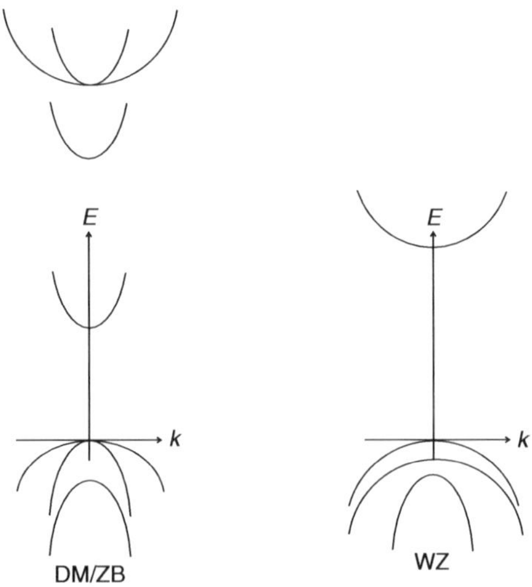

Fig. 1.1 Band structures to be studied in this book: diamond (DM), zincblende (ZB), wurtzite (WZ), and their nanostructures

and others are semimetallic (e.g., Sn [8] and HgTe [9] have a negative gap). For the semiconductors, the band gap tends to separate a valence band derived from atomic-like p orbitals and a conduction band derived from atomic-like s orbitals. The maximum of the valence band is, for most materials, at the zone center of the Brillouin zone ($\mathbf{k} = \mathbf{0}$). In the absence of spin-orbit coupling and spin degeneracy, the highest valence band is, therefore, three-fold degenerate while the lowest conduction band is nondegenerate. With spin-orbit coupling, the valence band consists of a four-fold degenerate band at the zone center (though the degeneracy is reduced for finite wave vectors into either two two-fold degenerate bands—so-called light and heavy-hole bands—or further splittings into nondegenerate states due to the phenomenon of spin splitting for zincblende) and a two-fold spin-hole split (also known as spin-orbit split) band.

Wurtzite has a primitive unit cell with four atoms and is the stable structure for many of the II-VI compounds (e.g., ZnO) and a few of the III-V compounds (e.g., GaN). A useful picture of the wurtzite structure is to envision it as a zincblende one strained along the [111] direction. Hence, the band structure can also be derived from the zincblende case by adding a crystal-field splitting. The main consequence is that the six highest valence states of zincblende are mixed and split into three three-fold degenerate states at the zone center.

The above band pictures and their modifications in external fields and in nanostructures are what we will be describing in this book.

1.3 Other Books

This book is about the $k \cdot p$ method and the resulting electronic structure of semiconductor bulk and nanostructure. Our approach differs significantly from other recent books and review articles where some exposition of the $k \cdot p$ theory has been given but with the emphasis on the application to, e.g., modeling devices or interpreting various types of experiments. While bits of $k \cdot p$ theory can be found in most textbooks [7, 10–19], we now give a brief outline of some of the more advanced presentations of the $k \cdot p$ theory and how they differ from ours.

First, the only books that place a heavy emphasis on developing the $k \cdot p$ theory appear to be those by Bir and Pikus [1], Ivchenko and Pikus [20] and Winkler [21]. They are all very useful books in their own rights and would be worth consulting together with the current one. For example, the book by Bir and Pikus [1] was a milestone in that it finally provided the first compendium in English of the seminal research by russian physicists in using symmetry to develop the $k \cdot p$ theory, particularly in two areas not previously considered by Luttinger: the impact of deformations and the study of hexagonal crystals. The book by Ivchenko and Pikus [20] follows along the same lines but the focus is on applications to heterostructures and their optical properties. Taken together, they cover a lot of the $k \cdot p$ theory and its applications. However, the book by Bir and Pikus is a little bit dated while the one by Ivchenko and Pikus only devotes a chapter to explaining $k \cdot p$ theory. Furthermore, both books are currently out of print. The book by Winkler [21] is rather more focused and is probably the most comprehensive exposition of the spin-splitting theory. Zeiger and Pratt [12] give a very detailed discussion of the Luttinger–Kohn theory, particularly as applied to the magnetic problem. A significant part of the modern use of the $k \cdot p$ theory is applied to semiconductor heterostructures so it is not surprising that the newer books focus on such applications [7, 14, 15, 17–22].

In many ways, the current book is a combination of all of the above work (and of the work cited throughout, of course). Thus, we present a comprehensive and, for that reason, both a historical and modern exposition of the $k \cdot p$ theory for both bulk crystals and nanostructures, taking into account deformations, impurities, and external static electric and magnetic fields. One can also use this book as an aid to reading the original literature. Certainly, the book is aimed at people who wishes to learn how to derive $k \cdot p$ Hamiltonians.

Part I
Homogeneous Crystals

Chapter 2
One-Band Model

2.1 Overview

Much of the physics of the $k \cdot p$ theory is displayed by considering a single isolated band. Such a band is relevant to the conduction band of many semiconductors and can even be applied to the valence band under certain conditions. We will illustrate using a number of derivations for a bulk crystal.

2.2 $k \cdot p$ Equation

The $k \cdot p$ equation is obtained from the one-electron Schrödinger equation

$$H\psi_{n\mathbf{k}}(\mathbf{r}) = E_n(\mathbf{k})\psi_{n\mathbf{k}}(\mathbf{r}), \tag{2.1}$$

upon representing the Bloch functions in terms of a set of periodic functions:

$$\psi_{n\mathbf{k}}(\mathbf{r}) = \mathrm{e}^{\mathrm{i}\mathbf{k}\cdot\mathbf{r}}u_{n\mathbf{k}}(\mathbf{r}). \tag{2.2}$$

The Bloch and cellular functions satisfy the following set of properties:

$$\langle\psi_{n\mathbf{k}}|\psi_{n'\mathbf{k}'}\rangle \equiv \int \mathrm{d}V\, \psi^*_{n\mathbf{k}}(\mathbf{r})\,\psi_{n'\mathbf{k}'}(\mathbf{r}) = \delta_{nn'}\delta(\mathbf{k}-\mathbf{k}'), \tag{2.3}$$

$$\langle u_{n\mathbf{k}}|u_{n'\mathbf{k}}\rangle \equiv \int \mathrm{d}\Omega\, u^*_{n\mathbf{k}}u_{n'\mathbf{k}} = \delta_{nn'}\frac{\Omega}{(2\pi)^3}, \tag{2.4}$$

where V (Ω) is the crystal (unit-cell) volume.

Let the Hamiltonian only consists of the kinetic-energy operator, a local periodic crystal potential, and the spin-orbit interaction term:

$$H = \frac{p^2}{2m_0} + V(\mathbf{r}) + \frac{\hbar}{4m_0^2c^2}(\sigma \times \nabla V)\cdot\mathbf{p}\,. \tag{2.5}$$

L.C. Lew Yan Voon, M. Willatzen, *The* k · p *Method*,
DOI 10.1007/978-3-540-92872-0_2,

Here, we only give the formal exact form for a periodic bulk crystal without external perturbations.

In terms of the cellular functions, Schrödinger's equation becomes

$$H(\mathbf{k})\,u_{n\mathbf{k}} = \mathcal{E}_n(\mathbf{k})\,u_{n\mathbf{k}}, \tag{2.6}$$

where

$$H(\mathbf{k}) \equiv H + H_{k\cdot p}, \tag{2.7}$$

$$H_{k\cdot p} = \frac{\hbar}{m_0}\mathbf{k}\cdot\boldsymbol{\pi}, \tag{2.8}$$

$$\boldsymbol{\pi} = \mathbf{p} + \frac{\hbar}{4m_0c^2}(\sigma \times \nabla V), \tag{2.9}$$

$$\mathcal{E}_n(\mathbf{k}) = E_n(\mathbf{k}) - \frac{\hbar^2k^2}{2m_0}. \tag{2.10}$$

Equation (2.6) is the $k \cdot p$ equation. If the states $u_{n\mathbf{k}}$ form a complete set of periodic functions, then a representation of $H(\mathbf{k})$ in this basis is exact; i.e., diagonalization of the infinite matrix

$$\langle u_{n\mathbf{k}}|H(\mathbf{k})|u_{m\mathbf{k}}\rangle$$

leads to the dispersion relation throughout the whole Brillouin zone. Note, in particular, that the off-diagonal terms are only linear in k. However, practical implementations only solve the problem in a finite subspace. This leads to approximate dispersion relations and/or applicability for only a finite range of k values. For GaAs and AlAs, the range of validity is of the order of 10% of the first Brillouin zone [7].

An even more extreme case is to only consider one $u_{n\mathbf{k}}$ function. This is then known as the one-band or effective-mass (the latter terminology will become clear below) model. Such an approximation is good if, indeed, the semiconductor under study has a fairly isolated band—at least, again, for a finite region in k space. This is typically true of the conduction band of most III–V and II–VI semiconductors. In such cases, one also considers a region in k space near the band extremum. This is partly driven by the fact that this is the region most likely populated by charge carriers in thermal equilibrium and also by the fact that linear terms in the energy dispersion vanish, i.e.,

$$\frac{\partial E_n(\mathbf{k}_0)}{\partial k_i} = 0.$$

A detailed discussion of the symmetry constraints on the locations of these extremum points was provided by Bir and Pikus [1]. In the rest of this chapter, we will discuss how to obtain the energy dispersion relation and analyze a few properties of the resulting band.

2.3 Perturbation Theory

One can apply nondegenerate perturbation theory to the $k \cdot p$ equation, Eq. (2.6), for an isolated band. Given the solutions at $\mathbf{k} = \mathbf{0}$, one can find the solutions for finite k via perturbation theory:

$$E_n(\mathbf{k}) = E_n(\mathbf{0}) + \frac{\hbar^2 k^2}{2m_0} + \frac{\hbar \mathbf{k}}{m_0} \cdot \langle n\mathbf{0}|\boldsymbol{\pi}|n\mathbf{0}\rangle + \frac{\hbar^2}{m_0^2} \sum_l{}' \frac{|\langle n\mathbf{0}|\boldsymbol{\pi}|l\mathbf{0}\rangle \cdot \mathbf{k}|^2}{E_n(\mathbf{0}) - E_l(\mathbf{0})} \tag{2.11}$$

to second order and where

$$\langle n\mathbf{0}|\boldsymbol{\pi}|l\mathbf{0}\rangle = \frac{(2\pi)^3}{\Omega} \int \mathrm{d}\Omega \, u_{n\mathbf{0}}^* \boldsymbol{\pi} u_{l\mathbf{0}}. \tag{2.12}$$

This is the basic effective-mass equation.

2.4 Canonical Transformation

A second technique for deriving the effective-mass equation is by the use of the canonical transformation introduced by Luttinger and Kohn in 1955 [6]. Here, one expands the cellular function in terms of a complete set of periodic functions:

$$u_{n\mathbf{k}}(\mathbf{r}) = \sum_{n'} A_{nn'}(\mathbf{k})\, u_{n'\mathbf{0}}(\mathbf{r}). \tag{2.13}$$

Then the $k \cdot p$ equation, Eq. (2.6), becomes

$$\begin{aligned}\sum_{n'} A_{nn'}(\mathbf{k}) \left[H + H_{k\cdot p}\right] u_{n'\mathbf{0}}(\mathbf{r}) &= \sum_{n'} A_{nn'}(\mathbf{k}) \left[E_{n'}(\mathbf{0}) + H_{k\cdot p}\right] u_{n'\mathbf{0}}(\mathbf{r}) \\ &= \mathcal{E}_n(\mathbf{k}) \sum_{n'} A_{nn'}(\mathbf{k})\, u_{n'\mathbf{0}}(\mathbf{r}).\end{aligned} \tag{2.14}$$

Multiplying by $(2\pi)^3/\Omega \int_\Omega \mathrm{d}^3\mathbf{r}\, u_{n\mathbf{0}}^*$ gives

$$E_n(\mathbf{0}) A_{nn} + \sum_{n'} \frac{\hbar \mathbf{k}}{m_0} \cdot \mathbf{p}_{nn'} A_{nn'}(\mathbf{k}) = \mathcal{E}_n(\mathbf{k}) A_{nn}, \tag{2.15}$$

where

$$\mathbf{p}_{nn'} \equiv \mathbf{p}_{nn'}(\mathbf{0}) = \frac{(2\pi)^3}{\Omega} \int \mathrm{d}\Omega \, u_{n\mathbf{0}}^* \mathbf{p} u_{n'\mathbf{0}}, \tag{2.16}$$

and we have left out the spin-orbit contribution to the momentum operator for simplicity. Now one can write (dropping one band index)

$$H(\mathbf{k})A = \mathcal{E}(\mathbf{k})A, \quad A = \begin{pmatrix} \vdots \\ A_n \\ \vdots \end{pmatrix}. \tag{2.17}$$

The linear equations are coupled. The solution involves uncoupling them. This can be achieved by a canonical transformation:

$$A = TB, \tag{2.18}$$

where T is unitary (in order to preserve normalization). Then

$$\overline{H}(\mathbf{k})B = \mathcal{E}(\mathbf{k})B, \tag{2.19}$$

where

$$\overline{H}(\mathbf{k}) = T^{-1}HT. \tag{2.20}$$

Writing $T = \mathrm{e}^S, \quad T^{-1} = \mathrm{e}^{-S} = T^\dagger$,

$$\begin{aligned}\overline{H} &= \left(1 - S + \frac{1}{2!}S^2 - \cdots\right) H(\mathbf{k}) \left(1 + S + \frac{1}{2!}S^2 + \cdots\right) \\ &= H(\mathbf{k}) + [H(\mathbf{k}), S] + \frac{1}{2!}[[H(\mathbf{k}), S], S] + \cdots \\ &= H + H_{k\cdot p} + [H, S] + [H_{k\cdot p}, S] \\ &\quad + \frac{1}{2!}[[H, S], S] + \frac{1}{2!}\left[[H_{k\cdot p}, S], S\right] + \cdots \end{aligned} \tag{2.21}$$

Since $H_{k\cdot p}$ induces the coupling, one would like to remove it to order S by

$$H_{k\cdot p} + [H, S] = 0, \tag{2.22}$$

or, with $|n\rangle \equiv |u_{n\mathbf{0}}\rangle$,

$$\begin{aligned}\langle n|H_{k\cdot p}|n'\rangle + \sum_{n''}\left[\langle n|H|n''\rangle\langle n''|S|n'\rangle - \langle n|S|n''\rangle\langle n''|H|n'\rangle\right] &= 0, \\ \frac{\hbar}{m_0}\mathbf{k}\cdot\mathbf{p}_{nn'} + E_n(\mathbf{0})\langle n|S|n'\rangle - \langle n|S|n'\rangle E_{n'}(\mathbf{0}) &= 0,\end{aligned}$$

giving, for $n \neq n'$,

$$\langle n|S|n'\rangle = -\frac{\hbar}{m_0}\frac{\mathbf{k}\cdot\mathbf{p}_{nn'}}{[E_n(\mathbf{0}) - E_{n'}(\mathbf{0})]}. \tag{2.23}$$

Now, Eq. (2.21) becomes

$$\overline{H}\,(\mathbf{k}) = H + \frac{1}{2}[H_{k\cdot p}, S] + \frac{1}{2}[[H_{k\cdot p}, S], S] + \cdots$$

and, to second order,

$$\begin{aligned}
\langle n|\ \overline{H}\ (\mathbf{k})|n'\rangle &\approx \langle n|H|n'\rangle + \frac{1}{2}\sum_{n''}\Big[\langle n|H_{k\cdot p}|n''\rangle\langle n''|S|n'\rangle - \langle n|S|n''\rangle\langle n''|H_{k\cdot p}|n'\rangle\Big] \\
&= E_n(\mathbf{0})\delta_{nn'} + \frac{\hbar^2}{2m_0^2}\sum_{n''}\left[\frac{\mathbf{k}\cdot\mathbf{p}_{nn''}\,\mathbf{k}\cdot\mathbf{p}_{n''n'}}{[E_{n'}(\mathbf{0}) - E_{n''}(\mathbf{0})]} + \frac{\mathbf{k}\cdot\mathbf{p}_{nn''}\,\mathbf{k}\cdot\mathbf{p}_{n''n'}}{[E_n(\mathbf{0}) - E_{n''}(\mathbf{0})]}\right] \\
&= \left[E_n(\mathbf{0}) + \frac{\hbar^2}{2}\sum_{\alpha\beta}k_\alpha\left(\frac{1}{m_n}\right)_{\alpha\beta}k_\beta\right]\delta_{nn'} + \text{interband terms of order } k^2,
\end{aligned}$$

which is, of course, the same as Eq. (2.11).

We now restrict ourselves to zincblende and diamond crystals for which $n = s = \Gamma_1$ (see Appendix B for the symmetry properties), $p_{nn} = 0$, and

$$E(\mathbf{k}) = E_{\Gamma_1} + \frac{\hbar^2k^2}{2m_0} + \frac{\hbar^2}{m_0^2}\sum_l{}' \frac{|\mathbf{p}_{\Gamma_1 l}\cdot\mathbf{k}|^2}{E_{\Gamma_1} - E_l}. \tag{2.24}$$

Note that, for conciseness, we are also only using the group notation for the electron states in a zincblende crystal. The standard state ordering for zincblende and diamond is given in Fig. 2.1. There are exceptions to these such as the inverted band structure of HgTe [9] and the inverted conduction band of Si. Thus, the interaction of the Γ_1 state with other states via $p_{\Gamma_1 l}$ changes the dispersion relation from that of a free-electron one. The new inverse effective-mass tensor is

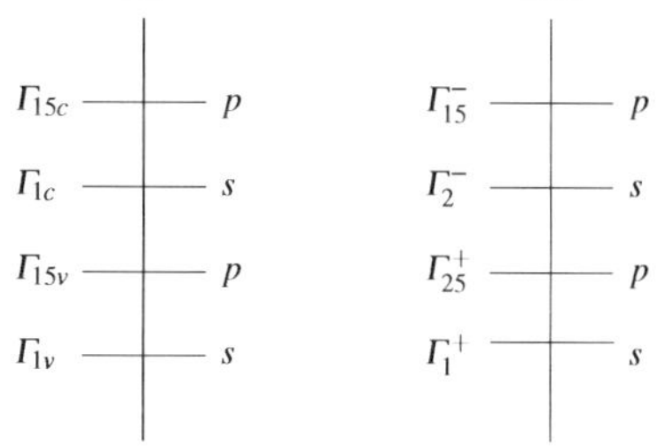

Fig. 2.1 Zone-center states for typical zincblende (ZB) and diamond (DM) crystals

$$\left(\frac{1}{m^*}\right)_{ij} = \frac{1}{m_0}\delta_{ij} + \frac{2}{m_0^2}\sum_l{}' \frac{p^i_{\Gamma_1 l} p^j_{l\Gamma_1}}{E_{\Gamma_1} - E_l}. \tag{2.25}$$

Equation (2.11) or Eqs. (2.24) and (2.25) define the one-band, effective-mass model. The band dispersion can be calculated given the momentum matrix elements and band gaps. Note that Eq. (2.24) is only approximate, giving the parabolic approximation. Constraints on the effective mass can now be written from Eq. (2.25).

2.5 Effective Masses

One can write down simple expressions for the effective masses of nondegenerate bands.

2.5.1 Electron

Because of the energy denominator, distant bands are expected to be less important. The two closest bands to the Γ_{1c} state for cubic semiconductors are the Γ_{15} states $\sim X, Y, Z$. Since $\Gamma_{1c} \sim S$, and

$$\langle S|p_x|X\rangle = \langle S|p_y|Y\rangle = \langle S|p_z|Z\rangle,$$

the conduction mass m_e is isotropic:

$$\begin{aligned}\frac{1}{m_e} &= \frac{1}{m_0} + \frac{2}{m_0^2}\frac{|\langle S|p_x|X_v\rangle|^2}{E_{\Gamma_{1c}} - E_{\Gamma_{15v}}} + \frac{2}{m_0^2}\frac{|\langle S|p_x|X_c\rangle|^2}{E_{\Gamma_{1c}} - E_{\Gamma_{15c}}} \\ &\equiv \frac{1}{m_0} + \frac{2P^2}{\hbar^2 E_0} - \frac{2P'^2}{\hbar^2 E_0'},\end{aligned} \tag{2.26}$$

where

$$P^2 = \frac{\hbar^2}{m_0^2}|\langle S|p_x|X_v\rangle|^2, \tag{2.27}$$

$$P'^2 = \frac{\hbar^2}{m_0^2}|\langle S|p_x|X_c\rangle|^2. \tag{2.28}$$

For diamond,

$$P' = 0 \implies 0 < m_e < m_0.$$

For zincblende, typically

$$\frac{P'^2}{E'_0} < \frac{P^2}{E_0} \quad \Longrightarrow \quad 0 < m_e < m_0.$$

Hence, the electron effective mass is usually smaller than the free-electron mass.

2.5.2 Light Hole

Of the three-fold degenerate Γ_{15v} states, only one couples with Γ_{1c} along a given Δ direction, giving rise to the light-hole (lh) mass. Consider $\mathbf{k} = (k_x, 0, 0)$. Then, since the lh state can now be assumed nondegenerate, again m_{lh} is isotropic (though a more accurate model will reveal them to be anisotropic):

$$\frac{1}{m_{lh}} = \frac{1}{m_0} + \frac{2}{m_0^2}\frac{|\langle S|p_x|X_v\rangle|^2}{E_{\Gamma_{15v}} - E_{\Gamma_{1c}}} = \frac{1}{m_0} - \frac{2P^2}{\hbar^2 E_0} \equiv \frac{1}{m_0}\left(1 - \frac{E_P}{E_0}\right), \tag{2.29}$$

with

$$E_P \equiv \frac{2m_0 P^2}{\hbar^2} \tag{2.30}$$

known as the Kane parameter. Typically, $E_p \sim 20\,\text{eV}$, $E_0 \sim 0$–$5\,\text{eV}$. Hence, $-m_0 < m_{lh} < 0$. Note that, contrary to the electron case, the lh mass does not contain the P' term.

To compare the lh and e masses,

$$\frac{1}{m_e} + \frac{1}{m_{lh}} = \frac{2}{m_0} - \frac{2P'^2}{\hbar^2 E'_0} = \frac{1}{m_0}\left(2 - \frac{E'_P}{E'_0}\right).$$

For diamond, $E'_P = 0$, giving

$$\frac{1}{m_e} + \frac{1}{m_{lh}} > 0 \quad \text{(always)}, \tag{2.31}$$

and

$$|m_{lh}| > m_e. \tag{2.32}$$

For zincblende, $E'_P \sim 1$–$10\,\text{eV}$, $E'_0 \sim 3$–$5\,\text{eV}$, and the masses are closer in magnitude. The qualitative effect of the e–lh interaction on the effective masses is sketched in Fig. 2.2. This is also known as a two-band model.

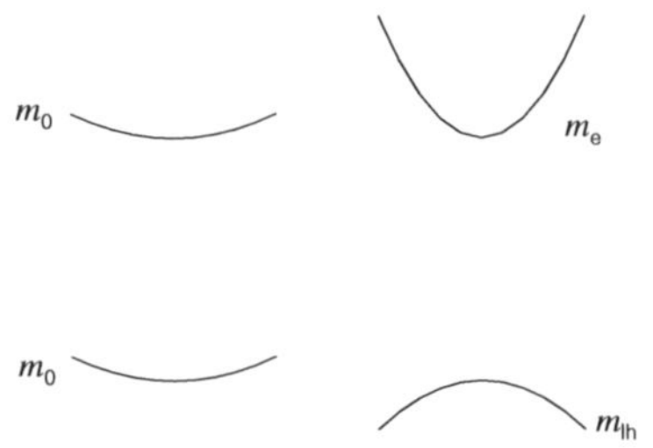

Fig. 2.2 Two-band model. The $k \cdot p$ interaction changes the curvatures

2.5.3 Heavy Hole

One may define the heavy-hole (hh) states as the partners in the Γ_{15v} representation which do not couple to the conduction s electron. In so far as the Γ_{15c} states are not considered, the hh state has the free-electron mass. Including the Γ_{15c} state and again assuming that the Γ_{15v} states are nondegenerate, the isotropic mass is

$$\frac{1}{m_{hh}} = \frac{1}{m_0} - \frac{2}{m_0^2}\frac{|\langle Y_v|p_x|Z_c\rangle|^2}{E_0 + E_0'} = \frac{1}{m_0}\left(1 - \frac{E_Q}{E_0 + E_0'}\right). \tag{2.33}$$

Typically, $E_Q \sim$20–25 eV, $E_0 + E_0' \sim 10$ eV, and $0 > m_{hh} > -m_0$.

We have seen how the simple one-band model can provide a semi-quantitative description of various bands for zincblende and diamond semiconductors, particularly the sign and relative magnitudes of the associated effective masses. The necessity of describing the band structure quantitatively and more accurately (such as nonparabolicity and anisotropy) leads to the consideration of multiband models.

2.6 Nonparabolicity

So far, we have presented the simplest one-band model in order to illustrate the theory; it does allow for anisotropy via an anisotropic effective mass. Still, a one-band model can be made to reproduce more detailed features of a real band including nonparabolicity, anisotropy and spin splitting. An example of such a model is the k^4 dispersion relation given by Rössler [23]:

$$\begin{aligned} E(\mathbf{k}) = \tfrac{\hbar^2 k^2}{2m^*} &+ \alpha k^4 + \beta\left(k_y^2 k_z^2 + k_z^2 k_x^2 + k_x^2 k_y^2\right) \\ &\pm \gamma\left\{k^2\left(k_y^2 k_z^2 + k_z^2 k_x^2 + k_x^2 k_y^2\right) - 9k_x^2 k_y^2 k_z^2\right\}^{1/2}. \end{aligned} \tag{2.34}$$

The first term on the right-hand side is the familiar isotropic and parabolic effective-mass term. The remaining terms give nonparabolicity, warping and spin splitting, respectively. We will derive them later in the book.

2.7 Summary

We have set up the fundamental $k \cdot p$ equation and shown, using a variety of techniques, how a one-band model (the so-called effective-mass model) can be obtained from it. This model was then used to derive a semi-quantitative understanding of the magnitude of the effective masses of band-edge states for cubic semiconductors. In particular, it was shown that the simplest effective-mass model for electrons and light holes gives isotropic masses.

Chapter 3
Perturbation Theory – Valence Band

3.1 Overview

Degenerate perturbation theory is presented in order to derive the valence-band Hamiltonian. This will be illustrated in detail for the Dresselhaus–Kip–Kittel Hamiltonian for diamond and for the valence-band Hamiltonian for wurtzite.

3.2 Dresselhaus–Kip–Kittel Model

We first give the derivation of the 3×3 (i.e., no spin) Dresselhaus–Kip–Kittel (DKK) Hamiltonian using the original second-order degenerate perturbation theory approach [2]. The theory applies to the valence band of diamond.

3.2.1 Hamiltonian

The starting equation is the $k \cdot p$ equation, Eq. (2.6), without the spin-orbit term:

$$\left[\frac{p^2}{2m_0} + V(\mathbf{r}) + \frac{\hbar}{m_0}\mathbf{k} \cdot \mathbf{p}\right] u_{n\mathbf{k}}(\mathbf{r}) = \mathcal{E}_n(\mathbf{k}) u_{n\mathbf{k}}(\mathbf{r}).$$

The unperturbed $u^r_{n\mathbf{0}}(\mathbf{r}) \equiv \varepsilon^+_r$ are the three-fold degenerate solutions at the zone center or Γ point. They transform according to the Γ^+_{25} irreducible representation (Fig. 3.1). The three states are: $\varepsilon^+_1 \sim yz$, $\varepsilon^+_2 \sim zx$, $\varepsilon^+_3 \sim xy$; they are even with respect to the inversion operator. An atomistic description of the transformation properties of some of the states of the DM structure is given in Table 3.1. Since the unperturbed states are degenerate, we have to use degenerate perturbation theory to find the solutions at finite $\mathbf{k}$.

The first-order correction is given by the matrix elements

$$\sim \langle \varepsilon^+_r | \mathbf{k} \cdot \mathbf{p} | \varepsilon^+_s \rangle = 0$$

L.C. Lew Yan Voon, M. Willatzen, *The* k · p *Method*,
DOI 10.1007/978-3-540-92872-0_3, © Springer-Verlag Berlin Heidelberg 2009

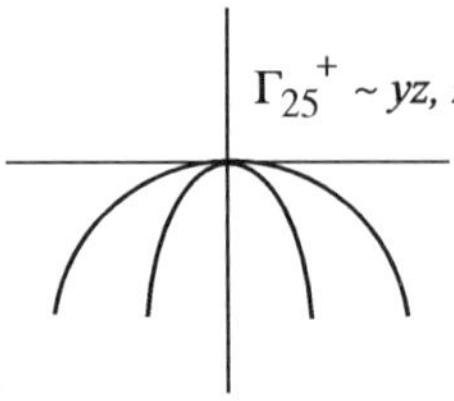

Fig. 3.1 Three-band model for diamond-type semiconductors

Table 3.1 Symmetries of states at the Γ point for diamond structure [5, 24]. The second column gives the orbitals on the two atoms in the basis. The far-right column gives the corresponding plane-wave states

		Cardona and Pollak [5]		
Γ_1^+	$s_a + s_a'$	Γ_1^l	s^+	[000]
Γ_{25}^+	$p_a - p_a' \sim yz, zx, xy$	$\Gamma_{25'}^l$	p^+	[111]
Γ_2^-	$s_a - s_a' \sim xyz$	$\Gamma_{2'}^l$	s^-	[111]
Γ_{15}^-	$p_a + p_a' \sim x, y, z$	Γ_{15}	p^-	[111]
Γ_{12}^-	$d_a - d_a' \sim \sqrt{3}\,(y^2 - z^2), 3x^2 - r^2$	$\Gamma_{12'}$	d^-	[200]
Γ_1^+	$s_a + s_a'$	Γ_1^u	s^+	[111]
Γ_{25}^+	$d_a + d_a' \sim yz, zx, xy$	$\Gamma_{25'}^u$	d^+	[200]
Γ_2^-	$s_a - s_a' \sim xyz$	$\Gamma_{2'}^u$	s^-	[200]

since the ε_r^+'s have the same parity and $\mathbf{p}$ is odd under inversion. In the language of group theory, one says that $\Gamma_{25}^+ \otimes \Gamma_{15}^- \otimes \Gamma_{25}^+$ does not contain Γ_1^+.

One, therefore, needs second-order degenerate perturbation theory. The corrections to the cellular functions and matrix elements are:

$$u_{n\mathbf{k}}^r = \varepsilon_r^+ + \frac{\hbar}{m_0}\mathbf{k} \cdot \sum_{l\alpha\nu}{}' \frac{|l\alpha\nu\rangle\langle l\alpha\nu|\mathbf{p}|r\rangle}{E_{\Gamma_{25}^+} - E_{l\alpha}}, \tag{3.1}$$

$$H_{rs} \equiv \langle r|H(\mathbf{k})|s\rangle = \frac{\hbar^2}{m_0^2}\sum_{l\alpha\nu}{}' \frac{\langle r|\mathbf{k}\cdot\mathbf{p}|l\alpha\nu\rangle\langle l\alpha\nu|\mathbf{k}\cdot\mathbf{p}|s\rangle}{E_{\Gamma_{25}^+} - E_{l\alpha}} \tag{3.2}$$

$$= \frac{\hbar^2}{m_0^2}k_i k_j \sum_{l\alpha\nu}{}' \frac{\langle r|p_i|l\alpha\nu\rangle\langle l\alpha\nu|p_j|s\rangle}{E_{\Gamma_{25}^+} - E_{l\alpha}},$$

where $|r\rangle = u_{n\mathbf{0}}^r$, $l\alpha\nu$ denotes the state ν (in case of degeneracy) belonging to the α representation in the band l; $E_{l\alpha}$ is the energy of that state at $\mathbf{k} = \mathbf{0}$. $E_{\Gamma_{25}^+}$ is the energy of the ε_r^+ states. The diagonal perturbation matrix elements are given by

$$H_{rr} = \frac{\hbar^2}{m_0^2}k_i k_j \sum_{l\alpha\nu}{}' \frac{\langle r|p_i|l\alpha\nu\rangle\langle l\alpha\nu|p_j|r\rangle}{E_{\Gamma_{25}^+} - E_{l\alpha}}.$$

In the group of DM, there are operators that invert all three coordinates or just one; taking, e.g., $r = xy$ (the others follow by permutation),

$$\langle xy|p_i|l\alpha\nu\rangle\langle l\alpha\nu|p_j|xy\rangle$$

is nonzero only if all coordinates appear pairwise. For example, in

$$\langle xy|p_x|l\alpha\nu\rangle\langle l\alpha\nu|p_j|xy\rangle,$$

using an operator that only changes the sign of x requires $p_j = p_x$ (equivalently for p_y), while in

$$\langle xy|p_z|l\alpha\nu\rangle\langle l\alpha\nu|p_j|xy\rangle,$$

using an operator that only changes the sign of z requires $p_j = p_z$. Thus, $p_i = p_j$ and two independent choices are $p_i = p_y (= p_x)$, p_z. Therefore,

$$H_{33} = \frac{\hbar^2}{m_0^2}\sum_{l\alpha\nu}{}' \left\{ k_z^2 \frac{|\langle xy|p_z|l\alpha\nu\rangle|^2}{E_{\Gamma_{25}^+} - E_{l\alpha}} + \left(k_x^2 + k_y^2\right) \frac{|\langle xy|p_y|l\alpha\nu\rangle|^2}{E_{\Gamma_{25}^+} - E_{l\alpha}} \right\}.$$

For the off-diagonal matrix elements, one again requires the coordinates to appear pairwise. For example, for $r = xy$ and $s = yz$,

$$\langle xy|p_x|l\alpha\nu\rangle\langle l\alpha\nu|p_j|yz\rangle,$$

using an operator that only changes the sign of x requires $p_j = p_y$ or p_z but, in addition, using an operator that only changes the sign of y requires $p_j = p_z$ only. One, therefore, gets

$$H_{31} = \frac{\hbar^2}{m_0^2} k_x k_z \sum_{l\alpha\nu}{}' \left\{ \frac{\langle xy|p_z|l\alpha\nu\rangle\langle l\alpha\nu|p_x|yz\rangle + \langle xy|p_x|l\alpha\nu\rangle\langle l\alpha\nu|p_z|yz\rangle}{E_{\Gamma_{25}^+} - E_{l\alpha}} \right\}.$$

Thus, one can introduce three independent parameters:

$$\begin{aligned} L &\equiv \frac{\hbar^2}{m_0^2}\sum_{l\alpha\nu}{}' \frac{|\langle xy|p_z|l\alpha\nu\rangle|^2}{E_{\Gamma_{25}^+} - E_{l\alpha}}, \\ M &\equiv \frac{\hbar^2}{m_0^2}\sum_{l\alpha\nu}{}' \frac{|\langle xy|p_y|l\alpha\nu\rangle|^2}{E_{\Gamma_{25}^+} - E_{l\alpha}}, \\ N &\equiv \frac{\hbar^2}{m_0^2}\sum_{l\alpha\nu}{}' \frac{\langle xy|p_z|l\alpha\nu\rangle\langle l\alpha\nu|p_x|yz\rangle + \langle xy|p_x|l\alpha\nu\rangle\langle l\alpha\nu|p_z|yz\rangle}{E_{\Gamma_{25}^+} - E_{l\alpha}}, \end{aligned} \tag{3.3}$$

giving rise to the DKK Hamiltonian:

$$H_{\mathrm{DKK}}(\mathbf{k}) = \begin{pmatrix} |yz\rangle & |zx\rangle & |xy\rangle \\ Lk_x^2 + M(k_y^2 + k_z^2) & Nk_xk_y & Nk_xk_z \\ Nk_xk_y & Lk_y^2 + M(k_z^2 + k_x^2) & Nk_yk_z \\ Nk_xk_z & Nk_yk_z & Lk_z^2 + M(k_x^2 + k_y^2) \end{pmatrix}. \quad (3.4)$$

The band energies are then given by

$$E_n(\mathbf{k}) = E_{\Gamma_{25}^+} + \frac{\hbar^2 k^2}{2m_0} + \lambda,$$

where λ are the eigenvalues of the DKK Hamiltonian. Luttinger and Kohn (LK) [6] came up with a slightly different notation for the Hamiltonian, which includes the free-electron term. They give the matrix as D with matrix elements

$$D_{mn} = D_{mn}^{ij} k_i k_j, \quad (3.5)$$

$$D_{mn}^{ij} = \frac{\hbar^2}{2m_0}\left[\delta_{mn}\delta_{jj'} + \frac{1}{m_0}\sum_l^B \frac{p_{ml}^i p_{ln}^j + p_{ml}^j p_{ln}^i}{E_{\Gamma_{25}^+} - E_l}\right]. \quad (3.6)$$

Then

$$D = \begin{pmatrix} |yz\rangle & |zx\rangle & |xy\rangle \\ A_L k_x^2 + B_L\left(k_y^2 + k_z^2\right) & C_L k_x k_y & C_L k_x k_z \\ C_L k_x k_y & A_L k_y^2 + B_L\left(k_z^2 + k_x^2\right) & C_L k_y k_z \\ C_L k_x k_z & C_L k_y k_z & A_L k_z^2 + B_L\left(k_x^2 + k_y^2\right) \end{pmatrix}, \quad (3.7)$$

with

$$\begin{aligned} A_L &= \frac{\hbar^2}{2m_0} + \frac{\hbar^2}{m_0^2}\sum_l \frac{p_{1l}^x p_{l1}^x}{E_{\Gamma_{25}^+} - E_l}, \\ B_L &= \frac{\hbar^2}{2m_0} + \frac{\hbar^2}{m_0^2}\sum_l \frac{p_{1l}^y p_{l1}^y}{E_{\Gamma_{25}^+} - E_l}, \\ C_L &= \frac{\hbar^2}{2m_0^2}\sum_l \frac{p_{1l}^x p_{l2}^y + p_{1l}^y p_{l2}^x}{E_{\Gamma_{25}^+} - E_l}. \end{aligned} \quad (3.8)$$

Thus, the LK parameters are related to the DKK ones via

$$A_L = L + \frac{\hbar^2}{2m_0},$$

$$B_L = M + \frac{\hbar^2}{2m_0},$$
$$C_L = N. \tag{3.9}$$

Terms of higher order, for example, k^4, have been considered by some authors [25] but will not be discussed in this book as they are not commonly used in the literature.

3.2.2 Eigenvalues

The DKK Hamiltonian cannot be diagonalized analytically throughout the first Brillouin zone (FBZ). However, analytical solutions exist when there is double degeneracy (e.g., along $L - \Gamma - X$). We now obtain those solutions:

$$\begin{vmatrix} Lk_x^2 + M\left(k_y^2 + k_z^2\right) - \lambda & Nk_xk_y & Nk_xk_z \\ Nk_xk_y & Lk_y^2 + M\left(k_z^2 + k_x^2\right) - \lambda & Nk_yk_z \\ Nk_xk_z & Nk_yk_z & Lk_z^2 + M\left(k_x^2 + k_y^2\right) - \lambda \end{vmatrix}$$
$$\begin{aligned} &= \left[Lk_x^2 + M\left(k_y^2 + k_z^2\right) - \lambda\right]\left[Lk_y^2 + M(k_x^2 + k_z^2) - \lambda\right]\left[Lk_z^2 + M\left(k_y^2 + k_x^2\right) - \lambda\right] \\ &\quad - \left[Lk_x^2 + M\left(k_y^2 + k_z^2\right) - \lambda\right]N^2k_y^2k_z^2 - \left[Lk_z^2 + M\left(k_x^2 + k_y^2\right) - \lambda\right]N^2k_x^2k_y^2 \\ &\quad - \left[Lk_y^2 + M\left(k_x^2 + k_z^2\right) - \lambda\right]N^2k_x^2k_z^2 + 2N^3k_x^2k_y^2k_z^2 = 0. \end{aligned}$$

Using

$$[a-\lambda][b-\lambda][c-\lambda] = abc - (ac + bc + ab)\lambda + (a + b + c)\lambda^2 - \lambda^3,$$

we have

$$\begin{aligned} &\lambda^3 - (L + 2M)k^2\lambda^2 \\ &\quad + \Big\{\left[Lk_x^2 + M\left(k_y^2 + k_z^2\right)\right]\left[Lk_y^2 + M\left(k_x^2 + k_z^2\right)\right] \\ &\quad + \left[Lk_x^2 + M\left(k_y^2 + k_z^2\right)\right]\left[Lk_z^2 + M\left(k_x^2 + k_y^2\right)\right] \\ &\quad + \left[Lk_y^2 + M\left(k_x^2 + k_z^2\right)\right]\left[Lk_z^2 + M\left(k_x^2 + k_y^2\right)\right] - N^2\left(k_x^2k_y^2 + k_x^2k_z^2 + k_y^2k_z^2\right)\Big\}\lambda \\ &\quad - \Big\{\left[Lk_x^2 + M\left(k_y^2 + k_z^2\right)\right]\left[Lk_y^2 + M\left(k_x^2 + k_z^2\right)\right]\left[Lk_z^2 + M\left(k_x^2 + k_y^2\right)\right] \\ &\quad + N^2\left[Lk_x^2 + M\left(k_y^2 + k_z^2\right)\right]k_y^2k_z^2 + N^2\left[Lk_y^2 + M\left(k_x^2 + k_z^2\right)\right]k_x^2k_z^2 \\ &\quad + N^2\left[Lk_z^2 + M\left(k_y^2 + k_x^2\right)\right]k_x^2k_y^2 - 2N^3k_x^2k_y^2k_z^2\Big\} = 0. \end{aligned} \tag{3.10}$$

This general characteristic equation is too complicated to be of much use. However, when there is a degenerate pair of eigenvalues,

$$(\lambda - a)^2(\lambda - b) = \lambda^3 - (2a + b)\lambda^2 + (2ab + a^2)\lambda - a^2 b = 0.$$

Comparing with Eq. (3.10), one has

$$2a + b = (L + 2M)k^2, \tag{3.11}$$

$$2ab + a^2 = \left(L^2 - N^2 + 3M^2 + 2ML\right) k_{xyz} + \left(M^2 + 2LM\right)\left(k_x^4 + k_y^4 + k_z^4\right) \equiv \alpha \tag{3.12}$$

where $k_{xyz} = k_x^2 k_y^2 + k_y^2 k_z^2 + k_z^2 k_x^2$. Eliminating b from Eqs. (3.11) and (3.12) gives

$$3a^2 - 2(L + 2M)k^2 a + \alpha = 0,$$

and

$$a = \frac{(L + 2M)}{3}k^2 \pm \left[\frac{(L + 2M)^2}{9}k^4 - \frac{\alpha}{3}\right]^{1/2}$$

$$= \frac{(L + 2M)}{3}k^2 \pm \left[\frac{(L - M)^2}{9}k^4 + \frac{1}{3}\left[N^2 - (L - M)^2\right]\left(k_x^2 k_y^2 + k_y^2 k_z^2 + k_z^2 k_x^2\right)\right]^{1/2}.$$

Hence,

$$E(\mathbf{k}) = Ak^2 \pm \left[B^2 k^4 + C^2\left(k_y^2 k_z^2 + k_z^2 k_x^2 + k_x^2 k_y^2\right)\right]^{1/2}, \tag{3.13}$$

where

$$A \equiv \frac{(L + 2M)}{3} + \frac{\hbar^2}{2m_0},\; B^2 \equiv \frac{(L - M)^2}{9},\; C^2 \equiv \frac{1}{3}\left[N^2 - (L - M)^2\right]. \tag{3.14}$$

This (restricted) result will turn out to be identical to the spin case. Note that the dispersion is both anisotropic (if $C \neq 0$) and nonparabolic (if $B, C \neq 0$); the first property is also known as warping. A careful study of warping was given in [26].

3.2.3 *L, M, N* Parameters

We next study the L, M, N parameters more closely. The matrix elements involved are

$$\langle r|\mathbf{p}|l\rangle \sim \Gamma_{25}^+ \otimes \Gamma_{15}^- \otimes \Gamma^l.$$

From Table 3.2, one deduces that

$$\Gamma_{25}^+ \otimes \Gamma_{15}^- \sim \Gamma_{12}^- \oplus \Gamma_{15}^- \oplus \Gamma_2^- \oplus \Gamma_{25}^-; \tag{3.15}$$

in effect, all the odd-parity representations except for Γ_1^-. In order, to be able to get Γ_1^+ in the decomposition, Γ^l must clearly have odd parity; however, Γ_1^- is not appropriate since the decomposition will then not include Γ_1^+. Otherwise, all of the four irreducible representations on the right-hand side of Eq. (3.15) can interact with the valence-band edge.

Table 3.2 Character table of the group O_h

O_h	E	$8C_3$	$3C_2$	$6C_4$	$6C_2'$	I	$8S_6 = IC_3$	$3\sigma_h = IC_4^2$	$6S_4 = IC_4$	$6\sigma_d = IC_2$
Γ_1^+	1	1	1	1	1	1	1	1	1	1
Γ_2^+	1	1	1	–1	–1	1	1	1	–1	–1
Γ_{12}^+	2	–1	2	0	0	2	–1	2	0	0
Γ_{15}^+	3	0	–1	1	–1	3	0	–1	1	–1
Γ_{25}^+	3	0	–1	–1	1	3	0	–1	–1	1
Γ_1^-	1	1	1	1	1	–1	–1	–1	–1	–1
Γ_2^-	1	1	1	–1	–1	–1	–1	–1	1	1
Γ_{12}^-	2	–1	2	0	0	–2	1	–2	0	0
Γ_{15}^-	3	0	–1	1	–1	–3	0	1	–1	1
Γ_{25}^-	3	0	–1	–1	1	–3	0	1	1	–1

The band-edge structure for diamond-type semiconductors is believed to be as given in Fig. 3.2. The state Γ_{12}^- was reported by von der Lage and Bethe [27] to have as smallest $l = 5$. However, Herman [28] showed that it originates from $\langle 200\rangle$ plane waves and as having the symmetry of d states. The discrepancy is likely due to the fact that von der Lage and Bethe were really studying the cubic group with a single atom (ion) per unit cell whereas Herman considered the case of two atoms per unit cell. Thus, one expects the maximum perturbation from the Γ_2^-, Γ_{12}^-, Γ_{15}^- bands; in particular, there is no perturbation from, e.g., Γ_1^+.

Similarly, not all states appear in the summation for the DKK parameters. Hence, instead of writing the DKK Hamiltonian in terms of the L, M, N, one can also write it in terms of interband parameters between states of given symmetries; furthermore, this will provide relations among the L, M, N parameters.

3.2.3.1 *L* Parameter

Starting with L, only Γ_2^- and Γ_{12}^- can contribute. This can be ascertained by looking at the reflection properties of the matrix element. Consider, e.g., $|r\rangle = |yz\rangle$. Then

$$p_x|yz\rangle \sim xyz \sim \Gamma_2^-.$$

Under an IC_4^2 reflection, Γ_{15}^- and Γ_{25}^- are even, while Γ_2^- is odd (see Table 3.2). This eliminates the former two representations from the matrix elements for L. This can be shown more explicitly, e.g., for the Γ_{15}^- representation ($\sim x, y, z$):

$$\langle yz|p_x|x, z\rangle = 0$$

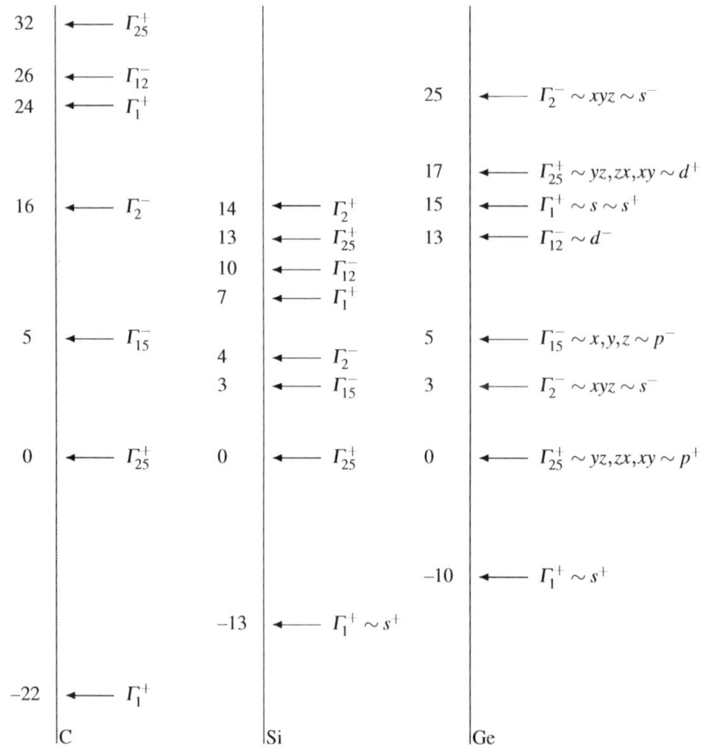

Fig. 3.2 Schematic of zone-center energy (in eV) ordering for diamond-structure semiconductors (not to scale; C from Willatzen et al. [29], Si from Cardona and Pollak [5], Ge from DKK [2])

using IC_{2y} and

$$\langle yz|p_x|y\rangle = 0$$

using IC_{2x}. We now wish to consider how many independent interband terms there are. Consider first

$$L = \frac{\hbar^2}{m_0^2} \sum_l{}' \left\{ \frac{|\langle yz|p_x|\beta_l^-\rangle|^2}{E_{\Gamma_{25}^+} - E_l} + \sum_\nu \frac{|\langle yz|p_x|\gamma_{\nu l}^-\rangle|^2}{E_{\Gamma_{25}^+} - E_l} \right\}, \tag{3.16}$$

where $\beta_l^- \in \Gamma_2^-$ and $\gamma_{\nu l}^- \in \Gamma_{12}^-$. DKK showed that each representation only contributes one matrix element per band, independent of its dimension. We will now obtain this result. Let

$$F \equiv \frac{\hbar^2}{m_0^2} \sum_{l \in \Gamma_2^-}{}' \frac{|\langle yz|p_x|\beta_l^-\rangle|^2}{E_{\Gamma_{25}^+} - E_l}, \tag{3.17}$$

$$G = \frac{\hbar^2}{m_0^2} \sum_{l \in \Gamma_{12}^-}' \frac{|\langle yz|p_x|\gamma_{1l}^-\rangle|^2}{E_{\Gamma_{25}^+} - E_l}. \tag{3.18}$$

Since Γ_2^- is one dimensional, there is only one such term. On the other hand, Γ_{12}^- is two dimensional; it turns out this still only contributes one distinct term. To establish this result, we need the basis functions. One could choose the d-like functions (as done by von der Lage and Bethe); however, the latter do not generate a unitary irreducible representation [2]. Hence, we follow DKK in choosing $\gamma_{\nu l}^-$ such that

$$\gamma_1^- = x^2 + \omega y^2 + \omega^2 z^2, \tag{3.19}$$

$$\gamma_2^- = x^2 + \omega^2 y^2 + \omega z^2, \tag{3.20}$$

where $\omega^3 = 1$ and $\gamma_\nu^- = \gamma_\nu - \gamma_\nu'$. To verify that they are basis functions for the Γ_{12}^- irreducible representation, one can generate the representation matrices and show that they have the right character vector. We use

$$\mathcal{O}(R) f_i(\mathbf{r}) = f(\mathsf{R}^{-1}\mathbf{r}) = \sum_j f_j(\mathbf{r}) \mathsf{R}_{ji},$$

where R is a rotation. For example (note that we only need the matrix for one element in each class):

$$\mathcal{O}(C_{4x})\gamma_1^- = C_{4x}^{-1}\gamma_1^- = C_{4x}^{-1}\left(x^2 + \omega y^2 + \omega^2 z^2\right)^- \overset{x\bar{z}y}{=} \left(x^2 + \omega z^2 + \omega^2 y^2\right)^- = \gamma_2^-,$$

$$\mathcal{O}(C_{4x})\gamma_2^- = \gamma_1^-,$$

$$\Longrightarrow \quad C_{4x} = \begin{pmatrix} 0 & 1 \\ 1 & 0 \end{pmatrix},$$

$$\mathcal{O}(C_{4y}^{-1})\gamma_1^- = C_{4y}\left(x^2 + \omega y^2 + \omega^2 z^2\right)^- \overset{\bar{z}yx}{=} \left(z^2 + \omega y^2 + \omega^2 x^2\right)^-$$

$$= \frac{1}{\omega}\left(x^2 + \omega^2 y^2 + \omega z^2\right)^- = \frac{1}{\omega}\gamma_2^-,$$

$$\mathcal{O}\left(C_{4y}^{-1}\right)\gamma_2^- = C_{4y}\left(x^2 + \omega z^2 + \omega^2 y^2\right)^- = \left(z^2 + \omega x^2 + \omega^2 y^2\right)^-$$

$$= \frac{1}{\omega^2}\left(x^2 + \omega y^2 + \omega^2 z^2\right)^- = \frac{1}{\omega^2}\gamma_1^-,$$

$$\Longrightarrow \quad C_{4y}^{-1} = \begin{pmatrix} 0 & \frac{1}{\omega} \\ \frac{1}{\omega^2} & 0 \end{pmatrix},$$

$$\mathcal{O}(C_2)\gamma_i^- = \gamma_i^-,$$

$$\Longrightarrow \quad C_2 = \begin{pmatrix} 1 & 0 \\ 0 & 1 \end{pmatrix},$$

$$\mathcal{O}\left(C_{3xyz}\right)\gamma_1^- = C_{3xyz}^{-1}\left(x^2+\omega y^2+\omega^2 z^2\right)^- \overset{zxy}{=} \left(z^2+\omega x^2+\omega^2 y^2\right)^-$$
$$= \frac{1}{\omega^2}\left(x^2+\omega y^2+\omega^2 z^2\right)^- = \frac{1}{\omega^2}\gamma_1^-,$$
$$\mathcal{O}\left(C_{3xyz}\right)\gamma_2^- = \left(z^2+\omega y^2+\omega^2 x^2\right)^- = \frac{1}{\omega}\gamma_2^-,$$
$$\Longrightarrow \quad C_{3xyz} = \begin{pmatrix} \frac{1}{\omega^2} & 0 \\ 0 & \frac{1}{\omega} \end{pmatrix} = \begin{pmatrix} \omega & 0 \\ 0 & \omega^2 \end{pmatrix}.$$

C_2' is similar to C_4, and so on. In order to get the characters, we have, for example for C_{3xyz}

$$\chi(C_{3xyz}) = \omega + \omega^2 = -1,$$

since $\omega + \omega^2 + \omega^3 = 0$ and $\omega^3 = 1$. Also, one should check that the representation is indeed unitary (i.e., $U^\dagger = U^{-1}$). For example, given C_{4y}^{-1} above and assuming unitarity,

$$C_{4y}^\dagger = C_{4y}^{-1} = \begin{pmatrix} 0 & \frac{1}{\omega^{*2}} \\ \frac{1}{\omega^*} & 0 \end{pmatrix} = \begin{pmatrix} 0 & \frac{1}{\omega} \\ \frac{1}{\omega^2} & 0 \end{pmatrix},$$

since

$$\omega^3 = 1 \Longrightarrow \quad \frac{1}{\omega^{*2}} = \omega^*,$$

and

$$\omega = e^{i\theta} \Longrightarrow \quad \omega^* = 1/\omega.$$

Also

$$C_{4y} = (C_{4y}^\dagger)^\dagger = \begin{pmatrix} 0 & \frac{1}{\omega} \\ \frac{1}{\omega^2} & 0 \end{pmatrix},$$

and

$$C_{4y}C_{4y}^{-1} = C_{4y}C_{4y}^\dagger = \begin{pmatrix} 0 & \frac{1}{\omega} \\ \frac{1}{\omega^2} & 0 \end{pmatrix}\begin{pmatrix} 0 & \frac{1}{\omega} \\ \frac{1}{\omega^2} & 0 \end{pmatrix} = \begin{pmatrix} \frac{1}{\omega^3} & 0 \\ 0 & \frac{1}{\omega^3} \end{pmatrix} = \begin{pmatrix} 1 & 0 \\ 0 & 1 \end{pmatrix}.$$

Thus $\gamma_{vl}^- \sim \Gamma_{12}^-$.

Now, following DKK, let

$$\langle yz|p_x|\gamma_1^-\rangle \equiv R. \tag{3.21}$$

We will show that $\langle yz|p_x|\gamma_2^-\rangle$ is related. To wit,

$$C_{4x}\langle yz|p_x|\gamma_2^-\rangle = C_{4x}\langle yz|p_x|(x^2+\omega^2y^2+\omega z^2)^-\rangle = -\langle yz|p_x|(x^2+\omega y^2+\omega^2z^2)\ \rangle$$
$$= -R.$$

The two matrix elements with respect to the two degenerate states are, therefore, the same up to a sign giving

$$L = F + 2G. \tag{3.22}$$

3.2.3.2 *M* Parameter

Similarly, one can show that M only has contributions from Γ_{15}^- and Γ_{25}^- (lowest $l = 3$ [27]). Again, taking the Γ_{15}^- representation as example:

$$\langle yz|p_y|x\rangle = 0$$

using IC_{2x},

$$\langle yz|p_y|y\rangle = 0$$

using IC_{2y}, but

$$\langle yz|p_y|z\rangle \neq 0\ \forall\ IC_2.$$

Then,

$$M = \frac{\hbar^2}{m_0^2}\sum_{lv}{}' \left\{ \frac{|\langle yz|p_y|lv\rangle|^2}{E_{\Gamma_{25}^+}-E_{\Gamma_{15}^-}} + \frac{|\langle yz|p_y|lv\rangle|^2}{E_{\Gamma_{25}^+}-E_{\Gamma_{25}^-}} \right\}. \tag{3.23}$$

Let

$$H_1 \equiv \frac{\hbar^2}{m_0^2}\sum_{l}{}' \frac{|\langle yz|p_y|\delta_{3l}^-\rangle|^2}{E_{\Gamma_{25}^+}-E_{\Gamma_{15}^-}}, \tag{3.24}$$

where $\delta_{3l}^- \in \Gamma_{15}^-$. There are no matrix elements with δ_{1l}^- and δ_{2l}^-, where $\delta_{1l}^-, \delta_{2l}^-, \delta_{3l}^- \sim x, y, z$:

$$\langle yz|p_y|y\rangle \overset{IC_{2y}}{=} 0,$$
$$\langle yz|p_y|x\rangle \overset{IC_{2z}}{=} 0,$$

but

$$\langle yz|p_y|z\rangle \neq 0.$$

Similarly, let

$$H_2 \equiv \frac{\hbar^2}{m_0^2} \sum_l{}' \frac{|\langle yz|p_y|\varepsilon_{3l}^-\rangle|^2}{E_{\Gamma_{25}^+} - E_{\Gamma_{25}^-}}. \tag{3.25}$$

Let $\Gamma_{25}^- \sim d_a - d_a'$ [e.g., $(yz)_a - (yz)_a'$]; except for operations with inversion, the matrix elements behave as for Γ_{25}^+. Then

$$\langle yz^+|p_y|yz^-\rangle = 0,$$

using C_{2x} and

$$\langle yz^+|p_y|\varepsilon_{2l}^-\rangle = 0,$$

using C_{2y} but

$$\langle yz^+|p_y|xy^-\rangle \neq 0.$$

Thus,

$$M = H_1 + H_2. \tag{3.26}$$

3.2.3.3 *N* Parameter

Finally, we study N:

$$\begin{aligned} N &= \frac{\hbar^2}{m_0^2} \sum_{l\nu\alpha}{}' \frac{\langle yz|p_x|l\nu\alpha\rangle\langle l\nu\alpha|p_y|yz\rangle + \langle yz|p_y|l\nu\alpha\rangle\langle l\nu\alpha|p_x|yz\rangle}{E_{\Gamma_{25}^+} - E_l} \\ &\sim \Gamma_2^- + \Gamma_{12}^- + \Gamma_{15}^- + \Gamma_{25}^-. \end{aligned} \tag{3.27}$$

We will consider the two matrix elements $\langle yz|p_x|l\rangle\langle l|p_y|yz\rangle$ and $\langle yz|p_y|l\rangle\langle l|p_x|yz\rangle$ separately. For Γ_2^-, we have first the matrix element $\langle yz|p_x|\beta_l^-\rangle\langle\beta_l^-|p_y|yz\rangle$, where $\beta_l^- \sim xyz$. Now,

$$\begin{aligned} \langle\beta_l^-|p_y|yz\rangle &= \langle xyz|p_y|zx\rangle \overset{IC_{2x\bar{y}}}{=} \langle xyz|p_x|yz\rangle = \langle\beta_l^-|p_x|yz\rangle, \\ \langle yz|p_y|l\alpha\rangle &= \langle yz|p_y|xyz\rangle = 0. \end{aligned}$$

Hence,

$$\frac{\hbar^2}{m_0^2} \sum_{l\in\Gamma_2^-}{}' \frac{\langle yz|p_x|l\rangle\langle l|p_y|yz\rangle + \langle yz|p_y|l\rangle\langle l|p_x|yz\rangle}{E_{\Gamma_{25}^+} - E_l} = \frac{\hbar^2}{m_0^2} \sum_{l\in\Gamma_2^-}{}' \frac{|\langle yz|p_x|\beta_l^-\rangle|^2}{E_{\Gamma_{25}^+} - E_l} \equiv F. \tag{3.28}$$

For Γ_{12}^-, we need the following:

$$\langle\gamma_2^-|p_y|yz\rangle = \left\langle\left(x^2+\omega^2y^2+\omega z^2\right)^-|p_y|zx\right\rangle \overset{C_{4z}(y\bar{x}z)}{=} -\left\langle\left(y^2+\omega^2x^2+\omega z^2\right)^-|p_x|zy\right\rangle$$

$$= -\frac{1}{\omega^*}\left\langle\left(x^2+\omega y^2+\omega^2z^2\right)^-|p_x|yz\right\rangle = -\omega\langle\gamma_1^-|p_x|yz\rangle,$$

$$\langle\gamma_1^-|p_y|yz\rangle = \frac{1}{\omega^2}\langle\gamma_2^-|p_x|yz\rangle = -\frac{R}{\omega^2} = -\omega R,$$

$$= \left\langle\left(x^2+\omega y^2+\omega^2z^2\right)^-|p_y|zx\right\rangle \overset{C_{2xz}(z\bar{y}x)}{=} -\left\langle\left(z^2+\omega y^2+\omega^2x^2\right)^-|p_y|xz\right\rangle$$

$$= -\frac{1}{\omega}\left\langle\left(x^2+\omega^2y^2+\omega z^2\right)^-|p_y|zx\right\rangle = -\omega^2\left\langle\gamma_2^-|p_y|yz\right\rangle,$$

$$\left\langle zx|p_y|\gamma_1^-\right\rangle = \left\langle zx|p_y|\left(x^2+\omega y^2+\omega^2z^2\right)^-\right\rangle \overset{IC_{2x\bar{y}}(yxz)}{=} -\left\langle zy|p_x|\left(y^2+\omega x^2+\omega^2z^2\right)^-\right\rangle$$

$$= -\frac{1}{\omega^2}\left\langle yz|p_x|(x^2+\omega^2y^2+\omega z^2)^-\right\rangle = -\frac{1}{\omega^2}\langle yz|p_x|\gamma_2^-\rangle,$$

or

$$\left\langle yz|p_x|\gamma_2^-\right\rangle = R = \left\langle yz|p_x|\gamma_1^-\right\rangle,$$

$$\langle\gamma_2^-|p_y|yz\rangle = \frac{1}{\omega}\left\langle\gamma_1^-|p_x|yz\right\rangle,$$

$$\Longrightarrow \quad \langle yz|p_x|\gamma_2^-\rangle\langle\gamma_2^-|p_y|yz\rangle = \frac{1}{\omega}\langle yz|p_x|\gamma_1^-\rangle\langle\gamma_1^-|p_x|yz\rangle = \frac{1}{\omega}|R|^2.$$

Next

$$\langle yz|p_x|\gamma_1^-\rangle\langle\gamma_1^-|p_y|yz\rangle = \langle yz|p_x|\gamma_2^-\rangle\frac{1}{\omega^2}\langle\gamma_2^-|p_x|yz\rangle = \frac{1}{\omega^2}|\langle yz|p_x|\gamma_2^-\rangle|^2 = \frac{1}{\omega^2}|R|^2.$$

Finally, in $\langle yz|p_y|l\rangle\langle l|p_x|yz\rangle$, e.g.,

$$\langle yz|p_y|(x^2+\omega y^2+\omega^2z^2)^-\rangle \overset{C_{2y}}{=} 0.$$

Hence,

$$\frac{\hbar^2}{m_0^2}\sum_\nu \frac{\langle yz|p_x|\gamma_\nu^-\rangle\langle\gamma_\nu^-|p_y|yz\rangle}{E_{\Gamma_{25}^+}-E_l}$$

$$= \frac{\hbar^2}{m_0^2}\left(\frac{1}{\omega}+\frac{1}{\omega^2}\right)\sum_\nu \frac{|R|^2}{E_{\Gamma_{25}^+}-E_l} = (\omega^2+\omega)\frac{\hbar^2}{m_0^2}\sum_\nu \frac{|R|^2}{E_{\Gamma_{25}^+}-E_l} = -\frac{\hbar^2}{m_0^2}\sum_\nu \frac{|R|^2}{E_{\Gamma_{25}^+}-E_l}$$

$$\equiv -G.$$

For $\Gamma_{15}^{-} \sim x, y, z$, we have

$$\langle yz|p_x|\delta_{\nu l}^{-}\rangle\langle\delta_{\nu l}^{-}|p_y|yz\rangle = \langle yz|p_x|x, y, z\rangle\langle x, y, z|p_y|zx\rangle = 0\ \forall\ \nu,$$
$$\langle yz|p_y|\delta_{\nu l}^{-}\rangle\langle\delta_{\nu l}^{-}|p_x|yz\rangle = \langle yz|p_y|z\rangle\langle z|p_x|zx\rangle \neq 0.$$

Indeed,

$$\langle yz|p_x|z\rangle = \langle yx|p_y|x\rangle = \langle xy|p_y|\delta_{1l}^{-}\rangle,$$
$$\langle z|p_y|zx\rangle = \langle x|p_y|xy\rangle = \langle\delta_{1l}^{-}|p_y|xy\rangle,$$

giving

$$\sum_{\Gamma_{15}^{-}} = H_1. \tag{3.29}$$

For $\Gamma_{25}^{-} \sim d_a - d_a'$, we have

$$\langle yz|p_x|\varepsilon_{\nu l}^{-}\rangle\langle\varepsilon_{\nu l}^{-}|p_y|zx\rangle = 0.$$

We already have that

$$\langle yz|p_y|\varepsilon_{1,2l}^{-}\rangle = 0.$$

Thus, the only non-zero element is

$$\langle yz|p_y|\varepsilon_{3l}^{-}\rangle\langle\varepsilon_{3l}^{-}|p_x|zx\rangle \overset{C_{4z}}{=} -\langle yz|p_y|xy^{-}\rangle\langle xy^{-}|p_y|yz\rangle = -H_2.$$

Finally,

$$N = F - G + H_1 - H_2. \tag{3.30}$$

3.2.4 Properties

Rewriting the DKK Hamiltonian, Eq. (3.4), as

$$Mk^2 + \begin{pmatrix} (L-M)k_x^2 & Nk_xk_y & Nk_xk_z \\ Nk_xk_y & (L-M)k_y^2 & Nk_yk_z \\ Nk_xk_z & Nk_yk_z & (L-M)k_z^2 \end{pmatrix},$$

one observes that if $(L - M) = N$, the energy dispersion is isotropic. This is known as the spherical approximation; it was first introduced by Baldereschi and Lipari in their study of acceptor states in cubic semiconductors [30].

Note that a minimal basis (sp^3) tight-binding model will give an isotropic dispersion for small wave vector [24]. In this respect, $k \cdot p$ theory gives a better representation of the symmetry group (if enough interactions are retained). Indeed, if one only retains the lowest interaction (assumed to be of symmetry $\Gamma_2^- \sim xyz$), and defines a Kane parameter analogously to the ZB case [Eq. (2.30)],

$$\frac{2}{m_0}|\langle xyz|p_z|xy\rangle|^2 \equiv E_P,$$

then

$$\begin{aligned} L &= \frac{\hbar^2}{m_0^2}\frac{|\langle xy|p_z|xyz\rangle|^2}{E_{\Gamma_{25}^+} - E_2^-} = \frac{\hbar^2}{2m_0}\left(\frac{E_P}{-E_0}\right), \\ M &= 0, \\ N &= L, \end{aligned}$$

giving an isotropic dispersion (and effective mass).

Table 3.3 DKK parameters in units of $\hbar^2/2m_0$ [2]

	L	M	N	F	G	H_1	H_2
Si	−1.9	−6.7	−7.5	−1.2	−0.4	−6.7	0
Ge	−31.8	−5.1	−32.1	−28.6	−1.4	−5.1	0

The band structures for Si and Ge along $L \to \Gamma \to X$ are plotted in Fig. 3.3, using the parameters from Table 3.3. H_2 was chosen zero since it involves interaction with Γ_{25}^- states and those are far from the valence band (Fig. 3.2). The warping is evident from the difference in dispersion along the two directions. Also, the significant failure of the $k \cdot p$ for large k is evident in the unnaturally large bandwidths.

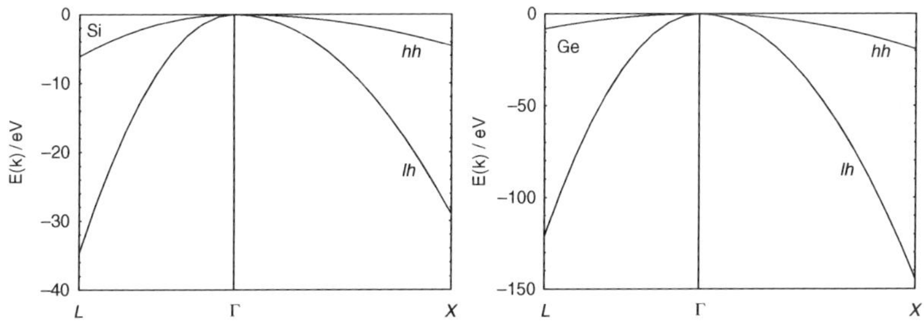

Fig. 3.3 Si and Ge band structures from DKK model

3.3 Six-Band Model for Diamond

We now add the spin-orbit coupling to the previous model.

3.3.1 Hamiltonian

First we recall that the $k \cdot p$ perturbation was present to second order. However, the spin-orbit interaction is nonzero even to first order. There are two contributions in the $k \cdot p$ equation:

$$H_{\mathrm{so}}(\mathbf{k}) = \frac{\hbar}{4m_0^2c^2}(\boldsymbol{\sigma} \times \nabla V) \cdot \mathbf{p} + \frac{\hbar^2}{4m_0^2c^2}(\boldsymbol{\sigma} \times \nabla V) \cdot \mathbf{k} = H_{\mathrm{so}} + H_{\mathrm{so},k}. \quad (3.31)$$

The second, k-dependent spin-orbit term is usually much smaller than the first one [3]; we will ignore it for now. Hence, we will be adding the matrix elements of H_{so} in the Γ_{25}^+ subspace and we rewrite the first term of Eq. (3.31) as

$$H_{\mathrm{so}} = \frac{\hbar}{4m_0^2c^2}(\boldsymbol{\sigma} \times \nabla V) \cdot \mathbf{p} = \frac{\hbar}{4m_0^2c^2}(\nabla V \times \mathbf{p}) \cdot \boldsymbol{\sigma} \equiv H_{\mathrm{s},i}\sigma_i. \quad (3.32)$$

One can show that there exists only one independent spin-orbit matrix element since $\nabla V \times \mathbf{p}$ is an axial vector. Then

$$\langle s|\nabla V \times \mathbf{p}|x\rangle = \langle x|\nabla V \times \mathbf{p}|x\rangle = 0,$$

and

$$\langle x|(\nabla V \times \mathbf{p})_z|y\rangle \neq 0.$$

Historically, one writes

$$\Delta_0 \equiv \frac{3\mathrm{i}\hbar}{4m_0^2c^2}\left\langle x \left| \frac{\partial V}{\partial x}p_y - \frac{\partial V}{\partial y}p_x \right| y \right\rangle. \quad (3.33)$$

Kane [31] wrote it differently, as

$$\Delta \equiv -\frac{3\mathrm{i}\hbar}{4m_0^2c^2}\langle x|(\nabla V \times \mathbf{p})_y|z\rangle,$$

but one can show both forms lead to equivalent H_{so} matrices.

It is necessary to include the spin functions to the basis states leading to a six-dimensional basis: $\varepsilon_1^+ \uparrow, \varepsilon_2^+ \uparrow, \varepsilon_3^+ \uparrow, \varepsilon_1^+ \downarrow, \varepsilon_2^+ \downarrow, \varepsilon_3^+ \downarrow$. In this LS representation, H_{so} is not diagonal. Equivalently, it mixes the states and can lead to a splitting of the degeneracy. Indeed, the six states transform according to (Table B.6)

$$\Gamma_{25}^+ \otimes \Gamma_6^+ = \Gamma_8^+ \oplus \Gamma_7^+,$$

and split into a quadruplet and a doublet. The double-group notation is that of Dresselhaus [32] as opposed to that of Koster et al. [33].

However, if one transforms to the JM_J basis, H_{so} is diagonal. That this is so can be seen from the fact that H_{so} behaves as

$$\mathbf{L}\cdot\mathbf{S}=\frac{1}{2}(J^2-L^2-S^2)=\frac{1}{2}\left[j(j+1)-l(l+1)-s(s+1)\right].$$

Combining the p-like orbitals ($l = 1$) with the spinors ($s = \frac{1}{2}$) leads to $j = \frac{3}{2}$ (four-fold degenerate) or $j = \frac{1}{2}$ (two-fold degenerate) states. We choose phases following Bastard [7] (Table 3.4). Our next task is to transform the Hamiltonian into the new JM_J basis and then attempt to find exact solutions. We do expect, from Kramer's degeneracy, that all the states will be doubly degenerate.

Table 3.4 $|J\ M_J\rangle$ states. The phase factors given agree with, e.g., Bastard [7] and Foreman [34]

$\left\vert\frac{3}{2}\ \frac{3}{2}\right\rangle$	=	$\frac{1}{\sqrt{2}}\vert(x+\mathrm{i}y)\uparrow\rangle$,
$\left\vert\frac{3}{2}\ \frac{1}{2}\right\rangle$	=	$\frac{1}{\sqrt{6}}\vert(x+\mathrm{i}y)\downarrow\rangle-\sqrt{\frac{2}{3}}\vert z\uparrow\rangle$,
$\left\vert\frac{3}{2}\ -\frac{1}{2}\right\rangle$	=	$-\frac{1}{\sqrt{6}}\vert(x-\mathrm{i}y)\uparrow\rangle-\sqrt{\frac{2}{3}}\vert z\downarrow\rangle$,
$\left\vert\frac{3}{2}\ -\frac{3}{2}\right\rangle$	=	$\frac{1}{\sqrt{2}}\vert(x-\mathrm{i}y)\downarrow\rangle$,
$\left\vert\frac{1}{2}\ \frac{1}{2}\right\rangle$	=	$\frac{1}{\sqrt{3}}\vert(x+\mathrm{i}y)\downarrow\rangle+\frac{1}{\sqrt{3}}\vert z\uparrow\rangle$,
$\left\vert\frac{1}{2}\ -\frac{1}{2}\right\rangle$	=	$-\frac{1}{\sqrt{3}}\vert(x-\mathrm{i}y)\uparrow\rangle+\frac{1}{\sqrt{3}}\vert z\downarrow\rangle$,
$\mathrm{i}\vert s\uparrow\rangle$,		
$\mathrm{i}\vert s\downarrow\rangle$.		

3.3.1.1 H_{so}

Consider first the spin-orbit term. We need the Pauli spin matrices:

$$\sigma_x=\begin{pmatrix}0&1\\1&0\end{pmatrix},\sigma_y=\begin{pmatrix}0&-\mathrm{i}\\\mathrm{i}&0\end{pmatrix},\sigma_z=\begin{pmatrix}1&0\\0&-1\end{pmatrix}. \tag{3.34}$$

In the JM_J basis, H_{so} is diagonal; furthermore, the matrix elements are only dependent upon J. For example,

$$\left\langle\frac{3}{2}\frac{3}{2}\left|H_{so}\right|\frac{3}{2}\frac{3}{2}\right\rangle=\left\langle\frac{3}{2}\frac{3}{2}\right|\left|\frac{\hbar}{4m_0^2c^2}\right|(\nabla V\times\mathbf{p})\cdot\boldsymbol{\sigma}\left|\frac{3}{2}\frac{3}{2}\right\rangle$$

$$=\frac{1}{2}\langle(x+\mathrm{i}y)\uparrow|\frac{\hbar}{4m_0^2c^2}(\nabla V\times\mathbf{p})\cdot\boldsymbol{\sigma}|(x+\mathrm{i}y)\uparrow\rangle$$

$$= \frac{1}{2}\langle(x+\mathrm{i}y)|\frac{\hbar}{4m_0^2c^2}(\nabla V \times \mathbf{p})\,|(x+\mathrm{i}y)\rangle \cdot \langle\uparrow|\boldsymbol{\sigma}|\uparrow\rangle$$

$$= \mathrm{i}\langle x|\frac{\hbar}{4m_0^2c^2}(\nabla V \times \mathbf{p})\,|y\rangle \cdot \langle\uparrow|\sigma|\uparrow\rangle$$

$$= \mathrm{i}\langle x|\frac{\hbar}{4m_0^2c^2}(\nabla V \times \mathbf{p})_z\,|y\rangle \equiv \frac{\Delta_0}{3}, \tag{3.35}$$

using Eq. (3.33). Similarly, one finds

$$\left\langle\frac{1}{2}\,\frac{1}{2}|H_{\mathrm{so}}|\frac{1}{2}\,\frac{1}{2}\right\rangle = -\frac{2\Delta_0}{3}.$$

A more formal procedure is to transform H_{so} in LS to H'_{so} in JM_J. In LS, we have

$$H_{\mathrm{so}} = \frac{1}{3}\Delta_0 \begin{pmatrix} \varepsilon_1^+\uparrow & \varepsilon_2^+\uparrow & \varepsilon_3^+\uparrow & \varepsilon_1^+\downarrow & \varepsilon_2^+\downarrow & \varepsilon_3^+\downarrow \\ 0 & -\mathrm{i} & 0 & 0 & 0 & 1 \\ \mathrm{i} & 0 & 0 & 0 & 0 & -\mathrm{i} \\ 0 & 0 & 0 & -1 & \mathrm{i} & 0 \\ 0 & 0 & -1 & 0 & \mathrm{i} & 0 \\ 0 & 0 & -\mathrm{i} & -\mathrm{i} & 0 & 0 \\ 1 & \mathrm{i} & 0 & 0 & 0 & 0 \end{pmatrix}. \tag{3.36}$$

We now need the transformation matrix S connecting the LS states to the JM_J ones. The equations in Table 3.4 can be rewritten in column vector form as

$$|JM_J\rangle^T = S^T \cdot |LS\rangle^T, \tag{3.37}$$

or in row vector form

$$\langle JM_J| = \langle LS| \cdot S, \tag{3.38}$$

which defines the matrix S as

$$S = \begin{pmatrix} & |\frac{3}{2}\,\frac{3}{2}\rangle & |\frac{3}{2}\,\frac{1}{2}\rangle & |\frac{3}{2}\,-\frac{1}{2}\rangle & |\frac{3}{2}\,-\frac{3}{2}\rangle & |\frac{1}{2}\,\frac{1}{2}\rangle & |\frac{1}{2}\,-\frac{1}{2}\rangle \\ |\varepsilon_1^+\uparrow\rangle & \frac{1}{\sqrt{2}} & 0 & -\frac{1}{\sqrt{6}} & 0 & 0 & -\frac{1}{\sqrt{3}} \\ |\varepsilon_2^+\uparrow\rangle & \frac{\mathrm{i}}{\sqrt{2}} & 0 & \frac{\mathrm{i}}{\sqrt{6}} & 0 & 0 & \frac{\mathrm{i}}{\sqrt{3}} \\ |\varepsilon_3^+\uparrow\rangle & 0 & -\sqrt{\frac{2}{3}} & 0 & 0 & \sqrt{\frac{1}{3}} & 0 \\ |\varepsilon_1^+\downarrow\rangle & 0 & \frac{1}{\sqrt{6}} & 0 & \frac{1}{\sqrt{2}} & \frac{1}{\sqrt{3}} & 0 \\ |\varepsilon_2^+\downarrow\rangle & 0 & \frac{\mathrm{i}}{\sqrt{6}} & 0 & -\frac{\mathrm{i}}{\sqrt{2}} & \frac{\mathrm{i}}{\sqrt{3}} & 0 \\ |\varepsilon_3^+\downarrow\rangle & 0 & 0 & -\sqrt{\frac{2}{3}} & 0 & 0 & \sqrt{\frac{1}{3}} \end{pmatrix}. \tag{3.39}$$

Note that this differs from Eq. (10) of Kane [3] in an overall sign and also in a different ordering of the JM_J basis. We can state that his $|JM_J\rangle$ states differ from ours by an overall sign (Table C.2). The spin-orbit Hamiltonian in the JM_J basis is

$$H'_{so} = S^{\dagger} H_{so} S = \frac{\Delta_0}{3} \begin{pmatrix} 1 & 0 & 0 & 0 & 0 & 0 \\ 0 & 1 & 0 & 0 & 0 & 0 \\ 0 & 0 & 1 & 0 & 0 & 0 \\ 0 & 0 & 0 & 1 & 0 & 0 \\ 0 & 0 & 0 & 0 & -2 & 0 \\ 0 & 0 & 0 & 0 & 0 & -2 \end{pmatrix}. \quad (3.40)$$

3.3.1.2 $H_{k \cdot p}$

We now consider the representation of the $k \cdot p$ perturbation in the JM_J basis:

$$H_{rs} = \sum_{l\alpha} \frac{\langle r|H_{k \cdot p}|l\alpha\rangle\langle l\alpha|H_{k \cdot p}|s\rangle}{E_0 - E_{l\alpha}}. \quad (3.41)$$

In principle, one would need to transform all the states. In Eq. (3.41), the combination $|l\alpha\rangle\langle l\alpha|$ is invariant under a unitary transformation; hence, they can be left unchanged. Their energies $E_{l\alpha}$ would, in general, experience spin-orbit splittings. Following DKK, we will neglect the latter effect. The Hamiltonian in the LS basis is

$$\begin{pmatrix} H_{\text{DKK}} & 0 \\ 0 & H_{\text{DKK}} \end{pmatrix}.$$

One can now obtain the Hamiltonian in the JM_J basis by performing a unitary transformation

$$S^{-1} H(\mathbf{k}) S,$$

The small size of the Hamiltonian together with hermiticity makes the direct transformation to the JM_J basis fairly straightforward. Furthermore, if left in terms of the matrix elements H_{ij} of H_{DKK}, the result reflects the choice of Clebsch-Gordan coefficients, i.e., the latter (together with hermiticity) govern the structure of the matrix. For example,

$$\begin{aligned}\left\langle \frac{3}{2}\ \frac{3}{2} |H(\mathbf{k})| \frac{3}{2}\ \frac{3}{2} \right\rangle &= \frac{1}{2} \langle (x + \mathrm{i}y) \uparrow |H(\mathbf{k})| (x + \mathrm{i}y) \uparrow \rangle \\ &= \frac{1}{2}(H_{11} + H_{22}) = \left\langle \frac{3}{2}\ -\frac{3}{2} |H(\mathbf{k})| \frac{3}{2}\ -\frac{3}{2} \right\rangle \equiv P',\end{aligned}$$

$$\left\langle \frac{3}{2}\ \frac{1}{2} |H(\mathbf{k})| \frac{3}{2}\ \frac{1}{2} \right\rangle = \left\langle \frac{1}{\sqrt{6}}(x + \mathrm{i}y) \downarrow -\sqrt{\frac{2}{3}} z \uparrow |H(\mathbf{k})| \frac{1}{\sqrt{6}}(x + \mathrm{i}y) \downarrow -\sqrt{\frac{2}{3}} z \uparrow \right\rangle$$

$$
\begin{aligned}
&= \frac{1}{6}\langle x+\mathrm{i}y|H(\mathbf{k})|x+\mathrm{i}y\rangle + \frac{2}{3}\langle z|H(\mathbf{k})|z\rangle \\
&= \frac{1}{6}(H_{11}+H_{22}) + \frac{2}{3}H_{33} = \left\langle \frac{3}{2}\ -\frac{1}{2}|H(\mathbf{k})|\frac{3}{2}\ -\frac{1}{2}\right\rangle \equiv P'',
\end{aligned}
$$

$$
\begin{aligned}
\left\langle \frac{1}{2}\ \frac{1}{2}|H(\mathbf{k})|\frac{1}{2}\ \frac{1}{2}\right\rangle &= \left\langle \frac{1}{\sqrt{3}}(x+\mathrm{i}y)\downarrow + \frac{1}{\sqrt{3}}z\uparrow |H(\mathbf{k})| \frac{1}{\sqrt{3}}(x+\mathrm{i}y)\downarrow + \frac{1}{\sqrt{3}}z\uparrow\right\rangle \\
&= \frac{1}{3}\langle x+\mathrm{i}y|H(\mathbf{k})|x+\mathrm{i}y\rangle + \frac{1}{3}\langle z|H(\mathbf{k})|z\rangle \\
&= \frac{1}{3}(H_{11}+H_{22}+H_{33}) = \left\langle \frac{1}{2}\ -\frac{1}{2}|H(\mathbf{k})|\frac{1}{2}\ -\frac{1}{2}\right\rangle \equiv P''',
\end{aligned}
$$

$$
\begin{aligned}
\left\langle \frac{3}{2}\ \frac{3}{2}|H(\mathbf{k})|\frac{3}{2}\ \frac{1}{2}\right\rangle &= \left\langle \frac{1}{\sqrt{2}}(x+\mathrm{i}y)\uparrow |H(\mathbf{k})| \frac{1}{\sqrt{6}}(x+\mathrm{i}y)\downarrow - \sqrt{\frac{2}{3}}z\uparrow\right\rangle \\
&= -\frac{1}{\sqrt{3}}\langle x+\mathrm{i}y|H(\mathbf{k})|z\rangle = -\frac{1}{\sqrt{3}}(H_{13}-\mathrm{i}H_{23}) \equiv S_-,
\end{aligned}
$$

$$
\begin{aligned}
\left\langle \frac{3}{2}\ \frac{3}{2}|H(\mathbf{k})|\frac{3}{2}\ -\frac{1}{2}\right\rangle &= \left\langle \frac{1}{\sqrt{2}}(x+\mathrm{i}y)\uparrow |H(\mathbf{k})| -\frac{1}{\sqrt{6}}(x+\mathrm{i}y)\uparrow - \sqrt{\frac{2}{3}}z\downarrow\right\rangle \\
&= -\frac{1}{2\sqrt{3}}\langle x+\mathrm{i}y|H(\mathbf{k})|x-\mathrm{i}y\rangle = -\frac{1}{2\sqrt{3}}(H_{11}-H_{22}-2\mathrm{i}H_{12}) \\
&\equiv -R,
\end{aligned}
$$

$$
\begin{aligned}
\left\langle \frac{3}{2}\ \frac{1}{2}|H(\mathbf{k})|\frac{1}{2}\ \frac{1}{2}\right\rangle &= \left\langle \frac{1}{\sqrt{6}}(x+\mathrm{i}y)\downarrow - \sqrt{\frac{2}{3}}z\uparrow |H(\mathbf{k})| \frac{1}{\sqrt{3}}(x+\mathrm{i}y)\downarrow + \frac{1}{\sqrt{3}}z\uparrow\right\rangle \\
&= \frac{1}{3\sqrt{2}}\langle x+\mathrm{i}y|H(\mathbf{k})|x+\mathrm{i}y\rangle - \frac{\sqrt{2}}{3}\langle z|H(\mathbf{k})|z\rangle \\
&= \frac{1}{3\sqrt{2}}(H_{11}+H_{22}-2H_{33}) \equiv -\sqrt{2}Q,
\end{aligned}
$$

$$
\begin{aligned}
\left\langle \frac{3}{2}\ -\frac{1}{2}|H(\mathbf{k})|\frac{1}{2}\ \frac{1}{2}\right\rangle &= -\frac{1}{3\sqrt{2}}\langle x-\mathrm{i}y|H(\mathbf{k})|z\rangle - \frac{\sqrt{2}}{3}\langle z|H(\mathbf{k})|x+\mathrm{i}y\rangle \\
&= -\frac{1}{\sqrt{2}}(H_{13}+\mathrm{i}H_{23}) = \sqrt{\frac{3}{2}}S_+.
\end{aligned}
$$

In the above derivations, we have used the fact that $H_{ij} = H_{ji}$.

3.3.1.3 $H(\mathbf{k}) + H_{so}$

The total Hamiltonian is given in Table 3.5. Note that there are a few sign differences between our Table 3.5 and Eq. (65) of DKK. These can be accounted for by differences in the phases of two of the basis functions (Table C.2), even though DKK did not give theirs explicitly. For example, in our case, the (1,5) and (2,6) matrix elements have the same sign; in DKK's case, they have opposite signs. Equation (4.3.14) in S. L. Chuang's book [15] (and Eq. (20) of Elliott [35] but there are other problems here as we will see below) also have the same structure as DKK's; in the former case, the basis functions are given and the sign difference in two basis functions is consistent with our interpretation of DKK's basis. Our Hamiltonian has the same structure as that of, e.g., Kane [3] (not surprisingly since our bases only differ by an overall negative sign) and, in fact, of Luttinger–Kohn [6] as well, as we will see later for both cases. Nevertheless, the subsequent evaluation of the secular determinant leads to the same characteristic equation. Our Hamiltonian does agree with that of Kane [3] upon re-ordering the basis. Before we treat the approximate eigenvalues, we give here the explicit and independent matrix elements of the 6×6 Hamiltonian given in Table 3.5, using the notation of Elliot [35]:

$$
\begin{aligned}
(1,1) = (4,4) &= \frac{H_{11} + H_{22}}{2} = \frac{1}{2}(L+M)(k_x^2 + k_y^2) + Mk_z^2 \equiv \frac{1}{2}P - \frac{\hbar^2 k^2}{2m_0},\\
(1,2) &= -\frac{H_{13} - \mathrm{i}H_{23}}{\sqrt{3}} = -\frac{N}{\sqrt{3}}(k_x - \mathrm{i}k_y)k_z \equiv R,\\
(1,3) &= -\frac{H_{11} - H_{22} - 2\mathrm{i}H_{12}}{2\sqrt{3}} = -\frac{1}{2\sqrt{3}}\left[(L-M)(k_x^2 - k_y^2) - 2\mathrm{i}Nk_xk_y\right] \equiv S,\\
(2,2) = (3,3) &= \frac{H_{11} + H_{22} + 4H_{33}}{6} = \frac{1}{6}\left[(L+5M)(k_x^2 + k_y^2) + 2(2L+M)k_z^2\right]\\
&\equiv \frac{1}{6}P + \frac{2}{3}Q - \frac{\hbar^2 k^2}{2m_0},\\
(2,5) &= \frac{H_{11} + H_{22} - 2H_{33}}{3\sqrt{2}} = \frac{1}{3\sqrt{2}}\left[(L-M)(k_x^2 + k_y^2) - 2(L-M)k_z^2\right]\\
&\equiv \frac{1}{3\sqrt{2}}(P - 2Q),\\
(5,5) = (6,6) &= \frac{H_{11} + H_{22} + H_{33}}{3} - \Delta_0 = \frac{L+2M}{3}k^2 - \Delta_0\\
&\equiv \frac{1}{3}(P+Q) - \Delta_0 - \frac{\hbar^2 k^2}{2m_0}.
\end{aligned}
$$

The above can be summarized into the Elliott–Luttinger–Kohn Hamiltonian in Table 3.6, after adding the diagonal free-electron energy term. Note, however, that our matrix differs slightly from the original Elliott [35] and LK [6] ones due to the different basis functions used. Indeed, if we correct for the difference in basis function, we do reproduce the LK Hamiltonian, Eq. (V.13); but we do not reproduce that of Elliott. Our P and Q are identical to the latter but not R and S. Since we have already used L, M, N as DKK parameters, we have introduced a new notation to replace the k-dependent L, M of LK; we, henceforth, refer to the LK k-dependent matrix elements as P_L, Q_L, L_L, M_L. The relationships between ours and LK's are:

Table 3.5 DKK Hamiltonian in JM_J basis. The top matrix is as given by DKK [2], the second emphasizes the structure due solely to the Clebsch-Gordan coefficients and hermiticity (thus the lower half of the matrix is not explicitly written out). H_{ij} are matrix elements of H_{DKK}

$$
\begin{pmatrix}
|\frac{3}{2}\,\frac{3}{2}\rangle & |\frac{3}{2}\,\frac{1}{2}\rangle & |\frac{3}{2}\,-\frac{1}{2}\rangle & |\frac{3}{2}\,-\frac{3}{2}\rangle & |\frac{1}{2}\,\frac{1}{2}\rangle & |\frac{1}{2}\,-\frac{1}{2}\rangle \\
\frac{H_{11}+H_{22}}{2} & -\frac{H_{13}-\mathrm{i}H_{23}}{\sqrt{3}} & -\frac{H_{11}-H_{22}-2\mathrm{i}H_{12}}{2\sqrt{3}} & 0 & \frac{H_{13}-\mathrm{i}H_{23}}{\sqrt{6}} & -\frac{H_{11}-H_{22}-2\mathrm{i}H_{12}}{\sqrt{6}} \\
-\frac{H_{13}+\mathrm{i}H_{23}}{\sqrt{3}} & \frac{4H_{33}+H_{11}+H_{22}}{6} & 0 & \frac{H_{11}-H_{22}-2\mathrm{i}H_{12}}{2\sqrt{3}} & \frac{H_{11}+H_{22}-2H_{33}}{3\sqrt{2}} & \frac{H_{13}-\mathrm{i}H_{23}}{\sqrt{2}} \\
-\frac{H_{11}-H_{22}+2\mathrm{i}H_{12}}{2\sqrt{3}} & 0 & \frac{4H_{33}+H_{11}+H_{22}}{6} & -\frac{H_{13}-\mathrm{i}H_{23}}{\sqrt{3}} & -\frac{H_{13}+\mathrm{i}H_{23}}{\sqrt{2}} & \frac{H_{11}+H_{22}-2H_{33}}{3\sqrt{2}} \\
0 & \frac{H_{11}-H_{22}+2\mathrm{i}H_{12}}{2\sqrt{3}} & -\frac{H_{13}+\mathrm{i}H_{23}}{\sqrt{3}} & \frac{H_{11}+H_{22}}{2} & \frac{H_{11}-H_{22}+2\mathrm{i}H_{12}}{\sqrt{6}} & \frac{H_{13}+\mathrm{i}H_{23}}{\sqrt{6}} \\
\frac{H_{13}+\mathrm{i}H_{23}}{\sqrt{6}} & \frac{H_{11}+H_{22}-2H_{33}}{3\sqrt{2}} & -\frac{H_{13}-\mathrm{i}H_{23}}{\sqrt{2}} & \frac{H_{11}-H_{22}-2\mathrm{i}H_{12}}{\sqrt{6}} & \frac{H_{11}+H_{22}+H_{33}}{3}-\Delta_0 & 0 \\
-\frac{H_{11}-H_{22}+2\mathrm{i}H_{12}}{\sqrt{6}} & \frac{H_{13}+\mathrm{i}H_{23}}{\sqrt{2}} & \frac{H_{11}+H_{22}-2H_{33}}{3\sqrt{2}} & \frac{H_{13}-\mathrm{i}H_{23}}{\sqrt{6}} & 0 & \frac{H_{11}+H_{22}+H_{33}}{3}-\Delta_0
\end{pmatrix}
$$

$$
\equiv
\begin{pmatrix}
|\frac{3}{2}\,\frac{3}{2}\rangle & |\frac{3}{2}\,\frac{1}{2}\rangle & |\frac{3}{2}\,-\frac{1}{2}\rangle & |\frac{3}{2}\,-\frac{3}{2}\rangle & |\frac{1}{2}\,\frac{1}{2}\rangle & |\frac{1}{2}\,-\frac{1}{2}\rangle \\
P' & S_- & -R & 0 & -\frac{1}{\sqrt{2}}S_- & -\sqrt{2}R \\
 & P'' & 0 & R & -\sqrt{2}Q & -\sqrt{\frac{3}{2}}S_- \\
 & & P'' & S_- & \sqrt{\frac{3}{2}}S_+ & -\sqrt{2}Q \\
 & \dagger & & P' & \sqrt{2}R^* & \frac{1}{\sqrt{2}}S_+ \\
 & & & & P'''-\Delta_0 & 0 \\
 & & & & & P'''-\Delta_0
\end{pmatrix}
$$

Table 3.6 DKK Hamiltonian in JM_J basis using Elliott-LK notation. Note differences compared to Eq. (20) of Elliott and Eq. (V.13) of LK. The free-electron energy was added to the DKK Hamiltonian

$$
\begin{pmatrix}
|\frac{3}{2}\,\frac{3}{2}\rangle & |\frac{3}{2}\,\frac{1}{2}\rangle & |\frac{3}{2}\,-\frac{1}{2}\rangle & |\frac{3}{2}\,-\frac{3}{2}\rangle & |\frac{1}{2}\,\frac{1}{2}\rangle & |\frac{1}{2}\,-\frac{1}{2}\rangle \\
\frac{1}{2}P & R & S & 0 & -\frac{1}{\sqrt{2}}R & \sqrt{2}S \\
R^* & \frac{1}{6}P+\frac{2}{3}Q & 0 & -S & \frac{1}{3\sqrt{2}}(P-2Q) & -\sqrt{\frac{3}{2}}R \\
S^* & 0 & \frac{1}{6}P+\frac{2}{3}Q & R & \sqrt{\frac{3}{2}}R^* & \frac{1}{3\sqrt{2}}(P-2Q) \\
0 & -S^* & R^* & \frac{1}{2}P & -\sqrt{2}S^* & -\frac{1}{\sqrt{2}}R^* \\
-\frac{1}{\sqrt{2}}R^* & \frac{1}{3\sqrt{2}}(P-2Q) & \sqrt{\frac{3}{2}}R & -\sqrt{2}S & \frac{1}{3}(P+Q)-\Delta_0 & 0 \\
\sqrt{2}S^* & -\sqrt{\frac{3}{2}}R^* & \frac{1}{3\sqrt{2}}(P-2Q) & -\frac{1}{\sqrt{2}}R & 0 & \frac{1}{3}(P+Q)-\Delta_0
\end{pmatrix}
$$

with

$$
\begin{aligned}
P &= (L+M)(k_x^2+k_y^2) + 2Mk_z^2 + \frac{\hbar^2k^2}{m_0}, \\
Q &= M(k_x^2+k_y^2) + Lk_z^2 + \frac{\hbar^2k^2}{2m_0}, \\
R &= -\frac{N}{\sqrt{3}}(k_x - \mathrm{i}k_y)k_z, \\
S &= -\frac{1}{2\sqrt{3}}\left[(L-M)(k_x^2-k_y^2) - 2\mathrm{i}Nk_xk_y\right].
\end{aligned}
$$

$$
\begin{aligned}
P &= (L+M)(k_x^2+k_y^2) + 2Mk_z^2 + \frac{\hbar^2k^2}{m_0} = (A_L+B_L)(k_x^2+k_y^2) + 2B_Lk_z^2 \equiv P_L, \\
Q &= M(k_x^2+k_y^2) + Lk_z^2 + \frac{\hbar^2k^2}{2m_0} = Q_L, \\
R &= -\frac{N}{\sqrt{3}}(k_x - \mathrm{i}k_y)k_z = -\mathrm{i}L_L, \\
S &= -\frac{1}{2\sqrt{3}}\left[(L-M)(k_x^2-k_y^2) - 2\mathrm{i}Nk_xk_y\right] = -M_L.
\end{aligned}
\tag{3.42}
$$

This form of the Hamiltonian is not widely used; exceptions include, e.g., in the paper by Sercel and Vahala [36].

While this Hamiltonian is often called the Luttinger–Kohn Hamiltonian, we note that it appears to have first been written down by Elliott in a related form [35]. There is also another form of this same Hamiltonian, e.g., in S. L. Chuang's book [15], in terms of the Luttinger parameters. However, since we have not yet introduced the Luttinger parameters, we will present this other form later. Note further that the quantities P, Q, R, S are different from the definitions in our canonical form in Table 3.5.

3.3.2 DKK Solution

The full 6×6 determinant cannot be solved analytically. If one assumes that the hh/lh states can be decoupled from the spin-split-off hole (sh) states (as they are for $\mathbf{k} = \mathbf{0}$), then one has a 2×2 and a 4×4 block. The spin-split-off block can obviously be diagonalized exactly. The 4×4 block, it turns out, can also be diagonalized exactly here due to the fact that all the states are doubly degenerate from time-reversal symmetry in an inversion-symmetric structure (Kramer's theorem).

We first obtain the spin-split-off dispersion. The 2×2 block is given by Eq. (3.43).

$$\begin{vmatrix} \frac{H_{11}+H_{22}+H_{33}}{3} - \lambda - \Delta_0 & 0 \\ 0 & \frac{H_{11}+H_{22}+H_{33}}{3} - \lambda - \Delta_0 \end{vmatrix} = 0. \tag{3.43}$$

Not surprisingly, the matrix is already diagonal, reflecting the double degeneracy expected. The eigenvalue is

$$\begin{aligned} \lambda &= -\Delta_0 + \frac{H_{11} + H_{22} + H_{33}}{3} \\ &= -\Delta_0 + \frac{1}{3}(L + 2M)k^2. \end{aligned}$$

The dispersion is

$$E_{sh}(\mathbf{k}) = \frac{\hbar^2 k^2}{2m_0} - \Delta_0 + \frac{1}{3}(L + 2M)k^2 \equiv -\Delta_0 + Ak^2, \tag{3.44}$$

where

$$A \equiv \frac{1}{3}(L + 2M) + \frac{\hbar^2}{2m_0}. \tag{3.45}$$

We now evaluate the 4×4 determinant in a straightforward manner:

$$\begin{vmatrix} \frac{H_{11}+H_{22}}{2} - \lambda & -\frac{H_{13}-\mathrm{i}H_{23}}{\sqrt{3}} & -\frac{H_{11}-H_{22}-2\mathrm{i}H_{12}}{2\sqrt{3}} & 0 \\ -\frac{H_{13}+\mathrm{i}H_{23}}{\sqrt{3}} & \frac{4H_{33}+H_{11}+H_{22}}{6} - \lambda & 0 & \frac{H_{11}-H_{22}-2\mathrm{i}H_{12}}{2\sqrt{3}} \\ -\frac{H_{11}-H_{22}+2\mathrm{i}H_{12}}{2\sqrt{3}} & 0 & \frac{4H_{33}+H_{11}+H_{22}}{6} - \lambda & -\frac{H_{13}-\mathrm{i}H_{23}}{\sqrt{3}} \\ 0 & \frac{H_{11}-H_{22}+2\mathrm{i}H_{12}}{2\sqrt{3}} & -\frac{H_{13}+\mathrm{i}H_{23}}{\sqrt{3}} & \frac{H_{11}+H_{22}}{2} - \lambda \end{vmatrix}$$

$$\begin{aligned} &= \left[\frac{1}{2}(H_{11} + H_{22}) - \lambda\right]^2 \left[\frac{1}{6}(4H_{33} + H_{11} + H_{22}) - \lambda\right]^2 \\ &\quad -\frac{1}{6}\left[\frac{1}{6}(4H_{33} + H_{11} + H_{22}) - \lambda\right]\left[\frac{1}{2}(H_{11} + H_{22}) - \lambda\right] \\ &\quad \times\left\{(H_{11} - H_{22})^2 + 4(H_{12}^2 + H_{13}^2 + H_{23}^2)\right\} \end{aligned}$$

$$+\frac{1}{18}(H_{13}^2 + H_{23}^2)\left[(H_{11} - H_{22})^2 + 4H_{12}^2\right] + \frac{1}{9}(H_{13}^2 + H_{23}^2)^2$$
$$+\frac{1}{144}\left[(H_{11} - H_{22})^2 + 4H_{12}^2\right]^2.$$

Making use of the double degeneracy of the states, we recognize that the secular equation can be rewritten as

$$(x - a)^2 = 0,$$

where

$$x = \left[\frac{1}{6}(4H_{33} + H_{11} + H_{22}) - \lambda\right]\left[\frac{1}{2}(H_{11} + H_{22}) - \lambda\right],$$
$$a = \frac{1}{12}\left[(H_{11} - H_{22})^2 + 4H_{12}^2 + 4(H_{13}^2 + H_{23}^2)\right].$$

Then, writing $x = a$, we have

$$\left[\frac{1}{6}(4H_{33} + H_{11} + H_{22}) - \lambda\right]\left[\frac{1}{2}(H_{11} + H_{22}) - \lambda\right]$$
$$= \frac{1}{12}\left[(H_{11} - H_{22})^2 + 4H_{12}^2 + 4(H_{13}^2 + H_{23}^2)\right],$$

or

$$\lambda^2 - \left[\frac{4}{3}(H_{11} + H_{22} + H_{33})\right]\lambda + \frac{1}{12}(4H_{33} + H_{11} + H_{22})(H_{11} + H_{22})$$
$$-\frac{1}{12}\left[(H_{11} - H_{22})^2 + 4(H_{12}^2 + H_{13}^2 + H_{23}^2)\right] = 0. \quad (3.46)$$

The eigenvalues are

$$\lambda = \frac{1}{3}(H_{11} + H_{22} + H_{33}) \pm \frac{1}{2}\left\{\frac{4}{9}(H_{11} + H_{22} + H_{33})^2\right.$$
$$\left.+\frac{1}{3}\left[(H_{11} - H_{22})^2 + 4(H_{12}^2 + H_{13}^2 + H_{23}^2) - (H_{11} + H_{22})(4H_{33} + H_{11} + H_{22})\right]\right\}^{1/2}$$
$$= \frac{1}{3}(L + 2M)k^2 \pm \left\{\frac{1}{9}(L + 2M)^2k^4 + \frac{N^2}{3}(k_x^2k_y^2 + k_y^2k_z^2 + k_z^2k_x^2)\right.$$
$$\left.+\frac{1}{12}\left[(H_{11} - H_{22})^2 - (H_{11} + H_{22})(4H_{33} + H_{11} + H_{22})\right]\right\}^{1/2}. \quad (3.47)$$

It remains to simplify the last term of Eq. (3.47). Note that

$$\frac{1}{12}\left[(H_{11} - H_{22})^2 - (H_{11} + H_{22})(4H_{33} + H_{11} + H_{22})\right]$$

$$
\begin{aligned}
&= \frac{1}{12}\left[(H_{11}-H_{22})^2-(H_{11}+H_{22})^2-4H_{33}(H_{11}+H_{22})\right] \\
&= -\frac{1}{3}\left[H_{11}H_{22}+H_{22}H_{33}+H_{33}H_{11}\right] \\
&= -\frac{1}{3}\left[(L^2+3M^2)(k_x^2k_y^2+k_y^2k_z^2+k_z^2k_x^2)+M^2(k_x^4+k_y^4+k_z^4)\right. \\
&\quad \left.+2ML(k_x^4+k_y^4+k_z^4+k_x^2k_y^2+k_y^2k_z^2+k_z^2k_x^2)\right], \qquad (3.48)
\end{aligned}
$$

while

$$
\begin{aligned}
\frac{1}{9}(L+2M)^2k^4 &= \frac{1}{9}(L-M)^2k^4+M^2k^4+\frac{2}{3}M(L-M)k^4 \\
&= \frac{1}{9}(L-M)^2k^4+\frac{M}{3}(2L+M)k^4. \qquad (3.49)
\end{aligned}
$$

Combining Eq. (3.48) with the second term on the right-hand side of Eq. (3.49) and using

$$
k^4=(k_x^2+k_y^2+k_z^2)^2=(k_x^4+k_y^4+k_z^4)+2(k_x^2k_y^2+k_y^2k_z^2+k_z^2k_x^2),
$$

we finally get

$$
\begin{aligned}
&\frac{1}{3}\Big\{M(2L+M)k^4-[(L^2+3M^2)(k_{xy}+k_{yz}+k_{zx})+M^2(k_x^4+k_y^4+k_z^4) \\
&+2ML(k_x^4+k_y^4+k_z^4+k_{xy}+k_{yz}+k_{zx})]\Big\} = \frac{1}{3}(L-M)^2(k_{xy}+k_{yz}+k_{zx}),
\end{aligned}
$$

where we have written

$$
k_{xy}+k_{yz}+k_{zx} \equiv k_x^2k_y^2+k_y^2k_z^2+k_z^2k_x^2.
$$

Hence, the dispersion is

$$
\begin{aligned}
E(\mathbf{k}) &= \frac{1}{3}(L+2M)k^2 \pm \left\{\frac{1}{9}(L-M)^2k^4+\frac{1}{3}\left[N^2-(L-M)^2\right](k_x^2k_y^2+k_y^2k_z^2+k_z^2k_x^2)\right\}^{1/2} \\
&\equiv Ak^2 \pm \left[B^2k^4+C^2(k_x^2k_y^2+k_y^2k_z^2+k_z^2k_x^2)\right]^{1/2}, \qquad (3.50)
\end{aligned}
$$

where

$$
\begin{aligned}
A &= \frac{1}{3}(L+2M)+\frac{\hbar^2}{2m_0}, \\
B &= \frac{1}{3}(L-M), \qquad (3.51)
\end{aligned}
$$

$$C^2 = \frac{1}{3}\left[N^2 - (L-M)^2\right].$$

Note that these solutions are identical to Eqs. (3.13) and (3.14).

3.3.3 Kane Solution

Kane [3] sought to block diagonalize the Hamiltonian as much as possible via unitary transformations before looking for analytic solutions. The key is to recognize the structure of the Hamiltonian matrix, here due to time-reversal symmetry. The Kramers operator for a crystal with inversion is

$$\mathcal{K} = -\mathrm{i}\sigma_y \mathcal{C} I, \tag{3.52}$$

where I is the inversion operator and $\mathcal{C}$ is complex conjugation. The basis can then be written as

$$\phi_1, \phi_2, \phi_3, \mathcal{K}\phi_1, \mathcal{K}\phi_2, \mathcal{K}\phi_3. \tag{3.53}$$

Here, they correspond to

$$\left|\frac{3}{2}\,\frac{3}{2}\right\rangle, \left|\frac{3}{2}\,\frac{1}{2}\right\rangle, \left|\frac{1}{2}\,\frac{1}{2}\right\rangle, \left|\frac{3}{2}\,-\frac{3}{2}\right\rangle, \left|\frac{3}{2}\,-\frac{1}{2}\right\rangle, \left|\frac{1}{2}\,-\frac{1}{2}\right\rangle.$$

Since $\mathcal{K}$ is equivalent to the time-reversed symmetry operator, the states ϕ_i and $\mathcal{K}\phi_i$ differ in their m values: $\pm m$, respectively. Luttinger and Kohn [6], on the other hand, defined the time-reversal operator as

$$\mathcal{K} = \sigma_y \mathcal{C}. \tag{3.54}$$

The $|J\ +|m|\rangle$ states of LK differ from ours; since $\mathcal{K}$ is also different, the $|J\ -|m|\rangle$ states differ as well. We have,

$$\begin{aligned}
\langle \mathcal{K}\phi_i | \mathcal{K}\phi_j \rangle &= \langle \phi_i | \phi_j \rangle^*, \\
\langle \mathcal{K}\phi_i | H | \mathcal{K}\phi_j \rangle &= \langle \phi_i | H | \phi_j \rangle^*, \\
\langle \phi_i | H | \mathcal{K}\phi_j \rangle &= -\langle \phi_j | H | \mathcal{K}\phi_i \rangle.
\end{aligned} \tag{3.55}$$

We, therefore, reorder the transformation matrix S from Eq. (3.39) into U defined by

$$U = \begin{pmatrix} & \left|\frac{3}{2}\ \frac{3}{2}\right\rangle & \left|\frac{3}{2}\ \frac{1}{2}\right\rangle & \left|\frac{1}{2}\ \frac{1}{2}\right\rangle & \left|\frac{3}{2}\ -\frac{3}{2}\right\rangle & \left|\frac{3}{2}\ -\frac{1}{2}\right\rangle & \left|\frac{1}{2}\ -\frac{1}{2}\right\rangle \\ |\varepsilon_1^+\uparrow\rangle & \frac{1}{\sqrt{2}} & 0 & 0 & 0 & -\frac{1}{\sqrt{6}} & -\frac{1}{\sqrt{3}} \\ |\varepsilon_2^+\uparrow\rangle & \frac{\mathrm{i}}{\sqrt{2}} & 0 & 0 & 0 & \frac{\mathrm{i}}{\sqrt{6}} & \frac{\mathrm{i}}{\sqrt{3}} \\ |\varepsilon_3^+\uparrow\rangle & 0 & -\sqrt{\frac{2}{3}} & \frac{1}{\sqrt{3}} & 0 & 0 & 0 \\ |\varepsilon_1^+\downarrow\rangle & 0 & \frac{1}{\sqrt{6}} & \frac{1}{\sqrt{3}} & \frac{1}{\sqrt{2}} & 0 & 0 \\ |\varepsilon_2^+\downarrow\rangle & 0 & \frac{\mathrm{i}}{\sqrt{6}} & \frac{\mathrm{i}}{\sqrt{3}} & -\frac{\mathrm{i}}{\sqrt{2}} & 0 & 0 \\ |\varepsilon_3^+\downarrow\rangle & 0 & 0 & 0 & 0 & -\sqrt{\frac{2}{3}} & \frac{1}{\sqrt{3}} \end{pmatrix}. \quad (3.56)$$

Our U differs from Kane's Eq. (10) [3] by an overall sign but otherwise has the same ordering. Thus, the Hamiltonian can be written as (using Kane's basis ordering)

$$H = \begin{pmatrix} G & \Gamma \\ -\Gamma^* & G^* \end{pmatrix}, \quad (3.57)$$

where $\Gamma^T = -\Gamma$. Note that our U in Eq. (3.56) is of this form too:

$$U = \begin{pmatrix} S & R \\ -R^* & S^* \end{pmatrix}.$$

Comparing Eq. (3.57) with Table 3.5 and accounting for the ordering difference, we find

$$G = \begin{pmatrix} \frac{H_{11}+H_{22}}{2} & \frac{-H_{13}+\mathrm{i}H_{23}}{\sqrt{3}} & \frac{H_{13}-\mathrm{i}H_{23}}{\sqrt{6}} \\ \frac{-H_{13}-\mathrm{i}H_{23}}{\sqrt{3}} & \frac{H_{11}+H_{22}+4H_{33}}{6} & \frac{H_{11}+H_{22}-2H_{33}}{3\sqrt{2}} \\ \frac{H_{13}+\mathrm{i}H_{23}}{\sqrt{6}} & \frac{H_{11}+H_{22}-2H_{33}}{3\sqrt{2}} & \frac{H_{11}+H_{22}+H_{33}}{3} - \Delta_0 \end{pmatrix}, \quad (3.58)$$

$$\Gamma = \begin{pmatrix} 0 & \frac{-H_{11}+H_{22}+2\mathrm{i}H_{12}}{2\sqrt{3}} & \frac{-H_{11}+H_{22}+2\mathrm{i}H_{12}}{\sqrt{6}} \\ \frac{H_{11}-H_{22}-2\mathrm{i}H_{12}}{2\sqrt{3}} & 0 & \frac{H_{13}-\mathrm{i}H_{23}}{\sqrt{2}} \\ \frac{H_{11}-H_{22}-2\mathrm{i}H_{12}}{\sqrt{6}} & \frac{-H_{13}+\mathrm{i}H_{23}}{\sqrt{2}} & 0 \end{pmatrix}. \quad (3.59)$$

Following Kane, we introduce

$$\begin{aligned} X\mathrm{e}^{\mathrm{i}\chi} &= -H_{11} + H_{22} + 2\mathrm{i}H_{12}, \\ Y\mathrm{e}^{\mathrm{i}\eta} &= H_{13} - \mathrm{i}H_{23}, \\ Z &= H_{11} + H_{22} - 2H_{33}. \end{aligned} \quad (3.60)$$

One can then show that the following transformation

$$U_1 = \begin{pmatrix} S_1 & R_1 \\ -R_1^* & S_1^* \end{pmatrix},$$

with

$$S_1 = \frac{1}{\sqrt{2}} \begin{pmatrix} e^{i\left(\frac{\chi+\eta}{2}-\frac{\pi}{4}\right)} & 0 & 0 \\ 0 & e^{i\left(\frac{\chi-\eta}{2}-\frac{\pi}{4}\right)} & 0 \\ 0 & 0 & e^{i\left(\frac{\chi-\eta}{2}-\frac{\pi}{4}\right)} \end{pmatrix}, \tag{3.61}$$

$$R_1 = \frac{1}{\sqrt{2}} \begin{pmatrix} -e^{i\left(\frac{\chi+\eta}{2}-\frac{\pi}{4}\right)} & 0 & 0 \\ 0 & -e^{i\left(\frac{\chi-\eta}{2}-\frac{\pi}{4}\right)} & 0 \\ 0 & 0 & -e^{i\left(\frac{\chi-\eta}{2}-\frac{\pi}{4}\right)} \end{pmatrix}, \tag{3.62}$$

transforms the Hamiltonian into

$$H' = \begin{pmatrix} G' & \Gamma' \\ -\Gamma'^* & G'^* \end{pmatrix}, \tag{3.63}$$

where

$$G' = \begin{pmatrix} \frac{H_{11}+H_{22}}{2} & \frac{-Y+iX/2}{\sqrt{3}} & \frac{Y+iX}{\sqrt{6}} \\ \frac{Y-iX/2}{\sqrt{3}} & \frac{H_{11}+H_{22}+4H_{33}}{6} & \frac{Z/3+iY\cos(\chi-2\eta)}{\sqrt{2}} \\ \frac{Y-iX}{\sqrt{6}} & \frac{Z/3-iY\cos(\chi-2\eta)}{\sqrt{2}} & \frac{H_{11}+H_{22}+H_{33}}{3} - \Delta_0 \end{pmatrix}, \tag{3.64}$$

$$\Gamma' = \begin{pmatrix} 0 & 0 & 0 \\ 0 & 0 & \frac{Y\sin(\chi-2\eta)}{\sqrt{2}} \\ 0 & \frac{-Y\sin(\chi-2\eta)}{\sqrt{2}} & 0 \end{pmatrix}. \tag{3.65}$$

Obviously, Γ' is zero for certain values of the wave vector; in that case, the Hamiltonian is block diagonal.

3.4 Wurtzite

Another important type of band structure is that of wurtzite-type materials.

3.4.1 Overview

The band structure and optical properties were first studied in the late 1950s [37–40]. Pikus [41] obtained the valence band using the method of invariants and provided a six-band model. Gutsche and Jahne [42] wrote down a 12-band model for the valence band at $\mathbf{k} = \mathbf{0}$. The bulk band structure regained attention in the 1990s due to the growth of high-quality epitaxial GaN and AlN. Thus, Sirenko and coworkers extended the study of the valence band [43–45] while Chuang and coworkers

[46, 47] introduced a Kane model and also used perturbation theory to compare to the invariant method. A coupled conduction-valence band model was introduced by Lew Yan Voon et al. [48], in which the conduction state was folded into the valence block and the latter diagonalized exactly. A treatment where they are treated on an equal footing was by Andreev and O'Reilly in 2000 [49] and used subsequently [50, 51]. Band parameters for a number of wurtzite (WZ) materials have now been obtained [45, 46, 52, 53]. There have also been a number of studies for nanostructures [43, 46, 47, 54–58].

Some basic differences compared to the ZB problem are:

- The symmetry is hexagonal [C^4_{6v} (6 mm)] instead of cubic.
- The six-band valence Hamiltonian now has 10 parameters (seven A_i and three Δ_i) instead of four.
- The bulk Hamiltonian with spin-orbit interaction cannot be diagonalized exactly at $\mathbf{k} = \mathbf{0}$ using only symmetry-adapted functions.

The WZ semiconductors are generally large-band-gap materials (except for InN with a band gap of ~ 0.7 eV [59, 60]) with a direct gap at the Γ point. The presence of both a crystal-field splitting (compared to a cubic structure) and spin-orbit interaction leads to a complex valence-band structure consisting of three doubly-degenerate bands at the Γ point; they are known as the A, B, and C bands.

Just as for the cubic semiconductors, the band structure near the Γ point will be of interest. One complication is that basis states that transforms as $s, p, d, ...$ will no longer diagonalize the Hamiltonian even at that point. This is clear from the irreducible basis functions given in Table B.10.

3.4.2 Basis States

Different models for the band structure of WZ materials have been considered, corresponding to different basis functions. In Table C.9, we list together the main choices of basis states. We have maintained the notation in the original papers, whereby the basis functions for, e.g., Chuang and Chang [46] are given in capital letters whereas the ones for Gutsche and Jahne [42] are in small letters. As noted by Chuang and Chang [46], other work have been less explicit in their basis functions and, as a result, discrepancies have arisen among the various Hamiltonians to be found in the literature. For example, Pikus [1, 61] and Sirenko et al. [43] only mention that they used the LS and JM_J basis without giving the full representations. As we did for ZB materials, we will provide as complete a picture as we can regarding the various Hamiltonians and the corresponding basis functions.

3.4.3 Chuang–Chang Hamiltonian

The presentation of the Hamiltonian within perturbation theory was done by Chuang and Chang (CC) [46].

Both spin-orbit and crystal-field splitting are important for WZ materials; hence, it is common to use basis functions that reflect this. The Hamiltonian consists of two terms–one with constant energies, and one k-dependent:

$$H(\mathbf{k}) = H_{k=0} + D_{6\times 6}(\mathbf{k}), \tag{3.66}$$

where the D matrix was defined in Eq. (3.6). As WZ differs from ZB by the inequivalence of the z axis from x and y, one can use a basis that resembles the JM_J states for cubic crystals but with the Z function separate. This is the so-called u basis and is given in Table 3.7.

Table 3.7 u basis states for the valence bands of wurtzite [46]

				J_z
Γ_5	$\lvert u_1\rangle$	$-\frac{1}{\sqrt{2}}\lvert(X+\mathrm{i}Y)\uparrow\rangle$	$Y_{1\ 1}\uparrow$	$\frac{3}{2}$
Γ_5	$\lvert u_2\rangle$	$\frac{1}{\sqrt{2}}\lvert(X-\mathrm{i}Y)\uparrow\rangle$	$Y_{1\ -1}\uparrow$	$-\frac{1}{2}$
Γ_1	$\lvert u_3\rangle$	$\lvert Z\uparrow\rangle$	$Y_{1\ 0}\uparrow$	$\frac{1}{2}$
Γ_5	$\lvert u_4\rangle$	$\frac{1}{\sqrt{2}}\lvert(X-\mathrm{i}Y)\downarrow\rangle$	$Y_{1\ -1}\downarrow$	$-\frac{3}{2}$
Γ_5	$\lvert u_5\rangle$	$-\frac{1}{\sqrt{2}}\lvert(X+\mathrm{i}Y)\downarrow\rangle$	$Y_{1\ 1}\downarrow$	$\frac{1}{2}$
Γ_1	$\lvert u_6\rangle$	$\lvert Z\downarrow\rangle$	$Y_{1\ 0}\downarrow$	$-\frac{1}{2}$

3.4.3.1 $H_{k=0}$

We have [using Eq. (3.32)],

$$H_{k=0} = H_0 + H_{\mathrm{so}} = H_0 + H_{\mathrm{s},i}\sigma_i. \tag{3.67}$$

In the u basis, for example,

$$\begin{aligned}
\langle u_1|H_{k=0}|u_1\rangle &= \langle -\frac{1}{\sqrt{2}}(X+\mathrm{i}Y)\uparrow|H|-\frac{1}{\sqrt{2}}(X+\mathrm{i}Y)\uparrow\rangle \\
&= \frac{1}{2}\left[\langle X\uparrow|H_0|X\uparrow\rangle + \langle Y\uparrow|H_0|Y\uparrow\rangle + \mathrm{i}\langle X|H_{\mathrm{s},z}|Y\rangle\langle\uparrow|\sigma_z|\uparrow\rangle - \mathrm{i}\langle Y|H_{\mathrm{s},z}|X\rangle\langle\uparrow|\sigma_z|\uparrow\rangle\right] \\
&= \langle X|H_0|X\rangle + \frac{\mathrm{i}}{2}\left[\langle X|H_{\mathrm{s},z}|X\rangle - \langle Y|H_{\mathrm{s},z}|X\rangle\right] \\
&\equiv E_v + \Delta_1 + \Delta_2,
\end{aligned}$$

$$\begin{aligned}
\langle u_2|H_{k=0}|u_6\rangle &= \left\langle \frac{1}{\sqrt{2}}(X-\mathrm{i}Y)\uparrow|H_{k=0}|Z\downarrow\right\rangle = \left\langle \frac{1}{\sqrt{2}}(X-\mathrm{i}Y)\uparrow|H_{\mathrm{so}}|Z\downarrow\right\rangle \\
&= \frac{1}{\sqrt{2}}\left[\langle X|H_{\mathrm{s},y}|Z\rangle\langle\uparrow|\sigma_y|\downarrow\rangle + \mathrm{i}\langle Y|H_{\mathrm{s},x}|Z\rangle\langle\uparrow|\sigma_x|\downarrow\rangle\right] \\
&= \frac{1}{\sqrt{2}}\left[-\mathrm{i}\langle X|H_{\mathrm{s},y}|Z\rangle + \mathrm{i}\langle Y|H_{\mathrm{s},x}|Z\rangle\right] \\
&\equiv \sqrt{2}\Delta_3.
\end{aligned}$$

Then $H_{k=0}$ in the u representation is

$$H_{k=0} = \begin{pmatrix} |u_1\rangle & |u_2\rangle & |u_3\rangle & |u_4\rangle & |u_5\rangle & |u_6\rangle \\ E_v + \Delta_1 + \Delta_2 & 0 & 0 & 0 & 0 & 0 \\ 0 & E_v + \Delta_1 - \Delta_2 & 0 & 0 & 0 & \sqrt{2}\Delta_3 \\ 0 & 0 & E_v & 0 & \sqrt{2}\Delta_3 & 0 \\ 0 & 0 & 0 & E_v + \Delta_1 + \Delta_2 & 0 & 0 \\ 0 & 0 & \sqrt{2}\Delta_3 & 0 & E_v + \Delta_1 - \Delta_2 & 0 \\ 0 & \sqrt{2}\Delta_3 & 0 & 0 & 0 & E_v \end{pmatrix}, \tag{3.68}$$

where

$$\begin{aligned} \langle X|H_0|X\rangle &= \langle Y|H_0|Y\rangle = E_v + \Delta_1, \\ \langle Z|H_0|Z\rangle &= E_v, \\ \langle X|H_{s,z}|Y\rangle &= -\mathrm{i}\Delta_2, \\ \langle Y|H_{s,x}|Z\rangle &= \langle Z|H_{s,y}|X\rangle = -\mathrm{i}\Delta_3, \end{aligned} \tag{3.69}$$

3.4.3.2 $D_{6\times 6}(\mathbf{k})$

Since $D_{6\times 6}(\mathbf{k})$ does not include the spin-orbit interaction, its matrix representation in the LS basis can be written as

$$D_{6\times 6} = \begin{pmatrix} \uparrow & \downarrow \\ D_{3\times 3} & 0 \\ 0 & D_{3\times 3} \end{pmatrix}. \tag{3.70}$$

For $D_{3\times 3}$, one can start with the Hamiltonian in the X, Y, Z basis:

$$D_{3\times 3} = \begin{pmatrix} |X\rangle & |Y\rangle & |Z\rangle \\ L_1k_x^2 + M_1k_y^2 + M_2k_z^2 & N_1k_xk_y & N_2k_xk_z \\ N_1k_xk_y & M_1k_x^2 + L_1k_y^2 + M_2k_z^2 & N_2k_yk_z \\ N_2k_zk_x & N_2k_zk_y & M_1(k_x^2 + k_y^2) + L_2k_z^2 \end{pmatrix}. \tag{3.71}$$

We have

$$\begin{aligned} L_1 &= \frac{\hbar^2}{2m_0}\left(1 + \sum_l^B \frac{2p^x_{Xl}p^x_{lX}}{m_0(E_0 - E_l)}\right) = \frac{\hbar^2}{2m_0}\left(1 + \sum_l^B \frac{2p^y_{Yl}p^y_{lY}}{m_0(E_0 - E_l)}\right), \\ L_2 &= \frac{\hbar^2}{2m_0}\left(1 + \sum_l^B \frac{2p^z_{Zl}p^z_{lZ}}{m_0(E_0 - E_l)}\right), \end{aligned}$$

$$
\begin{aligned}
M_1 &= \frac{\hbar^2}{2m_0}\left(1+\sum_l^B \frac{2p^y_{Xl}p^y_{lX}}{m_0(E_0-E_l)}\right) = \frac{\hbar^2}{2m_0}\left(1+\sum_l^B \frac{2p^x_{Yl}p^x_{lY}}{m_0(E_0-E_l)}\right),\\
M_2 &= \frac{\hbar^2}{2m_0}\left(1+\sum_l^B \frac{2p^z_{Xl}p^z_{lX}}{m_0(E_0-E_l)}\right) = \frac{\hbar^2}{2m_0}\left(1+\sum_l^B \frac{2p^z_{Yl}p^z_{lY}}{m_0(E_0-E_l)}\right),\\
M_3 &= \frac{\hbar^2}{2m_0}\left(1+\sum_l^B \frac{2p^x_{Zl}p^x_{lZ}}{m_0(E_0-E_l)}\right) = \frac{\hbar^2}{2m_0}\left(1+\sum_l^B \frac{2p^y_{Zl}p^y_{lZ}}{m_0(E_0-E_l)}\right),\\
N_1 &= \frac{\hbar^2}{m_0^2}\sum_l^B \frac{p^x_{Xl}p^y_{lY}+p^y_{Xl}p^x_{lY}}{(E_0-E_l)},\\
N_2 &= \frac{\hbar^2}{m_0}\left(1+\sum_l^B \frac{p^x_{Xl}p^z_{lZ}+p^z_{Xl}p^x_{lZ}}{m_0(E_0-E_l)}\right) = \frac{\hbar^2}{m_0}\left(1+\sum_l^B \frac{p^y_{Yl}p^z_{lZ}+p^z_{Yl}p^y_{lZ}}{m_0(E_0-E_l)}\right),
\end{aligned}
\tag{3.72}
$$

where $p^y_{Xl} = \langle X|p_y|l\rangle$, ...The matrix can now be written in the u basis. For example,

$$
\begin{aligned}
\langle u_1|D|u_1\rangle &= \frac{1}{2}\langle (X+\mathrm{i}Y)\uparrow|D|(X+\mathrm{i}Y)\uparrow\rangle\\
&= \frac{1}{2}\left[\langle X|D|X\rangle + \langle Y|D|Y\rangle + \mathrm{i}\left(\langle X|D|Y\rangle - \langle Y|D|X\rangle\right)\right]\\
&= \frac{1}{2}(L_1+M_1)(k_x^2+k_y^2) + M_2k_z^2 \equiv D_{11},\\
\langle u_3|D|u_3\rangle &= \langle Z\uparrow|D|Z\uparrow\rangle = M_3(k_x^2+k_y^2)+M_2k_z^2 \equiv D_{33},\\
\langle u_2|D|u_1\rangle &= \frac{1}{2}\langle (X-\mathrm{i}Y)\uparrow|D|(X+\mathrm{i}Y)\uparrow\rangle\\
&= \frac{1}{2}(L_1-M_1)(k_x^2-k_y^2) + \mathrm{i}N_1k_xk_y \equiv D_{21},\\
\langle u_2|D|u_3\rangle &= \frac{1}{\sqrt{2}}\langle (X-\mathrm{i}Y)\uparrow|D|Z\uparrow\rangle = \frac{1}{\sqrt{2}}N_2k_z(k_x+\mathrm{i}k_y).
\end{aligned}
$$

The D matrix is then

$$
\begin{pmatrix}
D_{11} & D_{21}^* & -D_{23}^* & & & \\
D_{21} & D_{11} & D_{23} & & 0 & \\
-D_{23} & D_{23}^* & D_{33} & & & \\
 & & & D_{11} & D_{21} & D_{23}\\
 & & & D_{21}^* & D_{11} & -D_{23}^*\\
 & & & D_{23}^* & -D_{23} & D_{33}
\end{pmatrix}.
\tag{3.73}
$$

The complete Hamiltonian matrix is given in Table 3.8. For WZ, we have [46]

$$L_1 - M_1 = N_1. \tag{3.74}$$

Table 3.8 Six-band Chuang–Chang Hamiltonian in u basis

$$H_{CC}(\mathbf{k}) = \begin{pmatrix} |u_1\rangle & |u_2\rangle & |u_3\rangle & |u_4\rangle & |u_5\rangle & |u_6\rangle \\ F & -K^* & -H^* & 0 & 0 & 0 \\ -K & G & H & 0 & 0 & \Delta \\ -H & H^* & \lambda & 0 & \Delta & 0 \\ 0 & 0 & 0 & F & -K & H \\ 0 & 0 & \Delta & -K^* & G & -H^* \\ 0 & \Delta & 0 & H^* & -H & \lambda \end{pmatrix},$$

where

$$\begin{aligned} F &= \Delta_1 + \Delta_2 + \lambda + \theta, \\ G &= \Delta_1 - \Delta_2 + \lambda + \theta, \\ \lambda &= L_2 k_z^2 + M_3(k_x^2 + k_y^2), \\ \theta &= (M_2 - L_2)k_z^2 + (\frac{L_1 + M_1}{2} - M_3)(k_x^2 + k_y^2), \\ K &= \frac{N_1}{2}(k_x + \mathrm{i}k_y)^2, \\ H &= \frac{N_2}{\sqrt{2}}(k_x + \mathrm{i}k_y)k_z, \\ \Delta &= \sqrt{2}\Delta_3. \end{aligned}$$

The model has an exact solution at $\mathbf{k} = \mathbf{0}$. The following states are decoupled: $|u_1\rangle$, and $|u_4\rangle$. There are two coupled pairs: $|u_2\rangle$ and $|u_6\rangle$, and $|u_3\rangle$ and $|u_5\rangle$. The energies are $E_1 = E_v + \Delta_1 + \Delta_2$, and

$$E_2 = E_v + \frac{(\Delta_1 - \Delta_2)}{2} + \sqrt{\left(\frac{(\Delta_1 - \Delta_2)}{2}\right)^2 + 2\Delta_3^2}, \tag{3.75}$$

$$E_3 = E_v + \frac{(\Delta_1 - \Delta_2)}{2} - \sqrt{\left(\frac{(\Delta_1 - \Delta_2)}{2}\right)^2 + 2\Delta_3^2}. \tag{3.76}$$

An example band structure for GaN is shown in Fig. 3.4. The band structure was obtained from an ab initio calculation and fitted to the RSP Hamiltonian. The latter is equivalent to the CC Hamiltonian and will be introduced in Chap. **??**. The bands are plotted along $\Gamma - A$ ($k_{||}$) and along two directions perpendicular to $k_{||}$. In-plane anisotropy is observed by the difference between the two $k_{\perp}$ directions.

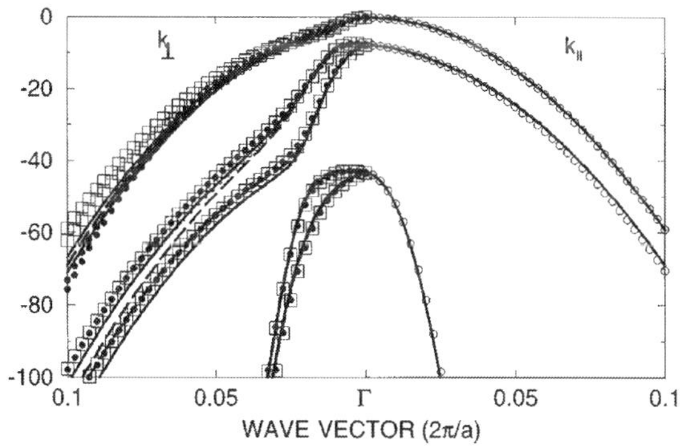

Fig. 3.4 GaN valence-band structure from an ab initio calculation with a fit to the RSP Hamiltonian(solid line). $k_{||}$ corresponds to $\Gamma - A$. *Open squares*: $\Gamma - K$ direction; *filled circles*: $\Gamma - M$ direction. *Dashed curve* is a fit without the A_7 term. Reprinted with permission from [62]. ©1997 by the American Physical Society

3.4.3.3 Block Diagonalization

A block diagonalization scheme can be written down. This was first carried out by Sirenko et al.[43] and Chuang and Chang [46]. The trick is to recognize that the vectors have azimuthal symmetry in the x-y plane. The choice

$$k_x + \mathrm{i}k_y = k_{||}\mathrm{e}^{\mathrm{i}\phi} \tag{3.77}$$

leads to the off-diagonal matrix elements being written as

$$K = K_{||}\mathrm{e}^{2\mathrm{i}\phi},\ K_{||} = \frac{N_1}{2}k_{||}^2, \tag{3.78}$$

$$H = H_{||}\mathrm{e}^{\mathrm{i}\phi},\ H_{||} = \frac{N_2}{\sqrt{2}}k_{||}k_z. \tag{3.79}$$

The explicit ϕ dependence can then be eliminated by rescaling the basis functions [15]:

$$u_1' = \mathrm{e}^{-\frac{3\mathrm{i}}{2}\phi}u_1,$$

$$u_2' = \mathrm{e}^{\frac{\mathrm{i}}{2}\phi}u_2,$$

$$u_3' = \mathrm{e}^{-\frac{\mathrm{i}}{2}\phi}u_3,$$

$$u_4' = e^{\frac{3i}{2}\phi} u_4, \tag{3.80}$$

$$u_5' = e^{-\frac{i}{2}\phi} u_5,$$

$$u_6' = e^{\frac{i}{2}\phi} u_6,$$

or $|u_i'\rangle = \sum_j T_{ij} |u_j\rangle$, where

$$T = \begin{pmatrix} \alpha^* & 0 & 0 & \alpha & 0 & 0 \\ 0 & \beta & 0 & 0 & \beta^* & 0 \\ 0 & 0 & \beta^* & 0 & 0 & \beta \\ \alpha^* & 0 & 0 & -\alpha & 0 & 0 \\ 0 & \beta & 0 & 0 & -\beta^* & 0 \\ 0 & 0 & -\beta^* & 0 & 0 & \beta \end{pmatrix}, \tag{3.81}$$

and

$$\alpha = \frac{1}{\sqrt{2}} e^{i(\frac{3}{4}\pi + \frac{3}{2}\phi)}, \beta = \frac{1}{\sqrt{2}} e^{i(\frac{1}{4}\pi + \frac{1}{2}\phi)}. \tag{3.82}$$

Finally, the Hamiltonian matrix becomes

$$H'(\mathbf{k}) = \begin{pmatrix} F & K_{||} & -iH_{||} & 0 & 0 & 0 \\ K_{||} & G & \Delta - iH_{||} & 0 & 0 & 0 \\ iH_{||} & \Delta + iH_{||} & \lambda & 0 & 0 & 0 \\ 0 & 0 & 0 & F & K_{||} & iH_{||} \\ 0 & 0 & 0 & K_{||} & G & \Delta + iH_{||} \\ 0 & 0 & 0 & -iH_{||} & \Delta - iH_{||} & \lambda \end{pmatrix}. \tag{3.83}$$

3.4.4 Gutsche–Jahne Hamiltonian

The basis for the above CC model consisted of the three Y_m^1 functions together with the two-component spin function. The spatial functions only transformed according to Γ_1 and Γ_5. In the Gutsche–Jahne (GJ) basis, spatial functions transforming according to Γ_1, Γ_5, Γ_3 and Γ_6 are used (Table 3.9).

Hence, the GJ model is a 12-band model with spin for the valence band of WZ. Note, however, that they only set up the Hamiltonian at the Γ point. In the GJ basis, $H_{k=0}$ actually block diagonalizes into two 6×6 blocks as the last six states are Kramers partners of the top six:

$$H(\mathbf{k}) = \begin{pmatrix} S & 0 \\ 0 & S^* \end{pmatrix}. \tag{3.84}$$

where

Table 3.9 Basis set for wurtzite: Gutsche–Jahne [42]

Γ_5	$\|u_5^* \uparrow\rangle$	$\|(x - \mathrm{i}y) \uparrow\rangle$
Γ_1	$\|u_1 \downarrow\rangle$	$\|z \downarrow\rangle$
Γ_6	$\|u_6 \uparrow\rangle$	$\|(x + \mathrm{i}y)^2 \uparrow\rangle$
Γ_3	$\|u_3 \downarrow\rangle$	$\|x(3y^2 - x^2) \downarrow\rangle$
Γ_5	$\|u_5 \uparrow\rangle$	$\|(x + \mathrm{i}y) \uparrow\rangle$
Γ_6	$\|u_6 \downarrow\rangle$	$\|(x + \mathrm{i}y)^2 \downarrow\rangle$
Γ_5	$\|u_5 \downarrow\rangle$	$\|(x + \mathrm{i}y) \downarrow\rangle$
Γ_1	$\|-u_1 \uparrow\rangle$	$\|-z \uparrow\rangle$
Γ_6	$\|u_6^* \downarrow\rangle$	$\|(x - \mathrm{i}y)^2 \downarrow\rangle$
Γ_3	$\|-u_3 \uparrow\rangle$	$\|x(x^2 - 3y^2) \uparrow\rangle$
Γ_5	$\|u_5^* \downarrow\rangle$	$\|(x - \mathrm{i}y) \downarrow\rangle$
Γ_6	$\|-u_6^* \uparrow\rangle$	$\|-(x - \mathrm{i}y)^2 \uparrow\rangle$

$$S = \begin{pmatrix} |u_5^* \uparrow\rangle & |u_1 \downarrow\rangle & |u_6 \uparrow\rangle & |u_3 \downarrow\rangle & |u_5 \uparrow\rangle & |u_6 \downarrow\rangle \\ E_5 - \frac{\Delta_5}{3} & \frac{\sqrt{2}}{3}\Delta_7^* & 0 & 0 & 0 & 0 \\ \frac{\sqrt{2}}{3}\Delta_7 & E_1 & 0 & 0 & 0 & 0 \\ & & E_6 + \frac{\Delta_6}{3} & \frac{\Delta_8^*}{3} & 0 & 0 \\ & & \frac{\Delta_8}{3} & E_3 & 0 & 0 \\ & & & & E_5 + \frac{\Delta_5}{3} & \frac{\sqrt{2}}{3}\Delta_9^* \\ & & & & \frac{\sqrt{2}}{3}\Delta_9 & E_6 - \frac{\Delta_6}{3} \end{pmatrix}, \tag{3.85}$$

$$\frac{\Delta_5}{3} = \langle u_5 | H_{\mathrm{s},z} | u_5 \rangle, \qquad \frac{\Delta_6}{3} = \langle u_6 | H_{\mathrm{s},z} | u_6 \rangle,$$
$$\frac{\Delta_7}{3} = \frac{1}{\sqrt{2}} \langle u_1 | H_{\mathrm{s},+} | u_5^* \rangle, \qquad \frac{\Delta_8}{3} = \frac{1}{\sqrt{2}} \langle u_3 | H_{\mathrm{s},+} | u_6 \rangle, \tag{3.86}$$
$$\frac{\Delta_9}{3} = \frac{1}{\sqrt{2}} \langle u_6 | H_{\mathrm{s},+} | u_5 \rangle,$$

with $H_{\mathrm{s},+} = H_{\mathrm{s},x} + \mathrm{i} H_{\mathrm{s},y}$.

The GJ Hamiltonian can be diagonalized exactly. The energies are [42]:

$$E_7 = \frac{1}{2} \left\{ (E_5 + E_1) - \frac{\Delta_5}{3} \pm \sqrt{\left[(E_5 - E_1) - \frac{\Delta_5}{3} \right]^2 + \frac{8}{9} |\Delta_7|^2} \right\},$$

$$E_8 = \frac{1}{2} \left\{ (E_6 + E_3) - \frac{\Delta_6}{3} \pm \sqrt{\left[(E_6 - E_3) - \frac{\Delta_6}{3} \right]^2 + \frac{8}{9} |\Delta_8|^2} \right\}, \tag{3.87}$$

$$E_9 = \frac{1}{2} \left\{ (E_6 + E_5) - \frac{\Delta_5 - \Delta_6}{3} \pm \sqrt{\left[(E_5 - E_6) - \frac{\Delta_5 + \Delta_6}{3} \right]^2 + \frac{8}{9} |\Delta_9|^2} \right\},$$

where each of the above energies are doubly-degenerate due to time-reversal symmetry and the corresponding linear combination of basis states is given by

$$\sqrt{1-q^2}|u'\uparrow\rangle + q|u''\downarrow\rangle, \tag{3.88}$$

with

$$\begin{aligned}
\left.\begin{matrix} q_7^2 \\ (1-q_7^2) \end{matrix}\right\} &= \frac{1}{2}\left[1 \mp \frac{E_5 - E_1 - \frac{\Delta_5}{3}}{\sqrt{\left[E_5 - E_1 - \frac{\Delta_5}{3}\right]^2 + \frac{8}{9}|\Delta_7|^2}}\right], \\
\left.\begin{matrix} q_8^2 \\ (1-q_8^2) \end{matrix}\right\} &= \frac{1}{2}\left[1 \mp \frac{E_6 - E_3 + \frac{\Delta_6}{3}}{\sqrt{\left[E_6 - E_3 + \frac{\Delta_6}{3}\right]^2 + \frac{8}{9}|\Delta_8|^2}}\right], \\
\left.\begin{matrix} q_9^2 \\ (1-q_9^2) \end{matrix}\right\} &= \frac{1}{2}\left[1 \mp \frac{E_5 - E_6 + \frac{\Delta_5+\Delta_6}{3}}{\sqrt{\left[E_5 - E_6 + \frac{\Delta_5+\Delta_6}{3}\right]^2 + \frac{8}{9}|\Delta_9|^2}}\right].
\end{aligned} \tag{3.89}$$

As discussed by Gutsche and Jahne [42], this model requires eight parameters (with a ninth set equal to zero) in order to describe the valence-band structure of WZ at $\mathbf{k} = \mathbf{0}$. This is to be contrasted to the $k \cdot p$ theory for ZB, with only one parameter (Δ_0). The structure of WZ can be viewed as starting from the ZB one and straining it along the [111] direction. The consequence in reciprocal space would be a zone folding of ZB bands along the L direction. This is consistent with the doubling of the number of valence bands for WZ compared to ZB. This effect leads to the Δ_5–Δ_8 parameters [42]. Δ_9 is an interaction between the two anions in the unit cell, an effect which is missing for ZB.

A closer comparison is to the CC model. Recall that the latter is only described by three parameters, Δ_1–Δ_3 at $\mathbf{k} = \mathbf{0}$. This reflects some of the shortcomings of the CC model–even though the latter is widely used.

3.5 Summary

It was shown how perturbation theory can be used to obtain the valence-band Hamiltonian for both cubic and hexagonal crystals. The cubic case was done specifically for the Dresselhaus–Kip–Kittel Hamiltonian and the latter was related to a few other Hamiltonians in the literature. It was found that the DKK Hamiltonian, in the absence of spin-orbit coupling and an external magnetic field, only has three band parameters. The case of a wurtzite semiconductor was then treated along similar lines; for the most widely used valence-band model, the Hamiltonian has 10 band parameters. Variations on the latter model were then discussed.

Chapter 4
Perturbation Theory – Kane Models

4.1 Overview

The models of Dresselhaus–Kip–Kittel and Luttinger–Kohn work well to describe the valence band of most cubic semiconductors. However, one might question the validity of degenerate perturbation theory for small band-gap semiconductors such as InSb ($E_0 \sim 0.5\,\mathrm{eV}$). Our next method is a form of quasi-degenerate perturbation theory applied to the multiband problem. There are two basic ways one can improve the band-structure model. In both cases, the basic idea is to replace the differential equation, Eq. (2.6), by the corresponding matrix representation. This is achieved by expanding the cellular function in terms of a complete set of cell-periodic states; to be exact, the representation should be infinite. In the first method, the problem will be approximated by diagonalization within an appropriate subspace. This corresponds to treating the interaction of the selected bands exactly. Any discrepancy with experiments can then be attributed to the neglect of interactions with the distant states. In the second method, one can include the distant bands via a downfolding procedure. The quasi-degenerate perturbation theory was introduced by Kane [3].

The approach taken by Kane was that, for small band-gap semiconductors, the valence-band model of Dresselhaus–Kip–Kittel and Luttinger–Kohn needed to be improved by the addition of the lowest conduction band. Subsequently, it has been found that such a model fails to reproduce the nonparabolicity of the lowest conduction band in materials such as GaAs, where the next conduction band is not any more distant. This has led to the consideration of so-called extended Kane models with as many as 14 bands [23]. When such details near an extremum point is not needed but rather an overall description over the whole Brillouin zone, the exact diagonalization within a finite subspace can work surprisingly well with as few as 15 bands [5].

4.2 First-Order Models

For conciseness, and partly for historical reasons, we illustrate the model for ZB.

L.C. Lew Yan Voon, M. Willatzen, *The* k · p *Method*,
DOI 10.1007/978-3-540-92872-0_4, © Springer-Verlag Berlin Heidelberg 2009

4.2.1 Four-Band Model

We first consider the subspace of Γ_{1c} and Γ_{15v} states (Fig. 4.1). The states are chosen to be $S, -\mathrm{i}X, -\mathrm{i}Y, -\mathrm{i}Z$; this choice of phases makes the Hamiltonian real. The exact

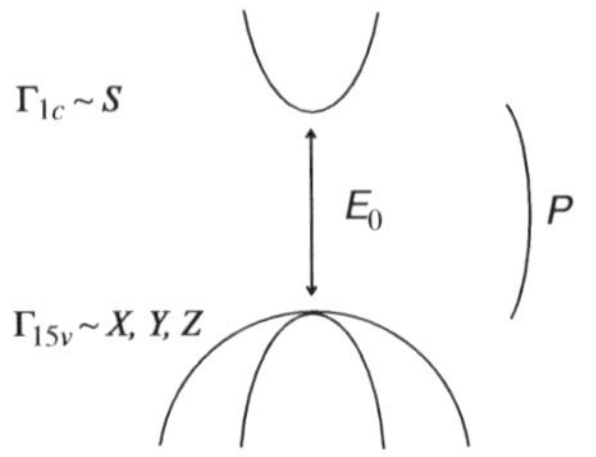

Fig. 4.1 Four-band model (zincblende symmetry)

Hamiltonian is

$$H(\mathbf{k}) = \begin{pmatrix} S & -\mathrm{i}X & -\mathrm{i}Y & -\mathrm{i}Z \\ \varepsilon(\mathbf{k}) + E_0 & k_x P & k_y P & k_z P \\ & \varepsilon(\mathbf{k}) & 0 & 0 \\ & & \varepsilon(\mathbf{k}) & 0 \\ & \dagger & & \varepsilon(\mathbf{k}) \end{pmatrix}, \tag{4.1}$$

where

$$\varepsilon((\mathbf{k}) = \frac{\hbar^2 k^2}{2m_0}, \quad P = -\mathrm{i}\frac{\hbar}{m_0}\langle S|p_x|X\rangle. \tag{4.2}$$

This Hamiltonian is actually equivalent to a two-band model for the conduction electron and light hole (but this is often referred to as a three-band model since the lh state is two-fold degenerate), with separate solutions for the heavy holes. The solutions are

$$E(\mathbf{k}) = \begin{cases} \frac{\hbar^2 k^2}{2m_0} \quad (\text{twice} — hh), \\ \varepsilon(k) + \frac{E_0}{2} \pm \sqrt{\left(\frac{E_0}{2}\right)^2 + k_x^2 P^2}. \end{cases} \tag{4.3}$$

The hh corresponds to a bare electron. All the dispersion relations are isotropic; but now nonparabolicity is present.

The effective masses are given by

$$\frac{1}{m^*} = \frac{1}{m_0}\left(1 \pm \frac{E_P}{E_0}\right), \tag{4.4}$$

similar to the two-band case (if P' is neglected for ZB), Eqs. (2.26) and (2.29). One disadvantage of the four-band model is that it does not differentiate between ZB and DM (since P' is absent).

4.2.2 Eight-Band Model

Before we study more accurate models that include remote contributions, we would like to see how the inclusion of spin transforms the four-band model. The appropriate bands are shown in Fig. 4.2. As we have already seen in previous chapters, one

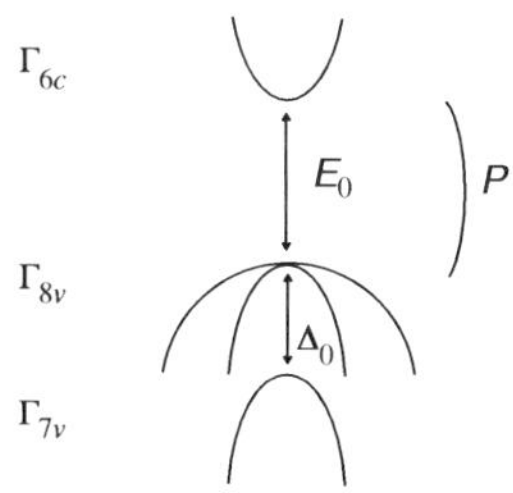

Fig. 4.2 Eight-band model (zincblende symmetry)

can diagonalize the spin-orbit perturbation by using the $|JM_J\rangle$ states (Table 3.4). With the notation,

$$k_\pm \equiv k_x \pm \mathrm{i}k_y, \tag{4.5}$$

and the basis ordering

$$\mathrm{i}|S\uparrow\rangle, \left|\frac{3}{2}\,\frac{1}{2}\right\rangle, \left|\frac{3}{2}\,\frac{3}{2}\right\rangle, \left|\frac{1}{2}\,\frac{1}{2}\right\rangle, \mathrm{i}|S\downarrow\rangle, \left|\frac{3}{2}\,-\frac{1}{2}\right\rangle, \left|\frac{3}{2}\,-\frac{3}{2}\right\rangle, \left|\frac{1}{2}\,-\frac{1}{2}\right\rangle,$$

the eight-band Hamiltonian in Table 4.1 is obtained. For example,

$$\left\langle \mathrm{i}S\uparrow|H|\frac{3}{2}\,\frac{3}{2}\right\rangle = \left\langle \mathrm{i}S\uparrow|H|\frac{1}{\sqrt{2}}(X+\mathrm{i}Y)\uparrow\right\rangle = \frac{1}{\sqrt{2}}P(k_x+\mathrm{i}k_y) \equiv \frac{1}{\sqrt{2}}Pk_+.$$

The latter cannot be completely diagonalized for arbitrary **k**. We will use intuition. First, note that the four-band problem is isotropic. We would, therefore, expect the present model to be too. Looking at the Hamiltonian, it is clear that only a 4×4 problem need be solved if $k = k_z$; such a choice is allowed by the isotropy and the Hamiltonian becomes

Table 4.1 First-order eight-band Kane Hamiltonian in JM_J basis

$$\begin{pmatrix}
\mathrm{i}|S\uparrow & \left|\frac{3}{2}\,\frac{1}{2}\right\rangle & \left|\frac{3}{2}\,\frac{3}{2}\right\rangle & \left|\frac{1}{2}\,\frac{1}{2}\right\rangle & \mathrm{i}|S\downarrow\rangle & \left|\frac{3}{2}\,-\frac{1}{2}\right\rangle & \left|\frac{3}{2}\,-\frac{3}{2}\right\rangle & \left|\frac{1}{2}\,-\frac{1}{2}\right\rangle \\
\varepsilon(\mathbf{k})+E_0 & -\sqrt{\frac{2}{3}}Pk_z & \frac{P}{\sqrt{2}}k_+ & \sqrt{\frac{1}{3}}Pk_z & 0 & -\sqrt{\frac{1}{6}}Pk_- & 0 & -\sqrt{\frac{1}{3}}Pk_- \\
-\sqrt{\frac{2}{3}}Pk_z & \varepsilon(\mathbf{k}) & 0 & 0 & \sqrt{\frac{1}{6}}Pk_- & 0 & 0 & 0 \\
\frac{P}{\sqrt{2}}k_- & 0 & \varepsilon(\mathbf{k}) & 0 & 0 & 0 & 0 & 0 \\
\sqrt{\frac{1}{3}}Pk_z & 0 & 0 & \varepsilon(\mathbf{k})-\Delta_0 & \sqrt{\frac{1}{3}}Pk_- & 0 & 0 & 0 \\
0 & \sqrt{\frac{1}{6}}Pk_+ & 0 & \sqrt{\frac{1}{3}}Pk_+ & \varepsilon(\mathbf{k})+E_0 & -\sqrt{\frac{2}{3}}Pk_z & \frac{P}{\sqrt{2}}k_- & \sqrt{\frac{1}{3}}Pk_z \\
-\sqrt{\frac{1}{6}}Pk_+ & 0 & 0 & 0 & -\sqrt{\frac{2}{3}}Pk_z & \varepsilon(\mathbf{k}) & 0 & 0 \\
0 & 0 & 0 & 0 & \frac{P}{\sqrt{2}}k_+ & 0 & \varepsilon(\mathbf{k}) & 0 \\
-\sqrt{\frac{1}{3}}Pk_+ & 0 & 0 & 0 & \sqrt{\frac{1}{3}}Pk_z & 0 & 0 & \varepsilon(\mathbf{k})-\Delta_0
\end{pmatrix},$$

where

$$\varepsilon(\mathbf{k}) = \frac{\hbar^2 k^2}{2m_0},$$

$$k_\pm \equiv k_x \pm \mathrm{i}k_y,$$

$$P = -\mathrm{i}\frac{\hbar}{m_0}\langle Sc|p_x|Xv\rangle.$$

$$H(\mathbf{k}) = \begin{bmatrix} H_{\frac{1}{2}} & 0 \\ 0 & H_{\frac{1}{2}} \end{bmatrix},$$

where

$$H_{\frac{1}{2}} = \begin{pmatrix} \varepsilon(\mathbf{k})+E_0 & -\sqrt{\frac{2}{3}}Pk & 0 & \sqrt{\frac{1}{3}}Pk \\ -\sqrt{\frac{2}{3}}Pk & \varepsilon(\mathbf{k}) & 0 & 0 \\ 0 & 0 & \varepsilon(\mathbf{k}) & 0 \\ \sqrt{\frac{1}{3}}Pk & 0 & 0 & \varepsilon(\mathbf{k})-\Delta_0 \end{pmatrix}. \tag{4.6}$$

All the bands are doubly degenerate. The determinantal equation is

$$[\varepsilon(\mathbf{k})-E(\mathbf{k})]\begin{vmatrix} \varepsilon(\mathbf{k})+E_0-E(\mathbf{k}) & -\sqrt{\frac{2}{3}}Pk & \sqrt{\frac{1}{3}}Pk \\ -\sqrt{\frac{2}{3}}Pk & \varepsilon(\mathbf{k})-E(\mathbf{k}) & 0 \\ \sqrt{\frac{1}{3}}Pk & 0 & \varepsilon(\mathbf{k})-\Delta_0-E(\mathbf{k}) \end{vmatrix} = 0. \tag{4.7}$$

Hence,

$$E(\mathbf{k}) = \frac{\hbar^2 k^2}{2m_0} \qquad (hh),$$

and

$$[E(\mathbf{k}) - \varepsilon(\mathbf{k}) - E_0]\,[E(\mathbf{k}) - \varepsilon(\mathbf{k})]\,[E(\mathbf{k}) - \varepsilon(\mathbf{k}) + \Delta_0]$$
$$= k^2 P^2 \left[E(\mathbf{k}) - \varepsilon(\mathbf{k}) + \frac{2}{3}\Delta_0 \right]. \tag{4.8}$$

Equation (4.8) contains the e, lh, and sh solutions. Formally, it can be seen that the $k \cdot p$ coupling only exists among the $M_J = \pm\frac{1}{2}$ states. Before we solve them for the effective masses, consider a general treatment of the Hamiltonian matrix. Since we expect the dispersion relation to be isotropic, one can rotate the axes. We have already seen that $k = k_z$ block diagonalizes H. Hence, one can rotate the basis functions such that the z-axis is along $\mathbf{k}$. If $\mathbf{k}$ is located at (θ, ϕ) with respect to the cubic axes, the transformation is

$$\begin{pmatrix} \uparrow' \\ \downarrow' \end{pmatrix} = \begin{pmatrix} e^{-i\phi/2}\cos\frac{\theta}{2} & e^{i\phi/2}\sin\frac{\theta}{2} \\ -e^{-i\phi/2}\sin\frac{\theta}{2} & e^{i\phi/2}\cos\frac{\theta}{2} \end{pmatrix} \begin{pmatrix} \uparrow \\ \downarrow \end{pmatrix}, \tag{4.9}$$

$$\begin{pmatrix} X' \\ Y' \\ Z' \end{pmatrix} = \begin{pmatrix} \cos\theta\cos\phi & \cos\theta\sin\phi & -\sin\theta \\ -\sin\phi & \cos\phi & 0 \\ \sin\theta\cos\phi & \sin\theta\sin\phi & \cos\theta \end{pmatrix} \begin{pmatrix} X \\ Y \\ Z \end{pmatrix}, \tag{4.10}$$

since (θ, ϕ) correspond to the first two Euler angles. In terms of the rotated basis functions, H is block diagonal. For example, using

$$\langle \uparrow' | \uparrow' \rangle - 1, \langle \uparrow' | \downarrow' \rangle = 0,$$

we have

$$\left\langle S' \uparrow' | H | \frac{1}{2}\, -\frac{1}{2}' \right\rangle = -\frac{1}{\sqrt{3}} \langle S | H | X' - iY' \rangle$$
$$= -\frac{1}{\sqrt{3}} \langle S | H | \cos\theta\cos\phi X + \cos\theta\sin\phi Y - \sin\theta Z + i\sin\phi X - i\cos\phi Y \rangle$$
$$= -\frac{i}{\sqrt{3}} P \left(\cos\theta\cos\phi k_x + \cos\theta\sin\phi k_y - \sin\theta k_z + i\sin\phi k_x - i\cos\phi k_y \right)$$
$$= 0,$$

since $\mathbf{k} = k(\cos\phi\sin\theta, \sin\phi\sin\theta, \cos\theta)$. The effective masses will now be extracted from the dispersion relation. The hh mass is obviously the free-electron mass.

4.2.2.1 Electron

For the electron,

$$E(k) \approx \frac{\hbar^2 k^2}{2m_e} + E_0.$$

To order k^2, Eq. (4.8) becomes

$$\left(\frac{\hbar^2k^2}{2m_e}-\frac{\hbar^2k^2}{2m_0}\right)E_0(E_0+\Delta_0)\approx\frac{\hbar^2k^2}{2m_0}E_P\left(E_0+\frac{2}{3}\Delta_0\right),$$

and

$$\frac{1}{m_e/m_0}=1+\frac{E_P(E_0+\frac{2}{3}\Delta_0)}{(E_0+\Delta_0)E_0}=1+\frac{2}{3}\frac{E_P}{E_0}+\frac{1}{3}\frac{E_P}{E_0+\Delta_0}. \tag{4.11}$$

4.2.2.2 Light Hole

For the light hole,

$$E(k)\approx\frac{\hbar^2k^2}{2m_{lh}}.$$

From Eq. (4.8),

$$\left(\frac{\hbar^2k^2}{2m_{lh}}-\frac{\hbar^2k^2}{2m_0}\right)(-E_0)\Delta_0\approx\frac{\hbar^2k^2}{2m_0}E_P\frac{2}{3}\Delta_0,$$

and

$$\frac{1}{m_{lh}/m_0}=1-\frac{2}{3}\frac{E_P}{E_0}. \tag{4.12}$$

4.2.2.3 Spin Hole

For the spin hole,

$$E(k)\approx\frac{\hbar^2k^2}{2m_{sh}}-\Delta_0.$$

Again, from Eq. (4.8),

$$\left(\frac{\hbar^2k^2}{2m_{sh}}-\frac{\hbar^2k^2}{2m_0}\right)(E_0+\Delta_0)\Delta_0\approx-\frac{\hbar^2k^2}{2m_0}E_P\frac{1}{3}\Delta_0,$$

and

$$\frac{1}{m_{sh}/m_0}=1-\frac{E_P}{3(E_0+\Delta_0)}. \tag{4.13}$$

4.3 Second-Order Kane Model

We are now ready to look at all the implications of the Kane model. Recall that we have studied the first-order Kane model, whereby only the interactions within the desired subspace are included. We will now include the second-order terms and consider DM and ZB.

4.3.1 Löwdin Perturbation

There are two approaches to including the remote bands. One is based upon the equations derived in Appendix A. The second-order renormalized Hamiltonian in the quasi-degenerate A class is given by [Eq. (A.13)]

$$U_{mn}^{A} \equiv H_{mn} + \sum_{\alpha \in B} \frac{H_{m\alpha} H_{\alpha n}}{(E - H_{\alpha\alpha})}$$

for $m, n \in A$. For $\alpha \in B$,

$$C_{\alpha} = \frac{1}{(E - H_{\alpha\alpha})} \sum_{n \in A} U_{\alpha n}^{A} C_{n}.$$

Note that, in this technique, the correct energy E appears in the definition of the Hamiltonian. Since it is to be determined, it is traditional to replace it by either the experimental energy (when the A set is degenerate) or by an average energy (when the A set has nondegenerate bands). This procedure was introduced by Kane [31].

A second expression was developed by Luttinger and Kohn within the framework of canonical transformation [6]:

$$\sum_{n \in A} \left\{ D_{mn}^{ij} k_i k_j + \pi_{mn}^{i} k_i + \frac{\hbar}{4m_0^2 c^2} [(\boldsymbol{\sigma} \times \nabla V) \cdot \mathbf{p}]_{mn} + E_n \delta_{mn} \right\} f_n(\mathbf{r}) \tag{4.14}$$
$$= E f_m(\mathbf{r}),$$

$$D_{mn}^{ij} = \frac{\hbar^2}{2m_0} \left\{ \delta_{mn}\delta_{ij} + \frac{1}{m_0} \sum_{\alpha \in B} \pi_{m\alpha}^{i} \pi_{\alpha n}^{j} \left[\frac{1}{(E_m - E_\alpha)} + \frac{1}{(E_n - E_\alpha)} \right] \right\}, \tag{4.15}$$

$$\pi_{m\alpha} = \frac{(2\pi)^3}{\Omega} \int_{\Omega} \mathrm{d}^3\mathbf{r}\, u_{m\mathbf{0}}(\mathbf{r}) \left[\mathbf{p} + \frac{\hbar^2}{4m_0 c^2} (\boldsymbol{\sigma} \times \nabla V) \right] u_{\alpha\mathbf{0}}(\mathbf{r}). \tag{4.16}$$

This formulation has been documented in Ivchenko and Pikus [20] and used by, e.g., Loehr [63] and Boujdaria et al. [64].

4.3.2 Four-Band Model

We start the study by leaving out the spin-orbit coupling. This is then a generalization of the four-band model studied earlier (Sect. 4.2.1). Recall that the previous model [Eq. (4.1)] only has diagonal elements and nondiagonal linear-in-k terms between s and p states. Inclusion of remote bands results in quadratic terms (Table 4.2). For example, the mass of the electron is related to the parameter A'. From the one-band model, we already know that the effective mass of the electron arises from interactions with neighboring bands. Since the top of the valence band, Γ_{25}^{+} (Γ_{15v}) for DM (ZB), is already part of class A, A' does not include those states and the next most important contribution comes from the p-like conduction band Γ_{25}^{+} (Γ_{15c}); note that, for C and Si, the latter state is actually the lowest conduction band (compare Figs. 3.2 and 4.3). As another example, consider the interaction of the s state with the p_x one. This is zero for DM but not for ZB. For the latter, the regular Hamiltonian leads to a linear-in-k term, as for the first-order Kane model. The second-order term from the Löwdin renormalization gives

$$\sum_{\alpha\in B} \frac{H_{s\alpha} H_{\alpha x}}{(E - H_{\alpha\alpha})}.$$

Instead of finding E self-consistently, one can approximate it by the average of the energies of the two states. This leads to the definition of B in Table 4.2. Should one use Eq. (4.15) instead, then [63]

$$B = \frac{\hbar^2}{m_0^2} \sum_{l\in\Gamma_{15}}{}' \langle S|p_x|u_l\rangle\langle u_l|p_y|Z\rangle \left[\frac{1}{E_c - E_l} + \frac{1}{E_v - E_l}\right]. \tag{4.17}$$

Due to time-reversal symmetry, the cellular functions are real, and all the parameters are too. The new parameters are given in Table 4.2. The symmetry symbols in the summation refer to the ZB and DM structures for the ones to the left and right of the semicolon, respectively. B is zero for DM. The symmetry states can be related to atomic-like states; this and a typical ordering is given in Fig. 4.3.

The primed parameters are related to DKK's unprimed ones by

$$\begin{aligned} A &= A' + \frac{P^2}{E_c - E_v}, \\ F &= F' + \frac{P^2}{E_v - E_c}. \end{aligned} \tag{4.18}$$

The reason is because some of the A bands of the Kane model were originally not in the DKK basis (in other words, they were part of the B states of the DKK model). Thus, in the summation over states, these A states of the Kane model should be removed from the summation over B states of the DKK model. If A' and F' are small, then $A \approx -F$; this was found to be the case for Ge.

Table 4.2 Second-order four-band Kane Hamiltonian [31]

$$H(\mathbf{k}) = \begin{bmatrix} S & X & Y & Z \\ A'k^2 + E_0 + \varepsilon(\mathbf{k}) & Bk_yk_z + \mathrm{i}k_xP & Bk_xk_z + \mathrm{i}k_yP & Bk_xk_y + \mathrm{i}k_zP \\ & L'k_x^2 + M(k_y^2 + k_z^2) + \varepsilon(\mathbf{k}) & N'k_xk_y & N'k_xk_z \\ & \dagger & L'k_y^2 + M(k_x^2 + k_z^2) + \varepsilon(\mathbf{k}) & N'k_yk_z \\ & & & L'k_z^2 + M(k_x^2 + k_y^2) + \varepsilon(\mathbf{k}) \end{bmatrix},$$

where

$$\varepsilon(\mathbf{k}) = \frac{\hbar^2 k^2}{2m_0}, \quad P = -\mathrm{i}\frac{\hbar}{m_0}\langle S|p_x|X\rangle,$$

$$L' = F' + 2G,$$

$$M = H_1 + H_2,$$

$$N' = F' - G + H_1 - H_2,$$

$$F' = \frac{\hbar^2}{m_0^2} \sum_l^{\Gamma_1;\Gamma_2^-}{}' \frac{|\langle X|p_x|u_l\rangle|^2}{E_v - E_l},$$

$$G \equiv \frac{\hbar^2}{2m_0^2} \sum_l^{\Gamma_{12};\Gamma_{12}^-}{}' \frac{|\langle X|p_x|u_l\rangle|^2}{E_v - E_l},$$

$$H_1 = \frac{\hbar^2}{m_0^2} \sum_l^{\Gamma_{15};\Gamma_{15}^-}{}' \frac{|\langle X|p_y|u_l\rangle|^2}{E_v - E_l},$$

$$H_2 = \frac{\hbar^2}{m_0^2} \sum_l^{\Gamma_{25};\Gamma_{25}^-}{}' \frac{|\langle X|p_y|u_l\rangle|^2}{E_v - E_l},$$

$$A' = \frac{\hbar^2}{m_0^2} \sum_l^{\Gamma_{15};\Gamma_{25}^+}{}' \frac{|\langle S|p_x|u_l\rangle|^2}{E_c - E_l},$$

$$B = \frac{2\hbar^2}{m_0^2} \sum_{l\in\Gamma_{15}}{}' \frac{\langle S|p_x|u_l\rangle\langle u_l|p_y|Z\rangle}{\frac{E_0}{2} - E_l}.$$

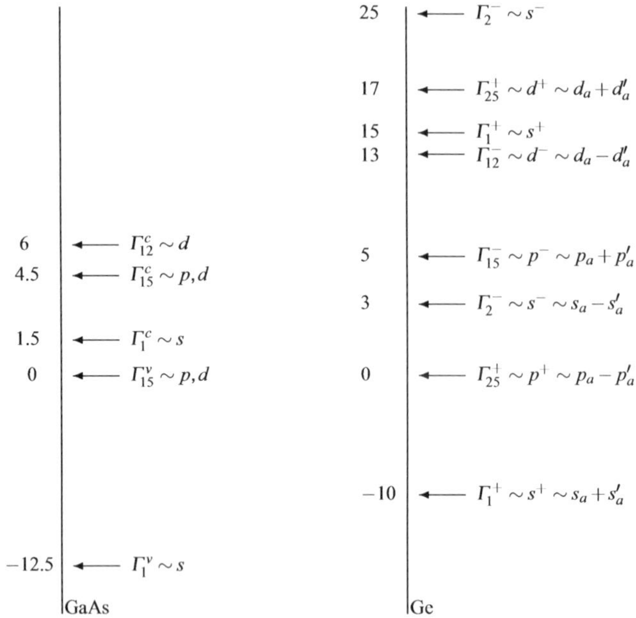

Fig. 4.3 Zone-center energy ordering for zincblende and diamond structure semiconductors. The energies on the left-hand side of the vertical lines are in eV's

4.4 Full-Zone $k \cdot p$ Model

The $k \cdot p$ theory has been extended to the full Brillouin zone for both bulk [5, 65–77] and heterostructures [78, 79]. The original objective was to reproduce a band structure that could be used to compute the optical properties.

The approach adopted so far has been to simply extend the first-order Kane models with a larger basis set. The rationale is that this should allow for a more complete representation of the band structure. Typically 15–30 bands are required in order to obtain a reasonable fit. The parameters are the energy gaps and momentum matrix elements at $\mathbf{k} = \mathbf{0}$, primarily obtained by fitting to first-principles calculations.

4.4.1 15-Band Model

The first attempt at a full-zone theory is the 1966 paper by Cardona and Pollak [5]. They considered a 15-band model for Si and Ge, these coming from the [000], $(2\pi/a)[111]$ and $(2\pi/a)[200]$ plane-wave states. This minimal-basis set appears adequate since the next set of $(2\pi/a)[220]$ states are much farther away in energy. There are 10 parameters; the choice of parameters was made using a combination of experimental data, free-electron values and pseudopotential fitting.

The basis states are given in Table 4.3. Cardona and Pollak premultiplied the odd-parity functions by i in order to generate real momentum matrix elements; for $\Gamma_{12'}$, we will only do so for $\Gamma_{12'}^{(2)}$. Thus, we will write the odd-parity states as $|\mathrm{i}\Gamma\rangle$ in order to make the presence of the complex number unambiguous. Next, we note that all matrix elements between even-parity and between odd-parity states are zero. The matrix elements are easily obtained. For example, for $\langle \Gamma_{25'}^{l}|\mathbf{k}\cdot\mathbf{p}|\mathrm{i}\Gamma_{15}\rangle$,

$$\begin{aligned}
\langle yz|\mathbf{k}\cdot\mathbf{p}|\mathrm{i}x\rangle &= 0,\\
\langle yz|\mathbf{k}\cdot\mathbf{p}|\mathrm{i}y\rangle &= \langle yz|p_z|\mathrm{i}y\rangle k_z = \frac{Q}{2}k_z,\\
\langle yz|\mathbf{k}\cdot\mathbf{p}|\mathrm{i}z\rangle &= \frac{Q}{2}k_y,
\end{aligned} \tag{4.19}$$

since $Q = 2\mathrm{i}\langle \Gamma_{25'}^{l}|p|\Gamma_{15}\rangle = 2\langle \Gamma_{25'}^{l}|p|\mathrm{i}\Gamma_{15}\rangle$, and we have left out a factor of $\hbar/m_0$ for the off-diagonal matrix elements for clarity. For $\langle \mathrm{i}\Gamma_{15}|\mathbf{k}\cdot\mathbf{p}|\Gamma_{25'}^{u}\rangle$,

$$\begin{aligned}
\langle \mathrm{i}y|\mathbf{k}\cdot\mathbf{p}|yz\rangle &= \langle \mathrm{i}y|p_z|yz\rangle k_z = -\mathrm{i}\langle y|p_z|yz\rangle k_z = -\mathrm{i}\langle yz|p_z|y\rangle^* k_z\\
&= \mathrm{i}\langle yz|p_z|y\rangle k_z = \langle yz|p_z|\mathrm{i}y\rangle k_z\\
&= \frac{Q'}{2}k_z.
\end{aligned} \tag{4.20}$$

For $\langle \Gamma_{25'}^{l}|\mathbf{k}\cdot\mathbf{p}|i\Gamma_{2'}^{l}\rangle$, using

$$\langle yz|p_x|\mathrm{i}s^-\rangle \overset{C_{3xyz}(yzx)}{=} \langle zx|p_y|\mathrm{i}s^-\rangle \overset{C_{3xyz}(yzx)}{=} \langle xy|p_z|\mathrm{i}s^-\rangle \equiv \frac{P}{2},$$

we have

$$\begin{aligned}
\langle yz|\mathbf{k}\cdot\mathbf{p}|\mathrm{i}s^-\rangle &= \frac{P}{2}k_x,\\
\langle zx|\mathbf{k}\cdot\mathbf{p}|\mathrm{i}s^-\rangle &= \frac{P}{2}k_y,\\
\langle xy|\mathbf{k}\cdot\mathbf{p}|\mathrm{i}s^-\rangle &= \frac{P}{2}k_z.
\end{aligned} \tag{4.21}$$

For $\langle \Gamma_{25'}^{l}|\mathbf{k}\cdot\mathbf{p}|\mathrm{i}\Gamma_{12'}\rangle$, we use the definition of Cardona and Pollak:

$$\Gamma_{12'}^{(1)} = \frac{1}{\sqrt{2}}(\gamma_1^- - \gamma_2^-), \tag{4.22}$$

$$\Gamma_{12'}^{(2)} = \frac{1}{\sqrt{2}}(\gamma_1^- + \gamma_2^-), \tag{4.23}$$

and, rewriting Eq. (3.20) here,

Table 4.3 15-band Cardona–Pollak Hamiltonian [5]

$$
\left[\begin{array}{ccccccccccccccc}
 & \Gamma^{l}_{25'} & & & \mathrm{i}\Gamma_{15} & & & \Gamma^{u}_{25'} & & \Gamma^{u}_{1} & \Gamma^{l}_{1} & \mathrm{i}\Gamma^{l}_{2'} & & \mathrm{i}\Gamma_{12'} & \mathrm{i}\Gamma^{u}_{2'} \\
yz & zx & xy & \mathrm{i}x & \mathrm{i}y & \mathrm{i}z & yz & zx & xy & s^{+} & s^{+} & \mathrm{i}s^{-} & d^{-}_{1} & \mathrm{i}d^{-}_{2} & \mathrm{i}s^{-} \\
0 & 0 & 0 & 0 & \frac{Q}{2}k_z & \frac{Q}{2}k_y & 0 & 0 & 0 & 0 & 0 & \frac{P}{2}k_x & \frac{R}{\sqrt{2}}k_x & 0 & \frac{P''}{2}k_x \\
0 & 0 & 0 & \frac{Q}{2}k_z & 0 & \frac{Q}{2}k_x & 0 & 0 & 0 & 0 & 0 & \frac{P}{2}k_y & -\frac{R}{2\sqrt{2}}k_y & -\frac{\sqrt{3}}{2\sqrt{2}}Rk_y & \frac{P''}{2}k_y \\
0 & 0 & 0 & \frac{Q}{2}k_y & \frac{Q}{2}k_x & 0 & 0 & 0 & 0 & 0 & 0 & \frac{P}{2}k_z & -\frac{R}{2\sqrt{2}}k_z & \frac{\sqrt{3}}{2\sqrt{2}}Rk_z & \frac{P''}{2}k_z \\
 & & & E(\Gamma_{15}) & 0 & 0 & 0 & \frac{Q'}{2}k_z & \frac{Q'}{2}k_y & \frac{T}{2}k_x & \frac{T'}{2}k_x & 0 & 0 & 0 & 0 \\
 & & & 0 & E(\Gamma_{15}) & 0 & \frac{Q'}{2}k_z & 0 & \frac{Q'}{2}k_x & \frac{T}{2}k_y & \frac{T'}{2}k_y & 0 & 0 & 0 & 0 \\
 & & & 0 & 0 & E(\Gamma_{15}) & \frac{Q'}{2}k_y & \frac{Q'}{2}k_x & 0 & \frac{T}{2}k_z & \frac{T'}{2}k_z & 0 & 0 & 0 & 0 \\
 & & & & & & E(\Gamma^{u}_{25'}) & 0 & 0 & 0 & 0 & \frac{P'}{2}k_x & \frac{R'}{\sqrt{2}}k_x & 0 & \frac{P'''}{2}k_x \\
 & & & & & & 0 & E(\Gamma^{u}_{25'}) & 0 & 0 & 0 & \frac{P'}{2}k_y & -\frac{R'}{2\sqrt{2}}k_y & -\frac{\sqrt{3}}{2\sqrt{2}}R'k_y & \frac{P'''}{2}k_y \\
 & & & & & & 0 & 0 & E(\Gamma^{u}_{25'}) & 0 & 0 & \frac{P'}{2}k_z & -\frac{1}{2\sqrt{2}}R'k_z & \frac{\sqrt{3}}{2\sqrt{2}}R'k_z & \frac{P'''}{2}k_z \\
 & & & & & & & & & E(\Gamma^{u}_{1}) & 0 & 0 & 0 & 0 & 0 \\
 & & & & & & & & & & E(\Gamma^{l}_{1}) & 0 & 0 & 0 & 0 \\
 & & & & & & & & & & & E(\Gamma^{l}_{2'}) & 0 & 0 & 0 \\
 & & & & & & & & & & & & E(\Gamma_{12'}) & 0 & 0 \\
 & & & & & & & & & & & & & E(\Gamma_{12'}) & 0 \\
 & & & & & & & & & & & & & & E(\Gamma^{u}_{2'})
\end{array}\right] + \frac{\hbar^2 k^2}{2m_0},
$$

where

$$P = 2\mathrm{i}\frac{\hbar}{m_0}\langle\Gamma^{l}_{25'}|p|\Gamma^{l}_{2'}\rangle,\ P' = 2\mathrm{i}\frac{\hbar}{m_0}\langle\Gamma^{l}_{25'}|p|\Gamma^{l}_{2'}\rangle,$$

$$P'' = 2\mathrm{i}\frac{\hbar}{m_0}\langle\Gamma^{l}_{25'}|p|\Gamma^{u}_{2'}\rangle,\ P''' = 2\mathrm{i}\frac{\hbar}{m_0}\langle\Gamma^{u}_{25'}|p|\Gamma^{u}_{2'}\rangle,$$

$$Q = 2\mathrm{i}\frac{\hbar}{m_0}\langle\Gamma^{l}_{25'}|p|\Gamma_{15}\rangle,\ Q' = 2\mathrm{i}\frac{\hbar}{m_0}\langle\Gamma^{u}_{25'}|p|\Gamma_{15}\rangle,$$

$$R = 2\frac{\hbar}{m_0}\langle(yz)^{l}|p_x|\gamma^{-}_{1}\rangle,\ R' = 2\frac{\hbar}{m_0}\langle(yz)^{u}|p_x|\gamma^{-}_{1}\rangle,$$

$$T = 2\mathrm{i}\frac{\hbar}{m_0}\langle\Gamma^{u}_{1}|p|\Gamma_{15}\rangle,\ T' = 2\mathrm{i}\frac{\hbar}{m_0}\langle\Gamma^{l}_{1}|p|\Gamma_{15}\rangle.$$

$$\gamma_1^- = x^2 + \omega y^2 + \omega^2 z^2,$$
$$\gamma_2^- = x^2 + \omega^2 y^2 + \omega z^2,$$

with $\gamma_i^- = \gamma_i - \gamma_i'$, $\omega^3 = 1$, $\omega + \omega^2 + \omega^3 = 0$. Then,

$$\Gamma_{12'}^{(1)} = (\omega - \omega^2)(y^2 - z^2) \sim (y^2 - z^2),$$
$$\Gamma_{12'}^{(2)} = 2x^2 + (\omega + \omega^2)y^2 + (\omega^2 + \omega)z^2 = 3x^2 - r^2.$$

Now, in order to reproduce the Hamiltonian of Cardona and Pollak along the Δ line, their Eq. (9) in particular, it will be necessary to give a different definition for R as given in [5]. Let (from Chap. 3)

$$\langle yz|p_x|\gamma_1^-\rangle = r,\ \langle yz|p_x|\gamma_2^-\rangle = -r,$$

and using

$$\langle zx|p_y|\gamma_1^-\rangle = \omega\langle yz|p_y|\gamma_1^-\rangle,$$
$$\langle zx|p_y|\gamma_2^-\rangle = -\omega^2\langle yz|p_y|\gamma_1^-\rangle,$$
$$\langle xy|p_z|\gamma_1^-\rangle = \omega^2\langle yz|p_y|\gamma_1^-\rangle,$$
$$\langle xy|p_z|\gamma_2^-\rangle = -\omega\langle yz|p_y|\gamma_1^-\rangle,$$

then

$$\langle yz|\mathbf{k}\cdot\mathbf{p}|\Gamma_{12'}^{(1)}\rangle = \frac{1}{\sqrt{2}}\langle yz|\mathbf{k}\cdot\mathbf{p}|(\gamma_1^- - \gamma_2^-)\rangle = \sqrt{2}r \equiv \frac{1}{\sqrt{2}}R, \tag{4.24}$$

if

$$R = 2r = \sqrt{2}\langle yz|\mathbf{k}\cdot\mathbf{p}|\Gamma_{12'}^{(1)}\rangle. \tag{4.25}$$

Also,

$$\langle yz|\mathbf{k}\cdot\mathbf{p}|\mathrm{i}\Gamma_{12'}^{(2)}\rangle = \frac{1}{\sqrt{2}}\langle yz|\mathbf{k}\cdot\mathbf{p}|\mathrm{i}(\gamma_1^- + \gamma_2^-)\rangle = 0, \tag{4.26}$$

which, as stated by Cardona and Pollak, says that the matrix element between the X state and $\Gamma_{12'}^{(2)}$ is zero. Also,

$$
\begin{aligned}
\langle zx|p_y|\Gamma_{12'}^{(1)}\rangle &= \frac{1}{\sqrt{2}}\langle zx|p_y|(\gamma_1^- - \gamma_2^-)\rangle = \frac{1}{\sqrt{2}}(\omega r + \omega^2 r) = -\frac{r}{\sqrt{2}} = -\frac{R}{2\sqrt{2}},\\
\langle zx|\mathbf{k}\cdot\mathbf{p}|\Gamma_{12'}^{(2)}\rangle &= \frac{1}{\sqrt{2}}\langle zx|\mathbf{k}\cdot\mathbf{p}|(\gamma_1^- + \gamma_2^-)\rangle = \frac{1}{\sqrt{2}}k_y(\omega r - \omega^2 r) = \mathrm{i}\frac{\sqrt{3}}{2\sqrt{2}}Rk_y,\\
\langle xy|\mathbf{k}\cdot\mathbf{p}|\Gamma_{12'}^{(1)}\rangle &= \frac{1}{\sqrt{2}}\langle xy|\mathbf{k}\cdot\mathbf{p}|(\gamma_1^- - \gamma_2^-)\rangle = \frac{1}{\sqrt{2}}k_z(\omega^2 r + \omega^2 r) = -\frac{R}{2\sqrt{2}}k_z,\\
\langle xy|\mathbf{k}\cdot\mathbf{p}|\Gamma_{12'}^{(2)}\rangle &= \frac{1}{\sqrt{2}}\langle xy|\mathbf{k}\cdot\mathbf{p}|(\gamma_1^- + \gamma_2^-)\rangle = \frac{1}{\sqrt{2}}k_z(\omega^2 r - \omega r) = -\mathrm{i}\frac{\sqrt{3}}{2\sqrt{2}}Rk_y.
\end{aligned}
$$

Hence, we premultiply $\Gamma_{12'}^{(2)}$ by i to make the matrix element real.

Along certain directions, the Hamiltonian block diagonalizes. For example, along Δ, one state remains free-electron-like ($\Gamma_{12'}^{(2)}$) and the others separate into the following three blocks:

$$
\begin{pmatrix}
\mathrm{i}\Gamma_{2'}^{(1)} & \Gamma_{25'}^{(1)} & \mathrm{i}\Gamma_{12'}^{(1)} & \Gamma_{25'}^{(u)} & \mathrm{i}\Gamma_{2'}^{(u)} \\
E(\Gamma_{2'}^{(1)}) + \frac{\hbar^2 k_x^2}{2m_0} & \frac{P}{2}k_x & 0 & \frac{P'}{2}k_x & 0 \\
 & \frac{\hbar^2 k_x^2}{2m_0} & \frac{R}{\sqrt{2}}k_x & 0 & \frac{P''}{2}k_x \\
 & & E(\Gamma_{12'}) + \frac{\hbar^2 k_x^2}{2m_0} & \frac{R'}{\sqrt{2}}k_x & 0 \\
 & & & E(\Gamma_{25'}^{(u)}) + \frac{\hbar^2 k_x^2}{2m_0} & \frac{P'''}{2}k_x \\
 & & & & E(\Gamma_{2'}^{(u)}) + \frac{\hbar^2 k_x^2}{2m_0}
\end{pmatrix}, \tag{4.27}
$$

$$
\begin{pmatrix}
\Gamma_{25'}^{(l)} & \mathrm{i}\Gamma_{15} & \Gamma_{25'}^{(u)} \\
\frac{\hbar^2 k_x^2}{2m_0} & \frac{Q}{2}k_x & 0 \\
 & E(\Gamma_{15}) + \frac{\hbar^2 k_x^2}{2m_0} & \frac{Q'}{2}k_x \\
 & & E(\Gamma_{25'}^{(u)}) + \frac{\hbar^2 k_x^2}{2m_0}
\end{pmatrix}, \tag{4.28}
$$

$$
\begin{pmatrix}
\mathrm{i}\Gamma_{15} & \Gamma_{1}^{(u)} & \Gamma_{1}^{(l)} \\
E(\Gamma_{15}) + \frac{\hbar^2 k_x^2}{2m_0} & \frac{T}{2}k_x & \frac{T'}{2}k_x \\
0 & E(\Gamma_{1}^{(u)}) + \frac{\hbar^2 k_x^2}{2m_0} & 0 \\
0 & 0 & E(\Gamma_{1}^{(l)}) + \frac{\hbar^2 k_x^2}{2m_0}
\end{pmatrix}. \tag{4.29}
$$

These blocks agree with those given by Cardona and Pollak [5], the last two being of symmetry Δ_5 (two-fold degenerate) and Δ_1 (three-fold degenerate).

The parameters obtained for Si and Ge are given in Table 4.4. The resulting band structures for Si and Ge are given in Fig. 4.4. We note that we have complete agreement with Cardona and Pollak.

Cardona and Pollak have discussed the fact that the X_3 band does not intersect the Brillouin zone edge perpendicularly; this being due to the neglect of higher states. We also note that the X_1 degeneracy, a characteristic of the DM lattice, is not

Table 4.1 Parameters for 15-band model in atomic units [5]

	Si	Ge
P	1.200	1.360
Q	1.070	1.050
R	0.830	0.8049
P''	0.100	0.100
P'	−0.090	0.1715
Q'	−0.807	−0.752
R'	1.210	1.4357
P'''	1.320	1.6231
T	1.080	1.2003
T'	0.206	0.5323

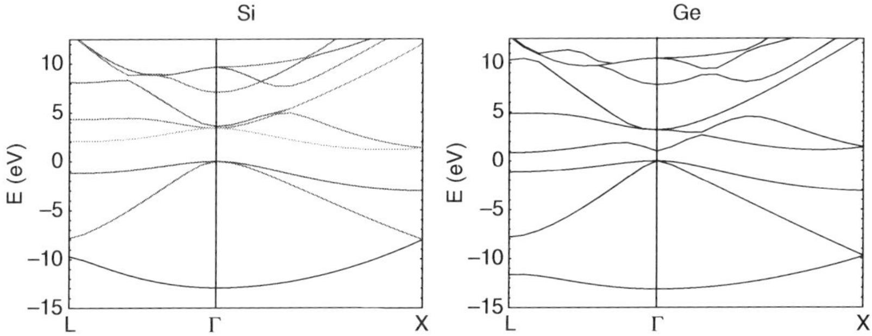

Fig. 4.4 Si and Ge band structures using 15-band Cardona–Pollak [5] $k \cdot p$ model

automatically enforced since the two states forming the degeneracy have different symmetries along Δ (Δ_1 and $\Delta_{2'}$, respectively). Indeed, the solutions come from two distinct subblocks. Hence, in this model, it appears that the degeneracy is not symmetry related but is enforced only by the appropriate choice of $k \cdot p$ parameters. Cardona and Pollak have also extended this work to ZB by adding six additional asymmetry parameters [65], and also to the band structure with spin-orbit [5, 67].

4.4.2 *Other Models*

Only in 2001 was the work of Cardona and Pollak 'modernized.' Cavassilas et al. [69] proposed a 20-band model for GaAs and Si. This consists of sp^3s^* bands and accounts for d states via Luttinger-like parameters. There are a total of 31 parameters in this model and the full Hamiltonian is given in [70]. The same group has subsequently extended their models into 24-band [71] and 30-band [72–77] ones.

4.5 Wurtzite

Various Kane-like models for wurtzite materials will now be presented.

4.5.1 Four-Band: Andreev-O'Reilly

The four-band model includes conduction-valence band coupling [49]. We label new $\widetilde{u}_i$ basis states defined in Table 4.5; $\widetilde{u}_2$ differs from u_1 by a sign.

Table 4.5 Basis states for the four-band model of wurtzite [49]

$\lvert\widetilde{u}_1\rangle$	$\lvert S\uparrow\rangle$
$\lvert\widetilde{u}_2\rangle$	$\frac{1}{\sqrt{2}}\lvert(X+\mathrm{i}Y)\uparrow\rangle$
$\lvert\widetilde{u}_3\rangle$	$\frac{1}{\sqrt{2}}\lvert(X-\mathrm{i}Y)\uparrow\rangle$
$\lvert\widetilde{u}_4\rangle$	$\lvert Z\uparrow\rangle$

Andreev and O'Reilly [49] gives

$$H_{\mathrm{AO}}(\mathbf{k}) = \begin{pmatrix} |\widetilde{u}_1\rangle & |\widetilde{u}_2\rangle & |\widetilde{u}_3\rangle & |\widetilde{u}_4\rangle \\ E_0 & \frac{1}{\sqrt{2}}P_\perp k_+ & \frac{1}{\sqrt{2}}P_\perp k_- & P_{||}k_z \\ \frac{1}{\sqrt{2}}P_\perp k_- & F & K^* & -H^* \\ \frac{1}{\sqrt{2}}P_\perp k_+ & K & F & H \\ P_{||}k_z & -H & H^* & \lambda \end{pmatrix}, \tag{4.30}$$

where

$$\begin{aligned} F &= \Delta_1 + \lambda + \theta, \\ K &= \widetilde{A}_5(k_x + \mathrm{i}k_y)^2, \\ H &= \mathrm{i}\widetilde{A}_6(k_x + \mathrm{i}k_y)k_z, \\ \theta &= \widetilde{A}_3 k_z^2 + \widetilde{A}_4(k_x^2 + k_y^2), \\ \lambda &= -\Delta_1 + \widetilde{A}_1 k_z^2 + \widetilde{A}_2(k_x^2 + k_y^2), \end{aligned} \tag{4.31}$$

and

$$\begin{aligned} \widetilde{A}_1 &= A_1 + \frac{2m_0}{\hbar^2}\frac{P_\perp^2}{E_0}, \\ \widetilde{A}_2 &= A_2, \\ \widetilde{A}_3 &= A_3 - \frac{2m_0}{\hbar^2}\frac{P_\perp^2}{E_0}, \\ \widetilde{A}_4 &= A_4 + \frac{m_0}{\hbar^2}\frac{P_{||}^2}{E_0}, \\ \widetilde{A}_5 &= A_5 + \frac{m_0}{\hbar^2}\frac{P_{||}^2}{E_0}, \end{aligned} \tag{4.32}$$

$$\widetilde{A}_6 = A_6 + \frac{\sqrt{2}m_0}{\hbar^2}\frac{P_{\|}P_{\perp}}{E_0},$$

$$P_{\|}^2 = \frac{\hbar^2}{2m_0}\left(\frac{m_0}{m_c^{\|}} - 1\right)\frac{(E_0 + \Delta_1 + \Delta_2)(E_0 + 2\Delta_2) - 2\Delta_3^2}{E_0 + 2\Delta_2},$$

$$P_{\perp}^2 = \frac{\hbar^2}{2m_0}\left(\frac{m_0}{m_c^{\perp}} - 1\right)\frac{\left[(E_0 + \Delta_1 + \Delta_2)(E_0 + 2\Delta_2) - 2\Delta_3^2\right]E_0}{(E_0 + \Delta_2)(E_0 + \Delta_1 + \Delta_2) - \Delta_3^2}.$$

Anisotropy between the z-direction and the x-y plane leads to two momentum matrix elements $P_{\|}$ and $P_{\perp}$, and to the anisotropic electron mass as well. A slightly different form of the four-band model has been used recently by Rinke and coworkers [60] to fit to ab initio calculations and, thereby, extract $k \cdot p$ parameters for AlN, GaN and InN.

4.5.2 Eight-Band: Chuang–Chang

Chuang and Chang [46] wrote down an eight-band model for wurtzite (Table 4.6). This differs from the previously given valence-band Hamiltonian $H_{k=0}$ [Eq. (3.68)] in having linear $k \cdot p$ coupling to the conduction state.

4.5.3 Eight-Band: Gutsche–Jahne

A new formulation of the eight-band model using the Gutsche–Jahne basis is now presented. A list of the upper diagonal of the 8×8 Hamiltonian in the lower conduction and upper valence states: $|e \uparrow\rangle, |e \downarrow\rangle, |7 \uparrow\rangle, |7 \downarrow\rangle, 7' \uparrow\rangle, |7' \downarrow\rangle, |9 \uparrow\rangle, |9 \downarrow\rangle$ will be given next using the GJ basis as given in Table 4.7.

Using the character table (Table B.9), it is easy to compute non-zero contributions from the various terms in the $k \cdot p$ equation. We shall assume that all other states are so far remote that they do not contribute significantly up to first-order in k in second-order or higher-order perturbation theory. Note that this assumption does not preclude first-order, second-order, and third-order in k contributions to the energies because we account exactly for all couplings among the eight states forming the basis. In fact, we expect the present Hamiltonian to include the most important contributions up to third order in k.

We split the wurtzite eight-band Hamiltonian contributions up to linear-in-k in two parts. The first part excludes the contribution from $\boldsymbol{\sigma} \cdot (\mathbf{k} \times \nabla V)$ to linear spin splittings while the second part (following subsection) is the contribution from the $\sigma \cdot (\mathbf{k} \times \nabla V)$ term to linear spin splittings.

Table 4.6 Chuang and Chang's eight-band model for wurtzite [46]

$$H_{\mathrm{CC8}}(\mathbf{k}) = \frac{\hbar^2 k^2}{2m_0} + \begin{pmatrix} |\mathrm{i}S\uparrow\rangle & |u_1\rangle & |u_2\rangle & |u_3\rangle & |\mathrm{i}S\downarrow\rangle & |u_4\rangle & |u_5\rangle & |u_6\rangle \\ E_c & -\frac{1}{\sqrt{2}}P_2k_+ & \frac{1}{\sqrt{2}}P_2k_- & P_1k_z & 0 & 0 & 0 & 0 \\ -\frac{1}{\sqrt{2}}P_2k_- & E_v+\Delta_1+\Delta_2 & 0 & 0 & 0 & 0 & 0 & 0 \\ \frac{1}{\sqrt{2}}P_2k_+ & 0 & E_v+\Delta_1-\Delta_2 & 0 & 0 & 0 & 0 & \sqrt{2}\Delta_3 \\ P_1k_z & 0 & 0 & E_v & 0 & 0 & \sqrt{2}\Delta_3 & 0 \\ 0 & 0 & 0 & 0 & E_c & \frac{1}{\sqrt{2}}P_2k_- & -\frac{1}{\sqrt{2}}P_2k_+ & P_1k_z \\ 0 & 0 & 0 & 0 & \frac{1}{\sqrt{2}}P_2k_+ & E_v+\Delta_1+\Delta_2 & 0 & 0 \\ 0 & 0 & 0 & \sqrt{2}\Delta_3 & -\frac{1}{\sqrt{2}}P_2k_- & 0 & E_v+\Delta_1-\Delta_2 & 0 \\ 0 & 0 & \sqrt{2}\Delta_3 & 0 & P_1k_z & 0 & 0 & E_v \end{pmatrix},$$

where

$$\begin{aligned} k_\pm &= k_x \pm \mathrm{i}k_y, \\ E_c &= \langle S|H_0|S\rangle, \\ E_v+\Delta_1 &= \langle X|H_0|X\rangle = \langle Y|H_0|Y\rangle, \\ E_v &= \langle Z|H_0|Z\rangle, \\ -\mathrm{i}\Delta_2 &= \langle X|H_{\mathrm{s},z}|Y\rangle, \\ -\mathrm{i}\Delta_3 &= \langle Y|H_{\mathrm{s},x}|Z\rangle = \langle Z|H_{\mathrm{s},y}|X\rangle, \\ P_1 &= \frac{\hbar}{m_0}\langle \mathrm{i}S|p_z|Z\rangle, \\ P_2 &= \frac{\hbar}{m_0}\langle \mathrm{i}S|p_x|X\rangle = \frac{\hbar}{m_0}\langle \mathrm{i}S|p_y|Y\rangle. \end{aligned}$$

Table 4.7 Basis states for the eight-band model of wurtzite

$\|e\uparrow\rangle = q_s\|s\uparrow\rangle + q_z\|z\uparrow\rangle,$
$\|e\downarrow\rangle = q_s\|s\downarrow\rangle + q_z\|z\downarrow\rangle,$
$\|7\uparrow\rangle = \sqrt{1-q_7^2}\|u_5\downarrow\rangle - q_7\|u_1\uparrow\rangle,$
$\|9\uparrow\rangle = \sqrt{1-q_9^2}\|u_5\uparrow\rangle + q_9\|u_6\downarrow\rangle,$
$\|7\downarrow\rangle = -\sqrt{1-q_7^2}\|u_5^*\uparrow\rangle - q_7\|u_1\downarrow\rangle,$
$\|9\downarrow\rangle = \sqrt{1-q_9^2}\|u_5^*\downarrow\rangle - q_9\|u_6\uparrow\rangle,$
$\|7'\uparrow\rangle = q_7\|u_5\downarrow\rangle + \sqrt{1-q_7^2}\|u_1\uparrow\rangle,$
$\|7'\downarrow\rangle = -q_7\|u_5^*\downarrow\rangle + \sqrt{1-q_7^2}\|u_1\downarrow\rangle.$

4.5.3.1 Hamiltonian

The non-zero matrix elements to the upper diagonal as described above are

$$
\begin{aligned}
\langle e\uparrow|H|e\uparrow\rangle &= \langle e\downarrow|H|e\downarrow\rangle = E_e + \frac{\hbar^2k^2}{2m_0},\\
\langle 7\uparrow|H|7\uparrow\rangle &= \langle 7\downarrow|H|7\downarrow\rangle = E_7 + \frac{\hbar^2k^2}{2m_0},\\
\langle 7'\uparrow|H|7'\uparrow\rangle &= \langle 7'\downarrow|H|7'\downarrow\rangle = E_{7'} + \frac{\hbar^2k^2}{2m_0},\\
\langle 9\uparrow|H|9\uparrow\rangle &= \langle 9\downarrow|H|9\downarrow\rangle = E_9 + \frac{\hbar^2k^2}{2m_0},\\
\langle e\uparrow|H|7\uparrow\rangle &= -\sqrt{1-q_7^2}q_z\langle z|H_{s,+}|u_5^*\rangle - q_7q_s\frac{\hbar}{m_0}k_z\langle s|p_z|u_1\rangle,\\
\langle e\uparrow|H|7\downarrow\rangle &= -\sqrt{1-q_7^2}q_s\frac{1}{2}\frac{\hbar}{m_0}k_-\langle s|p_+|u_5^*\rangle,\\
\langle e\uparrow|H|7'\uparrow\rangle &= -q_7q_z\langle z|H_{s,+}|u_5^*\rangle + \sqrt{1-q_7^2}q_s\frac{\hbar}{m_0}k_z\langle s|p_z|u_1\rangle,\\
\langle e\uparrow|H|7'\downarrow\rangle &= -q_7q_s\frac{1}{2}\frac{\hbar}{m_0}k_-\langle s|p_+|u_5^*\rangle,\\
\langle e\uparrow|H|9\uparrow\rangle &= \sqrt{1-q_9^2}q_s\frac{1}{2}\frac{\hbar}{m_0}k_+\langle s|p_-|u_5\rangle,\\
\langle e\downarrow|H|7\uparrow\rangle &= \sqrt{1-q_7^2}q_s\frac{1}{2}\frac{\hbar}{m_0}k_+\langle s|p_-|u_5\rangle,\\
\langle e\downarrow|H|7\downarrow\rangle &= -\sqrt{1-q_7^2}q_z\langle z|H_{s,+}|u_5^*\rangle - q_7q_s\frac{\hbar}{m_0}k_z\langle s|p_z|u_1\rangle,
\end{aligned}
\tag{4.33}
$$

$$\langle e\downarrow|H|7'\uparrow\rangle = q_7 q_s \frac{1}{2}\frac{\hbar}{m_0}k_+\langle s|p_-|u_5\rangle,$$

$$\langle e\downarrow|H|7'\downarrow\rangle = -q_7 q_z \langle z|H_{\mathrm{s},+}|u_5^*\rangle + \sqrt{1-q_7^2}\,q_s\frac{\hbar}{m_0}k_z\langle s|p_z|u_1\rangle,$$

$$\langle e\downarrow|H|9\downarrow\rangle = \sqrt{1-q_9^2}\,q_s\frac{1}{2}\frac{\hbar}{m_0}k_-\langle s|p_+|u_5^*\rangle,$$

$$\langle 7\uparrow|H|7'\downarrow\rangle = \frac{1}{2}\frac{\hbar}{m_0}k_-\langle u_5|p_+|u_1\rangle,$$

$$\langle 7\uparrow|H|9\uparrow\rangle = \frac{1}{2}\left(q_9\sqrt{1-q_7^2}\langle u_5|p_-|u_6\rangle - q_7\sqrt{1-q_9^2}\langle u_1|p_-|u_5\rangle\right)\frac{\hbar}{m_0}k_+,$$

$$\langle 7\downarrow|H|7'\uparrow\rangle = -\frac{1}{2}\frac{\hbar}{m_0}k_+\langle u_1|p_-|u_5\rangle,$$

$$\langle 7\downarrow|H|9\downarrow\rangle = \frac{1}{2}\frac{\hbar}{m_0}k_-\left(q_9\sqrt{1-q_7^2}\langle u_5^*|p_+|u_6^*\rangle - q_7\sqrt{1-q_9^2}\langle u_1|p_+|u_5^*\rangle\right),$$

$$\langle 7'\uparrow|H|9\uparrow\rangle = \frac{1}{2}\frac{\hbar}{m_0}k_+\left(q_7q_9\langle u_5|p_-|u_6\rangle + \sqrt{1-q_7^2}\sqrt{1-q_9^2}\langle u_1|p_-|u_5\rangle\right),$$

$$\langle 7'\downarrow|H|9\downarrow\rangle = \frac{1}{2}\frac{\hbar}{m_0}k_-\left(q_7q_9\langle u_5^*|p_+|u_6^*\rangle + \sqrt{1-q_7^2}\sqrt{1-q_9^2}\langle u_1|p_+|u_5^*\rangle\right).$$

Here

$$\begin{aligned} H_{\mathrm{s},\pm} &= H_{\mathrm{s},x} \pm \mathrm{i}H_{\mathrm{s},y}, \\ k_\pm &= k_x \pm \mathrm{i}k_y, \\ p_\pm &= p_x \pm \mathrm{i}p_y. \end{aligned} \tag{4.34}$$

In deriving the above matrix, we have several times made use of the relation

$$\langle u_i|p_z|u_i\rangle \propto \langle u_i|[H_0, z]|u_i\rangle = 0, \tag{4.35}$$

where H_0 is the Hamiltonian in the absence of spin-orbit interaction with eigenstates u_i at the Γ point.

We note that the energy E_e denotes the Γ-point energy of conduction e states in the bulk case in the absence of interband (electron-hole) spin-orbit couplings and in the absence of external fields (a similar decoupling of spin-orbit couplings between electron and hole states was assumed in [48]). The same conclusion applies for, e.g., the valence-band energies E_7 but observe that intraband (hole-hole) spin-orbit couplings are accounted for in the GJ basis. Nevertheless, it is an easy matter to obtain the Γ-point energies taking all spin-orbit couplings among the eight basis states into account by diagonalizing the present Hamiltonian at $\mathbf{k} = \mathbf{0}$ as we shall show below. This is because the only off-diagonal terms are spin-orbit terms such as $\sqrt{1-q_7^2}\,q_z\langle z|H_{\mathrm{s},+}|u_5\rangle$.

4.5.3.2 Linear-in-k Term

This subsection describes the additional contribution to the eight-band Hamiltonian model from the linear-in-k spin-orbit term, $\hbar^2\boldsymbol{\sigma}\cdot(\mathbf{k}\times\nabla V)/(4m_0^2c^2)$ (matrix elements left out are zero):

$$\begin{aligned}
\langle e\uparrow|H_{\mathrm{so},k}|7\uparrow\rangle &= \frac{\hbar^2}{4m_0^2c^2}\sqrt{1-q_7^2}q_s\mathrm{i}k_z\langle s|V_-|u_5\rangle,\\
\langle e\uparrow|H_{\mathrm{so},k}|7\downarrow\rangle &= \frac{\hbar^2}{4m_0^2c^2}\left(-\frac{1}{2}\sqrt{1-q_7^2}q_s\mathrm{i}k_-\langle s|V_+|u_5^*\rangle+\mathrm{i}k_-q_7\langle s|V_z|u_1\rangle\right),\\
\langle e\uparrow|H_{\mathrm{so},k}|7'\uparrow\rangle &= \frac{\hbar^2}{4m_0^2c^2}q_7q_s\mathrm{i}k_z\langle s|V_-|u_5\rangle,\\
\langle e\uparrow|H_{\mathrm{so},k}|7'\downarrow\rangle &= \frac{\hbar^2}{4m_0^2c^2}\left(-q_7q_s\frac{1}{2}\mathrm{i}k_-\langle s|V_+|u_5^*\rangle-\mathrm{i}k_-\sqrt{1-q_7^2}q_s\langle s|V_z|u_1\rangle\right),\\
\langle e\uparrow|H_{\mathrm{so},k}|9\uparrow\rangle &= -\frac{\hbar^2}{4m_0^2c^2}\frac{1}{2}\sqrt{1-q_9^2}q_s\mathrm{i}k_+\langle s|V_-|u_5\rangle,\\
\langle e\downarrow|H_{\mathrm{so},k}|7\uparrow\rangle &= \frac{\hbar^2}{4m_0^2c^2}\left(\frac{1}{2}\sqrt{1-q_7^2}q_s\mathrm{i}k_+\langle s|V_-|u_5\rangle-\mathrm{i}k_+q_7q_s\langle s|V_z|u_1\rangle\right),\\
\langle e\downarrow|H_{\mathrm{so},k}|7\downarrow\rangle &= \frac{\hbar^2}{4m_0^2c^2}\sqrt{1-q_7^2}q_s\mathrm{i}k_z\langle s|V_+|u_5^*\rangle,\\
\langle e\downarrow|H_{\mathrm{so},k}|7'\uparrow\rangle &= \frac{\hbar^2}{4m_0^2c^2}\left(q_7q_s\frac{1}{2}\mathrm{i}k_+\langle s|V_-|u_5\rangle+\mathrm{i}k_+\sqrt{1-q_7^2}q_s\langle s|V_z|u_1\rangle\right),\\
\langle e\downarrow|H_{\mathrm{so},k}|7'\downarrow\rangle &= \frac{\hbar^2}{4m_0^2c^2}q_7q_s\mathrm{i}k_z\langle s|V_+|u_5^*\rangle,\\
\langle e\downarrow|H_{\mathrm{so},k}|9\downarrow\rangle &= -\frac{\hbar^2}{4m_0^2c^2}\frac{1}{2}\sqrt{1-q_9^2}q_s\mathrm{i}k_-\langle s|V_+|u_5^*\rangle, \qquad (4.36)\\
\langle 7\uparrow|H_{\mathrm{so},k}|7'\uparrow\rangle &= -\frac{\hbar^2}{4m_0^2c^2}\mathrm{i}k_z\langle u_5|V_+|u_1\rangle,\\
\langle 7\uparrow|H_{\mathrm{so},k}|7'\downarrow\rangle &= -\frac{\hbar^2}{4m_0^2c^2}\mathrm{i}k_-(1-2q_7^2)\frac{1}{2}\langle u_5|V_+|u_1\rangle,\\
\langle 7\uparrow|H_{\mathrm{so},k}|9\uparrow\rangle &= \frac{\hbar^2}{4m_0^2c^2}\mathrm{i}\frac{1}{2}k_+\left(q_9\sqrt{1-q_7^2}\langle u_5|V_-|u_6\rangle+q_7\sqrt{1-q_9^2}\langle u_1|V_-|u_5\rangle\right),\\
\langle 7\downarrow|H_{\mathrm{so},k}|7'\uparrow\rangle &= -\frac{\hbar^2}{4m_0^2c^2}(2q_7^2-1)\frac{1}{2}\mathrm{i}k_+\langle u_1|V_-|u_5\rangle,\\
\langle 7\downarrow|H_{\mathrm{so},k}|7'\downarrow\rangle &= -\frac{\hbar^2}{4m_0^2c^2}\mathrm{i}k_z\langle u_1|V_-|u_5\rangle,\\
\langle 7\downarrow|H_{\mathrm{so},k}|9\downarrow\rangle &= \frac{\hbar^2}{4m_0^2c^2}\mathrm{i}\frac{1}{2}k_-\left(q_9\sqrt{1-q_7^2}(\langle u_5|V_-|u_6\rangle)^*+q_7\sqrt{1-q_9^2}\langle u_1|V_+|u_5^*\rangle\right),\\
\langle 7'\uparrow|H_{\mathrm{so},k}|7'\downarrow\rangle &= -\frac{\hbar^2}{4m_0^2c^2}q_7\sqrt{1-q_7^2}\mathrm{i}k_-\langle u_5|V_+|u_1\rangle,
\end{aligned}$$

$$\langle 7' \uparrow |H_{\text{so},k}|9 \uparrow\rangle = \frac{\hbar^2}{4m_0^2c^2}\mathrm{i}\frac{1}{2}k_+\left(q_7q_9\langle u_5|V_-|u_6\rangle - \sqrt{1-q_7^2}\sqrt{1-q_9^2}\langle u_1|V_-|u_5\rangle\right),$$
$$\langle 7' \downarrow |H_{\text{so},k}|9 \downarrow\rangle = \frac{\hbar^2}{4m_0^2c^2}\mathrm{i}\frac{1}{2}k_-\left(q_7q_9\langle u_5|V_-|u_6\rangle^* - \sqrt{1-q_7^2}\sqrt{1-q_9^2}\langle u_1|V_-|u_5\rangle\right),$$

where we introduced the notation:

$$V_z = \frac{\partial V}{\partial z},$$
$$V_\pm = \left(\frac{\partial V}{\partial x} \pm \mathrm{i}\frac{\partial V}{\partial y}\right). \tag{4.37}$$

We point out that matrix elements of V_z, V_+, V_- are given in terms of the corresponding matrix elements of p_z, p_+, p_- by use of the following simple relation

$$\nabla V = \frac{\mathrm{i}}{\hbar}[\mathbf{p}, H_0]. \tag{4.38}$$

Thus,

$$\langle u_i|\nabla V|u_j\rangle = \frac{\mathrm{i}(E_j - E_i)}{\hbar}\langle u_i|\mathbf{p}|u_j\rangle, \tag{4.39}$$

where E_i, E_j denote Γ-point energy eigenvalues in the absence of spin-orbit interaction. Employing the latter equation allows us to significantly reduce the number of independent parameters in the eight-band wurtzite model.

4.5.3.3 Energies and States at the Γ Point

It is instructive to obtain the energies and associated states at the Γ point based on the Hamiltonian derived in Eq. (4.33) and compare with the CC and GJ versions. Note that there is a difference between the GJ results and the present result since the electron state $|e\rangle$ is included among the interesting states (A states in Löwdin perturbation theory) in the latter case. Immediate inspection shows that at the Γ point, the Γ_9 states decouple from other states while $|e\rangle$ states mix with the $|7\rangle$ and $|7'\rangle$ states only due to spin-orbit interaction. Two identical hermitian 3×3 blocks in the $|e \uparrow\rangle$, $|7 \uparrow\rangle$, $|7' \uparrow\rangle$ and the $|e \downarrow\rangle$, $|7 \downarrow\rangle$, $|7' \downarrow\rangle$ bases result (only upper diagonal is given):

$$\begin{pmatrix} E_e & -\sqrt{(1-q_7^2)}q_z\langle z|H_{\text{s},+}|u_5^*\rangle & -q_7q_z\langle z|H_{\text{s},+}|u_5^*\rangle \\ & E_7 & 0 \\ & & E_{7'} \end{pmatrix}. \tag{4.40}$$

The secular equation for the matrix in Eq. (4.40) is

$$x^3 + x^2(E_{7'} + E_7 - 2E_e) + x\left[(E_7 - E_e)(E_{7'} - E_e) - q_z^2|\langle z|H_{s,+}|u_5^*\rangle|^2\right] \\ - q_z^2|\langle z|H_{s,+}|u_5^*\rangle|^2\left[(1 - q_7^2)(E_{7'} - E_e) + q_7^2(E_7 - E_e)\right] = 0, \quad (4.41)$$

where $x = E_e - E$ and E denotes the Γ-point energies. The three solutions E_i ($i = 1, 2, 3$) to this third-order polynomial are the Γ-point energies accounting for mixing among the e, 7, and 7′ states and not just the mixing between 7 and 7′ states found by GJ. Evidently, if $\langle z|H_{s,+}|u_5^*\rangle \neq 0$, the three eigenstates are (apart from normalization)

$$|u_i \uparrow\rangle = |e\uparrow\rangle + \frac{\sqrt{(1-q_7^2)}q_z\langle z|H_{s,+}|u_5^*\rangle^*}{E_7 - E_i}|7\uparrow\rangle + \frac{q_7 q_z\langle z|H_{s,+}|u_5^*\rangle^*}{E_{7'} - E_i}|7'\uparrow\rangle, i = 1, 2, 3. \quad (4.42)$$

If, on the other hand, $\langle z|H_{s,+}|u_5^*\rangle = 0$, the GJ energies and states are obtained. For the Kramers degenerate set, we find

$$|u_i \downarrow\rangle = |e\downarrow\rangle + \frac{\sqrt{(1-q_7^2)}q_z\langle z|H_{s,+}|u_5^*\rangle^*}{E_7 - E_i}|7\downarrow\rangle + \frac{q_7 q_z\langle z|H_{s,+}|u_5^*\rangle^*}{E_{7'} - E_i}|7'\downarrow\rangle, i = 1, 2, 3, \quad (4.43)$$

if $\langle z|H_{s,+}|u_5^*\rangle \neq 0$, otherwise the GJ energies and states are obtained.

4.6 Summary

The method associated with Kane is the unified treatment of quasi-degenerate states such as the conduction and valence states of semiconductors. The interaction among the states in this quasi-degenerate group are then treated exactly. Should remote states (i.e., outside the quasi-degenerate group) be considered then the method is known as Löwdin perturbation. The method is illustrated for both zincblende and wurtzite. It is also shown how $k \cdot p$ can reproduce a band structure over a full Brillouin zone. The chapter concludes with an eight-band model of wurtzite in the Gutsche–Jahne basis, a previously unpublished result.

Chapter 5
Method of Invariants

5.1 Overview

We had, in earlier chapters, described various perturbation methods for obtaining the $k \cdot p$ Hamiltonian. We note that, intrinsically, one still used symmetry arguments in order to determine the number of independent parameters in each of the Hamiltonian matrix elements. One could, therefore, argue whether the application of symmetry arguments alone, and from the beginning, could determine uniquely the basic structure of the Hamiltonian matrix. This is the method of invariants introduced by Luttinger in 1956 [4].

It was subsequently generalized into a theory of invariants by Pikus and coworkers [41, 61] and detailed in the book by Bir and Pikus [1], repeated in the more recent book by Ivchenko and Pikus [20]. Since then, there have been applications to, for example, the valence band of diamond in the presence of strain and a magnetic field [80], the valence band of zincblende in the presence of strain and a magnetic field [81], the spin-splitting problem in semiconductors [21], and to the internal structure of excitons [82, 83]. The theory of invariants is equivalent to the effective Hamiltonian method used in other areas in physics.

To ease the transition into the formal theory of invariants, we first present a hybrid discussion that combines elements of both the perturbation and invariant methods. Following that, the formal theory of invariants will be explained and then various examples are given. The application to the eight-band model of wurtzite has not been previously published.

5.2 DKK Hamiltonian – Hybrid Method

The basic idea is that the Hamiltonian must be an invariant with respect to the symmetry group of the crystal and within the standard approximation of being quadratic in the wave vector. While this is sufficient for a general expression, one also knows that the individual terms originate from expressions of the form $k \cdot p$ (to second order). Indeed, following DKK, one can write down the valence-band Hamiltonian to second order in k, the perturbation matrix element being

L.C. Lew Yan Voon, M. Willatzen, *The* k · p *Method*,
DOI 10.1007/978-3-540-92872-0_5, © Springer-Verlag Berlin Heidelberg 2009

$$\sum_l{}' \frac{\langle\xi|H_{k\cdot p}|l\rangle\langle l|H_{k\cdot p}|\eta\rangle}{E_0 - E_l} = \frac{\hbar^2}{m_0^2}\sum_l{}' \frac{1}{E_0 - E_l}\sum_{ij} k_i k_j \langle\xi|p_i|l\rangle\langle l|p_j|\eta\rangle$$
$$= \frac{\hbar^2}{m_0^2}\sum_{ij} k_i k_j \left\langle \xi|\frac{p_i p_j}{E_0 - H}|\eta\right\rangle.$$

Under a coordinate transformation, both the wave vector and the momentum operator change (e.g., $k \cdot p$ is invariant); hence, we need to study the transformation properties of both. Note that, in the following, we will use the group notation of DKK instead of Luttinger.

The wave vector **k** transforms under the diamond group O_h as a polar vector (i.e., just like a position vector); hence, as $\Gamma_{15}^- \sim T_2$ in Luttinger's notation (note that Luttinger wrongly assigns it to T_1, though that will be of no consequence as we will see later). The product $k_i k_j$ (being symmetric) then generates a six-dimensional reducible representation $[\Gamma_{15}^- \otimes \Gamma_{15}^-]_S$. One can find the irreducible representations using the standard techniques. We will do one case in detail. First, we need the characters of the symmetric-product representation [Eq. (A.21)]:

$$\chi[\Gamma \otimes \Gamma]_S = \frac{1}{2}\left(\chi^2(\Gamma) + \chi\left(\Gamma^2\right)\right).$$

The product of two odd representations will be an even one (whether the product is symmetric or antisymmetric); thus, we only need to write down the characters corresponding to half the group. Here, we find

	E	$8C_3$	$3C_2$	$6S_4$	$6C_2'$
$[\Gamma_{15}^- \otimes \Gamma_{15}^-]_S$	6	0	2	0	2

We now use the decomposition formula to find the irreducible representations [Eq. (A.23)]:

$$n_\gamma = \frac{1}{l_g}\sum_l c_l \chi^{(\gamma)*}\chi,$$

where l_g is the order of the group, and c_l is the number of elements in class l. Note that, since the reducible representation is even, only even irreducible representations need be considered (as is obvious from the character equation); hence, we also only need to work with the subgroup of 24 elements given above. Thus,

$$n_1 = \frac{1}{24}(6 + 6 + 12) = 1,$$
$$n_2 = \frac{1}{24}(6 + 6 - 12) = 0,$$
$$n_{12} = \frac{1}{24}(18 + 6 + 0) = 1,$$

$$n_{15} = \frac{1}{24}(18 - 6 - 12) = 0,$$
$$n_{25} = \frac{1}{24}(18 - 6 + 12) = 1,$$

using Table B.5 for finding the square of elements [e.g., $(I\delta_{4x})^2 = \delta_{2x}$], and using the character table given in Table B.6. Therefore,

$$[\Gamma_{15}^{-} \otimes \Gamma_{15}^{-}]_S = \Gamma_1^{+} \oplus \Gamma_{12}^{+} \oplus \Gamma_{25}^{+}. \tag{5.1}$$

Combinations of $k_i k_j$ that transform according to those irreducible representations are

$$\sim \left(k_x^2 + k_y^2 + k_z^2\right), \left\{2k_z^2 - k_x^2 - k_y^2, \sqrt{3}\left(k_x^2 - k_y^2\right)\right\}, \left\{k_y k_z, k_z k_x, k_x k_y\right\}.$$

A similar decomposition exists for the product of the momentum operator except that, for the Γ_{25}^{+} representation, one requires symmetrized products

$$\left\{p_y p_z\right\}, \left\{p_z p_x\right\}, \left\{p_x p_y\right\},$$

where

$$\{PQ\} \equiv \frac{1}{2}(PQ + QP). \tag{5.2}$$

We must now construct, out of the k's and p's, combinations that are invariant, i.e., that transform according to Γ_1^{+}. Another well-known result from group theory is that, one requires the product of a representation with itself in order for the decomposition to include the invariant representation. Thus, the invariant $k \cdot p$ operators are then

$$k^2 P^2,$$
$$\left(2k_z^2 - k_x^2 - k_y^2\right)\left(2P_z^2 - P_x^2 - P_y^2\right) + 3\left(k_x^2 - k_y^2\right)\left(P_x^2 - P_y^2\right),$$
$$\sum_{i<j} k_i k_j \left\{P_i P_j\right\}.$$

Since each involves summing over intermediate states and dividing by the appropriate energy denominator, the three terms will, in general, be premultiplied by three different constants. That each term involves only one independent constant can be seen from the decomposition of the matrix element with respect to the valence-band states:

$$\langle \Gamma_{25}^{+} | \Gamma_{\gamma}^{+} | \Gamma_{25}^{+} \rangle = \Gamma_1^{+} \oplus \cdots,$$

where Γ_γ^+ is one of Γ_1^+, Γ_{12}^+, Γ_{25}^+ and with the identity representation only appearing once.

So far, we have given the irreducible invariant form of the Hamiltonian. In order to relate to the DKK Hamiltonian, one needs the representation of the Hamiltonian in terms of the valence-band states. Given the three-fold degeneracy (without spin), the Hamiltonian becomes a 3×3 matrix. However, the terms should retain the same symmetry properties; i.e., the goal is to replace the matrix elements of $P_i P_j$ by matrices. While P_i transforms according to Γ_{15}^-, it turns out that the symmetrized product of axial vectors ($\sim \Gamma_{15}^+$) has the *same decomposition*:

$$[\Gamma_{15}^+ \otimes \Gamma_{15}^+]_S = \Gamma_1^+ \oplus \Gamma_{12}^+ \oplus \Gamma_{25}^+.$$

This is useful because angular momentum is an axial vector and its representations can be used. Indeed, for the valence band, one would use a three-dimensional representation of the angular momentum.

Using Eqs. (A.27) and writing $\mathbf{I} = \mathbf{L}/\hbar$, one easily finds I_i in the $|l\ m\rangle$ basis for $l = 1$ ($|1\ 1\rangle$, $|1\ 0\rangle$, $|1\ -1\rangle$):

$$I_x = \frac{1}{\sqrt{2}} \begin{pmatrix} 0 & 1 & 0 \\ 1 & 0 & 1 \\ 0 & 1 & 0 \end{pmatrix},\ I_y = \frac{\mathrm{i}}{\sqrt{2}} \begin{pmatrix} 0 & -1 & 0 \\ 1 & 0 & -1 \\ 0 & 1 & 0 \end{pmatrix},\ I_z = \begin{pmatrix} 1 & 0 & 0 \\ 0 & 0 & 0 \\ 0 & 0 & -1 \end{pmatrix}. \tag{5.3}$$

We need a representation of I_i with respect to the valence states $\varepsilon_1^+, \varepsilon_2^+, \varepsilon_3^+$ (Sect. 3.2.1). Using Eq. (A.28), we have

$$|1 \pm 1\rangle = \mp \frac{1}{\sqrt{2}} (\varepsilon_1^+ \pm \mathrm{i}\varepsilon_2^+),$$
$$|1\ 0\rangle = \varepsilon_3^+,$$

giving

$$I_x = \begin{pmatrix} 0 & 0 & 0 \\ 0 & 0 & -\mathrm{i} \\ 0 & \mathrm{i} & 0 \end{pmatrix},\ I_y = \begin{pmatrix} 0 & 0 & \mathrm{i} \\ 0 & 0 & 0 \\ -\mathrm{i} & 0 & 0 \end{pmatrix},\ I_z = \begin{pmatrix} 0 & -\mathrm{i} & 0 \\ \mathrm{i} & 0 & 0 \\ 0 & 0 & 0 \end{pmatrix}, \tag{5.4}$$

$$I^2 = I_x^2 + I_y^2 + I_z^2 = \begin{pmatrix} 2 & 0 & 0 \\ 0 & 2 & 0 \\ 0 & 0 & 2 \end{pmatrix} = l(l+1)\mathbf{1}. \tag{5.5}$$

Our choice of matrices agrees with Luttinger [4].

We can now write

$$
\begin{aligned}
&k^2 P^2 \sim \frac{1}{2} A k^2 I^2 \equiv A k^2, \\
&\left(2k_z^2 - k_x^2 - k_y^2\right)\left(2P_z^2 - P_x^2 - P_y^2\right) + 3\left(k_x^2 - k_y^2\right)\left(P_x^2 - P_y^2\right) \\
&\sim \frac{\overline{B}}{2}\left(2k_z^2 - k_x^2 - k_y^2\right)\left(2I_z^2 - I_x^2 - I_y^2\right) + 3\left(k_x^2 - k_y^2\right)\left(I_x^2 - I_y^2\right) = 3\overline{B}\sum_i k_i^2\left(I_i^2 - \frac{1}{3}I^2\right), \\
&\sum_{i<j} k_i k_j \{P_i P_j\} \sim \frac{2\overline{D}}{\sqrt{3}} \sum_{i<j} k_i k_j \{I_i I_j\}.
\end{aligned}
$$

Hence,

$$
H_{\mathrm{mL}} = \overline{A}k^2 + 3\overline{B}\sum_i k_i^2\left(I_i^2 - \frac{1}{3}I^2\right) + 2\overline{D}\sqrt{3}\sum_{i<j} k_i k_j \{I_i I_j\}. \tag{5.6}
$$

We call this the modified Luttinger Hamiltonian without spin and without magnetic field, since it is in a slightly different form from the one in Luttinger's paper. We now show the connection of the new parameters ($\overline{A}$, $\overline{B}$, $\overline{D}$) to the DKK A, B, C by comparing the new Hamiltonian, Eq. (5.6), to the DKK one, Eq. (3.4). Since

$$
I_x^2 = \begin{pmatrix} 0 & 0 & 0 \\ 0 & 1 & 0 \\ 0 & 0 & 1 \end{pmatrix},\ I_y^2 = \begin{pmatrix} 1 & 0 & 0 \\ 0 & 0 & 0 \\ 0 & 0 & 1 \end{pmatrix},\ I_z^2 = \begin{pmatrix} 1 & 0 & 0 \\ 0 & 1 & 0 \\ 0 & 0 & 0 \end{pmatrix},
$$

$$
\{I_x I_y\} = \begin{pmatrix} 0 & -\frac{1}{2} & 0 \\ -\frac{1}{2} & 0 & 0 \\ 0 & 0 & 0 \end{pmatrix},\ \{I_y I_z\} = \begin{pmatrix} 0 & 0 & 0 \\ 0 & 0 & -\frac{1}{2} \\ 0 & -\frac{1}{2} & 0 \end{pmatrix},\ \{I_z I_x\} = \begin{pmatrix} 0 & 0 & -\frac{1}{2} \\ 0 & 0 & 0 \\ -\frac{1}{2} & 0 & 0 \end{pmatrix},
$$

we have

$$
\begin{aligned}
H_{\mathrm{mL}} &= \begin{pmatrix}
\begin{matrix}\left(\overline{A} - 2\overline{B}\right)k_x^2 \\ + \left(\overline{A} + \overline{B}\right)\left(k_y^2 + k_z^2\right)\end{matrix} & -\overline{D}\sqrt{3}k_x k_y & -\overline{D}\sqrt{3}k_x k_z \\
-\overline{D}\sqrt{3}k_x k_y & \begin{matrix}\left(\overline{A} - 2\overline{B}\right)k_x^2 \\ + \left(\overline{A} + \overline{B}\right)\left(k_x^2 + k_z^2\right)\end{matrix} & -\overline{D}\sqrt{3}k_y k_z \\
-\overline{D}\sqrt{3}k_x k_z & -\overline{D}\sqrt{3}k_y k_z & \begin{matrix}\left(\overline{A} - 2\overline{B}\right)k_x^2 \\ + \left(\overline{A} + \overline{B}\right)\left(k_x^2 + k_y^2\right)\end{matrix}
\end{pmatrix} \\
&\equiv \begin{pmatrix}
Lk_x^2 + M\left(k_y^2 + k_z^2\right) & Nk_x k_y & Nk_x k_z \\
Nk_x k_y & Lk_y^2 + M\left(k_x^2 + k_z^2\right) & Nk_y k_z \\
Nk_x k_z & Nk_y k_z & Lk_z^2 + M\left(k_x^2 + k_y^2\right)
\end{pmatrix},
\end{aligned}
$$

where

$$L = \overline{A} - 2\overline{B},\ M = \overline{A} + \overline{B},\ N = -\overline{D}\sqrt{3}. \tag{5.7}$$

Note that the $\overline{A}$, $\overline{B}$ here are slightly different from the DKK A, B parameters:

$$\begin{aligned}\overline{A} &= A - \frac{\hbar^2}{2m_0},\\ \overline{B} &= -B,\\ \overline{D} &= -\left(3B^2 + C^2\right)^{\frac{1}{2}}.\end{aligned} \tag{5.8}$$

5.3 Formalism

A general theory of invariants was first written down by Pikus [41, 61]. We reproduce here the discussions of Bir and Pikus [1] and Trebin et al. [81].

5.3.1 Introduction

The Hamiltonian is an $n \times n$ matrix in a multiband $k \cdot p$ theory. In principle, the n^2 matrix elements can be described using n^2 independent matrices; the latter can be chosen to transform according to the irreducible representations of the symmetry group. When multiplied by appropriate irreducible combinations of a tensor (e.g., the wave vector, magnetic field, strain, electric field) invariants are formed.

5.3.2 Spatial Symmetries

Let the Hamiltonian H be represented in terms of a set of basis functions $\psi_i (i = 1, n)$, the latter transforming according to an n-dimensional representation $D(G)$ of the group $\mathcal{G}$. Let the Hamiltonian be a function of a tensorial parameter $\mathcal{K}$, where $\mathcal{K}$, in this chapter, will be the wave vector $\mathbf{k}$ but will later be chosen as the electric field $\mathcal{E}$, magnetic field $\mathbf{B}$, or the strain ε. Then

$$\psi_i'(\mathbf{r}) = \psi_i(G^{-1}\mathbf{r}) = \sum_j \psi_j(\mathbf{r}) D_{ji}(G), \tag{5.9}$$

and $H(\mathcal{K})$ becomes

$$D^{-1}(G)H(\mathcal{K})D(G).$$

If the Hamiltonian is invariant under transformation G, then

$$H'(\mathcal{K}') = D(G)H(G^{-1}\mathcal{K})D^{-1}(G) = H(\mathcal{K}). \tag{5.10}$$

This leads to n^2 equations from which the matrix $H(\mathcal{K})$ can be deduced. Note that only the group generators are needed. Instead of using the full representation in Eq. (5.10), only the characters will be needed to solve for the Hamiltonian matrix elements.

The Hamiltonian will be written in terms of the components $\mathcal{K}_i$ and the n^2 linearly-independent X_i matrices. We call the X_i basis matrices. They are such that

$$X_i' = G^{-1}X_i = D(G)X_iD^{-1}(G) = \sum_j X_j D_{ji}^{(X)}(G). \tag{5.11}$$

The $D^{(X)}(G)$ is a representation of the group of dimension n^4. One can derive an expression for it in terms of the original $D(G)$ matrices by first relabeling the X_i matrices as X^{lk} such that

$$X_{l'k'}^{lk} = \delta_{ll'}\delta_{kk'}. \tag{5.12}$$

Substituting into Eq. (5.11),

$$D_{im}X_{mn}^{i'j'}D_{nj}^{-1} = \sum_{m'n'} X_{ij}^{m'n'}D_{m'n'i'j'}^{(X)},$$

and using Eq. (5.12),

$$D_{iji'j'}^{(X)} = D_{ii'}D_{jj'}^{-1} = D_{ii'}D_{jj'}^{*}. \tag{5.13}$$

Hence, $D^{(X)}$ is the direct product $D \otimes D^*$, with character

$$\chi^{(X)}(G) = |\chi(G)|^2. \tag{5.14}$$

Summarizing what we have so far, the set of n^2 X_i matrices form a basis for the generation of a representation D^X of the group. In general, this representation is reducible. The process of finding the irreducible representations is equivalent to finding a linear transformation of the X_i into a set of so-called irreducible basis operators. The Hamiltonian will then, instead, be expressed in terms of such an irreducible basis. Note that, since we are still dealing with the group $\mathcal{G}$, the maximum number of such irreducible basis operators is the same as the number of irreducible representations of $\mathcal{G}$. However, not all need show up in $D \otimes D^*$.

Example: T_d. The 3×3 valence band without spin-orbit coupling requires nine 3×3 matrices which can be combined into five irreducible components. D^X is nine-dimensional.

Labeling the irreducible representation of $\mathcal{G}$ by γ, this implies Eq. (5.11) splits into a set of disjoint equations

$$G^{-1}X_i^{(\gamma)} = D(G)X_i^{(\gamma)}D^{-1}(G) = \sum_{i'} X_{i'}^{(\gamma)} D_{i'i}^{(\gamma)}(G), \tag{5.15}$$

for each γ. Similarly, the tensorial variables $\mathcal{K}_i$ can be expressed in terms of irreducible components $\mathcal{K}_i^{(\gamma)}$ transforming as

$$G^{-1}\mathcal{K}_i^{(\gamma)*} = \sum_{i'} \mathcal{K}_{i'}^{(\gamma)*} D_{i'i}^{(\gamma)*}(G), \tag{5.16}$$

and, therefore, invariant combinations are given by

$$H(\mathcal{K}) = \sum_{\gamma} a_\gamma \sum_{l} X_l^{(\gamma)} \mathcal{K}_l^{(\gamma)*}. \tag{5.17}$$

Since the various a_γ are not connected by symmetry, they are arbitrary except that the Hamiltonian must be hermitian.

Should the original representation D be reducible (e.g., by using s and p functions for a cubic crystal), then the above formalism carries through by generalizing with the irreducible-representation index. Using two irreducible representations as an example, consider D to consist of the two irreducible representations D_{I} and D_{II}. Then, when diagonalized,

$$D = \begin{pmatrix} D_{\mathrm{I}} & 0 \\ 0 & D_{\mathrm{II}} \end{pmatrix}. \tag{5.18}$$

The Hamiltonian can be written into four blocks:

$$H(\mathcal{K}) = \begin{pmatrix} H^{\mathrm{I\,I}} & H^{\mathrm{I\,II}} \\ H^{\mathrm{II\,I}} & H^{\mathrm{II\,II}} \end{pmatrix}. \tag{5.19}$$

Equation (5.10) then becomes

$$D_{\mathrm{I}}(G)H^{\mathrm{I\,I}}(G^{-1}\mathcal{K})D_{\mathrm{I}}^{-1}(G) = H^{\mathrm{I\,I}}(\mathcal{K}), \tag{5.20}$$

$$D_{\mathrm{II}}(G)H^{\mathrm{II\,II}}(G^{-1}\mathcal{K})D_{\mathrm{II}}^{-1}(G) = H^{\mathrm{II\,II}}(\mathcal{K}), \tag{5.21}$$

$$D_{\mathrm{I}}(G)H^{\mathrm{I\,II}}(G^{-1}\mathcal{K})D_{\mathrm{II}}^{-1}(G) = H^{\mathrm{I\,II}}(\mathcal{K}), \tag{5.22}$$

$$D_{\mathrm{II}}(G)H^{\mathrm{II\,I}}(G^{-1}\mathcal{K})D_{\mathrm{I}}^{-1}(G) = H^{\mathrm{II\,I}}(\mathcal{K}). \tag{5.23}$$

Equations (5.20) and (5.21) are similar to the earlier case. For Eqs. (5.22) and (5.23), the basis matrices X are, in general, rectangular. More generally, for a multiband Hamiltonian that has a block structure with a block formed from irreducible representations α and β, the above equations are modified such that

$$H^{ij}(\mathcal{K}) = \sum_{\gamma} a_{\gamma} \sum_{l} X_l^{(\alpha)} \mathcal{K}_l^{(\beta)*}, \tag{5.24}$$

$$G^{-1} X_i^{(\gamma)} = D^{(\alpha)}(G) X_i^{(\gamma)} D^{(\beta)}(G)^{-1} = \sum_{i'} X_{i'}^{(\gamma)} D_{i'i}^{(\gamma)}(G), \tag{5.25}$$

where the irreducible representations Γ_{γ} are those contained in $\Gamma_{\alpha} \otimes \Gamma_{\beta}^{*}$.

We will choose the $X_i^{(\gamma)}$ basis matrices to transform according to irreducible representations of the point group. One such procedure has been outlined by Trebin et al. [81]. Thus, the symmetrized basis matrices $X_i^{(\gamma)}$ can be constructed from cartesian components using the basis functions of coordinates and angular momenta operators found in standard group theory texts [33] and reproduced in Appendix B. The parity is equal to the product of the parities of $D^{(\alpha)}$ and $D^{(\beta)}$. For diagonal blocks the remaining procedure is straightforward. For off-diagonal blocks (i.e., coupling different representations), each cartesian operator can be re-expressed in terms of spherical ones of the appropriate rank and the matrix elements of the latter can be written in terms of a reduced set by using the Wigner-Eckart theorem [Eq. (A.30)]. For instance, in combining a $j = \frac{1}{2}$ subspace (e.g., conduction electrons) with a $j' = \frac{3}{2}$ one (e.g., holes), the irreducible tensor matrices T present are of rank 1 and 2. Rank 1 operators are vectors, with cartesian components T_i. The representation of these matrices in the jj' subblock is simplified by re-expressing the cartesian vectors in terms of spherical vectors T_{μ}. An explicit calculation will be presented later for the 14-band model of ZB.

5.3.2.1 Time-Reversal Symmetry

The diagonal blocks are invariant with respect to time-reversal symmetry while there is no such constraint for the off-diagonal blocks [1]. We note that the Pauli Hamiltonian operator is invariant; however, the matrix representation need not be. For diagonal blocks, this is the case because the two representations are the same. But, for the off-diagonal blocks, the two representations are linearly independent; hence, no definite time-reversal property is required. One consequence is that, when one builds invariants, it will not be necessary to match the time-reversal property of the symmetrized quantities for the off-diagonal blocks.

5.3.2.2 Example

We follow Bir and Pikus [1] in clarifying the above theory as it applies to generating the Hamiltonian $H_{\mathbf{k}_0}(\mathcal{K})$ at some point $\mathbf{k}_0$. The appropriate group is the little group $G_{\mathbf{k}_0}$ and $D = D_{\mathbf{k}_0}^{(\mu)}$ with character $\chi_{\mathbf{k}_0}^{(\mu)}$. The character of $D^{(X)}$ is

$$\chi_{\mathbf{k}_0}^{(X)} = \left|\chi_{\mathbf{k}_0}^{(\mu)}\right|^2. \tag{5.26}$$

$D^{(X)}$ is determined by the rotational elements of the point group $F_{\mathbf{k}_0}$. Thus, $D^{(X)}$ is a vector representation of the point group and can be expanded in terms of other vector representations. Therefore, the number of sets of $X^{(\gamma)}$ which transform according to the representation D^γ of the point group is given by

$$n_\gamma = \frac{1}{l_g} \sum_{g \in G'_{\mathbf{k}_0}} \chi^{(X)}_{\mathbf{k}_0} \chi^{(\gamma)}(g) = \frac{1}{l_g} \sum_{g \in G'_{\mathbf{k}_0}} \left| \chi^{(\mu)}_{\mathbf{k}_0} \right|^2 \chi^{(\gamma)}(g). \tag{5.27}$$

The summation is over all the elements of the little group except those containing primitive translations. For projective representations of the point group, $D(R) = D^{(\mu)}_{\mathbf{k}_0}(g) \mathrm{e}^{\mathrm{i}\mathbf{k}_0 R}$, and

$$n_\gamma = \frac{1}{h} \sum_{R \in F_{\mathbf{k}_0}} \left| \chi^{(\mu)}_{\mathbf{k}_0}(R) \right|^2 \chi^{(\gamma)}(R). \tag{5.28}$$

Finally, for two combined representations D_μ and D_ν, the expression for the off-diagonal submatrices is

$$n_\gamma = \frac{1}{l_g} \sum_{g \in G'_{\mathbf{k}_0}} \chi^{(\mu)*}_{\mathbf{k}_0}(g) \chi^{(\nu)}_{\mathbf{k}_0}(g) \chi^{(\gamma)}(g). \tag{5.29}$$

5.3.3 Spinor Representation

There are two methods for constructing $H(\mathcal{K})$. The first involves using the spinor representation of $D(G)$ from the start, which means the spin properties is built into the formalism. The simplification is that the formalism for constructing the basis matrices X_i explained above carries through. The disadvantage is that there is no procedure for determining the relative strength of the spin-orbit coupling.

The second approach involves using product basis functions $\psi(\mathbf{r}, \sigma) = \psi(\mathbf{r})\alpha$, where α is a spinor, which leads to direct-product representations $D_l \otimes D_{\frac{1}{2}}$. The basis matrices may then be expressed in terms of products of spatial matrices $X^{(\gamma)}$ and spin operators σ_i.

5.4 Valence Band of Diamond

We describe again, in summary form, the procedure explained above for obtaining the Hamiltonian using the theory of invariants. The first step is to identify a set of basis matrices. The latter are then symmetrized according to the irreducible representations present in the reduction of the direct product of the basis states. Finally, invariants are formed with the correspondingly symmetrized functions. In practice, the invariant equation is next recast into a matrix form and usually compared to the perturbation theory in order to relate the parameters in the two methods.

5.4.1 No Spin

The problem statement is: given that the valence band of diamond is three-fold degenerate at $\mathbf{k} = \mathbf{0}$, how can one construct the most general Hamiltonian for finite $\mathbf{k}$ in the absence of external fields? Using the theory of invariants, we note first that any 3×3 matrix can be written in terms of the following nine linearly-independent ones:

$$I_x, I_y, I_z, I_x^2, I_y^2, \{I_x I_y\}, \{I_y I_z\}, \{I_z I_x\}.$$

Restricting oneselves to these matrices and up to $k_i k_j$ terms, one obtains the irreducible basis functions and matrices given in Table 5.1.

These are obtained by noting that

$$\mathbf{k} \sim \Gamma_{15}^-, \mathbf{I} \sim \Gamma_{15}^+.$$

Then,

$$\Gamma_{15}^- \otimes \Gamma_{15}^- = \Gamma_1^+ \oplus \Gamma_{12}^+ \oplus \Gamma_{15}^+ \oplus \Gamma_{25}^+, \tag{5.30}$$

$$\Gamma_{15}^+ \otimes \Gamma_{15}^+ = \Gamma_1^+ \oplus \Gamma_{12}^+ \oplus \Gamma_{15}^+ \oplus \Gamma_{25}^+. \tag{5.31}$$

Table 5.1 Irreducible basis functions for O_h in 3×3 subspace. $\omega = e^{2\pi i/3}$, $\{AB\} = \frac{1}{2}(AB + BA)$, $[A, B] = (AB - BA)$

	Time reversal	
	Even	Odd
Γ_1^+	k^2, I^2	
Γ_{12}^+	$k_x^2 + \omega k_y^2 + \omega^2 k_z^2, k_x^2 + \omega^2 k_y^2 + \omega k_z^2$ $I_x^2 + \omega I_y^2 + \omega^2 I_z^2, I_x^2 + \omega^2 I_y^2 + \omega I_z^2$	
Γ_{15}^+		I_x, I_y, I_z $[k_y, k_z], [k_z, k_x], [k_x, k_y]$ $\sigma_x, \sigma_y, \sigma_z$
Γ_{25}^+	$\{k_y k_z\}, \{k_z k_x\}, \{k_x k_y\}$ $\{I_y I_z\}, \{I_z I_x\}, \{I_x I_y\}$	
Γ_{15}^-		k_x, k_y, k_z

Note that the antisymmetric combination $[k_i, k_j]$ will be shown later to be related to the magnetic field and the linear-in-k terms (the Γ_{15}^- representation) do not participate since there are no corresponding I_i matrices of the same symmetry (and vice versa). Finally, there are three popular forms for the basis functions of Γ_{12}^+.

We have chosen the complex conjugate pair of Bir and Pikus [1]. The other two are: $I_x^2 - I_y^2, I_y^2 - I_z^2$ [80] and $3I_z^2 - I^2, \sqrt{3}(I_x^2 - I_y^2)$ (e.g., Koster et al. [33]).

The next step is to form invariants with combinations of the $k_i k_j$ terms together with the corresponding $I_i I_j$ factors. This then leads to [4]

$$\begin{aligned} H(\mathbf{k}) &= \alpha_1 k^2 + \alpha_2 \left(k_x^2 I_x^2 + k_y^2 I_y^2 + k_z^2 I_z^2\right) \\ &\quad + \alpha_3 \left(\{k_x k_y\}\{I_x I_y\} + \{k_y k_z\}\{I_y I_z\} + \{k_z k_x\}\{I_z I_x\}\right) \qquad (5.32) \\ &= Ak^2 - (A - B)\left(k_x^2 I_x^2 + k_y^2 I_y^2 + k_z^2 I_z^2\right) \\ &\quad - 2C\left(\{k_x k_y\}\{I_x I_y\} + \{k_y k_z\}\{I_y I_z\} + \{k_z k_x\}\{I_z I_x\}\right). \qquad (5.33) \end{aligned}$$

The matrices can be expanded giving

$$H(\mathbf{k}) = \begin{pmatrix} \alpha_1 k^2 + \alpha_2 \left(k_y^2 + k_z^2\right) & -\frac{\alpha_3}{2} k_x k_y & -\frac{\alpha_3}{2} k_x k_z \\ -\frac{\alpha_3}{2} k_x k_y & \alpha_1 k^2 + \alpha_2 \left(k_x^2 + k_z^2\right) & -\frac{\alpha_3}{2} k_y k_z \\ -\frac{\alpha_3}{2} k_x k_z & -\frac{\alpha_3}{2} k_y k_z & \alpha_1 k^2 + \alpha_2 \left(k_x^2 + k_y^2\right) \end{pmatrix}. \qquad (5.34)$$

Comparing to the DKK (noting that the Luttinger Hamiltonians include the free-electron term, the DKK does not), we have

$$\begin{aligned} \alpha_1 &= A_L = L + \frac{\hbar^2}{2m_0}, \\ \alpha_2 &= B_L - A_L = M - L, \qquad (5.35) \\ \alpha_3 &= -2C_L = -2N. \end{aligned}$$

5.4.2 Magnetic Field

In the discussion of Luttinger's Hamiltonian with spin later, it will be useful for completeness to include the interaction with an external magnetic field. Hence, we present a derivation of the magnetic Hamiltonian at this point, leaving the complete treatment of the magnetic problem for Chap. 9.

5.4.2.1 DKK Hamiltonian

One can look at the influence of an externally applied homogeneous magnetic field on the DKK Hamiltonian. While the general theory will be treated next using the method of invariants, we repeat here a discussion of Luttinger [4] in terms of the antisymmetric components of the D matrix since this issue of antisymmetry is an important one. We are following the work of Luttinger and will use his D matrix rather than the DKK one for illustration.

The key lies in the observation that D^{ij}_{nm} is not necessarily symmetric in the cartesian indices, as is clear in its definition, Eq. (3.7). Let us, therefore, introduce symmetrized and antisymmetrized components:

$$\begin{aligned} k_i k_j &\equiv \frac{1}{2}(k_i k_j - k_j k_i) + \frac{1}{2}(k_i k_j + k_j k_i) \equiv \frac{1}{2}[k_i, k_j] + \{k_i k_j\}, \\ D^{ij}_{nm} &\equiv \frac{1}{2}(D^{ij}_{nm} - D^{ji}_{nm}) + \frac{1}{2}(D^{ij}_{nm} + D^{ji}_{nm}), \\ D^{ij}_{nm} k_i k_j &= \frac{1}{4}(D^{ij}_{nm} - D^{ji}_{nm})[k_i, k_j] + \frac{1}{2}(D^{ij}_{nm} + D^{ji}_{nm})\{k_i k_j\}. \end{aligned} \tag{5.36}$$

Now

$$(D^{ij}_{nm} - D^{ji}_{nm})\{k_i k_j\} = 0,$$

so, in the absence of a magnetic field (i.e., $[k_i, k_j] = 0$), we have

$$D^{ij}_{nm} k_i k_j = \frac{1}{2}\left(D^{ij}_{nm} + D^{ji}_{nm}\right)\left\{k_i k_j\right\}. \tag{5.37}$$

In the presence of a magnetic field, the components of $\mathbf{k}$ no longer commute. To show this, we assume that the magnetic field is included into the effective Hamiltonian via minimal coupling [4] (with $e > 0$):

$$\mathbf{k} = \frac{1}{\hbar}(\mathbf{p} + e\mathbf{A}).$$

This will be derived later in the chapter on magnetic field (Chap. 9). Then,

$$\begin{aligned} \left[k_y, k_z\right] &= k_y k_z - k_z k_y = \frac{1}{\hbar^2}\left[\left(p_y + eA_y\right)(p_z + eA_z) - (p_z + eA_z)\left(p_y + eA_y\right)\right] \\ &= \frac{1}{i\hbar} e B_x, \\ [k_z, k_x] &= \frac{1}{i\hbar} e B_y, \\ \left[k_x, k_y\right] &= \frac{1}{i\hbar} e B_z. \end{aligned} \tag{5.38}$$

Thus, there are additional contributions to the Hamiltonian. Recall that, in the absence of the magnetic field, the D^{ij}_{nm} is symmetric and has three independent real constants (A_L, B_L, C_L). We need to correct this by adding antisymmetric terms. The question is how many new constants there are.

Note that D must be hermitian for real eigenvalues. In particular, for the diagonal elements,

$$D^*_{nn} = D_{nn} = D^{ij}_{nn} k_i k_j = \frac{1}{2}\left[D^{ij}_{nn}, D^{ji}_{nn}\right]\left[k_i, k_j\right] + \left\{D^{ij}_{nn}, D^{ji}_{nn}\right\}\left\{k_i k_j\right\}.$$

But, since D^{ij}_{nn} is real and $[k_i, k_j]$ pure imaginary, then there cannot be an antisymmetric contribution in the diagonal terms.

Consider, therefore, the off-diagonal ones. For example,

$$D_{XY} = D^{xy}_{XY} k_x k_y + D^{yx}_{XY} k_y k_x.$$

Let

$$K \equiv D^{xy}_{XY} - D^{yx}_{XY}. \tag{5.39}$$

Then

$$\begin{aligned} D_{XY} &= \left(D^{xy}_{XY} + D^{yx}_{XY}\right)\left\{k_x k_y\right\} + \frac{1}{2}\left(D^{xy}_{XY} - D^{yx}_{XY}\right)\left[k_x, k_y\right] \\ &= C_L\left\{k_x k_y\right\} + \frac{K}{2}\left[k_x, k_y\right] = C_L\left\{k_x k_y\right\} + \frac{\mathrm{i}e}{2\hbar} K B_z, \\ D_{YX} &= \left(D^{xy}_{YX} + D^{yx}_{YX}\right)\left\{k_x k_y\right\} + \frac{1}{2}\left(D^{xy}_{YX} - D^{yx}_{YX}\right)\left[k_x, k_y\right] \\ &= C_L\left\{k_x k_y\right\} + \frac{\mathrm{i}e}{2\hbar} K B_z. \end{aligned}$$

Thus, we can write

$$D = D^{(S)} + D^{(A)}, \tag{5.40}$$

where

$$D^{(A)} = \frac{eK}{2\hbar}\begin{pmatrix} 0 & -\mathrm{i}B_z & \mathrm{i}B_y \\ \mathrm{i}B_z & 0 & -\mathrm{i}B_x \\ -\mathrm{i}B_y & \mathrm{i}B_x & 0 \end{pmatrix}. \tag{5.41}$$

There is, therefore, one new parameter, K.

5.4.2.2 Method of Invariants

We now consider the valence band of diamond in a magnetic field using the method of invariants [4]. The components of **k** no longer commute. It is, therefore, necessary to keep the antisymmetrized products of the wave vector. The latter has the following character table:

	E	$8C_3$	$3C_2$	$6S_4$	$6C_2'$
$[\Gamma_{15}^- \otimes \Gamma_{15}^-]_A$	3	0	−1	1	−1

which is none other than that of Γ_{15}^+. Since the angular momentum also transforms according to the same irreducible representation, one can get an invariant by combining the two:

$$[k_y, k_z]\, I_x + [k_z, k_x]\, I_y + [k_x, k_y]\, I_z.$$

The total Hamiltonian can, therefore, be written as

$$\begin{aligned} H_{\mathrm{aL}} &= \alpha_1 k^2 + \alpha_2 \left(k_x^2 I_x^2 + k_y^2 I_y^2 + k_z^2 I_z^2\right) + \alpha_3\left(\{k_x k_y\}\{I_x I_y\} + \{k_y k_z\}\{I_y I_z\} + \{k_z k_x\}\{I_z I_x\}\right) \\ &\quad + \alpha_4\left([k_y, k_z]\, I_x + [k_z, k_x]\, I_y + [k_x, k_y]\, I_z\right), \end{aligned} \tag{5.42}$$

which is Eq. (32) of Luttinger [4]. Comparing with Eq. (5.41), one also finds that the last term can be written as

$$D^{(A)} = \frac{e}{2\hbar} K \mathbf{I} \cdot \mathbf{B}.$$

Note that, in the limit of large mass, the energy of a tightly-bound electron is given by the Zeeman term

$$\Delta E = \mu_B \mathbf{I} \cdot \mathbf{B} = \frac{e\hbar}{2m_0} \mathbf{I} \cdot \mathbf{B}.$$

Hence, $K \sim \hbar^2/m_0$.

5.4.3 Spin-Orbit Interaction

The spin-orbit Hamiltonian had been studied earlier in Chap. 3. We now wish to obtain a form of the spin-orbit matrix using the method of invariants.

5.4.3.1 *LS* Basis

We first note that $\nabla V \times \mathbf{p}$ transforms like an axial vector which, as we have already seen, leads to only one independent matrix element in a basis of Γ_{25}^+ states. We thus write

$$H_{\mathrm{so}} = \frac{1}{3} \Delta_0 \mathbf{I} \cdot \boldsymbol{\sigma}, \tag{5.43}$$

where the constant Δ_0 is chosen to parametrize the spin-orbit energy. To agree with Eq. (3.33), we write

$$\left\langle \varepsilon_1^+ \left| \frac{\hbar}{4m_0^2c^2}(\nabla V \times \mathbf{p})_z \right| \varepsilon_2^+ \right\rangle \equiv -\frac{\mathrm{i}}{3}\Delta_0.$$

We also already know that the spin-orbit Hamiltonian is not diagonal in the LS basis. The proper basis (or a good quantum number) is the total angular momentum:

$$\mathbf{J} = \mathbf{I} + \frac{1}{2}\boldsymbol{\sigma}. \tag{5.44}$$

Then

$$J^2 = \left(\mathbf{I} + \frac{1}{2}\boldsymbol{\sigma}\right)^2 = I^2 + \frac{1}{4}\sigma^2 + \mathbf{I}\cdot\boldsymbol{\sigma},$$

$$\Longrightarrow \quad \mathbf{I}\cdot\boldsymbol{\sigma} = J^2 - I^2 - \frac{1}{4}\sigma^2 = j(j+1) - i(i+1) - \frac{1}{4}\sigma(\sigma+1).$$

Now, $i = 1, \sigma = 1$ and $j = \frac{5}{2}, \frac{3}{2}$. Hence,

$$\mathbf{I}\cdot\boldsymbol{\sigma} = j(j+1) - \frac{11}{4},$$

$$H_{\text{so}} = \frac{\Delta_0}{3}\left[j(j+1) - \frac{11}{4}\right] = \frac{\Delta_0}{3}, -\frac{2}{3}\Delta_0.$$

The full Hamiltonian in the LS basis is obtained from Eq. (5.6) by adding the spin-orbit interaction:

$$H(\mathbf{k}) = E_v + \frac{1}{3}\Delta_0\mathbf{I}\cdot\boldsymbol{\sigma} + \overline{A}k^2 + 3\overline{B}\sum_i k_i^2\left(I_i^2 - \frac{1}{3}I^2\right) + 2\overline{D}\sqrt{3}\sum_{i<j} k_ik_j\left\{I_iI_j\right\}. \tag{5.45}$$

It can be diagonalized exactly at $\mathbf{k} = \mathbf{0}$. Letting $E_v = -\frac{1}{3}\Delta_0$,

$$\begin{aligned} H(\mathbf{k}=\mathbf{0}) &= E_v + \frac{1}{3}\Delta_0\mathbf{I}\cdot\boldsymbol{\sigma} = -\frac{1}{3}\Delta_0\left[\frac{15}{4} - j(j+1)\right] \\ &= 0, -\Delta_0. \end{aligned}$$

5.4.3.2 JM_J Basis – First Version

The above Hamiltonian was arrived at by first considering the 3×3 Hamiltonian without spin-orbit coupling, then adding the latter. In many cases, Δ_0 is large and the hh and lh states can be understood to be derived from a 4×4 Hamiltonian. In other words, we have a four-dimensional space. A convenient basis set is expected to be the angular momentum matrices for $j = \frac{3}{2}$. Since $\mathbf{J}$ hasthe same properties as $\mathbf{I}$, the terms in the above Hamiltonian still apply. Additional terms might be expected since the number of independent basis matrices is now 16, but Luttinger showed them not to contribute in the absence of a magnetic field and in the presence of time-reversal symmetry. We will explore this later.

Thus, the Hamiltonian for the hh and lh is

$$H(\mathbf{k}) = \overline{A}k^2 + \overline{B}\sum_i k_i^2\left(J_i^2 - \frac{1}{3}J^2\right) + \frac{2\overline{D}}{\sqrt{3}}\sum_{i<j} k_i k_j \left\{J_i J_j\right\}. \tag{5.46}$$

We will now show that the $\overline{A}$, $\overline{B}$, $\overline{D}$ are the same as before. The proof is based upon the Wigner-Eckart theorem, Eq. (A.30).

If one applies the Wigner-Eckart theorem to two irreducible tensors T and S,

$$\left\langle njm\left|T_\kappa^{(k)}\right|n'j'm'\right\rangle = \frac{\left\langle nj||T^{(k)}||n'j'\right\rangle}{\sqrt{2j+1}}\left\langle j'k;m'\kappa|j'k;jm\right\rangle,$$

$$\left\langle njm\left|S_\kappa^{(k)}\right|n'j'm'\right\rangle = \frac{\left\langle nj||S^{(k)}||n'j'\right\rangle}{\sqrt{2j+1}}\left\langle j'k;m'\kappa|j'k;jm\right\rangle,$$

then one can write

$$\begin{aligned}\left\langle njm\left|T_\kappa^{(k)}\right|n'j'm'\right\rangle &= \frac{\left\langle nj||T^{(k)}||n'j'\right\rangle}{\left\langle nj||S_\kappa^{(k)}||n'j'\right\rangle}\left\langle njm\left|S^{(k)}\right|n'j'm'\right\rangle \\ &= \gamma\left\langle njm\left|S_\kappa^{(k)}\right|n'j'm'\right\rangle,\end{aligned} \tag{5.47}$$

where $\gamma \neq f(m, m', \kappa)$. This is known as the equivalent-operator method [84].

In Eqs. (5.6) and (5.46), the angular momenta can be cast into irreducible tensors. Indeed, note that all three of 1, $J_i^2 - \frac{1}{3}J^2$, and $\{J_i J_j\}$ in Eq. (5.46) can be obtained by working with the traceless symmetric tensor

$$J_{ij} \equiv \{J_i J_j\} - \frac{1}{3}J^2\delta_{ij}. \tag{5.48}$$

For example, $J_i^2 - \frac{1}{3}J^2 = J_{ii}$. J_{ij} is a spherical tensor of rank 2 and transforms under the rotation group according to the D_2 representation; hence, we can use it in the Wigner-Eckart theorem. For completeness, note that J_{ij} has only five independent components. They can be chosen as follows:

$$J_0^{(2)} = \sqrt{\frac{3}{2}}\left[J_{zz} - \frac{1}{3}\left(J_{xx} + J_{yy} + J_{zz}\right)\right] = \sqrt{\frac{3}{2}}\left[J_z^2 - \frac{1}{3}J^2\right],$$

$$J_{\pm1}^{(2)} = \mp\left(J_{xz} \pm \mathrm{i}J_{yz}\right) = \mp\frac{1}{2}\left[\left(J_x \pm \mathrm{i}J_y\right), J_z\right], \tag{5.49}$$
$$J_{\pm2}^{(2)} = \frac{1}{2}\left(J_{xx} - J_{yy} \pm 2\mathrm{i}J_{xy}\right) = \frac{1}{2}\left(J_x \pm \mathrm{i}J_y\right)^2.$$

We wish to find γ such that

$$\left\langle njm|J_{ij}|n'j'm'\right\rangle = \gamma\left\langle njm|I_{ij}|n'j'm'\right\rangle. \tag{5.50}$$

In our case, $n = n'$, $j = j' = \frac{3}{2}$; m, m', κ arbitrary. Let $m = m' = \frac{3}{2}, i = j = z$. Then,

$$\text{LHS} = \langle njm|J_{zz}|njm\rangle = \left\langle\frac{3}{2}\frac{3}{2}\middle|J_z^2 - \frac{1}{3}J^2\middle|\frac{3}{2}\frac{3}{2}\right\rangle = \frac{9}{4} - \frac{1}{3}j(j+1) = 1,$$
$$\text{RHS} = \gamma\,\langle njm|I_{zz}|njm\rangle = \gamma\left\langle\frac{3}{2}\frac{3}{2}\middle|I_z^2 - \frac{1}{3}I^2\middle|\frac{3}{2}\frac{3}{2}\right\rangle = \left[1 - \frac{1}{3}l(l+1)\right]\gamma = \frac{1}{3}\gamma,$$
$$\implies \quad \gamma = 3.$$

This difference was made explicit in Eqs. (5.6) and (5.46).

If we now apply the same procedure to the sh band, $j = \frac{1}{2}$, and

$$\left\langle\frac{1}{2}\frac{1}{2}\middle|J_z^2 - \frac{1}{3}J^2\middle|\frac{1}{2}\frac{1}{2}\right\rangle = 0,$$
$$\implies \quad \gamma = 0.$$

Hence,

$$H_{\text{so}} = -\Delta_0 + Ak^2. \tag{5.51}$$

5.4.3.3 Eigenvalues

An analytical solution to the 4×4 Hamiltonian exists; we have already obtained it from the DKK matrix. We now do so starting from Eq. (5.46). The first term on the RHS is already diagonal, hence we focus on the remaining terms. Let

$$\Omega = \sum_i \alpha_{ii}\left(J_i^2 - \frac{1}{3}J^2\right) + \frac{2}{\sqrt{3}}\sum_{i<j}\alpha_{ij}\{J_iJ_j\}, \tag{5.52}$$

where

$$\alpha_{ii} = \overline{B}k_i^2, \alpha_{ij} = \overline{D}k_ik_j\ (i \neq j). \tag{5.53}$$

Each solution is doubly degenerate (Kramer's theorem). Writing them as ω_1 and ω_2, we have

$$(\omega_1 - \omega)^2(\omega_2 - \omega)^2 = \left[\omega^2 - (\omega_1 + \omega_2) + \omega_1\omega_2\right]^2 = 0.$$

But, by construction, Ω is traceless,

$$\mathrm{Tr}\Omega = 2(\omega_1 + \omega_2) = 0, \tag{5.54}$$

$$\Longrightarrow \quad \omega_1 + \omega_2 = 0. \tag{5.55}$$

The fact that Ω is traceless can also be obtained directly. For this, we need the matrix representation of the J_i matrices and the result

$$\mathrm{Tr}(\alpha A + \beta B) = d\alpha \mathrm{Tr}A + d\beta \mathrm{Tr}B,$$

where d is the dimension of the representation. We know that

$$\mathrm{Tr}J_i = 0;\, J_i^2 = 1.$$

Also, using Eqs. (A.27) and with the basis ordering $M_J = \frac{3}{2}, \frac{1}{2}, -\frac{1}{2}, -\frac{3}{2}$,

$$J_+ = \begin{pmatrix} 0 & \sqrt{3} & 0 & 0 \\ 0 & 0 & 2 & 0 \\ 0 & 0 & 0 & \sqrt{3} \\ 0 & 0 & 0 & 0 \end{pmatrix}, \; J_- = \begin{pmatrix} 0 & 0 & 0 & 0 \\ \sqrt{3} & 0 & 0 & 0 \\ 0 & 2 & 0 & 0 \\ 0 & 0 & \sqrt{3} & 0 \end{pmatrix},$$

$$J_x = \begin{pmatrix} 0 & \frac{\sqrt{3}}{2} & 0 & 0 \\ \frac{\sqrt{3}}{2} & 0 & 1 & 0 \\ 0 & 1 & 0 & \frac{\sqrt{3}}{2} \\ 0 & 0 & \frac{\sqrt{3}}{2} & 0 \end{pmatrix}, \; J_y = \begin{pmatrix} 0 & -\mathrm{i}\frac{\sqrt{3}}{2} & 0 & 0 \\ \mathrm{i}\frac{\sqrt{3}}{2} & 0 & -\mathrm{i} & 0 \\ 0 & \mathrm{i} & 0 & -\mathrm{i}\frac{\sqrt{3}}{2} \\ 0 & 0 & \mathrm{i}\frac{\sqrt{3}}{2} & 0 \end{pmatrix}, \; J_z = \begin{pmatrix} \frac{3}{2} & 0 & 0 & 0 \\ 0 & \frac{1}{2} & 0 & 0 \\ 0 & 0 & -\frac{1}{2} & 0 \\ 0 & 0 & 0 & -\frac{3}{2} \end{pmatrix},$$

$$J_x^2 = \begin{pmatrix} \frac{3}{4} & 0 & \frac{\sqrt{3}}{2} & 0 \\ 0 & \frac{7}{4} & 0 & \frac{\sqrt{3}}{2} \\ \frac{\sqrt{3}}{2} & 0 & \frac{7}{4} & 0 \\ 0 & \frac{\sqrt{3}}{2} & 0 & \frac{3}{4} \end{pmatrix}, \; J_y^2 = \begin{pmatrix} \frac{3}{4} & 0 & -\frac{\sqrt{3}}{2} & 0 \\ 0 & \frac{7}{4} & 0 & -\frac{\sqrt{3}}{2} \\ -\frac{\sqrt{3}}{2} & 0 & \frac{7}{4} & 0 \\ 0 & -\frac{\sqrt{3}}{2} & 0 & \frac{3}{4} \end{pmatrix}, \; J_z^2 = \begin{pmatrix} \frac{9}{4} & 0 & 0 & 0 \\ 0 & \frac{1}{4} & 0 & 0 \\ 0 & 0 & \frac{1}{4} & 0 \\ 0 & 0 & 0 & \frac{9}{4} \end{pmatrix},$$

$$J^2 = \frac{15}{4}\mathbf{1}, \; \mathrm{Tr}J^2 = j(j+1) \times (2j+1) = 15, \tag{5.56}$$

giving

$$\mathrm{Tr}\left(J_i^2 - \frac{1}{3}J^2\right) = \mathrm{Tr}\left(J_i^2\right) - \mathrm{Tr}\left(\frac{1}{3}J^2\right) = 0.$$

Our J_i's agree with Fishman. Luttinger's ones [4] differ from ours by a factor of i and complex conjugate:

$$\mathrm{i}J_{x,y}^* = J_{x,y}^L.$$

Now, for $\mathrm{Tr}\{J_i J_j\}$, one can either evaluate directly or first note that

$$\begin{aligned}
\{J_i J_j\} &= \frac{1}{2}(J_i J_j + J_j J_i),\\
[J_i, J_j] &= \mathrm{i}\varepsilon_{ijk} J_k = J_i J_j - J_j J_i,\\
\Longrightarrow \quad \{J_i J_j\} &= J_j J_i + \frac{\mathrm{i}}{2}\varepsilon_{ijk} J_k,\\
\mathrm{Tr}\{J_i J_j\} &= \mathrm{Tr}(J_j J_i) + \frac{\mathrm{i}d}{2}\varepsilon_{ijk}\mathrm{Tr}J_k = \mathrm{Tr}(J_j J_i),
\end{aligned}$$

and the latter can be shown to be zero by explicit calculation. Hence, $\mathrm{Tr}\Omega = 0$. Also, when Ω is diagonalized (say Λ), the diagonal elements are the eigenvalues: $\omega_1, \omega_1, \omega_2, \omega_2$. Then the diagonal elements of Λ^2 are $\omega_1^2, \omega_1^2, \omega_2^2, \omega_2^2$. Since the trace is invariant,

$$\mathrm{Tr}\Omega^2 = \mathrm{Tr}\Lambda^2 = 2\left(\omega_1^2 + \omega_2^2\right) = 4\omega_1^2, \tag{5.57}$$

using Eq. (5.55). In computing $\mathrm{Tr}\Omega^2$, we use

$$\mathrm{Tr}(A+B)^2 = \mathrm{Tr}A^2 + \mathrm{Tr}B^2 + 2\mathrm{Tr}AB.$$

The first term gives

$$\mathrm{Tr}\left\{\left[\sum_{i=1}^{3}\alpha_{ii}\left(J_i^2 - \frac{1}{3}J^2\right)\right]^2\right\} = 2\left[\left(\alpha_{yy} - \alpha_{zz}\right)^2 + \left(\alpha_{zz} - \alpha_{xx}\right)^2 + \left(\alpha_{xx} - \alpha_{yy}\right)^2\right],$$

the second

$$\mathrm{Tr}\left[\sum_{i<j}^{3}\alpha_{ij}\{J_i J_j\}\right]^2 = 4\left(\alpha_{yz}^2 + \alpha_{zx}^2 + \alpha_{xy}^2\right),$$

and the product of terms gives zero. Hence,

$$\omega_{1,2} = \pm\left[\alpha_{xx}^2 + \alpha_{yy}^2 + \alpha_{zz}^2 - \alpha_{yy}\alpha_{zz} - \alpha_{zz}\alpha_{xx} - \alpha_{xx}\alpha_{yy} + \alpha_{yz}^2 + \alpha_{zx}^2 + \alpha_{xy}^2\right]^{\frac{1}{2}}, \tag{5.58}$$

and

$$\begin{aligned} E_{1,2}(\mathbf{k}) &= \overline{A}k^2 + \omega_{1,2} \\ &= \overline{A}k^2 \pm \left[\overline{B}^2\left(k_x^4 + k_y^4 + k_z^4\right) - B^2\left(k_y^2k_z^2 + k_z^2k_x^2 + k_x^2k_y^2\right)\right. \\ &\quad \left. + \overline{D}\left(k_y^2k_z^2 + k_z^2k_x^2 + k_x^2k_y^2\right)\right]^{\frac{1}{2}} \\ &= \overline{A}k^2 \pm \left[\overline{B}^2k^4 + \overline{C}^2\left(k_y^2k_z^2 + k_z^2k_x^2 + k_x^2k_y^2\right)\right]^{\frac{1}{2}}, \end{aligned} \tag{5.59}$$

where

$$\overline{C} \equiv \overline{D}^2 - 3\overline{B}^2. \tag{5.60}$$

The parameters are consistent with Eq. (5.8).

5.4.3.4 *JM$_J$* Basis – Luttinger Model

For cases when the spin-orbit energy Δ_0 is large, it is often appropriate to study the four-dimensional Γ_8^+ subspace. First, we note that there are 16 linearly independent 4×4 matrices [4]:

$$\begin{aligned} &1,\ J_x,\ J_y,\ J_z,\ J_x^2,\ J_y^2,\ \{J_xJ_y\},\ \{J_yJ_z\},\ \{J_zJ_x\}, \\ &\left\{(J_y^2 - J_z^2)J_x\right\} \equiv V_x,\ \left\{(J_z^2 - J_x^2)J_y\right\} \equiv V_y,\ \left\{(J_x^2 - J_y^2)J_z\right\} \equiv V_z, \\ &J_x^3,\ J_y^3,\ J_z^3,\ J_xJ_yJ_z + J_zJ_yJ_x, \end{aligned} \tag{5.61}$$

and the table of irreducible basis functions and matrices is given in Table 5.2. They are found by decomposing $\Gamma_8^+ \otimes \Gamma_8^{+*}$ into irreducible components.

Since Γ_2^- cannot be formed from k_ik_j, there are only two new invariants:

$$V_x\{k_yk_z\} + V_y\{k_zk_x\} + V_z\{k_xk_y\}, \tag{5.62}$$

and

$$J_x^3B_x + J_y^3B_y + J_z^3B_z. \tag{5.63}$$

On the basis of time-reversal symmetry, Eq. (5.62) cannot occur because it consists of three factors of J_i and is odd with respect to time reversal. On the other hand,

Table 5.2 Irreducible basis functions, operators, and tensors for O_h' in 4×4 subspace [1]. $\omega = e^{2\pi i/3}$, $\{AB\} = \frac{1}{2}(AB + BA)$

	Time reversal	
	Even	Odd
Γ_1^+	k^2, J^2	
Γ_2^+		$J_x J_y J_z + J_z J_y J_x$
Γ_{12}^+	$k_x^2 + \omega k_y^2 + \omega^2 k_z^2, k_x^2 + \omega^2 k_y^2 + \omega k_z^2$ $J_x^2 + \omega J_y^2 + \omega^2 J_z^2, J_x^2 + \omega^2 J_y^2 + \omega J_z^2$	
Γ_{15}^+		$J_x, J_y, J_z; J_x^3, J_y^3, J_z^3; \sigma_x, \sigma_y, \sigma_z$ $[k_y, k_z], [k_z, k_x], [k_x, k_y]$
Γ_{25}^+	$\{k_y k_z\}, \{k_z k_x\}, \{k_x k_y\}$ $\{J_y J_z\}, \{J_z J_x\}, \{J_x J_y\}$	V_x, V_y, V_z
Γ_{15}^-		k_x, k_y, k_z
Γ_1^-		$k_x\sigma_x + k_y\sigma_y + k_z\sigma_z$
Γ_{12}^-		$K_1 = k_x\sigma_x + \omega k_y\sigma_y + \omega^2 k_z\sigma_z, K_2 = K_1^\dagger$
Γ_{25}^-		$k_x\sigma_y + k_y\sigma_x, k_y\sigma_z + k_z\sigma_y, k_x\sigma_z + k_z\sigma_x$

Eq. (5.63) is permissible since the magnetic field also changes sign. Hence, the most general Hamiltonian with spin and magnetic field is Eq. (38) of Luttinger [4]:

$$\begin{aligned} H_{\mathrm{bL}}(\mathbf{k}) = {} & \beta_1 k^2 + \beta_2 \left(k_x^2 J_x^2 + k_y^2 J_y^2 + k_z^2 J_z^2\right) \\ & + \beta_3 \left(\{k_x k_y\}\{J_x J_y\} + \{k_y k_z\}\{J_y J_z\} + \{k_z k_x\}\{J_z J_x\}\right) \\ & + \beta_4 \left(B_x J_x + B_y J_y + B_z J_z\right) + \beta_5 \left(B_x J_x^3 + B_y J_y^3 + B_z J_z^3\right). \end{aligned} \tag{5.64}$$

5.4.3.5 Luttinger Parameters

In his paper, Luttinger introduced the now popular dimensionless Luttinger parameters. Furthermore, the parameters in his Hamiltonians with and without spin, Eqs. (5.33) and (5.64), were related. We first establish the latter by rewriting Eq. (5.33) with the inclusion of the magnetic-field dependent terms:

$$H(\mathbf{k}) = Ak^2 - (A-B)\sum_i k_i^2 I_i^2 - 2C\sum_{i<j}\{k_i k_j\}\{I_i I_j\} + \frac{e}{2\hbar}K\mathbf{I}\cdot\mathbf{B} + \mu_B \boldsymbol{\sigma}\cdot\mathbf{B}. \tag{5.65}$$

Using the Wigner-Eckart theorem, he showed that

$$\sigma \to \frac{2}{3}\mathbf{J}, \mathbf{I} \to \frac{2}{3}\mathbf{J},$$

$$\{I_\alpha I_\beta\} - \frac{1}{3}\delta_{ij} I^2 \to \frac{1}{3}\left(\{J_\alpha J_\beta\} - \frac{1}{3}\delta_{ij} J^2\right),$$

which give

$$\{I_\alpha I_\beta\} \to \frac{1}{3}\{J_\alpha J_\beta\} + \frac{1}{4}\delta_{ij},$$

$$I_\alpha^2 = \{I_\alpha I_\alpha\} \to \frac{1}{3}J_\alpha^2 + \frac{1}{4}.$$

Then,

$$\sum_i k_i^2 I_i^2 = -\frac{1}{3}(A - B)\sum_i k_i^2 J_i^2 - \frac{1}{4}(A - B)k^2,$$

and Eq. (5.65) becomes

$$\begin{aligned} H(\mathbf{k}) &= Ak^2 - \frac{1}{3}(A - B)\sum_i k_i^2 J_i^2 - \frac{1}{4}(A - B)k^2 - \frac{2}{3}C\sum_{i<j}\{k_i k_j\}\{J_i J_j\} \\ &\quad + \frac{eK}{2\hbar}\frac{2}{3}\mathbf{J}\cdot\mathbf{B} + \frac{2}{3}\mu_B \mathbf{J}\cdot\mathbf{B} \\ &= \frac{1}{4}(3A + B)k^2 - \frac{1}{3}(A - B)\sum_i k_i^2 J_i^2 - \frac{2}{3}C\sum_{i<j}\{k_i, k_j\}\{J_i, J_j\} \\ &\quad + \frac{e\hbar}{3m_0}(1 + \frac{Km_0}{\hbar^2})\mathbf{J}\cdot\mathbf{B}, \end{aligned} \tag{5.66}$$

where we have replaced $\mu_B = e\hbar/(2m_0)$. Comparing with Eq. (5.64) gives

$$\begin{aligned} \beta_1 &= \frac{1}{4}(3A + B), \\ \beta_2 &= -\frac{1}{3}(A - B), \\ \beta_3 &= -\frac{2}{3}C, \\ \beta_4 &= \frac{e\hbar}{3m_0}\left(1 + \frac{Km_0}{\hbar^2}\right), \\ \beta_5 &= 0. \end{aligned} \tag{5.67}$$

This shows that β_5 is zero in the limit of no spin-orbit. Luttinger now introduced the dimensionless parameters

$$\begin{aligned}
\frac{\hbar^2}{2m_0}\gamma_1 &= -\frac{1}{3}(A+2B),\\
\frac{\hbar^2}{2m_0}\gamma_2 &= -\frac{1}{6}(A-B),\\
\frac{\hbar^2}{2m_0}\gamma_3 &= -\frac{1}{6}C,\\
\frac{\hbar^2}{2m_0}\kappa &= \frac{\hbar}{3m_0}\left(1+\frac{Km_0}{\hbar^2}\right),\\
\frac{e\hbar}{m_0}q &= -\beta_5,
\end{aligned} \tag{5.68}$$

so that

$$\begin{aligned}
\beta_1 &= -\frac{\hbar^2}{2m_0}\left(\gamma_1+\frac{5}{2}\gamma_2\right),\\
\beta_2 &= \frac{\hbar^2}{m_0}\gamma_2,\\
\beta_3 &= \frac{2\hbar^2}{m_0}\gamma_3,\\
\beta_4 &= -\frac{e\hbar}{m_0}\kappa,
\end{aligned} \tag{5.69}$$

and

$$\begin{aligned}
H_{\mathrm{L}}(\mathbf{k}) \equiv -\frac{\hbar^2}{m_0}\Bigg\{&\left(\gamma_1+\frac{5}{2}\gamma_2\right)\frac{k^2}{2} - \gamma_2\left(k_x^2J_x^2+k_y^2J_y^2+k_z^2J_z^2\right)\\
&-2\gamma_3\left(\{k_xk_y\}\{J_xJ_y\}+\{k_yk_z\}\{J_yJ_z\}+\{k_zk_x\}\{J_zJ_x\}\right)\\
&+\frac{e}{\hbar}\kappa\left(B_xJ_x+B_yJ_y+B_zJ_z\right)+\frac{e}{\hbar}q\left(B_xJ_x^3+B_yJ_y^3+B_zJ_z^3\right)\Bigg\}.
\end{aligned} \tag{5.70}$$

Note that this work was done for a basis of four degenerate states. When nondegenerate states are included, the Luttinger parameters are no longer uniquely defined. A recent discussion is that of Fishman and coworkers [64]. The first detailed study of the valence-band parameters for ZB semiconductors was carried by Lawaetz [85].

Example Luttinger parameters are given in Table 5.5. It is worth pointing out that κ is not determined by classical (i.e., large quantum number) cyclotron resonance even though it affects magnetic levels [25]. q was first measured by Pidgeon and Brown [86] and an expression for it derived by Hensel and Suzuki [87].

5.4.3.6 Kane's Version: Zero Field

We will now write out the Luttinger Hamiltonian in the absence of a magnetic field explicitly using the angular momentum matrices in Eq. (5.4). Following Kane [88] and Fishman [24], we write

$$H(\mathbf{k}, \mathbf{B} = \mathbf{0}) = -\frac{\hbar^2}{2m_0}\left[\gamma_1 k^2 + h_{23}(\mathbf{k})\right], \tag{5.71}$$

with

$$\begin{aligned} h_{23}(\mathbf{k}) = {} & \frac{5}{2}\gamma_2 k^2 - 2\gamma_2\left(k_x^2 J_x^2 + k_y^2 J_y^2 + k_z^2 J_z^2\right) \\ & -4\gamma_3\left(\{k_x k_y\}\{J_x J_y\} + \{k_y k_z\}\{J_y J_z\} + \{k_z k_x\}\{J_z J_x\}\right). \end{aligned} \tag{5.72}$$

Now,

$$\begin{aligned} \{J_x J_y\} &= \frac{\mathrm{i}}{2}\begin{pmatrix} 0 & 0 & -\sqrt{3} & 0 \\ 0 & 0 & 0 & -\sqrt{3} \\ \sqrt{3} & 0 & 0 & 0 \\ 0 & \sqrt{3} & 0 & 0 \end{pmatrix}, \\ \{J_y J_z\} &= \frac{\mathrm{i}}{2}\begin{pmatrix} 0 & -\sqrt{3} & 0 & 0 \\ \sqrt{3} & 0 & 0 & 0 \\ 0 & 0 & 0 & \sqrt{3} \\ 0 & 0 & -\sqrt{3} & 0 \end{pmatrix}, \\ \{J_x J_z\} &= \frac{1}{2}\begin{pmatrix} 0 & \sqrt{3} & 0 & 0 \\ \sqrt{3} & 0 & 0 & 0 \\ 0 & 0 & 0 & -\sqrt{3} \\ 0 & 0 & -\sqrt{3} & 0 \end{pmatrix}, \end{aligned} \tag{5.73}$$

giving

$$\{k_xk_y\}\{J_xJ_y\}+\{k_yk_z\}\{J_yJ_z\}+\{k_zk_x\}\{J_zJ_x\}$$
$$=\frac{1}{2}\begin{pmatrix} 0 & \sqrt{3}(k_xk_z-\mathrm{i}k_yk_z) & -\sqrt{3}\mathrm{i}k_xk_y & 0 \\ \sqrt{3}(k_xk_z+\mathrm{i}k_yk_z) & 0 & 0 & -\sqrt{3}\mathrm{i}k_xk_y \\ \sqrt{3}\mathrm{i}k_xk_y & 0 & 0 & -\sqrt{3}(k_xk_z-\mathrm{i}k_yk_z) \\ 0 & \sqrt{3}\mathrm{i}k_xk_y & -\sqrt{3}(k_xk_z+\mathrm{i}k_yk_z) & 0 \end{pmatrix}.$$

Next

$$k_x^2J_x^2+k_y^2J_y^2+k_z^2J_z^2$$
$$=\frac{1}{4}\begin{pmatrix} 3(k_x^2+k_y^2+3k_z^2) & 0 & 2\sqrt{3}(k_x^2-k_y^2) & 0 \\ 0 & 7(k_x^2+k_y^2)+k_z^2 & 0 & 2\sqrt{3}(k_x^2-k_y^2) \\ 2\sqrt{3}(k_x^2-k_y^2) & 0 & 7(k_x^2+k_y^2)+k_z^2 & 0 \\ 0 & 2\sqrt{3}(k_x^2-k_y^2) & 0 & 3(k_x^2+k_y^2+3k_z^2) \end{pmatrix}.$$

Thus,

$$h_{23}(\mathbf{k})$$
$$=\begin{pmatrix} \gamma_2(k_x^2+k_y^2-2k_z^2) & -2\sqrt{3}\gamma_3(k_x-\mathrm{i}k_y)k_z & \sqrt{3}[-\gamma_2(k_x^2-k_y^2)+2\mathrm{i}\gamma_3k_xk_y] & 0 \\ -2\sqrt{3}\gamma_3(k_x+\mathrm{i}k_y)k_z & -\gamma_2(k_x^2+k_y^2-2k_z^2) & 0 & \sqrt{3}[-\gamma_2(k_x^2-k_y^2)+2\mathrm{i}\gamma_3k_xk_y] \\ \sqrt{3}[-\gamma_2(k_x^2-k_y^2)-2\mathrm{i}\gamma_3k_xk_y] & 0 & -\gamma_2(k_x^2+k_y^2-2k_z^2) & 2\sqrt{3}\gamma_3(k_x-\mathrm{i}k_y)k_z \\ 0 & \sqrt{3}[\gamma_2(k_x^2-k_y^2)-2\mathrm{i}\gamma_3k_xk_y] & 2\sqrt{3}\gamma_3(k_x+\mathrm{i}k_y)k_z & \gamma_2(k_x^2+k_y^2-2k_z^2) \end{pmatrix}$$

$$\equiv\begin{pmatrix} A & B & C & 0 \\ B^* & -A & 0 & C \\ C^* & 0 & -A & -B \\ 0 & C^* & -B^* & A \end{pmatrix}, \tag{5.74}$$

with

$$
\begin{aligned}
A &= \gamma_2 \left(k_x^2 + k_y^2 - 2k_z^2\right), \\
B &= -2\sqrt{3}\gamma_3 \left(k_x - \mathrm{i}k_y\right) k_z, \\
C &= -\sqrt{3}\gamma_2 \left(k_x^2 - k_y^2\right) + 2\mathrm{i}\sqrt{3}\gamma_3 k_x k_y.
\end{aligned} \tag{5.75}
$$

This form of the Hamiltonian was given by Kane [88]; however, with the same definition of A, B, C as Kane, we get a slightly different matrix. Indeed, the structure of his matrix agrees with the 4×4 subblock of the six-band DKK Hamiltonian we had obtained earlier, Tables 3.5 and 3.6. On the other hand, the structure of the matrix in Eq. (5.74) is similar to that of Elliott [35], DKK [2], Bir and Pikus [1] (after adding the diagonal term), and S. L. Chuang [15].

Finally

$$
H(\mathbf{k}, \mathbf{B} = \mathbf{0}) = -\frac{\hbar^2}{2m_0} \left[\gamma_1 k^2 + \begin{pmatrix} A & B & C & 0 \\ B^* & -A & 0 & C \\ C^* & 0 & -A & -B \\ 0 & C^* & -B^* & A \end{pmatrix} \right]
$$

$$
= -\frac{\hbar^2}{2m_0} \begin{pmatrix} \left|\frac{3}{2}\,\frac{3}{2}\right\rangle & \left|\frac{3}{2}\,\frac{1}{2}\right\rangle & \left|\frac{3}{2}\,-\frac{1}{2}\right\rangle & \left|\frac{3}{2}\,-\frac{3}{2}\right\rangle \\ A + \gamma_1 k^2 & B & C & 0 \\ B^* & -A + \gamma_1 k^2 & 0 & C \\ C^* & 0 & -A + \gamma_1 k^2 & -B \\ 0 & C^* & -B^* & A + \gamma_1 k^2 \end{pmatrix}. \tag{5.76}
$$

By comparing Table 3.6 and Eq. (5.76), one can relate the Luttinger parameters to the DKK parameters. Thus, taking the (1,1) element,

$$
\frac{1}{2}P = -\frac{\hbar^2}{2m_0}\left(A + \gamma_1 k^2\right),
$$

$$
\Longrightarrow \quad \frac{\hbar^2 k^2}{2m_0} + \frac{(L+M)}{2}\left(k_x^2 + k_y^2\right) + M k_z^2 = -\frac{\hbar^2}{2m_0}\gamma_1 k^2 - \frac{\hbar^2}{2m_0}\gamma_2\left(k_x^2 + k_y^2 - 2k_z^2\right),
$$

giving

$$\frac{\hbar^2}{2m_0}\gamma_1 = -\frac{1}{3}(L+2M) - \frac{\hbar^2}{2m_0},$$
$$\frac{\hbar^2}{2m_0}\gamma_2 = -\frac{1}{6}(L-M). \tag{5.77}$$

Similarly, comparing the (1,2) element,

$$R = -\frac{\hbar^2}{2m_0}B,$$
$$\Longrightarrow \quad -\frac{N}{\sqrt{3}}\left(k_x - \mathrm{i}k_y\right)k_z = -\frac{\hbar^2}{2m_0}\left(-2\sqrt{3}\gamma_3(k_x - \mathrm{i}k_y)k_z\right),$$

giving

$$\frac{\hbar^2}{2m_0}\gamma_3 = -\frac{1}{6}N. \tag{5.78}$$

Finally,

$$\frac{\hbar^2}{2m_0}2(3\kappa+1) = -K,$$
$$\frac{e\hbar}{m_0}q = -\beta_5.$$

5.4.3.7 Chuang's Version

We now compare Kane's four-band to Chuang's six-band model [15]. Let the matrix elements in Chuang's model be

$$\begin{aligned}
P_C &= \frac{\hbar^2\gamma_1}{2m_0}\left(k_x^2+k_y^2+k_z^2\right) = \frac{\hbar^2\gamma_1}{2m_0}k^2,\\
Q_C &= \frac{\hbar^2\gamma_2}{2m_0}\left(k_x^2+k_y^2-2k_z^2\right) = \frac{\hbar^2 A}{2m_0},\\
S_C &= \frac{\hbar^2}{2m_0}2\sqrt{3}\gamma_3\left(k_x-\mathrm{i}k_y\right)k_z = -\frac{\hbar^2 B}{2m_0},\\
R_C &= \frac{\hbar^2}{2m_0}\left[-\sqrt{3}\gamma_2\left(k_x^2-k_y^2\right)+2\mathrm{i}\sqrt{3}\gamma_3 k_x k_y\right] = \frac{\hbar^2 C}{2m_0}\\
&= \sqrt{3}\mu k_+^2 - \sqrt{3}\overline{\gamma}k_-^2,
\end{aligned} \tag{5.79}$$

where $k_\pm = k_x \pm \mathrm{i}k_y$, $\overline{\gamma} = \frac{1}{2}(\gamma_3 + \gamma_2)$, $\mu = \frac{1}{2}(\gamma_3 - \gamma_2)$. Then Eq. (5.76) becomes

$$-\begin{pmatrix} \left|\frac{3}{2}\,\frac{3}{2}\right\rangle & \left|\frac{3}{2}\,\frac{1}{2}\right\rangle & \left|\frac{3}{2}\,-\frac{1}{2}\right\rangle & \left|\frac{3}{2}\,-\frac{3}{2}\right\rangle \\ P_C + Q_C & -S_C & R_C & 0 \\ -S_C^* & P_C - Q_C & 0 & R_C \\ R_C^* & 0 & P_C - Q_C & S_C \\ 0 & R_C^* & S_C^* & P_C + Q_C \end{pmatrix}. \tag{5.80}$$

The rest of the 6×6 matrix can be similarly obtained, e.g., starting from Table 3.6. (Note that this will give our phase convention.) We need the DKK parameters in terms of the Luttinger ones. Inverting Eqs. (5.77) and (5.78), we have

$$\begin{aligned} L &= -\frac{\hbar^2}{2m_0}(\gamma_1 + 4\gamma_2 + 1), \\ M &= -\frac{\hbar^2}{2m_0}(\gamma_1 - 2\gamma_2 + 1), \\ N &= -\frac{\hbar^2}{2m_0}6\gamma_3. \end{aligned} \tag{5.81}$$

Then, the matrix elements of Table 3.6 are

$$\begin{aligned} P &= (L+M)\left(k_x^2 + k_y^2\right) + 2Mk_z^2 + \frac{\hbar^2 k^2}{m_0} \\ &= \frac{\hbar^2}{2m_0}\left[-2(\gamma_1 + \gamma_2)\left(k_x^2 + k_y^2\right) - 2(\gamma_1 - 2\gamma_2)k_z^2\right] = -2(P_C + Q_C), \\ Q &= M\left(k_x^2 + k_y^2\right) + Lk_z^2 + \frac{\hbar^2 k^2}{2m_0} \\ &= \frac{\hbar^2}{2m_0}\left[-(\gamma_1 - 2\gamma_2)\left(k_x^2 + k_y^2\right) - (\gamma_1 + 4\gamma_2)k_z^2\right] = -(P_C - 2Q_C), \\ R &= -\frac{N}{\sqrt{3}}\left(k_x - \mathrm{i}k_y\right)k_z = \frac{\hbar^2}{2m_0}\left[2\sqrt{3}\gamma_3(k_x - \mathrm{i}k_y)k_z\right] = S_C, \\ S &= -\frac{1}{2\sqrt{3}}\left[(L-M)(k_x^2 - k_y^2) - 2\mathrm{i}Nk_xk_y\right] = \frac{\hbar^2}{2m_0}\left[\sqrt{3}(\gamma_2\left(k_x^2 - k_y^2\right) - 2\mathrm{i}\gamma_3 k_xk_y)\right] \\ &= -R_c, \\ (1,1) &= \frac{1}{2}P = -(P_C + Q_C), \\ (1,2) &= R = S_C, \end{aligned}$$

$$
\begin{aligned}
(1,3) &= S = -R_C, \\
(2,2) &= \frac{1}{6}P + \frac{2}{3}Q = -(P_C - Q_C), \\
(2,5) &= \frac{1}{3\sqrt{2}}(P - 2Q) = -\sqrt{2}Q_C, \\
(3,5) &= \frac{1}{3}(P + Q) - \Delta_0 = -P_C - \Delta_0.
\end{aligned}
$$

Therefore, we get the Hamiltonian in Table 5.3. We have dropped the C subscript on the matrix elements for conciseness. Changing the sign of the $\left|\frac{3}{2}\ -\frac{3}{2}\right\rangle$ and $\left|\frac{1}{2}\ \frac{1}{2}\right\rangle$ states leads to a matrix identical to Chuang's.

Of course, this matrix is the same as the canonical one in Table 3.5. Indeed, we have defined the quantities Q, R, S to be identical in both cases and

$$P' \equiv -(P + Q), \; P'' \equiv -(P - Q), \; P''' \equiv -P. \tag{5.82}$$

The form in Table 5.3 is often known as the Luttinger–Kohn Hamiltonian though we already clarified in Chap. 3 that this is not the form to be found in their paper [6].

Table 5.3 Six-band Hamiltonian in JM_J basis using our phase convention. Note differences compared to Eq. (4.3.14) of Chuang [15]. We have dropped the C subscript on the matrix elements used in the text. Also note the conventional negative sign outside the matrix

$$
H(\mathbf{k}) = -\begin{pmatrix}
\left|\frac{3}{2}\ \frac{3}{2}\right\rangle & \left|\frac{3}{2}\ \frac{1}{2}\right\rangle & \left|\frac{3}{2}\ -\frac{1}{2}\right\rangle & \left|\frac{3}{2}\ -\frac{3}{2}\right\rangle & \left|\frac{1}{2}\ \frac{1}{2}\right\rangle & \left|\frac{1}{2}\ -\frac{1}{2}\right\rangle \\
P+Q & -S & R & 0 & \frac{1}{\sqrt{2}}S & \sqrt{2}R \\
-S^* & P-Q & 0 & -R & \sqrt{2}Q & \sqrt{\frac{3}{2}}S \\
R^* & 0 & P-Q & -S & -\sqrt{\frac{3}{2}}S^* & \sqrt{2}Q \\
0 & -R^* & -S^* & P+Q & -\sqrt{2}R^* & \frac{1}{\sqrt{2}}S^* \\
\frac{1}{\sqrt{2}}S^* & \sqrt{2}Q & -\sqrt{\frac{3}{2}}S & -\sqrt{2}R & P+\Delta_0 & 0 \\
\sqrt{2}R^* & \sqrt{\frac{3}{2}}S^* & \sqrt{2}Q & \frac{1}{\sqrt{2}}S & 0 & P+\Delta_0
\end{pmatrix},
$$

with

$$
\begin{aligned}
P &= \frac{\hbar^2}{2m_0}\gamma_1\left(k_x^2 + k_y^2 + k_z^2\right), \\
Q &= \frac{\hbar^2}{2m_0}\gamma_2\left(k_x^2 + k_y^2 - 2k_z^2\right), \\
S &= \frac{\hbar^2}{2m_0}2\sqrt{3}\gamma_3\left(k_x - \mathrm{i}k_y\right)k_z, \\
R &= \frac{\hbar^2}{2m_0}\left[-\sqrt{3}\gamma_2\left(k_x^2 - k_y^2\right) + 2\mathrm{i}\sqrt{3}\gamma_3 k_x k_y\right].
\end{aligned}
$$

The earliest we have been able to trace it is in a 1985 paper by Broido and Sham in a four-band version [89].

If inversion asymmetry is neglected, this six-band model applies to ZB as well. An example of the six-band calculation for the valence band of GaAs is given in Fig. 5.1.

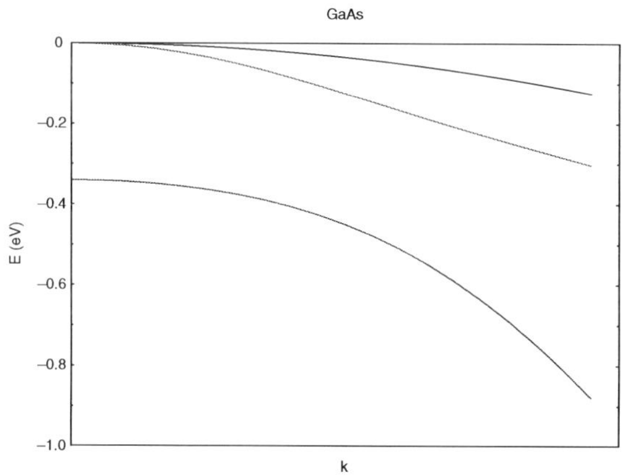

Fig. 5.1 Valence band of GaAs using six-band model. Parameters used are given in Table 5.5

5.4.3.8 Broido-Sham Transformation

The four-band Hamiltonian can be block diagonalized. The idea actually originated from Kane [3]. However, the work of Broido and Sham [89] is often quoted and we will develop it here. The procedure is to apply a unitary transformation to the Hamiltonian in, e.g., Eq. (5.80). The Broido-Sham transformation is given by the matrix

$$U = \begin{pmatrix} \frac{1}{\sqrt{2}}e^{-i\phi} & 0 & 0 & -\frac{1}{\sqrt{2}}e^{i\phi} \\ 0 & \frac{1}{\sqrt{2}}e^{-i\eta} & -\frac{1}{\sqrt{2}}e^{i\eta} & 0 \\ 0 & \frac{1}{\sqrt{2}}e^{-i\eta} & \frac{1}{\sqrt{2}}e^{i\eta} & 0 \\ \frac{1}{\sqrt{2}}e^{-i\phi} & 0 & 0 & \frac{1}{\sqrt{2}}e^{i\phi} \end{pmatrix}. \tag{5.83}$$

Then [note that the basis has been re-ordered from Eq. (5.80)]

$$UH = \begin{pmatrix} \frac{1}{\sqrt{2}}e^{-i\phi} & 0 & 0 & -\frac{1}{\sqrt{2}}e^{i\phi} \\ 0 & \frac{1}{\sqrt{2}}e^{-i\eta} & -\frac{1}{\sqrt{2}}e^{i\eta} & 0 \\ 0 & \frac{1}{\sqrt{2}}e^{-i\eta} & \frac{1}{\sqrt{2}}e^{i\eta} & 0 \\ \frac{1}{\sqrt{2}}e^{-i\phi} & 0 & 0 & \frac{1}{\sqrt{2}}e^{i\phi} \end{pmatrix} \begin{pmatrix} P+Q & R & -S & 0 \\ R^* & P-Q & 0 & S \\ -S^* & 0 & P-Q & R \\ 0 & S^* & R^* & P+Q \end{pmatrix}$$

$$= \frac{1}{\sqrt{2}} \begin{pmatrix} e^{-i\phi}(P+Q) & e^{-i\phi}R - e^{i\phi}S^* & -e^{-i\phi}S - e^{i\phi}R^* & -e^{i\phi}(P+Q) \\ e^{-i\eta}R^* + e^{i\eta}S^* & e^{-i\eta}(P-Q) & -e^{i\eta}(P-Q) & e^{-i\eta}S - e^{i\eta}R \\ e^{-i\eta}R^* - e^{i\eta}S^* & e^{-i\eta}(P-Q) & e^{i\eta}(P-Q) & e^{-i\eta}S + e^{i\eta}R \\ e^{-i\phi}(P+Q) & e^{-i\phi}R + e^{i\phi}S^* & -e^{-i\phi}S + e^{i\phi}R^* & e^{i\phi}(P+Q) \end{pmatrix}.$$

Finally, UHU^T is given in Table 5.4. The matrix is block diagonal if the following element is zero:

$$\begin{aligned} \widetilde{S} &\equiv \frac{1}{2}\left[e^{i(\eta-\phi)}R - e^{-i(\eta-\phi)}R^* - e^{i(\eta+\phi)}S^* - e^{-i(\eta+\phi)}S\right] \\ &= \frac{1}{2}\left[e^{i\alpha}R - e^{-i\alpha}R^* - e^{i(\frac{\pi}{2}+\beta)}S^* - e^{-i(\frac{\pi}{2}+\beta)}S\right]. \end{aligned} \tag{5.84}$$

Writing

$$e^{i\alpha}R - e^{-i\alpha}R^* = \left(e^{i(\alpha+\theta)} - e^{-i(\alpha+\theta)}\right)|R|,$$

then, the term is zero if $\alpha = -\theta$ or

$$\tan\alpha = -\tan\theta = \frac{2\gamma_3 k_x k_y}{\gamma_2\left(k_x^2 - k_y^2\right)}. \tag{5.85}$$

Similarly, writing

$$e^{i(\frac{\pi}{2}+\beta)}S^* + e^{-i(\frac{\pi}{2}+\beta)}S = \left(e^{-i(\frac{\pi}{2}+\beta-\theta')} + e^{i(\frac{\pi}{2}+\beta-\theta')}\right)|S|,$$

then, the term is zero if $\beta = \theta'$ or

$$\tan\beta = \tan\theta' = \frac{k_x}{k_y}. \tag{5.86}$$

A discussion in the presence of strain was given by Lee and Vassell [90].

5.4.3.9 Eigenvalues

Recall that the four-band DKK model could be diagonalized exactly. Since the Luttinger Hamiltonian is equivalent, one should expect an exact diagonalization too. This is straightforward to demonstrate. The problem is equivalent to diagonalizing h_{23}. The eigenvalues are given by

$$||h_{23} - \lambda|| = 0$$

Table 5.4 Broido-Sham Hamiltonian [89]

$$
UHU^T = \begin{pmatrix}
P+Q & \begin{matrix}\frac{1}{2}\Big[\mathrm{e}^{\mathrm{i}(\eta-\phi)}R + \mathrm{e}^{-\mathrm{i}(\eta-\phi)}R^* \\ -\mathrm{e}^{\mathrm{i}(\eta+\phi)}S^* + \mathrm{e}^{-\mathrm{i}(\eta+\phi)}S\Big]\end{matrix} & \begin{matrix}\frac{1}{2}\Big[\mathrm{e}^{\mathrm{i}(\eta-\phi)}R - \mathrm{e}^{-\mathrm{i}(\eta-\phi)}R^* \\ -\mathrm{e}^{\mathrm{i}(\eta+\phi)}S^* - \mathrm{e}^{-\mathrm{i}(\eta+\phi)}S\Big]\end{matrix} & 0 \\
\begin{matrix}\frac{1}{2}\Big[\mathrm{e}^{-\mathrm{i}(\eta-\phi)}R^* + \mathrm{e}^{\mathrm{i}(\eta-\phi)}R \\ -\mathrm{e}^{-\mathrm{i}(\eta+\phi)}S + \mathrm{e}^{\mathrm{i}(\eta+\phi)}S^*\Big]\end{matrix} & P-Q & 0 & \begin{matrix}\frac{1}{2}\Big[\mathrm{e}^{-\mathrm{i}(\eta-\phi)}R^* - \mathrm{e}^{\mathrm{i}(\eta-\phi)}R \\ +\mathrm{e}^{\mathrm{i}(\eta+\phi)}S^* + \mathrm{e}^{-\mathrm{i}(\eta+\phi)}S\Big]\end{matrix} \\
\begin{matrix}\frac{1}{2}\Big[\mathrm{e}^{-\mathrm{i}(\eta-\phi)}R^* - \mathrm{e}^{\mathrm{i}(\eta-\phi)}R \\ -\mathrm{e}^{-\mathrm{i}(\eta+\phi)}S - \mathrm{e}^{\mathrm{i}(\eta+\phi)}S^*\Big]\end{matrix} & 0 & P-Q & \begin{matrix}\frac{1}{2}\Big[\mathrm{e}^{-\mathrm{i}(\eta-\phi)}R^* + \mathrm{e}^{\mathrm{i}(\eta-\phi)}R \\ -\mathrm{e}^{\mathrm{i}(\eta+\phi)}S^* + \mathrm{e}^{-\mathrm{i}(\eta+\phi)}S\Big]\end{matrix} \\
0 & \begin{matrix}\frac{1}{2}\Big[\mathrm{e}^{\mathrm{i}(\eta-\phi)}R - \mathrm{e}^{-\mathrm{i}(\eta-\phi)}R^* \\ +\mathrm{e}^{\mathrm{i}(\eta+\phi)}S^* + \mathrm{e}^{-\mathrm{i}(\eta+\phi)}S\Big]\end{matrix} & \begin{matrix}\frac{1}{2}\Big[\mathrm{e}^{\mathrm{i}(\eta-\phi)}R + \mathrm{e}^{-\mathrm{i}(\eta-\phi)}R^* \\ +\mathrm{e}^{\mathrm{i}(\eta+\phi)}S^* - \mathrm{e}^{-\mathrm{i}(\eta+\phi)}S\Big]\end{matrix} & P+Q
\end{pmatrix}.
$$

Table 5.5 Examples of Luttinger parameters [96]

	γ_1	γ_2	γ_3	κ	q
AlAs	3.45	0.68	1.29	0.12	0.03
GaAs	6.85	2.10	2.90	1.20	0.01
GaSb	11.80	4.03	5.26	3.18	0.13
InP	5.04	1.56	1.73	0.97	0.02
InAs	19.67	8.37	9.29	7.68	0.04

$$= (A-\lambda)\begin{vmatrix} -A-\lambda & 0 & C \\ 0 & -A-\lambda & -B \\ C^* & -B^* & A-\lambda \end{vmatrix} - B\begin{vmatrix} B^* & 0 & C \\ C^* & -A-\lambda & -B \\ 0 & -B^* & A-\lambda \end{vmatrix} + C\begin{vmatrix} B^* & -A-\lambda & C \\ C^* & 0 & -B \\ 0 & C^* & A-\lambda \end{vmatrix}$$

$$= \left[\left(A^2-\lambda^2\right)+|B|^2+|C|^2\right]^2.$$

Therefore,

$$\lambda = \pm\sqrt{A^2+|B|^2+|C|^2}$$

$$= \pm\left\{\gamma_2^2\left(k_x^2+k_y^2-2k_z^2\right)^2+12\gamma_3^2k_z^2\left(k_x^2+k_y^2\right)+3\gamma_2^2\left(k_x^2-k_y^2\right)^2+12\gamma_3^2k_x^2k_y^2\right\}^{\frac{1}{2}}$$

$$= \pm 2k^2\left\{\gamma_2^2+3\left(\gamma_3^2-\gamma_2^2\right)\frac{k_x^2k_y^2+k_y^2k_z^2+k_z^2k_x^2}{k^4}\right\}^{\frac{1}{2}}. \tag{5.87}$$

We can now write down the band structure along various directions. For $\mathbf{k} = [00k]$,

$$E(00k) = -\frac{\hbar^2k^2}{2m_0}(\gamma_1 \pm 2\gamma_2), \tag{5.88}$$

and the effective masses are given by

$$m_{lh,hh} = \frac{1}{\gamma_1 \pm 2\gamma_2}. \tag{5.89}$$

Note that $\mathbf{B} = \mathbf{0}$ in Eq. (5.74). Thus, the $m = \frac{3}{2}$ and $m = \frac{1}{2}$ states are decoupled along this direction and we can say that the heavy-hole (light-hole) states are pure $m = \pm\frac{3}{2}$ ($m = \pm\frac{1}{2}$). For all other cases, those states are mixed.

For $\mathbf{k} = [kkk]/\sqrt{3}$,

$$E(kkk) = -\frac{\hbar^2k^2}{2m_0}(\gamma_1 \pm 2\gamma_3), \tag{5.90}$$

and the effective masses are given by

$$m_{lh,hh} = \frac{1}{\gamma_1 \pm 2\gamma_3}. \tag{5.91}$$

For $\mathbf{k} = [kk0]/\sqrt{2}$,

$$E(kk0) = -\frac{\hbar^2 k^2}{2m_0}\left[\gamma_1 \pm \left(\gamma_2^2 + 3\gamma_3^2\right)^{\frac{1}{2}}\right], \tag{5.92}$$

and the effective masses are given by

$$m_{lh,hh} = \frac{1}{\gamma_1 \pm \left(\gamma_2^2 + 3\gamma_3^2\right)^{\frac{1}{2}}}. \tag{5.93}$$

Sometimes it is useful to quantize the spin along the wave vector $\mathbf{k}$. One advantage is seen by rewriting

$$\begin{aligned} h_{23} &= \frac{5}{2}\gamma_2 k^2 - 2\gamma_2\left(k_x^2 J_x^2 + k_y^2 J_y^2 + k_z^2 J_z^2\right) \\ &\quad -4\gamma_3\left(\{k_x k_y\}\{J_x J_y\} + \{k_y k_z\}\{J_y J_z\} + \{k_z k_x\}\{J_z J_x\}\right) \\ &= \frac{5}{2}\gamma_2 k^2 - 2\gamma_2(\mathbf{k}\cdot\mathbf{J})^2 \\ &\quad +4(\gamma_2 - \gamma_3)\left(\{k_x k_y\}\{J_x J_y\} + \{k_y k_z\}\{J_y J_z\} + \{k_z k_x\}\{J_z J_x\}\right). \end{aligned} \tag{5.94}$$

For example, if $\gamma_2 = \gamma_3$, then h_{23} is diagonal. But there are other instances when the latter is diagonal. The simplest is when both the spin and the wave vector are along a cubic direction. For example, take $\mathbf{k} = [k00]$. If $\mathbf{J}$ is quantized along $\mathbf{k}$, then it must be diagonal in that direction; hence, $(\mathbf{k}\cdot\mathbf{J})^2$ becomes $k^2 J^2$. The diagonal form of J^2 is the same as for J_z^2 for z-axis quantization. Thus,

$$h_{23}(k00) = \frac{5}{2}\gamma_2 k^2 - 2\gamma_2 k^2 \begin{pmatrix} \frac{9}{4} & 0 & 0 & 0 \\ 0 & \frac{1}{4} & 0 & 0 \\ 0 & 0 & \frac{1}{4} & 0 \\ 0 & 0 & 0 & \frac{9}{4} \end{pmatrix} = \gamma_2 k^2 \begin{pmatrix} -2 & 0 & 0 & 0 \\ 0 & 2 & 0 & 0 \\ 0 & 0 & 2 & 0 \\ 0 & 0 & 0 & -2 \end{pmatrix}, \tag{5.95}$$

which is diagonal. It is, of course, the same result as for z-axis quantization with $\mathbf{k} = [00k]$.

5.4.3.10 Spherical and Axial Approximations

The condition

$$\gamma_2 = \gamma_3, \tag{5.96}$$

is known as the spherical approximation since the valence band is then isotropic. This was already discussed in terms of the DKK parameters in Sect. 3.2.4. Ekenberg and Altarelli [91] showed that the spherical approximation is also obtained if one replaces γ_2 and γ_3 by

$$\overline{\gamma} = \frac{1}{5}(2\gamma_2 + 3\gamma_3). \tag{5.97}$$

Then, the Hamiltonian is obtained by combining Eqs. (5.71) and (5.72) and (5.94):

$$H(\mathbf{k}) = -\frac{\hbar^2}{2m_0}\left[\left(\gamma_1 + \frac{5}{2}\overline{\gamma}\right)k^2 - 2\overline{\gamma}\,(\mathbf{k}\cdot\mathbf{J})^2\right]. \tag{5.98}$$

The above is to be contrasted to the axial approximation whereby one term of cubic symmetry is retained so that full spherical symmetry is not recovered but rather axial symmetry in the x-y plane [92], i.e., the Hamiltonian becomes cylindrically symmetric. This corresponds to replacing γ_2 and γ_3 by [91, 93, 94]

$$\widetilde{\gamma} = \frac{1}{2}(\gamma_2 + \gamma_3). \tag{5.99}$$

This only affects the R matrix element (e.g., in Table 5.3). Indeed, it is easy to see that the valence-band Hamiltonian then conserves angular momentum. For example, in Table 5.3,

$$\left\langle \frac{3}{2}\,\frac{3}{2}|H|\frac{3}{2}\,\frac{1}{2}\right\rangle = S \propto k_z k_-,$$

and the increase in M_J from $\frac{1}{2}$ to $\frac{3}{2}$ is offset by the decrease in the orbital angular momentum via k_- [95].

5.5 Six-Band Model for Diamond

This work was first carried out by Suzuki and Hensel [80]. It was a generalization of the work of Luttinger by treating all six valence states (heavy hole, light hole and spin-orbit split off).

The representation obtained from $\Gamma_{25}^+ \otimes D_{\frac{1}{2}}$ is six-dimensional; hence, 36 independent matrices are needed. They will be formed from combinations of the angular-momenta $\mathbf{I}$ matrices with $I = 1$ and of the Pauli matrices $\boldsymbol{\sigma}$ [Eq. (3.34)]. The Hamiltonian will be at most second order in $\mathbf{I}$ and first order in $\boldsymbol{\sigma}$. We extend Table 5.1 into Table 5.6, with the basis choices of Suzuki and Hensel [80]. One can now write down the lowest-order $k \cdot p$ Hamiltonian for the $\Gamma_{25}^+ \otimes \Gamma_{\frac{1}{2}}$ subspace. Following Suzuki and Hensel [80], one can separate it into two components.

Table 5.6 Irreducible basis functions, operators and tensors for O_h' in 6×6 subspace [80]. c.p. stands for cyclic permutation and the partners in Γ_{12}^+, Γ_{15}^+, Γ_{25}^+, and Γ_{15}^- are implied. $\{AB\} = \frac{1}{2}(AB + BA)$

	Time reversal	
	Even	Odd
Γ_1^+	k^2, I^2	
	$\mathbf{I} \cdot \boldsymbol{\sigma}$	
Γ_2^+	$\{I_x I_z\} \sigma_x +$ c.p.	
Γ_{12}^+	$k_x^2 - k_y^2, I_x^2 - I_y^2$	
	$I_x \sigma_x - I_y \sigma_y$	
	$\{I_x (I_y \sigma_z - I_z \sigma_y)\} - \{I_y (I_z \sigma_x - I_x \sigma_z)\}$	
Γ_{15}^+		I_x
		σ_x
	$I_y \sigma_z - I_z \sigma_y$	
		$\left(I_x^2 - \frac{1}{3} I^2\right) \sigma_x + \{I_x I_y\} \sigma_y + \{I_z I_x\} \sigma_z$
		$\left(I_x^2 - \frac{1}{3} I^2\right) \sigma_x - \frac{2}{3} \left(\{I_x I_y\} \sigma_y + \{I_z I_x\} \sigma_z\right)$
Γ_{25}^+	$\{k_y k_z\}, \{I_y I_z\}$	
	$I_y \sigma_z + I_z \sigma_y$	
		$\left(I_y^2 - I_z^2\right) \sigma_x - \left(\{I_x I_y\} \sigma_y - \{I_z I_x\} \sigma_z\right)$
		$\left(I_y^2 - I_z^2\right) \sigma_x + 2 \left(\{I_x I_y\} \sigma_y - \{I_z I_x\} \sigma_z\right)$
Γ_{15}^-		k_x

5.5.1 Spin-Orbit Interaction

From Table 5.6, the only invariant (Γ_1^+) involving both orbital and spin operators is $\mathbf{I} \cdot \boldsymbol{\sigma}$ and

$$H_{\text{so}} = \frac{1}{3} \Delta_0 \mathbf{I} \cdot \boldsymbol{\sigma}. \tag{5.100}$$

5.5.2 k-Dependent Part

Again from Table 5.6, the quadratic Hamiltonian is given by

$$\begin{aligned} H_{\text{HS}}(\mathbf{k}) = \frac{\hbar^2}{2m_0} \Bigg\{ & \Big(A_1 + B_1(\mathbf{I} \cdot \boldsymbol{\sigma})\Big] k^2 + \Big[A_2(I_x^2 - \frac{1}{3} I^2) + B_2(I_x \sigma_x - \frac{1}{3} \mathbf{I} \cdot \boldsymbol{\sigma})\Big] k_x^2 + \text{c.p.} \\ & + \Big[A_3(I_x I_y + I_y I_x) + B_3(I_x \sigma_y + I_y \sigma_x)\Big) \{k_x k_y\} + \text{c.p.} \Bigg\}, \end{aligned} \tag{5.101}$$

where c.p. stands for cyclic permutation. These parameters can be compared to the DKK ones. Thus, comparing to a perturbative expansion [80], one finds, e.g.,

$$
\begin{aligned}
A_1 &= 1 + \frac{1}{3}(F + 2G + 2H_1 + 2H_2), \\
A_2 &= -F - 2G + H_1 + H_2, \\
A_3 &= -F + G - H_1 + H_2.
\end{aligned}
\tag{5.102}
$$

Suzuki and Hensel have redefined the parameters on the RHS to be dimensionless compared to the equivalent ones by DKK [2]. The new B_i parameters involve spin-orbit correction in the k^2 terms and were absent in the DKK model; hence, they are typically dropped. Since this six-band model is identical to the DKK one, no further analysis is presented here.

5.6 Four-Band Model for Zincblende

The most direct extension of the work of Luttinger to ZB was given in the paper by Pidgeon and Groves [97], whereby they extended the Luttinger Hamiltonian by including an odd part to the invariant form of the Hamiltonian in a four-dimensional representation:

$$
\begin{aligned}
H_{\mathrm{PG}}(\mathbf{k}) = \frac{\hbar^2}{m_0}\Big[&\left(\gamma_1 + \frac{5}{2}\gamma_2\right)\frac{k^2}{2} - \gamma_2\left(k_x^2 J_x^2 + k_y^2 J_y^2 + k_z^2 J_z^2\right) \\
&-2\gamma_3\left(\{k_x k_y\}\{J_x J_y\} + \{k_y k_z\}\{J_y J_z\} + \{k_z k_x\}\{J_z J_x\}\right) \\
&+\frac{e}{\hbar}\kappa\,(\mathbf{B}\cdot\mathbf{J}) + \frac{e}{\hbar}q\left(B_x J_x^3 + B_y J_y^3 + B_z J_z^3\right)\Big] \\
&-\frac{2C_k}{\sqrt{3}}\left[k_x\{J_x V_x\} + k_y\{J_y V_y\} + k_z\{J_z V_z\}\right].
\end{aligned}
\tag{5.103}
$$

Note the overall sign difference compared to Luttinger for the even terms. The current model introduces one additional parameter: C_k, which is obviously related to the inversion asymmetry and gives a term linear in k. This contribution will be discussed in detail in Chap. 6. Representative values of the band parameters obtained by Pidgeon and Grove for InSb are given in Table 5.7.

Table 5.7 Parameters of Pidgeon-Grove model: InSb [97]

γ_1	γ_2	γ_3	κ	q	C_k
3.60	−0.47	0.70	−1.48	0.39	9.3×10^{-11} eV cm

5.7 Eight-Band Model for Zincblende

This work was first carried out by Weiler [98–100] and restudied by Trebin et al. [81]. We give the version by Weiler here as the formalism of Trebin et al. is closely related to the 14-band of Winkler [21] to be presented next.

5.7.1 Weiler Hamiltonian

Weiler generalized the eight-band model of Pidgeon and Brown [86] to include inversion-asymmetry terms. Her basis is given in Table 5.8. All the parameters of the theory are given in Table 5.9 and the Hamiltonian in Table 5.10.

Table 5.8 Weiler's eight-band basis set for zincblende [98, 99]

a set

$$|1\rangle = S\uparrow,$$

$$|3\rangle = \left|\frac{3}{2}\ \frac{3}{2}\right\rangle = -\frac{\mathrm{i}}{\sqrt{2}}|(X+\mathrm{i}Y)\uparrow\rangle,$$

$$|5\rangle = \left|\frac{3}{2}\ -\frac{1}{2}\right\rangle = \frac{\mathrm{i}}{\sqrt{6}}[|(X-\mathrm{i}Y)\uparrow\rangle + 2|Z\downarrow\rangle],$$

$$|7\rangle = \left|\frac{1}{2}\ -\frac{1}{2}\right\rangle = -\frac{\mathrm{i}}{\sqrt{3}}[|(X-\mathrm{i}Y)\uparrow\rangle - |Z\downarrow\rangle],$$

b set

$$|2\rangle = |S\downarrow\rangle,$$

$$|6\rangle = \left|\frac{3}{2}\ \frac{1}{2}\right\rangle = -\frac{\mathrm{i}}{\sqrt{6}}[|(X+\mathrm{i}Y)\downarrow\rangle - 2|Z\uparrow\rangle],$$

$$|4\rangle = \left|\frac{3}{2}\ -\frac{3}{2}\right\rangle = \frac{\mathrm{i}}{\sqrt{2}}|(X-\mathrm{i}Y)\downarrow\rangle,$$

$$|8\rangle = \left|\frac{1}{2}\ \frac{1}{2}\right\rangle = -\frac{\mathrm{i}}{\sqrt{3}}[|(X+\mathrm{i}Y)\downarrow\rangle + |Z\uparrow\rangle].$$

The following definitions are used (but where we have interchanged $\Gamma_4' \leftrightarrow \Gamma_5'$ since Weiler followed the Dresselhaus notation [32] and we follow Koster et al. [33]):

$$k^2 = k_x^2 + k_y^2 + k_z^2,\ k_\pm = k_x \pm \mathrm{i}k_y,$$

$$F_3^1 = 2k_z^2 - k_x^2 - k_y^2,\ F_3^2 = \sqrt{3}\left(k_x^2 - k_y^2\right),$$

$$F_4^\pm = \{k_z, k_\pm\} = (k_z k_\pm + k_\pm k_z),\ F_4^z = \{k_x, k_y\}, \tag{5.104}$$

Table 5.9 Weiler's eight-band Hamiltonian for zincblende: parameters. Adapted with permission from [98]. ©1978 by the American Physical Society

	Energy	k^2	$(2k_z^2 - k_x^2 - k_y^2)$, $\sqrt{3}(k_x^2 - k_y^2)$	$\{k_y, k_z\}$, $\{k_z, k_x\}$, $\{k_x, k_y\}$	k_i	B_i
$H^{6c\,6c}$	E_0	F				N_1
$H^{8v\,8v}$		γ_1	γ_2	γ_3	C	κ, q
$H^{7v\,7v}$	$-\Delta_0$	γ_1'				κ'
$H^{6c\,8v}$			N_2	G	P	N_3
$H^{6c\,7v}$				G'	P'	
$H^{7v\,8v}$			γ_2'	γ_3'	C'	κ''

$$B_z = \mathrm{i}[k_x, k_y],\ B_\pm = \pm[k_\pm, k_z],$$

$$P = -\mathrm{i}\frac{\hbar}{m_0}\langle S|p_x|X\rangle,\ P' = -\mathrm{i}\frac{\hbar}{m_0}\langle S|p_x|X'\rangle,\ Q = -\mathrm{i}\frac{\hbar}{m_0}\langle X|p_y|Z'\rangle,$$

$$F = \frac{1}{m_0}\sum_{\Gamma_5'}\frac{|\langle S|p_x|\Gamma_5'\rangle|^2}{E_0 - E_{\Gamma_5'}},\ G = \frac{1}{m_0}\sum_{\Gamma_5'}\frac{\langle S\uparrow|p_x|\Gamma_5'\rangle\langle\Gamma_5'|p_y|Z\rangle}{E_0 - E(\Gamma_5')},$$

$$q = \frac{8}{27}\frac{\mathrm{i}}{m_0}\sum_{\Gamma_8'}\frac{\langle\frac{3}{2}\,\frac{3}{2}|p_x|\Gamma_8'\rangle\langle\Gamma_8'|p_y|\frac{3}{2}\,\frac{3}{2}\rangle}{E_0 - E(\Gamma_8')}, \tag{5.105}$$

$$N_1 = -\frac{\mathrm{i}}{m_0}\sum_{\Gamma_8'}\frac{\langle S\uparrow|p_x|\Gamma_8'\rangle\langle\Gamma_8'|p_x|S\uparrow\rangle}{E(\Gamma_6) - E(\Gamma_8')},$$

$$N_2 = \frac{1}{m_0}\sum_{\Gamma_8'}\frac{\langle S\uparrow|p_x|\Gamma_8'\rangle\langle\Gamma_8'|p_x|\frac{3}{2}\ -\frac{3}{2}\rangle}{E(\Gamma_{6,8}) - E(\Gamma_8')},$$

$$N_3 = -\frac{\mathrm{i}}{m_0}\sum_{\Gamma_8'}\frac{\langle S\uparrow|p_x|\Gamma_8'\rangle\langle\Gamma_8'|p_y|\frac{3}{2}\ -\frac{3}{2}\rangle}{E(\Gamma_{6,8}) - E(\Gamma_8')}.$$

As defined, P, P' and Q are real. Note that there are different sets of "Luttinger" parameters for the different subspaces, for a total of 20 parameters plus two energies. In the double-group picture, these parameters are independent. However, it is possible to derive relations connecting some of them by starting from the single-group picture and adding spin. Indeed, all the primed and unprimed parameters are then the same. Weiler et al. makes the point that there are only three new parameters beyond Luttinger's model: the N_i's. For a physical interpretation of the parameters in the Weiler model, we note that F is related to the conduction-band effective mass, N_1 to the corresponding g factor, G gives the quadratic coupling between the

Table 5.10 Weiler's eight-band Hamiltonian for zincblende. Adapted with permission from [98]. ©1978 by the American Physical Society

	$\|1\rangle$	$\|3\rangle$	$\|5\rangle$	$\|7\rangle$	$\|2\rangle$	$\|6\rangle$	$\|4\rangle$	$\|8\rangle$
$\|1\rangle$	$E_0+(F+\frac{1}{2})k^2$ $+(N_1+\frac{1}{2})B_z$	$\frac{1}{\sqrt{2}}Pk_+ + \frac{1}{\sqrt{2}}GF_4^-$ $+\frac{1}{2}N_3B_-$	$-\frac{1}{\sqrt{6}}Pk_- + \frac{1}{\sqrt{6}}GF_4^+$ $\frac{\sqrt{3}}{2}N_3B_+$	$\frac{1}{\sqrt{3}}P'k_- - \frac{1}{\sqrt{3}}G'F_4^+$	$(N_1+\frac{1}{2})B_-$ $+N_2F_3^2$	$-\sqrt{\frac{2}{3}}Pk_z + \mathrm{i}\sqrt{\frac{2}{3}}GF_4^z$	$-N_2F_3^1 + N_3B_z$	$\frac{1}{\sqrt{3}}P'k_z - \frac{\mathrm{i}}{\sqrt{3}}G'F_4^z$
$\|3\rangle$		$-\frac{1}{2}\gamma_1k^2 + \frac{1}{2}\gamma_2F_3^1$ $-\frac{3}{2}(\kappa+\frac{9}{4}q)B_z$	$\frac{1}{2}\gamma_2F_3^2 - \mathrm{i}\frac{\sqrt{3}}{2}\gamma_3F_4^z$ $+Ck_z$	$-\frac{1}{\sqrt{2}}\gamma_2'F_3^2 + \mathrm{i}\frac{\sqrt{6}}{2}\gamma_3'F_4^z$ $+\frac{1}{\sqrt{2}}C'k_z$	$N_2F_3^1 + N_3B_z$	$\frac{\sqrt{3}}{2}\gamma_3F_4^- - \frac{1}{2}Ck_+$ $-\frac{\sqrt{3}}{2}(\kappa+\frac{7}{4}q)B_-$	$-\frac{\sqrt{3}}{2}Ck_- - \frac{3}{4}qH_+$	$-\sqrt{\frac{3}{8}}\gamma_3'F_4^- + \frac{1}{\sqrt{8}}C'k_+$ $+\sqrt{\frac{3}{8}}(\kappa''+1)B_-$
$\|5\rangle$			$-\frac{1}{2}\gamma_1k^2 - \frac{1}{2}\gamma_2F_3^1$ $+\frac{1}{2}(\kappa+\frac{1}{4}q)B_z$	$-\frac{1}{\sqrt{2}}\gamma_2'F_3^1$ $-\frac{1}{\sqrt{2}}(\kappa''+1)B_z$	$-\sqrt{\frac{2}{3}}Pk_z - \mathrm{i}\sqrt{\frac{2}{3}}GF_4^z$ $-N_2F_3^2$	$\frac{\sqrt{3}}{2}Ck_-$ $-(\kappa+\frac{5}{2}q)B_+$	$-\frac{\sqrt{3}}{2}\gamma_3F_4^- - \frac{1}{2}Ck_+$ $-\frac{\sqrt{3}}{2}(\kappa+\frac{7}{4}q)B_-$	$\frac{3}{\sqrt{8}}\gamma_3'F_4^+ + \sqrt{\frac{3}{8}}C'k_-$ $-\frac{1}{\sqrt{8}}(\kappa''+1)B_+$
$\|7\rangle$				$-\Delta_0 - \frac{1}{2}\gamma_1'k^2$ $+(\kappa'+\frac{1}{2})B_z$	$-\frac{1}{\sqrt{3}}P'k_z - \frac{\mathrm{i}}{\sqrt{3}}G'F_4^z$	$\frac{3}{\sqrt{8}}\gamma_3'F_4^+ - \frac{\sqrt{3}}{8}C'k_-$ $+\frac{1}{\sqrt{8}}(\kappa''+1)B_+$	$-\sqrt{\frac{3}{8}}\gamma_3'F_4^- - \frac{1}{\sqrt{8}}C'k_+$ $-\sqrt{\frac{3}{8}}(\kappa''+1)B_-$	$-(\kappa'+\frac{1}{2})B_+$
$\|2\rangle$					$E_0+(F+\frac{1}{2})k^2$ $-(N_1+\frac{1}{2})B_z$	$\frac{1}{\sqrt{6}}Pk_+ + \frac{1}{\sqrt{6}}GF_4^-$ $-\frac{\sqrt{3}}{2}N_3H_-$	$-\frac{1}{\sqrt{2}}Pk_- + \frac{1}{\sqrt{2}}GF_4^+$ $-\frac{1}{2}N_3H_+$	$\frac{1}{\sqrt{3}}P'k_+ + \frac{1}{\sqrt{3}}G'F_4^-$
$\|6\rangle$						$-\frac{1}{2}\gamma_1k^2 - \frac{1}{2}\gamma_2F_3^1$ $-\frac{1}{2}(\kappa+\frac{1}{4}q)B_z$	$\frac{1}{2}\gamma_2F_3^2 - \mathrm{i}\frac{\sqrt{3}}{2}\gamma_3F_4^z$ $-Ck_z$	$\frac{1}{\sqrt{2}}\gamma_2'F_3^1$ $-\frac{1}{\sqrt{2}}(\kappa''+1)B_z$
$\|4\rangle$							$-\frac{1}{2}\gamma_1k^2 + \frac{1}{2}\gamma_2F_3^1$ $+\frac{3}{2}(\kappa+\frac{9}{4}q)B_z$	$\frac{1}{\sqrt{2}}\gamma_2'F_3^2 + \mathrm{i}\frac{\sqrt{6}}{2}\gamma_3'F_4^z$ $+\frac{1}{\sqrt{2}}C'k_z$
$\|8\rangle$								$-\Delta_0 - \frac{1}{2}\gamma_1'k^2$ $-(\kappa'+\frac{1}{2})B_z$

conduction and valence bands (i.e., must be related to Kane's B parameter), and N_2 and N_3 are additional coupling between the conduction and valence bands absent in the original Kane model. The fact that the N_i's are zero in the absence of spin-orbit splitting of the Γ_8' bands explains why N_2 does not appear in the model of Kane (Table 4.2). Conceptually, even when the Kane model is expressed in double-group notation and the spin-orbit splitting of the valence band (Δ_0) is included, N_2 is still absent because the latter is due to the spin-orbit splitting of the remote states.

Within the approximation of including only one set of remote (conduction) bands, one can derive simplified expressions for many of the parameters. For example,

$$G \approx -\frac{m_0 P' Q}{E(\Gamma_8') - \frac{1}{2}\Delta' - E(\Gamma_{6,8})}, \tag{5.106}$$

where Δ' is the spin-orbit splitting of the Γ_8' bands.

We will carry out a further study of all the parameters in comparison with the next model.

5.8 14-Band Model for Zincblende

The eight-band model of Trebin et al. [81] was extended by Rössler [23, 101] into a 14-band one by including the second set of conduction bands as part of the A set (Fig. 5.2). Thus, a 14-band model consists of the spin-split valence band (Γ_{8v} and Γ_{7v} for ZB), the lowest conduction band (Γ_{6c}) and the second set of conduction bands (Γ_{8c} and Γ_{7c}). This is sometimes known as a five-level model [102, 103]. It was considered by Rössler [23, 101] in his study of nonparabolicity and spin splitting of the conduction band of GaAs. It is also known as an extended Kane model [21]. A variation on this was used by Cardona et al. for the study of spin splitting in ZB by the addition of a set of conduction s states, giving a 16-band model [104].

This model is interesting for a number of reasons. It is certainly one of the most complex models set up for ZB. It will also allow us to gain further insight into the Weiler Hamiltonian.

The structure of the 14-band Hamiltonian is

$$H(\mathbf{k}) = \begin{pmatrix} H^{8c8c} & H^{8c7c} & H^{8c6c} & H^{8c8v} & H^{8c7v} \\ H^{7c8c} & H^{7c7c} & H^{7c6c} & H^{7c8v} & H^{7c7v} \\ H^{6c8c} & H^{6c7c} & H^{6c6c} & H^{6c8v} & H^{6c7v} \\ H^{8v8c} & H^{8v7c} & H^{8v6c} & H^{8v8v} & H^{8v7v} \\ H^{7v8c} & H^{7v7c} & H^{7v6c} & H^{7v8v} & H^{7v7v} \end{pmatrix}. \tag{5.107}$$

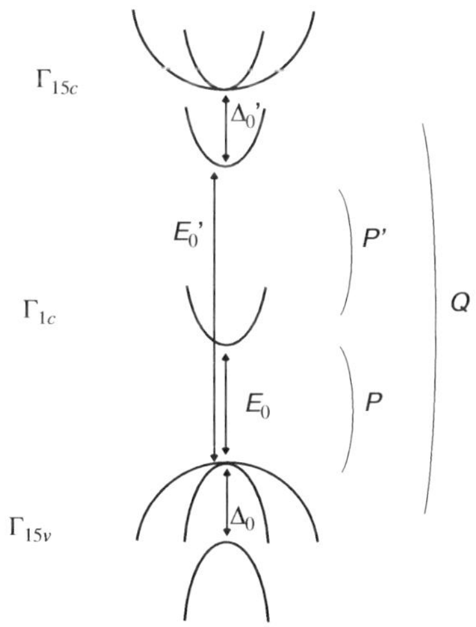

Fig. 5.2 14-band model for zincblende in single-group notation

5.8.1 Symmetrized Matrices

The irreducible representations and symmetrized matrices needed for the Hamiltonian are given in Table 5.11. It can be seen that the Hamiltonian will be build out of the following matrices: 1_σ, σ_i, J_i, 1_ρ, ρ_i, T_i, T_{ij}, U_i, and U_{ij}. We have already come across σ_i and J_i. The H^{cc} block is two-dimensional and, thus, the corresponding invariant is the 2×2 identity matrix (1_σ) and an axial vector is the set of Pauli matrices. The T and U matrices couple states in off-diagonal blocks. In this particular case, the subspaces of different angular momenta have $j = \frac{1}{2}$ and $j = \frac{3}{2}$. Hence, the allowed irreducible components are $\gamma = 1, 2$; in cartesian form, they are first and second-rank tensors. Finally, in constructing H^{6c7v}, it will be necessary to have 2×2 matrices that have odd parity; hence, we define 1_ρ and ρ_i to have the same matrix representation as 1_σ and σ_i but are otherwise odd.

The properties of ρ_i and U deserve additional discussion. Even though inversion is not a symmetry operator for T_d, it will be useful to define the spatial parity of the operators and representations since consistency with the $O(3)$ group is important as a limiting case [i.e., T_d is a subgroup of $O(3)$]. For instance, the lowest conduction band for ZB is s-like (i.e., double group $D_{\frac{1}{2}}$) which has even parity for $O(3)$ (i.e., double group $D^+_{\frac{1}{2}}$). However, the p-like valence-band states are derived from the odd parity states of $O(3)$ (double groups $D^-_{\frac{3}{2}}$ and $D^-_{\frac{1}{2}}$). Thus, the σ's are axial vectors and transform like Γ_{25}. The resulting desired parity of the X matrix will be given by the product of the parities of the two D representations. Thus, for diagonal blocks, the X matrices will need to have positive parities (like the Pauli spin matrices) but not so for, for example, the cs block. One can, therefore, similarly define a set of matrices that have the same matrix elements as the Pauli spin matrices but whose

Table 5.11 Symmetrized matrices for 14-band model of zincblende. $\{AB\} = \frac{1}{2}(AB + BA)$

Block	Representations	Symmetrized matrices	Time reversal
H^{66}	$\Gamma_6 \otimes \Gamma_6^* = \Gamma_1 \oplus \Gamma_{25}$	$\Gamma_1 : 1_\sigma$	+
		$\Gamma_{25} : \sigma_x, \sigma_y, \sigma_z$	−
H^{88}	$\Gamma_8 \otimes \Gamma_8^*$		
	$= \Gamma_1 \oplus \Gamma_2 \oplus \Gamma_{12}$	$\Gamma_1 : 1_{4\times 4}, J^2$	+
	$\oplus 2\Gamma_{15} \oplus 2\Gamma_{25}$	$\Gamma_2 : J_x J_y J_z + J_z J_y J_z$	−
		$\Gamma_{12} : \sqrt{3}(J_z^2 - \frac{1}{3}J^2), J_x^2 - J_y^2$	+
		$\Gamma_{15} : \{J_x J_y\}, \{J_y J_z\}, \{J_z J_x\};$	+
		$\{(J_y^2 - J_z^2)J_x\}, \{(J_z^2 - J_x^2)J_y\}, \{(J_x^2 - J_y^2)J_z\}$	−
		$\Gamma_{25} : J_x, J_y, J_z; J_x^3, J_y^3, J_z^3$	−
H^{77}	$\Gamma_7 \otimes \Gamma_7^* = \Gamma_1 \oplus \Gamma_{25}$	$\Gamma_1 : 1_\sigma$	+
		$\Gamma_{25} : \sigma_x, \sigma_y, \sigma_z$	−
H^{68}	$\Gamma_6 \otimes \Gamma_8^*$	$\Gamma_{12} : T_{xx} - T_{yy}, -\sqrt{3}T_{zz}$	
	$= \Gamma_{12} \oplus \Gamma_{15} \oplus \Gamma_{25}$	$\Gamma_{15} : T_x, T_y, T_z$	
		$\Gamma_{25} : T_{yz}, T_{zx}, T_{xy}$	
H^{67}	$\Gamma_6 \otimes \Gamma_7^* = \Gamma_2 \oplus \Gamma_{15}$	$\Gamma_1 : 1_\sigma$	
		$\Gamma_{15} : \rho_x, \rho_y, \rho_z$	
H^{87}	$\Gamma_8 \otimes \Gamma_7^*$	$\Gamma_{12} : \sqrt{3}U_{zz}, U_{xx} - U_{yy},$	
	$= \Gamma_{12} \oplus \Gamma_{15} \oplus \Gamma_{25}$	$\Gamma_{15} : U_{yz}, U_{zx}, U_{xy}$	
		$\Gamma_{25} : U_x, U_y, U_z$	

Adapted with permission from [21, 81]. © 1979 by American Physical Society

cartesian components transform like a polar vector. These will be called ρ matrices and transform like Γ_{15}. Similarly, one defines the U matrices as

$$(U_i)_{mm'} = (T_i)^*_{m'm}, \tag{5.108}$$

but also with opposite parity to T.

The properly symmetrized interaction terms are given in Table 5.12. Terms involving the magnetic field and strain tensor are included here only for completeness and in order to illustrate the generality of the method; they will be discussed in later chapters.

Table 5.12 Irreducible tensors for 14-band model of zincblende. They are separated into time-reversal even and odd combinations. $\{AB\} = \frac{1}{2}(AB + BA)$

		Time reversal
Terms from wave vector:		
Γ_1	$1, k^2$	+
Γ_{12}	$\sqrt{3}\left(k_z^2 - \frac{1}{3}k^2\right), k_x^2 - k_y^2$	+
Γ_{15}	$\{k_x k_y\}, \{k_y k_z\}, \{k_z k_x\}$	+
	k_x, k_y, k_z	−
Terms from magnetic field:		
Γ_1	B^2	+
Γ_{12}	B_x^2, B_y^2	+
Γ_{15}	$B_x B_y, B_y B_z, B_z B_x$	+
Γ_{25}	B_x, B_y, B_z	−
Terms from strain interaction:		
Γ_1	$\varepsilon_{xx} + \varepsilon_{yy} + \varepsilon_{zz}$	+
Γ_{12}	$\sqrt{3}\left(\varepsilon_{zz} - \frac{1}{3}\mathrm{Tr}\varepsilon\right), \varepsilon_{xx} - \varepsilon_{yy}$	+
Γ_{15}	$\varepsilon_{yz}, \varepsilon_{zx}, \varepsilon_{xy}$	+
Mixed terms:		
Γ_1	$k_x\varepsilon_{yz} + k_y\varepsilon_{zx} + k_z\varepsilon_{xy}$	−
Γ_{12}	$\frac{1}{\sqrt{3}}\left(2k_z\varepsilon_{xy} - k_x\varepsilon_{yz} - k_y\varepsilon_{zx}\right), \left(k_x\varepsilon_{yz} - k_y\varepsilon_{zx}\right)$	−
Γ_{15}	$k_y\varepsilon_{xy} + k_z\varepsilon_{xz}, k_z\varepsilon_{yz} + k_x\varepsilon_{yx}, k_x\varepsilon_{zx} + k_y\varepsilon_{zy};$	−
	$k_x\left(\varepsilon_{xx} - \frac{1}{3}\mathrm{Tr}\varepsilon\right), k_y\left(\varepsilon_{yy} - \frac{1}{3}\mathrm{Tr}\varepsilon\right), k_z\left(\varepsilon_{zz} - \frac{1}{3}\mathrm{Tr}\varepsilon\right);$	−
	$k_x\mathrm{Tr}\varepsilon, k_y\mathrm{Tr}\varepsilon, k_z\mathrm{Tr}\varepsilon$	−
Γ_{25}	$\left(k_y\varepsilon_{xy} - k_z\varepsilon_{xz}\right), \left(k_z\varepsilon_{yz} - k_x\varepsilon_{yx}\right), \left(k_x\varepsilon_{zx} - k_y\varepsilon_{zy}\right);$	−
	$k_x\left(\varepsilon_{yy} - \varepsilon_{zz}\right), k_y\left(\varepsilon_{zz} - \varepsilon_{xx}\right), k_z\left(\varepsilon_{xx} - \varepsilon_{yy}\right)$	−

Adapted with permission from [81]. ©1979 by the American Physical Society

5.8.2 Invariant Hamiltonian

Combining the symmetrized matrices with the interaction terms gives the Hamiltonian of Table 5.13, where the blocks belong to Eq. (5.107). It is equivalent to a second-order Kane model. The form given in Table 5.13 has been compared to the one obtained from perturbation theory in order to express the invariant coefficients in terms of more standard $k \cdot p$ parameters.

Table 5.13 Invariant expansion of Hamiltonian for 14-band model of zincblende. $\{AB\} = \frac{1}{2}(AB + BA)$, c.p. stands for cyclic permutation

$$H^{8c8c} = E_0' + \Delta_0',$$

$$H^{7c7c} = E_0',$$

$$H^{6c6c} = E_0 + \frac{\hbar^2 k^2}{2m_0} + A'k^2,$$

$$H^{8v8v} = -\frac{\hbar^2}{2m_0}\left\{\gamma_1' k^2 - 2\gamma_2'\left[\left(J_x^2 - \frac{1}{3}J^2\right)k_x^2 + \text{c.p.}\right] - 4\gamma_3'\left[\{J_x J_y\}\{k_x k_y\} + \text{c.p.}\right]\right\}$$

$$+\frac{2C}{\sqrt{3}}\left[\{J_x\left(J_y^2 - J_z^2\right)\}k_x + \text{c.p.}\right],$$

$$H^{7v7v} = -\left(\Delta_0 + \frac{\hbar^2}{2m_0}\gamma_1' k^2\right),$$

$$H^{8c7c} = 0,$$

$$H^{8c6c} = -\sqrt{3}P'(U_x k_x + \text{c.p.}),$$

$$H^{8c8v} = -\frac{2}{3}Q(\{J_y J_z\}k_x + \text{c.p.}) + \frac{1}{3}\Delta^-,$$

$$H^{8c7v} = -2Q(U_{yz}k_x + \text{c.p.}),$$

$$H^{7c6c} = \frac{1}{\sqrt{3}}P'(\mathbf{k}\cdot\boldsymbol{\sigma}),$$

$$H^{7c8v} = -2Q(T_{yz}k_x + \text{c.p.}),$$

$$H^{7c7v} = -\frac{2}{3}\Delta^-,$$

$$H^{6c8v} = \sqrt{3}\Big\{P\mathbf{k}\cdot\mathbf{T} + \mathrm{i}B_{8v}^+(T_x\{k_y k_z\} + \text{c.p.})$$

$$+\frac{1}{2}B_{8v}^-\left[\left(T_{xx} - T_{yy}\right)\left(k_z^2 - \frac{1}{3}k^2\right) - T_{zz}\left(k_x^2 - k_y^2\right)\right]\Big\},$$

$$H^{6c7v} = -\frac{1}{\sqrt{3}}\left[P(\mathbf{k}\cdot\boldsymbol{\rho}) + \mathrm{i}B_{7v}(\rho_x\{k_y k_z\} + \text{c.p.})\right],$$

$$H^{8v7v} = -\frac{\hbar^2}{2m_0}\left[-3\gamma_2'\left(U_{xx}k_x^2 + \text{c.p.}\right) - 6\gamma_3'\left(U_{xy}\{k_x k_y\} + \text{c.p.}\right)\right] - \mathrm{i}\sqrt{3}C_k\left(U_{yz}k_x + \text{c.p.}\right).$$

5.8.3 T Basis Matrices

While new matrices such as T_i and T_{ij} have been introduced in the invariant expression, their explicit form is only needed when expressing the Hamiltonian in matrix representation. Therefore, we now look at how these matrices are obtained and, in the process, we will also be comparing to perturbation theory.

For a detailed example, let us consider the H^{6c8v} block and we will leave out external fields for simplicity. The theory of invariant, prior to comparing to perturbation theory within the second-order Kane model (Table 4.2), gives

$$H^{6c8v} = \alpha_1 \mathbf{k}\cdot\mathbf{T} + \alpha_2\left(T_x\{k_y k_z\} + \text{c.p.}\right). \tag{5.109}$$

We thus need a matrix representation for the T_i. Note that this leaves out the terms proportional to T_{ij} as they are absent in this Kane model.

The technique is to relate the cartesian tensors to the spherical tensors of rank γ (Sect. A.3.2). As an illustration, we will obtain the T_z matrix given in Table 5.14. Combining Eq. (A.28) with the Wigner-Eckart theorem [Eq. (A.30)], we have

$$\begin{aligned}\left\langle \frac{1}{2}m|T_z|\frac{3}{2}m'\right\rangle &= \left\langle \frac{1}{2}m|T_0^{(1)}|\frac{3}{2}m'\right\rangle = (-)^{\frac{1}{2}-m}\left\langle \alpha\frac{1}{2}||T^{(1)}||\beta\frac{3}{2}\right\rangle \begin{pmatrix} \frac{1}{2} & 1 & \frac{3}{2} \\ -m & 0 & m' \end{pmatrix} \\ &= (-1)^{\frac{1}{2}-m} T^{(1)} \begin{pmatrix} \frac{1}{2} & 1 & \frac{3}{2} \\ -m & 0 & m' \end{pmatrix} \delta_{m,m'},\end{aligned} \tag{5.110}$$

where the reduced matrix element $\langle \alpha\frac{1}{2}||T^{(1)}||\beta\frac{3}{2}\rangle$ has been written as $T^{(1)}$.

Thus, the only nonzero elements for T_z are when $m = m' = \pm\frac{1}{2}$. and

$$a^{(\frac{1}{2}1)}_{\frac{3}{2}\mp\frac{1}{2}\pm\frac{1}{2}} = \frac{\sqrt{2}}{\sqrt{3}}.$$

Finally

$$\begin{pmatrix} \frac{1}{2} & 1 & \frac{3}{2} \\ \mp\frac{1}{2} & 0 & \pm\frac{1}{2} \end{pmatrix} = \mp\frac{1}{2} a^{(\frac{1}{2}1)}_{\frac{3}{2}\mp\frac{1}{2}\pm\frac{1}{2}} = \mp\frac{1}{\sqrt{6}}.$$

Putting everything together gives

$$\left\langle \frac{1}{2}m|T_z|\frac{3}{2}m'\right\rangle = \frac{T^{(1)}}{\sqrt{6}}\begin{pmatrix} 0 & -1 & 0 & 0 \\ 0 & 0 & -1 & 0 \end{pmatrix}. \tag{5.111}$$

Similarly, we find

$$\left\langle \frac{1}{2}m|T_x|\frac{3}{2}m'\right\rangle = -\frac{T^{(1)}}{2\sqrt{6}}\begin{pmatrix} -\sqrt{3} & 0 & 1 & 0 \\ 0 & -1 & 0 & \sqrt{3} \end{pmatrix}, \tag{5.112}$$

$$\left\langle \frac{1}{2}m|T_y|\frac{3}{2}m'\right\rangle = \frac{\mathrm{i}T^{(1)}}{2\sqrt{6}}\begin{pmatrix} \sqrt{3} & 0 & 1 & 0 \\ 0 & 1 & 0 & \sqrt{3} \end{pmatrix}. \tag{5.113}$$

Then, the matrix representation of Eq. (5.109) is

$$\begin{aligned} H^{6c8v} &= \alpha_1 \frac{T^{(1)}}{2\sqrt{6}}\begin{pmatrix} \sqrt{3}k_+ & -2k_z & -k_- & 0 \\ 0 & k_+ & -2k_z & -\sqrt{3}k_- \end{pmatrix} \\ &+\alpha_2 \frac{T^{(1)}}{2\sqrt{6}}\begin{pmatrix} \sqrt{3}k_z(k_y+\mathrm{i}k_x) & -2k_xk_y & -k_z(k_y-\mathrm{i}k_x) & 0 \\ 0 & k_z(k_y+\mathrm{i}k_x) & -2k_xk_y & -\sqrt{3}k_z(k_y-\mathrm{i}k_x) \end{pmatrix}. \end{aligned} \tag{5.114}$$

In perturbation theory, we had the Hamiltonian in Table 4.2. Note, however, that if one were to use the $|JM_J\rangle$ states from Table 3.4 and compare to Eq. (5.114), then, one cannot obtain a consistent definition of $T^{(1)}$; this is due to the inconsistency in the phase definition in the Clebsch-Gordan decomposition and in the $3j$ symbol. Thus, for this section, we will use the same $|JM_J\rangle$ states as Winkler [21]. Converting to a JM_J basis we find, for example,

$$\left\langle \frac{1}{2}\,\frac{1}{2}|H|\frac{3}{2}\,\frac{3}{2}\right\rangle = -\frac{1}{\sqrt{2}}Pk_+ + \frac{\mathrm{i}}{\sqrt{2}}Bk_z(k_y+\mathrm{i}k_x). \tag{5.115}$$

Comparing to the corresponding matrix element in Eq. (5.114), we have

$$\alpha_1 T^{(1)} = -2P,$$

$$\alpha_2 T^{(1)} = 2\mathrm{i}B,$$

and, if we choose $\alpha_1 = \sqrt{3}P$, then the first equation gives

$$T^{(1)} = \left\langle c\frac{1}{2}||T^{(1)}||v\frac{3}{2}\right\rangle = -\frac{2}{\sqrt{3}}, \tag{5.116}$$

and, from the second one, $\alpha_2 = -\mathrm{i}\sqrt{3}B$. The T_i matrices then agree with the literature [21, 81].

To find the T_{ij}, one carries out a similar calculation. First, one needs the second-rank spherical tensors:

$$T_2^{(2)} = \frac{T^{(2)}}{\sqrt{5}} \begin{pmatrix} 0 & 0 & 0 & -1 \\ 0 & 0 & 0 & 0 \end{pmatrix}, \qquad T_{-2}^{(2)} = \frac{T^{(2)}}{\sqrt{5}} \begin{pmatrix} 0 & 0 & 0 & 0 \\ 1 & 0 & 0 & 0 \end{pmatrix},$$

$$T_1^{(2)} = \frac{T^{(2)}}{2\sqrt{5}} \begin{pmatrix} 0 & 0 & \sqrt{3} & 0 \\ 0 & 0 & 0 & -1 \end{pmatrix}, \qquad T_1^{(2)} = \frac{T^{(2)}}{2\sqrt{5}} \begin{pmatrix} 1 & 0 & 0 & 0 \\ 0 & -\sqrt{3} & 0 & 0 \end{pmatrix},$$

$$T_0^{(2)} = -\frac{T^{(2)}}{\sqrt{10}} \begin{pmatrix} 0 & 1 & 0 & 0 \\ 0 & 0 & -1 & 0 \end{pmatrix},$$

Then

$$T_{xx} = \frac{1}{2}\left(T_2^{(2)} + T_{-2}^{(2)}\right) - \frac{1}{\sqrt{6}} T_0^{(2)} = \frac{T^{(2)}}{2\sqrt{15}} \begin{pmatrix} 0 & 1 & 0 & -\sqrt{3} \\ \sqrt{3} & 0 & -1 & 0 \end{pmatrix}, \tag{5.117}$$

$$T_{yy} = -\frac{1}{2}\left(T_2^{(2)} + T_{-2}^{(2)}\right) - \frac{1}{\sqrt{6}} T_0^{(2)} = \frac{T^{(2)}}{2\sqrt{15}} \begin{pmatrix} 0 & 1 & 0 & \sqrt{3} \\ -\sqrt{3} & 0 & -1 & 0 \end{pmatrix}, \tag{5.118}$$

$$T_{zz} = \sqrt{\frac{2}{3}} T_0^{(2)} = -\frac{T^{(2)}}{\sqrt{15}} \begin{pmatrix} 0 & 1 & 0 & 0 \\ 0 & 0 & -1 & 0 \end{pmatrix}. \tag{5.119}$$

One can now choose a matrix element, for example, in the H^{7v8v} block to compare. Thus, from the second-order Kane Hamiltonian and using the Winkler basis states,

$$\begin{aligned} \langle sh \downarrow |H| lh \downarrow \rangle &= \left\langle \frac{1}{2} - \frac{1}{2} \right| H \left| \frac{3}{2} - \frac{1}{2} \right\rangle = -\frac{1}{3\sqrt{2}} \left(H_{11} + H_{22} - 2H_{33}\right) \\ &= \frac{\hbar^2}{2m_0} \frac{\gamma_2'}{\sqrt{2}} \left(k_x^2 + k_y^2 - 2k_z^2\right). \end{aligned} \tag{5.120}$$

On the other hand, from Table 5.13,

$$H^{7v8v} = \frac{\hbar^2}{2m_0} \left[-3\gamma_2' \left(T_{xx} k_x^2 + \text{c.p.}\right)\right] \tag{5.121}$$

if the T_{ii} matrices are real. Comparing, e.g., the term proportional to k_z^2 (which only requires the T_{zz} matrix), one finds

$$\left\langle c\frac{1}{2} || T^{(2)} || v\frac{3}{2} \right\rangle = -\sqrt{\frac{10}{3}}. \tag{5.122}$$

All the needed symmetrized matrices are collected together in Table 5.11. Note that the T_{yz} matrix given by Winkler [21] has a sign mistake.

Table 5.14 Cartesian matrices T for the $j = \frac{1}{2}$ and $j = \frac{3}{2}$ cross-space for zincblende [81]

$$T_x = \frac{1}{3\sqrt{2}} \begin{pmatrix} -\sqrt{3} & 0 & 1 & 0 \\ 0 & -1 & 0 & \sqrt{3} \end{pmatrix}, \quad T_y = \frac{-\mathrm{i}}{3\sqrt{2}} \begin{pmatrix} \sqrt{3} & 0 & 1 & 0 \\ 0 & 1 & 0 & \sqrt{3} \end{pmatrix},$$

$$T_z = \frac{\sqrt{2}}{3} \begin{pmatrix} 0 & 1 & 0 & 0 \\ 0 & 0 & 1 & 0 \end{pmatrix},$$

$$T_{xx} = \frac{1}{3\sqrt{2}} \begin{pmatrix} 0 & -1 & 0 & \sqrt{3} \\ -\sqrt{3} & 0 & 1 & 0 \end{pmatrix}, \quad T_{yy} = \frac{1}{3\sqrt{2}} \begin{pmatrix} 0 & -1 & 0 & -\sqrt{3} \\ \sqrt{3} & 0 & 1 & 0 \end{pmatrix},$$

$$T_{zz} = \frac{\sqrt{2}}{3} \begin{pmatrix} 0 & 1 & 0 & 0 \\ 0 & 0 & -1 & 0 \end{pmatrix}, \quad T_{yz} = \frac{\mathrm{i}}{2\sqrt{6}} \begin{pmatrix} -1 & 0 & -\sqrt{3} & 0 \\ 0 & \sqrt{3} & 0 & 1 \end{pmatrix},$$

$$T_{zx} = \frac{1}{2\sqrt{6}} \begin{pmatrix} -1 & 0 & \sqrt{3} & 0 \\ 0 & \sqrt{3} & 0 & -1 \end{pmatrix}, \quad T_{xy} = \frac{\mathrm{i}}{\sqrt{6}} \begin{pmatrix} 0 & 0 & 0 & -1 \\ -1 & 0 & 0 & 0 \end{pmatrix}.$$

5.8.4 Parameters

Having presented the Hamiltonian, we will now spend some time understanding some of the parameters. One of the reasons is because it is possible to relate the parameters from different models. Indeed, it will also be important to establish any connection among parameters in different subblocks of the same model. Finally, there are some inconsistencies in the literature. For example, Winkler [21] has three B parameters to the single one in the paper by Trebin et al. [81]. The other two models to be compared to here are the one by Weiler [98] and by Kane [31].

First consider the γ_i' parameters of Weiler. We can use the DKK Hamiltonian as reference. From Table 3.5,

$$\begin{aligned} \left\langle \frac{3}{2}\ \frac{3}{2} | H(\mathbf{k}) | \frac{1}{2}\ -\frac{1}{2} \right\rangle &= -\frac{1}{\sqrt{6}} (H_{11} - H_{22} - 2\mathrm{i} H_{12}) \\ &= \frac{1}{\sqrt{6}} \left[(L - M)^K \left(k_x^2 - k_y^2\right) - 2\mathrm{i} N^K k_x k_y \right] \\ &= \frac{\hbar^2}{2m_0} \left[\sqrt{6}\gamma_2^K \left(k_x^2 - k_y^2\right) - 2\mathrm{i}\gamma_3^K k_x k_y \right]. \end{aligned}$$

We have labeled the parameters as γ_i^K to reflect the fact that, while Table 3.5 is for the valence band only, the comparison to the Weiler parameters requires a Kane-like model. Thus, in the Weiler model, we have the same matrix element being

$$-\frac{1}{\sqrt{2}}\gamma_2' F_3^2 - \mathrm{i}\frac{\sqrt{6}}{2}\gamma_3' F_4^z = -\sqrt{\frac{3}{2}}\gamma_2'\left(k_x^2 - k_y^2\right) - \mathrm{i}\sqrt{6}\gamma_3' k_x k_y.$$

Comparing the two expressions give

$$\gamma_2' = \gamma_2^K, \gamma_3' = \gamma_3^K.$$

Similarly, we find

$$\gamma_1' = \gamma_1^K.$$

The γ_i^K can be related to the Luttinger parameters as follows. Using Eqs. (5.77) and (5.78), one has

$$\begin{aligned}
\gamma_1 &= -\frac{2m_0}{3\hbar^2}(L+2M) - 1 = \gamma_1^K + \frac{2P^2}{3E_0},\\
\gamma_2 &= -\frac{m_0}{6\hbar^2}(L-M) = \gamma_2^K + \frac{P^2}{3E_0},\\
\gamma_3 &= -\frac{2m_0}{6\hbar^2}N = \gamma_3^K + \frac{P^2}{3E_0}.
\end{aligned}$$

For example, comparing the $\left\langle \frac{3}{2}\,\frac{3}{2}|H|S\uparrow\right\rangle$, Weiler has

$$\frac{1}{\sqrt{2}}Pk_- + \frac{1}{\sqrt{2}}GF_4^+ + \frac{1}{2}N_3 H^+ = \frac{1}{\sqrt{2}}Pk^- + \sqrt{2}Gk_+k_z + \frac{1}{2}N_3\frac{e}{\hbar}B_+,$$

while Winkler has

$$\begin{aligned}
&-\frac{1}{\sqrt{2}}Pk^- + \frac{1}{\sqrt{2}}B_{8v}^+ k_+ k_z + \frac{1}{\sqrt{3}}\mu_B D'\left\langle\frac{3}{2}\,\frac{3}{2}|\left(T_{yz}B_x + \text{c.p.}\right)|S\uparrow\right\rangle\\
&= -\frac{1}{\sqrt{2}}Pk^- + \frac{1}{\sqrt{2}}B_{8v}^+ k_+ k_z + \frac{\mathrm{i}}{6\sqrt{2}}\mu_B D' B_+,
\end{aligned}$$

and Kane has

$$\langle S|H|X\rangle = \mathrm{i}Pk_x + Bk_y k_z.$$

Note that all three use bases that differ in phase. Taking this into account gives us the relations among some of the parameters presented in Table 5.15.

Table 5.15 Relationship among some zincblende parameters. Differences in the expansions are not made explicit

Winkler [21]	Weiler [98]	Kane [31]
P	$P = P'$	P
B_{8v}^{+}	$-2G$	B
B_{8v}^{-}	$-3\sqrt{2}N_2$	0
B_{7v}	$-2G'$	$-B$
D'	$3\sqrt{2}\mathrm{i}m_0 N_3/\hbar^2$	

In Table 5.15, differences in contribution from remote bands are not made explicit for simplicity. However, the differences can, of course, be important and we illustrate a few cases. We recall that, in the Kane model, the valence-band parameters L' and N' differed from those of Luttinger due to the difference in F. In the extended Kane model, M' also differs due to the difference in H_1:

$$H_1 = H_1' + \frac{Q^2}{(E_v - E_{\Gamma_{15c}})}, \tag{5.123}$$

where Q is the momentum matrix element between the Γ_{15v} and Γ_{15c} states (Fig. 5.2). For instance, for γ_1, the relationship among the Winkler (Wi), Kane (primed) and Luttinger (unprimed) parameters are as follows:

$$\gamma_1^{\mathrm{Wi}} = -\frac{2m_0}{3\hbar^2}(L' + 2M') - 1 = \gamma_1 + \frac{2m_0 P^2}{3\hbar^2(E_v - E_{\Gamma_{1c}})} + \frac{4m_0 Q^2}{3\hbar^2(E_v - E_{\Gamma_{15c}})} - 1. \tag{5.124}$$

Typical values of the parameters are given in Table 5.16. Most of the band parameters have been illustrated in Fig. 5.2. Δ^- is a spin-orbit energy between a conduction state and a valence state. C_k characterizes the spin splitting of bands to be discussed in Chap. 6. g^* is the g-factor to be derived in the magnetic-field chapter (Chap. 9). The cellular functions $|X'\rangle$, $|Y'\rangle$ and $|Z'\rangle$ forming the $\Gamma_{7,8c}$ states were chosen to be pure imaginary (and so was the Γ_{6c} $|S\rangle$ state, while $|X\rangle$, $|Y\rangle$ and $|Z\rangle$ forming the $\Gamma_{7,8v}$ states were chosen to be pure real) so that some of the matrix elements are imaginary. For diamond-type semiconductors, P', Δ^- and C_k all vanish due to inversion symmetry.

An example of a 14-band model band structure is given in Fig. 5.3. One can clearly see anisotropy and nonparabolicity in the valence bands. The additional splitting of the hh and lh bands along the [011] direction will be discussed in the next chapter.

Table 5.16 Example band parameters for the 14-band extended Kane model [21]

	GaAs	InSb
E_0 (eV)	1.519	0.237
Δ_0 (eV)	0.341	0.810
E'_0 (eV)	4.488	3.160
Δ'_0 (eV)	0.171	0.330
Δ^- (eV)	−0.050i	
P (eVÅ)	10.493	9.641
P' (eVÅ)	4.780i	6.325i
Q (eVÅ)	8.165	8.130
C_k (eVÅ)	−0.0034	−0.0082
m_e (m_0)	0.0665	0.0139
g^*	−0.44	−51.56
γ_1	6.85	37.10
γ_2	2.10	16.50
γ_3	2.90	17.70
κ	1.20	15.60
q	0.01	0.39

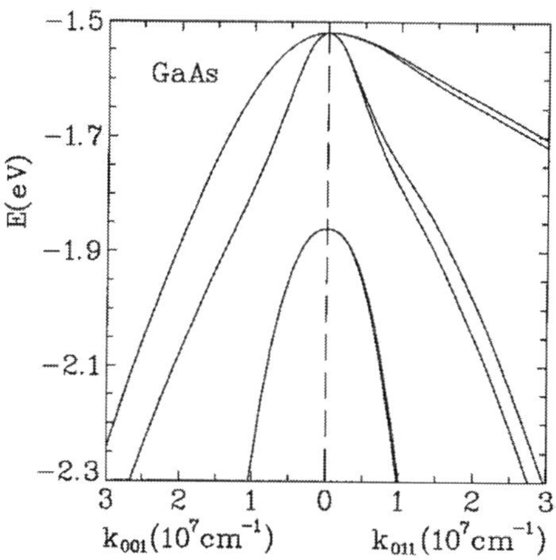

Fig. 5.3 14-band model of GaAs valence band. Reprinted with permission from [103]. ©1996 by the American Physical Society

5.9 Wurtzite

The valence band of WZ semiconductors has been extensively studied using the method of invariants.

5.9.1 Six-Band Model

The six-band model had been originally derived by Rashba, Sheka and Pikus [1, 41, 61] and we will refer to the latest form in the book by Bir and Pikus. There are now at least two other forms, by Sirenko and coworkers [43] and Chuang and Chang [46]. We will compare all three here.

5.9.1.1 *LS* Basis

Near the Γ point, Bir and Pikus [1] gave the following result using the method of invariants:

$$
\begin{aligned}
H_{RSP}(\mathbf{k}) = {} & \Delta_1 J_z^2 + \Delta_2 J_z\sigma_z + \sqrt{2}\Delta_3(J_+\sigma_- + J_-\sigma_+) \\
& + \left(A_1 + A_3 J_z^2\right) k_z^2 + (A_2 + A_4 J_z^2) k_\perp^2 - A_5\left(J_+^2 k_-^2 + J_-^2 k_+^2\right) \\
& -2\mathrm{i} A_6 k_z \left(\{J_z J_+\} k_- - \{J_z, J_+\} k_+\right) + A_7(k_- J_+ + k_+ J_-) \\
& + \left(\alpha_1 + \alpha_3 J_z^2\right)(\sigma_+ k_- + \sigma_- k_+) + \alpha_2\left(J_+^2 k_- \sigma_- + J_-^2 k_+ \sigma_+\right) \\
& +2\alpha_4 \sigma_z \left(\{J_z J_+\} k_- + \{J_z J_-\} k_+\right) + 2\mathrm{i}\alpha_5 k_z \left(\{J_z J_+\}\sigma_- - \{J_z J_-\}\sigma_+\right),
\end{aligned}
\tag{5.125}
$$

where $\{AB\} = (AB + BA)/2$ and the individual terms in the above equation were constructed from the following irreducible representations: Γ_1, Γ_2, Γ_5, Γ_1, Γ_1, Γ_6, Γ_5, Γ_5, Γ_1, Γ_6, Γ_5, and Γ_5, respectively. They wrote down the six-band form given in Table 5.17.

In comparing with the work of Sirenko and coworkers [43] and Chuang and Chang [46], it is realized that the RSP Hamiltonian does not correspond to either of the latter if the J matrices are chosen real, as given in Table 31.7 of Ref. [1] and reproduced in Appendix C [Eq. (C.4)]. Indeed, real $J_\pm$ are inconsistent with the definitions given in Table 5.18. Thus, Sirenko and coworkers [43] give

$$
\begin{aligned}
H_{\mathrm{SJKLS}}(\mathbf{k}) = {} & -I\left(\Delta_1 + \Delta_2 + B_1 k_z^2 + B_3 k_\perp^2\right) + \Delta_2 J_z \sigma_z \\
& +\sqrt{2}\Delta_3 (J_+\sigma_- + J_-\sigma_+) - J_z^2\left(\Delta_1 - B_2 k_z^2 + B_4 k_\perp^2\right) \\
& -B_5\left(J_+^2 k_-^2 + J_-^2 k_+^2\right) - 2B_6 k_z \left(\{J_z J_+\} k_- + \{J_z J_+\} k_-\right) \\
& -\mathrm{i}\frac{\hbar^2 \mathcal{K}}{2m_0}(k_- J_+ - k_+ J_-)
\end{aligned}
\tag{5.126}
$$

Table 5.17 Six-band RSP Hamiltonian in LS basis

$$H_{\mathrm{RSP}}(\mathbf{k}) = \begin{pmatrix} |Y_{1\;1}\uparrow\rangle & |Y_{1\;1}\downarrow\rangle & |Y_{1\;0}\uparrow\rangle & |Y_{1\;0}\downarrow\rangle & |Y_{1\;-1}\uparrow\rangle & |Y_{1\;-1}\downarrow\rangle \\ F & 0 & -H^* & 0 & k^* & 0 \\ 0 & G & \Delta & -H^* & 0 & k^* \\ -H & \Delta & \lambda & 0 & I^* & 0 \\ 0 & -H & 0 & \lambda & \Delta & I^* \\ k & 0 & I & \Delta & G & 0 \\ 0 & k & 0 & I & 0 & F \end{pmatrix},$$

where

$$\begin{aligned} \Delta &= \sqrt{2}\Delta_3, \\ F &= \Delta_1 + \Delta_2 + \lambda + \theta, \\ G &= \Delta_1 - \Delta_2 + \lambda + \theta, \\ k &= A_5 k_+^2, \\ H &= \mathrm{i}(A_6 k_+ k_z + A_7 k_+), \\ I &= \mathrm{i}(A_6 k_+ k_z - A_7 k_+), \\ \lambda &= A_1 k_z^2 + A_2 k_\perp^2, \\ \theta &= A_3 k_z^2 + A_4 k_\perp^2. \end{aligned}$$

Table 5.18 Notations for irreducible representations, functions, and matrices for wurtzite

KDWS [33]	BP75 [1]	P62 [61] SJKLS [43] MU99 [55]	CC96 [46]
Γ_5	Γ_5	Γ_6	
Γ_6	Γ_6	Γ_5	
	$k_\pm = k_x \pm \mathrm{i}k_y$	$k_\pm = k_x \pm \mathrm{i}k_y$	
	$J_\pm = \pm\frac{\mathrm{i}}{\sqrt{2}}\left(J_x \pm \mathrm{i}J_y\right)$	$J_\pm = \frac{1}{\sqrt{2}}\left(J_x \pm \mathrm{i}J_y\right)$	
	$\sigma_\pm = \pm\frac{\mathrm{i}}{2}\left(\sigma_x \pm \mathrm{i}\sigma_y\right)$	$\sigma_\pm = \frac{1}{2}\left(\sigma_x \pm \mathrm{i}\sigma_y\right)$	

$$\begin{aligned} &= -PI - QJ_z^2 + \Delta_2 J_z \sigma_z + \sqrt{2}\Delta_3(J_+\sigma_- + J_-\sigma_+) \\ &\quad -RJ_+^2 - R^* J_-^2 - 2S\{J_z J_+\} - 2S^*\{J_z J_-\} - TJ_+ - T^* J_-, \end{aligned} \tag{5.127}$$

where the quantities in the second equation are defined in Table 5.19, together with the matrix form of the Hamiltonian.

Table 5.19 Six-band Hamiltonian in LS basis according to Sirenko and coworkers [43]

$$H_{SJKLS}(\mathbf{k}) = -\begin{pmatrix} |Y_{1\ 1}\uparrow\rangle & |Y_{1\ 1}\downarrow\rangle & |Y_{1\ 0}\uparrow\rangle & |Y_{1\ 0}\downarrow\rangle & |Y_{1\ -1}\uparrow\rangle & |Y_{1\ -1}\downarrow\rangle \\ P+Q-\Delta_2 & 0 & T+S & 0 & R & 0 \\ 0 & P+Q+\Delta_2 & -\sqrt{2}\Delta_3 & T+S & 0 & R \\ T^*+S^* & -\sqrt{2}\Delta_3 & P & 0 & T-S & 0 \\ 0 & T^*+S^* & 0 & P & -\sqrt{2}\Delta_3 & T-S \\ R^* & 0 & T^*-S^* & -\sqrt{2}\Delta_3 & P+Q+\Delta_2 & 0 \\ 0 & R^* & 0 & T^*-S^* & 0 & P+Q-\Delta_2 \end{pmatrix},$$

where

$$P = \Delta_1 + \Delta_2 + B_1 k_z^2 + B_3 k_\perp^2,$$

$$Q = -\Delta_1 + B_2 k_z^2 + B_4 k_\perp^2,$$

$$R = B_5 k_-^2,$$

$$S = B_6 k_z k_-,$$

$$T = \mathrm{i}\frac{\hbar^2 K}{2m_0} k_-.$$

Finally, the third form we would like to present is the one by Chuang and Chang [46].

$$\begin{aligned} H_{\mathrm{CC}}(\mathbf{k}) &= \Delta_1 J_z^2 + \Delta_2 J_z \sigma_z + \Delta(J_+\sigma_- + J_-\sigma_+) \\ &+ \frac{\hbar^2}{2m_0}\left[\left(A_1 + A_3 J_z^2\right)k_z^2 + \left(A_2 + A_4 J_z^2\right)k_\perp^2 - A_5\left(J_+^2 k_-^2 + J_-^2 k_+^2\right)\right. \\ &\left. - 2A_6 k_z\left(\{J_z J_+\}k_- + \{J_z J_-\}k_+\right)\right]. \end{aligned} \tag{5.128}$$

In the above, terms linear-in-k have been dropped. They wrote down the six-band form given in Table 5.20. We now compare the above three Hamiltonians. The CC Hamiltonian is the same as the SJKLS one, Table 5.19, with a slightly different ordering. Comparison of the different constants is given in Table 5.21. The basis of RSP appears to be the same as that of SJKLS. As Chuang and Chang [46] stated, there are phase differences between their Hamiltonian and the RSP one, in addition to the different basis ordering. The two are made almost identical by proposing the following basis for RSP: $\{-u_1, u_2, \mathrm{i}u_3, u_4, -u_5, \mathrm{i}u_6\}$. In the CC Hamiltonian, this changes the sign of K and changes their H and Δ to $\mathrm{i}H$ and $\mathrm{i}\Delta$. The first two changes are needed to bring the RSP Hamiltonian to look like the CC one. However, the discrepancy that remains in Δ means it is only real in one of the two cases, and pure imaginary in the other. We follow Chuang and Chang [46] in taking theirs to be real. Indeed, we will use the Hamiltonian of Chuang and Chang [46] as the six-band Hamiltonian for WZ in the LS basis.

Table 5.20 Six-band wurtzite Chuang–Chang Hamiltonian in LS basis [46]

$$H_{\mathrm{CC}}(\mathbf{k}) = \begin{pmatrix} |u_1\rangle & |u_2\rangle & |u_3\rangle & |u_4\rangle & |u_5\rangle & |u_6\rangle \\ F & -K^* & -H^* & 0 & 0 & 0 \\ -K & G & H & 0 & 0 & \Delta \\ -H & H^* & \lambda & 0 & \Delta & 0 \\ 0 & 0 & 0 & F & -K & H \\ 0 & 0 & \Delta & -K^* & G & -H^* \\ 0 & \Delta & 0 & H^* & -H & \lambda \end{pmatrix},$$

where

$$\begin{aligned} F &= \Delta_1 + \Delta_2 + \lambda + \theta, \\ G &= \Delta_1 - \Delta_2 + \lambda + \theta, \\ \lambda &= \frac{\hbar^2}{2m_0}\left[A_1 k_z^2 + A_2\left(k_x^2 + k_y^2\right)\right], \\ \theta &= \frac{\hbar^2}{2m_0}\left[A_3 k_z^2 + A_4\left(k_x^2 + k_y^2\right)\right], \\ K &= \frac{\hbar^2}{2m_0} A_5\left(k_x + \mathrm{i}k_y\right)^2, \\ H &= \frac{\hbar^2}{2m_0} A_6\left(k_x + \mathrm{i}k_y\right)k_z, \\ \Delta &= \sqrt{2}\Delta_3. \end{aligned}$$

Table 5.21 Parameters for six-band Hamiltonian in LS basis for wurtzite

CC96 [46]	BP75 [1]	SJKLS [43]
$\frac{\hbar^2}{2m_0}A_1$	A_1	$-B_1$
$\frac{\hbar^2}{2m_0}A_2$	A_2	$-B_3$
$\frac{\hbar^2}{2m_0}A_3$	A_3	$-B_2$
$\frac{\hbar^2}{2m_0}A_4$	A_4	$-B_4$
$\frac{\hbar^2}{2m_0}A_5$	$-A_5$	B_5
$\frac{\hbar^2}{2m_0}A_6$	$-A_6$	B_6
	A_7	$\frac{\hbar^2 K}{2m_0}$

5.9.2 Quasi-Cubic Approximation

The six-band Hamiltonian is often simplified using the so-called quasi-cubic approximation [1, 105]:

$$\begin{aligned} \Delta_1 &= \Delta_{\rm cr}, \Delta_2 = \Delta_3 = \frac{1}{3}\Delta_{\rm so}, \\ A_3 &= 4A_5 - \sqrt{2}A_6, \\ 2A_4 &= -A_3 = A_1 - A_2, \\ A_7 &= 0. \end{aligned} \tag{5.129}$$

In the GJ model, the equivalence is to set $q_7^2 = 2/3$. Within this approximation, the WZ structure is viewed as a ZB one, strained along the [111] direction. According to Bir and Pikus [1], the difference between Δ_2 and Δ_3 is at most about 2 % for CdS, CdSe, ZnS and ZnSe, while $A_7 \sim 10^{-9}$ eV/cm, i.e., reasonably small. These would indicate that the quasi-cubic approximation is excellent for these materials. Gutsche and Jahne [42] have reported $q_7^2 \sim 0.43$, 0.6 and 0.01 for CdS, CdSe, and ZnO. Thus, one might question the applicability of this approximation for ZnO; the same conclusion was reached by Langer et al. [106] by fitting excitonic lines under a uniaxial stress.

5.9.2.1 JM_J Basis

Sirenko et al. [43] stated explicitly their choice of the JM_J states:

$$\begin{aligned} \left|\frac{3}{2}\, m\right\rangle &= \sqrt{\frac{3-2m}{6}}\left|1\; m+\frac{1}{2}\right\rangle|\downarrow\rangle + \sqrt{\frac{3+2m}{6}}\left|1\; m-\frac{1}{2}\right\rangle|\uparrow\rangle, \\ \left|\frac{1}{2}\, m\right\rangle &= \sqrt{\frac{3+2m}{6}}\left|1\; m+\frac{1}{2}\right\rangle|\downarrow\rangle - \sqrt{\frac{3-2m}{6}}\left|1\; m-\frac{1}{2}\right\rangle|\uparrow\rangle. \end{aligned} \tag{5.130}$$

This then gives

$$\begin{aligned} \left|\frac{3}{2}\;\frac{3}{2}\right\rangle &= |1\;1\rangle|\uparrow\rangle, \\ \left|\frac{3}{2}\;\frac{1}{2}\right\rangle &= \frac{1}{\sqrt{3}}|1\;1\rangle|\downarrow\rangle + \sqrt{\frac{2}{3}}|1\;0\rangle|\uparrow\rangle, \\ \left|\frac{3}{2}\;-\frac{1}{2}\right\rangle &= \sqrt{\frac{2}{3}}|1\;0\rangle|\downarrow\rangle + \frac{1}{\sqrt{3}}|1\;-1\rangle|\uparrow\rangle, \end{aligned} \tag{5.131}$$

$$\left|\frac{3}{2}\ -\frac{3}{2}\right\rangle = \ |1\ -1\rangle|\downarrow\rangle,$$

$$\left|\frac{1}{2}\ \frac{1}{2}\right\rangle = \ \sqrt{\frac{2}{3}}|1\ 1\rangle|\downarrow\rangle - \frac{1}{\sqrt{3}}|1\ 0\rangle|\uparrow\rangle,$$

$$\left|\frac{1}{2}\ -\frac{1}{2}\right\rangle = \frac{1}{\sqrt{3}}|1\ 0\rangle|\downarrow\rangle - \sqrt{\frac{2}{3}}|1\ -1\rangle|\uparrow\rangle.$$

One can then rewrite the LS Hamiltonian in this basis.

5.9.3 Eight-Band Model

In Table 5.22, we give the functions and matrices transforming according to the different representations. Since we are about to construct the general $k \cdot p$ Hamiltonian near the Γ point, time-reversal symmetry imposes additional restrictions namely that all terms in the Hamiltonian must be even subject to time reversal. For that reason, we partition the representation functions into even or odd components. In Table 5.22, we have introduced the notation:

$$\begin{aligned}\mathbf{S} &= \mathrm{i}\mathbf{k} \times \mathbf{k} = \frac{e}{\hbar}\mathbf{B},\\ S_\pm &= \frac{\mathrm{i}}{2}\left(S_x \pm \mathrm{i}S_y\right),\\ \sigma_\pm &= \frac{\mathrm{i}}{2}\left(\sigma_x \pm \mathrm{i}\sigma_y\right),\\ k_\pm &= \left(k_x \pm \mathrm{i}k_y\right),\end{aligned} \tag{5.132}$$

and $\mathbf{B}$ and $\mathcal{E}$ denote the external magnetic field and electric field, respectively.

Table 5.22 Basis functions and matrices for wurtzite C_{6v}. $\{AB\} = \frac{1}{2}(AB + BA)$

Representations	Matrices	Odd functions	Even functions
Γ_1	I	$k_z, k_z\mathcal{E}_z, k_x\mathcal{E}_x + k_y\mathcal{E}_y$	$S_+k_- + S_-k_+, \mathcal{E}_z$
			$k_z^2; \{k_+k_-\}$
Γ_2	σ_z	$(\mathbf{k} \times \mathcal{E})_z$	$S_z, S_zk_z, S_+k_- - S_-k_+$
Γ_3			$k_x^3 - 3\{k_xk_y^2\}$
Γ_4			$k_y^3 - 3\{k_yk_x^2\}$
Γ_5	σ_+, σ_-	S_+, S_-, k_+, k_-	$S_+k_z, S_-k_z, S_zk_+, S_zk_-, \mathcal{E}_+, \mathcal{E}_-$
		$(\mathbf{k} \times \mathcal{E})_+, (\mathbf{k} \times \mathcal{E})_-$	$\{k_+k_z\}, \{k_-k_z\}$
Γ_6			S_+k_+, S_-k_-
			k_+^2, k_-^2

Before continuing, we also need the double-group product representations of C_{6v}:

$$\Gamma_7 \otimes \Gamma_7 = \Gamma_1 \oplus \Gamma_2 \oplus \Gamma_5,$$
$$\Gamma_8 \otimes \Gamma_8 = \Gamma_1 \oplus \Gamma_2 \oplus \Gamma_5,$$
$$\Gamma_9 \otimes \Gamma_9 = \Gamma_1 \oplus \Gamma_2 \oplus \Gamma_3 \oplus \Gamma_4,$$
$$\Gamma_7 \otimes \Gamma_9 = \Gamma_5 \oplus \Gamma_6. \tag{5.133}$$

5.9.3.1 Spin-Splitting Terms

Discarding first external fields and collecting spin-splitting terms linear-in-k and cubic-in-k only, we find the following general result for the Γ_7 and Γ_8 bands using Table 5.22:

$$\begin{aligned} H_{7,8}(\mathbf{k}) = {} & \gamma_1 (\sigma_- k_+ + \sigma_+ k_-) + \gamma_2 k_z^2 (\sigma_- k_+ + \sigma_+ k_-) \\ & + \gamma_3 k_\perp^2 (\sigma_- k_+ + \sigma_+ k_-), \end{aligned} \tag{5.134}$$

where $\gamma_1, \gamma_2, \gamma_3$ are constants to be determined using perturbation theory, independent theoretical results (e.g., ab initio), experimental data, or combinations hereof.

For the Γ_9 bands, we find the following spin-splitting invariants up to third order in k in the absence of external fields:

$$H_9(\mathbf{k}) = \gamma_1 \left(k_x^3 - 3k_x k_y^2\right) \sigma_x + \gamma_2 \left(k_y^3 - 3k_y k_x^2\right) \sigma_y, \tag{5.135}$$

where, again, γ_1, γ_2 are constants to be determined using perturbation theory, independent theoretical results.

5.9.3.2 Linear-in-k Terms

One can now give the general 8×8 Hamiltonian matrix structure using the method of invariants. We shall consider only up to linear-in-k invariants but include external electric and magnetic fields. Note that couplings between the eight states will allow for higher order in k contributions.

For the 2×2 $\Gamma_7 \times \Gamma_7$ matrix part, we have, using Table 5.22,

$$\begin{aligned} H_{7\times7}(\mathbf{k}) = {} & \alpha_0^7 + \alpha_1^7 (\sigma_- k_+ + \sigma_+ k_-) + \alpha_2^7 \mathcal{E}_z + \alpha_3^7 (S_- k_+ + S_+ k_-) + \alpha_4^7 S_z \sigma_z \\ & + \alpha_5^7 (S_- \sigma_+ + S_+ \sigma_-) + \alpha_6^7 (\mathbf{k} \times \mathcal{E})_z \sigma_z + \alpha_7^7 \left[(\mathbf{k} \times \mathcal{E})_- \sigma_+ + (\mathbf{k} \times \mathcal{E})_+ \sigma_-\right] \\ & + \beta_1^7 k_z + \beta_2^7 (\sigma_+ \mathcal{E}_- + \sigma_- \mathcal{E}_+) + \beta_3^7 k_z \mathcal{E}_z + \beta_4^7 (k_+ \mathcal{E}_- + k_- \mathcal{E}_+) \\ & + \beta_5^7 k_z S_z \sigma_z + \beta_6^7 (S_- k_+ - S_+ k_-) \sigma_z + \beta_7^7 (\mathcal{E}_- k_+ - \mathcal{E}_+ k_-) \sigma_z \\ & + \beta_8^7 k_z (S_- \sigma_+ + S_+ \sigma_-) + \beta_9^7 S_z (\sigma_- k_+ + \sigma_+ k_-), \end{aligned} \tag{5.136}$$

where α (β) coefficients distinguish between terms fulfilling (not fulfilling) time-reversal symmetry. This is relevant for the $H_{7\times7}$ blocks only since they appear in the total Hamiltonian as both on- and off-diagonal blocks (on-diagonal blocks are required to satisfy time-reversal symmetry while off-diagonal blocks are not).

For the 2×2 $\Gamma_9 \times \Gamma_9$ matrix part, we have

$$H_{9\times9}(\mathbf{k}) = \alpha_0^9 + \alpha_1^9 \mathcal{E}_z + \alpha_2^9 (S_- k_+ + S_+ k_-) + \alpha_3^9 S_z \sigma_z + \alpha_4^9 (\mathbf{k} \times \mathcal{E})_z \sigma_z, \quad (5.137)$$

while for the 2×2 $\Gamma_7 \times \Gamma_9$ matrix part, we find:

$$\begin{aligned} H_{7\times9}(\mathbf{k}) = {} & \gamma_1^{79} (\sigma_- k_+ + \sigma_+ k_-) + \gamma_2^{79} (S_- \sigma_+ + S_+ \sigma_-) + \gamma_3^{79} \left(S_- k_- \sigma_+^2 + S_+ k_+ \sigma_-^2\right) \\ & + \gamma_4^{79} \left[(\mathbf{k} \times \mathcal{E})_- \sigma_+ + (\mathbf{k} \times \mathcal{E})_+ \sigma_-\right] + \gamma_5^{79} k_z (S_- \sigma_+ + S_+ \sigma_-) \\ & + \gamma_6^{79} S_z (\sigma_- k_+ + \sigma_+ k_-) + \gamma_7^{79} k_z (\mathcal{E}_+ \sigma_- + \mathcal{E}_- \sigma_+) \\ & + \gamma_8^{79} \mathcal{E}_z (k_+ \sigma_- + k_- \sigma_+) + \gamma_9^{79} \left(k_+ \mathcal{E}_+ \sigma_-^2 + k_- \mathcal{E}_- \sigma_+^2\right). \end{aligned} \quad (5.138)$$

The general 8×8 matrix then takes the following form:

$$H_{8\times8}(\mathbf{k}) = \begin{pmatrix} H_{7\times7} & H'_{7\times7} & H'_{7\times7} & H_{7\times9} \\ H'_{7\times7} & H_{7\times7} & H'_{7\times7} & H_{7\times9} \\ H'_{7\times7} & H'_{7\times7} & H_{7\times7} & H_{7\times9} \\ H_{9\times7} & H_{9\times7} & H_{9\times7} & H_{9\times9} \end{pmatrix}, \quad (5.139)$$

where each of the 16 blocks are 2×2 matrices of the form given in Eqs. (5.136), (5.137), and (5.138). The primes on the $H'_{7\times7}$ indicate that time-reversal symmetry is not required such that terms with coefficients β_i^7 are allowed. For the diagonal blocks $H_{7\times7}$ (without primes), however, the coefficients β_i^7 must be zero so as to fulfill time-reversal symmetry. Obviously, since the Hamiltonian is hermitian, there are only 10 independent blocks. Note that the constants (α's and β's) are different for each of the 10 independent blocks. These must be determined using independent information.

5.9.3.3 Effective-Mass Terms

Next, we use the method of invariants to obtain the general structure of the 8×8 Hamiltonian including linear-in-k and effective-mass terms in the absence of external fields. At the Γ point, we find

$$\begin{aligned} H_{7\times7}(\mathbf{k}) = {} & \alpha_0^7 + \alpha_1^7 (\sigma_- k_+ + \sigma_+ k_-) + \alpha_8^7 k_z^2 + \alpha_9^7 \{k_+ k_-\} + \beta_1^7 k_z \\ & + \beta_{10}^7 (\sigma_- \{k_z k_+\} + \sigma_+ \{k_z k_-\}), \end{aligned} \quad (5.140)$$

where α (β) refer to terms fulfilling (not fulfilling) time-reversal symmetry. The β terms are allowed for the off-diagonal blocks only, i.e., the $H'_{7\times7}$ blocks.

Similarly, we find

$$H_{7\times9}(\mathbf{k}) = \gamma_1^{79}\left(\sigma_- k_+ + \sigma_+ k_-\right) + \gamma_{10}^{79}\left(\sigma_-\{k_z k_+\} + \sigma_+\{k_z k_-\}\right), \quad (5.141)$$

and

$$H_{9\times9}(\mathbf{k}) = \alpha_0^9 + \alpha_5^9 k_z^2 + \alpha_6^9\{k_+ k_-\}. \quad (5.142)$$

The total Hamiltonian structure is the same as in Eq. (5.139) and the text following Eq. (5.139) still applies.

5.10 Method of Invariants Revisited

Cho [82] has provided another formulation of the method of invariants. This is now described.

5.10.1 *Zincblende*

The irreducible tensors present in the effective Hamiltonian are given in Table 5.23. Note that l_x is the x component of an angular vector, e.g., the three-dimensional orbital-angular matrix in the valence-band states [Eq. (5.148)] or the Pauli spin matrices.

Table 5.23 Irreducible tensors for zincblende

Γ_1	S [1]
Γ_2	T $[xl_x + yl_y + zl_z]$
Γ_{12}	(U, V) $[\sqrt{3}\,(x^2 - y^2), (2z^2 - x^2 - y^2)]$
Γ_{25}	(P, Q, R) $[l_x, l_y, l_z]$
Γ_{15}	(X, Y, Z) $[x, y, z]$

In Table 5.24, a list of relevant physical variables are given transforming according to the various irreducible representations separated according to their time-reversal transformation properties. While the transformation of the wave vector and strain were already given in Table 5.12, they are repeated here in order to relate to the notation in Table 5.23.

For example, the only independent components of the product of l_i with itself can be obtained from Table 5.25 by noting that $(P, Q, R) = (P', Q', R') = (l_x, l_y, l_z)$ and only combinations of the following symmetries are present in the T_d group:

Table 5.24 Examples of fields and transformation properties for zincblende. k_i, $\mathcal{E}_i$, B_i, and ε_{ij} denote wave vector, electric field, magnetic field, and strain field, respectively. A field transforming according to K_+ (K_-) is invariant (changes sign) under time reversal

		K_-	K_+
Γ_1	S		$k^2 = k_x^2 + k_y^2 + k_z^2$
			$\mathcal{E}^2 = \mathcal{E}_x^2 + \mathcal{E}_y^2 + \mathcal{E}_z^2$
			$\varepsilon_{xx} + \varepsilon_{yy} + \varepsilon_{zz}$
Γ_2	T		
Γ_{12}	U		$\mathcal{E}_x^2 - \mathcal{E}_y^2, \varepsilon_{xx} - \varepsilon_{yy}$
	V		$\mathcal{E}_z^2 - \mathcal{E}^2, 2\varepsilon_{zz} - \varepsilon_{xx} - \varepsilon_{yy}$
Γ_{25}	P	B_x	
	Q	B_y	
	R	B_z	
Γ_{15}	X	k_x	$\mathcal{E}_x, k_y k_z, \varepsilon_{yz}$
	Y	k_y	$\mathcal{E}_y, k_x k_z, \varepsilon_{xz}$
	Z	k_z	$\mathcal{E}_z, k_x k_y, \varepsilon_{xy}$

$$
\begin{aligned}
S &= l_x^2 + l_y^2 + l_z^2, \\
(U, V) &= \sqrt{3}\left(l_x^2 - l_y^2\right), 2l_z^2 - l_x^2 - l_y^2, \\
(X, Y, Z) &= (\{l_y l_z\}, \{l_z l_x\}, \{l_x l_y\}).
\end{aligned}
$$

In forming the invariant Hamiltonian, we have for the conduction band,

$$H_{\mathrm{S}}^{(c)} = \tau S, \tag{5.143}$$

$$H_{\mathrm{AS}}^{(c)} = 2\overline{\lambda}(P\sigma_{ex} + Q\sigma_{ey} + R\sigma_{ez}). \tag{5.144}$$

For the valence band,

$$
\begin{aligned}
H_{\mathrm{S}}^{(v)} &= (a_1 + 2a_2 \boldsymbol{\sigma} \cdot \mathbf{I})S + c_1\left[\sqrt{3}U\left(l_x^2 - l_y^2\right) + V(3l_z^2 - 2)\right] \\
&+ \sqrt{2}c_2\left[\sqrt{3}U(l_x\sigma_x - l_y\sigma_y) + V(3l_z\sigma_z - \boldsymbol{\sigma} \cdot \mathbf{I})\right]
\end{aligned}
$$

Table 5.25 Multiplication of irreducible components of representations for zincblende

Γ_1	S	SS'	TT'	$UU' + VV'$		$PP' + QQ' + RR'$		$XX' + YY' + ZZ'$
Γ_2	T	ST'		$UV' - VU'$			$PX' + QY' + RZ'$	
Γ_{12}	U	SU'	$-TV'$	$UV' + VU'$		$\sqrt{3}(PP' - QQ')$	$2RZ' - PX' - QY'$	$\sqrt{3}(XX' - YY')$
	V	SV'	$-TU'$	$UU' - VV'$		$2RR' - PP' - QQ'$	$\sqrt{3}(QY' - PX')$	$2ZZ' - XX' - YY'$
Γ_{25}	P	SP'	TX'	$(\sqrt{3}U - V)P'$	$-(\sqrt{3}V + U)X'$	$QR' - RQ'$	$QZ' + RY'$	$YZ' - ZY'$
	Q	SQ'	TY'	$-(\sqrt{3}U + V)Q'$	$(\sqrt{3}V - U)Y'$	$RP' - PR'$	$RX' + PZ'$	$ZX' - XZ'$
	R	SR'	TZ'	$2VR'$	$2UZ'$	$PQ' - QP'$	$PY' + QX'$	$XY' - YX'$
Γ_{15}	X	SX'	TP'	$-(\sqrt{3}V + U)P'$	$(\sqrt{3}U - V)X'$	$QR' + RQ'$	$QZ' - RY'$	$YZ' + ZY'$
	Y	SY'	TQ'	$(\sqrt{3}V - U)Q'$	$-(\sqrt{3}U + V)Y'$	$RP' + PR'$	$RX' - PZ'$	$ZX' + XZ'$
	Z	SZ'	TR'	$2UR'$	$2VZ'$	$PQ' + QP''$	$PY' - QX'$	$XY' + YX'$

Reprinted with permission from [82] using the coupling-constant tables in [33].

$$+\sqrt{8}d_1\left[P(l_y\sigma_z - l_z\sigma_y) + \text{c.p.}\right] + \sqrt{8}e_1\left[X\{l_y l_z\} + \text{c.p.}\right]$$

$$+\sqrt{8}e_2\left[X(l_y\sigma_z + l_z\sigma_y) + \text{c.p.}\right], \tag{5.145}$$

$$H_{\text{AS}}^{(v)} = 4\bar{b}_1(\sigma_x\{l_y l_z\} + \sigma_y\{l_z l_x\} + \sigma_z\{l_x l_y\})T$$

$$+\sqrt{8}\bar{c}_1\left[U(2\{l_x l_y\}\sigma_z - \{l_y l_z\}\sigma_x - \{l_z l_x\}\sigma_y) + \sqrt{3}V(\{l_z l_x\}\sigma_y - \{l_y l_z\}\sigma_x)\right]$$

$$+\sqrt{2}\bar{d}_1(Pl_x + Ql_y + Rl_z) + 2\bar{d}_2(P\sigma_x + Q\sigma_y + R\sigma_z)$$

$$+4\bar{d}_3[P(\{l_x l_z\}\sigma_z + \{l_y l_x\}\sigma_y) + \text{c.p.}] + 4\bar{d}_4\left(P\sigma_x l_x^2 + \text{c.p.}\right)$$

$$+\sqrt{8}\bar{e}_1\left[X\left(\{l_z l_x\}\sigma_z - \{l_y l_x\}\sigma_y\right) + \text{c.p.}\right] + 4\bar{e}_2\left[X(l_y^2 - l_x^2)\sigma_x + \text{c.p.}\right]. \tag{5.146}$$

The above expressions for the valence-band Hamiltonian components $H_{\text{S}}^{(v)}$ and $H_{\text{AS}}^{(v)}$ can be used next to find a 6×6 Hamiltonian in the one-particle valence-band states [82]:

$$\begin{aligned}
\phi_1 &= -\frac{1}{\sqrt{2}}[(X + \text{i}Y)\uparrow], \\
\phi_2 &= \frac{1}{\sqrt{6}}[2Z\uparrow -(X + \text{i}Y)\downarrow], \\
\phi_3 &= \frac{1}{\sqrt{6}}[2Z\downarrow +(X - \text{i}Y)\uparrow], \\
\phi_4 &= \frac{1}{\sqrt{2}}[(X - \text{i}Y)\downarrow], \\
\phi_5 &= \frac{1}{\sqrt{3}}[Z\uparrow +(X + \text{i}Y)\downarrow], \\
\phi_6 &= \frac{1}{\sqrt{3}}[-Z\downarrow +(X - \text{i}Y)\uparrow].
\end{aligned} \tag{5.147}$$

Note that these differ from our choice in Table 3.4 by a sign change for the ϕ_1, ϕ_2, ϕ_3 and ϕ_6 states. We make use of 2×2 spin matrices σ_i $(i = x, y, z)$ and the orbital angular momentum matrices,

$$l_x = \frac{\hbar}{\sqrt{2}}\begin{pmatrix} 0 & 1 & 0 \\ 1 & 0 & 1 \\ 0 & 1 & 0 \end{pmatrix}, \; l_y = \frac{\text{i}\hbar}{\sqrt{2}}\begin{pmatrix} 0 & -1 & 0 \\ 1 & 0 & -1 \\ 0 & 1 & 0 \end{pmatrix}, l_z = \hbar\begin{pmatrix} 1 & 0 & 0 \\ 0 & 0 & 0 \\ 0 & 0 & -1 \end{pmatrix}, \tag{5.148}$$

in the basis set consisting of the three states: $|1\rangle = -(X + \mathrm{i}Y)/\sqrt{2}$, $|0\rangle = Z$, $|-1\rangle = (X - \mathrm{i}Y)/\sqrt{2}$. This is consistent with our earlier choice of **I** matrices for diamond (Sect. 5.2).

The matrix for $H_{\mathrm{S}}^{(c)}$ and $H_{\mathrm{AS}}^{(c)}$ can be found immediately:

$$\begin{pmatrix} \tau S + \bar{\lambda} R & \bar{\lambda}(P - \mathrm{i}Q) \\ \bar{\lambda}(P + \mathrm{i}Q) & \tau S - \bar{\lambda} R \end{pmatrix}. \tag{5.149}$$

A matrix expression for $H_{\mathrm{S}}^{(v)}$ in the valence-band states defined by Eq. (5.147) is then as given in Table 5.26, where, e.g., the matrix entry $H_{\mathrm{S}}^{(v)}(2,3)$ is $\langle\phi_2|H_{\mathrm{S}}^{(v)}|\phi_3\rangle$ and so on. The matrix coefficients are

$$\begin{aligned}
a &= (a_1 + a_2)S + (c_1 + \sqrt{2}c_2)V, & f &= \sqrt{3}c_1 U - \mathrm{i}\sqrt{2}e_1 Z, \\
b &= a_1 S - 2c_1 V, & g &= \sqrt{2}a_2 S - c_2 V - 2\mathrm{i}d_1 R, \\
c &= (a_1 - a_2)S + (c_1 - \sqrt{2}c_2)V, & h &= \sqrt{2}d_1(Q + \mathrm{i}P) + \sqrt{2}e_2(Y - \mathrm{i}X), \\
d &= -d_1(Q + \mathrm{i}P)S + (e_1 + e_2)(Y - \mathrm{i}X), & k &= \sqrt{3}c_2 U - 2\mathrm{i}e_2 Z, \\
e &= -d_1(Q + \mathrm{i}P)S - (e_1 - e_2)(Y - \mathrm{i}X), & g_+ &= g + g^* = 2\sqrt{2}a_2 S - 2c_2 V.
\end{aligned} \tag{5.150}$$

As an example, note that the only terms in $H_{\mathrm{S}}^{(v)}$ contributing to entry $H_{\mathrm{S}}^{(v)}(1,1)$ are the terms where none of the operators $\sigma_x, \sigma_y, l_x, l_y$ appear. Thus,

$$\begin{aligned}
\left\langle \frac{3}{2}\frac{3}{2}\right| H_{\mathrm{S}}^{(v)} \left| \frac{3}{2}\frac{3}{2}\right\rangle &= \left\langle \frac{3}{2}\frac{3}{2}\right| S\,(a_1 + 2a_2\boldsymbol{\sigma}\cdot\mathbf{l}) + c_1 V\,(3l_z^2 - 2) + \sqrt{2}c_2 V\,(3l_z\sigma_z - \boldsymbol{\sigma}\cdot\mathbf{l}) \left| \frac{3}{2}\frac{3}{2}\right\rangle \\
&= \left\langle \frac{3}{2}\frac{3}{2}\right| S\,(a_1 + 2a_2\sigma_z l_z) + c_1 V\,(3l_z^2 - 2) + \sqrt{2}c_2 V\,(3l_z\sigma_z - l_z\sigma_z) \left| \frac{3}{2}\frac{3}{2}\right\rangle \\
&= \left\langle \frac{3}{2}\frac{3}{2}\right| S\left(a_1 + 2a_2\cdot\frac{1}{2}\cdot 1\right) + c_1 V\,(3\cdot 1\cdot 1 - 2) + \sqrt{2}c_2 V\left(3\cdot 1\cdot\frac{1}{2} - 1\cdot\frac{1}{2}\right) \left| \frac{3}{2}\frac{3}{2}\right\rangle \\
&= S\,(a_1 + a_2) + V\left(c_1 + \sqrt{2}c_2\right).
\end{aligned} \tag{5.151}$$

In analogy, we compute the matrix expression for $H_{\mathrm{AS}}^{(v)}$ in the same valence-band states obtaining Table 5.27. As above, we give the derivation of the matrix entry $H_{\mathrm{AS}}^{(v)}(1,1)$:

$$\begin{aligned}
\left\langle \frac{3}{2}\frac{3}{2}\right| H_{\mathrm{AS}}^{(v)} \left| \frac{3}{2}\frac{3}{2}\right\rangle &= \left\langle \frac{3}{2}\frac{3}{2}\right| \left(\sqrt{2}\bar{d}_1 R l_z + 2\bar{d}_2 R\sigma_z + 4\bar{d}_4 R\sigma_z l_z^2\right) \left| \frac{3}{2}\frac{3}{2}\right\rangle \\
&= \sqrt{2}\bar{d}_1 R + \bar{d}_2 R + 2\bar{d}_4 R.
\end{aligned}$$

Table 5.26 $H_S^{(v)}$

$$
\frac{1}{3}\begin{pmatrix}
3a & \sqrt{6}d+\sqrt{3}h & \sqrt{3}f+\sqrt{6}k & 0 & \sqrt{3}d-\sqrt{6}h & \sqrt{6}f-\sqrt{3}k \\
 & 2b+c+\sqrt{2}g & 0 & \sqrt{5}k+\sqrt{3}f & \sqrt{2}(b-c)+g^*-2g & 3e \\
 & & 2b+c+\sqrt{2}g_+ & -\sqrt{6}d-\sqrt{3}h & 3e^* & \sqrt{2}(c-b)+2g^*-g \\
 & & & 3a & \sqrt{3}k^*-\sqrt{6}f^* & -\sqrt{6}h^*+\sqrt{3}d^* \\
 & & & & b+2c-\sqrt{2}g_+ & 0 \\
 & & & & & b+2c-\sqrt{2}g_+
\end{pmatrix},
$$

Other matrix elements can be obtained in a similar fashion. We will now show the equivalence of these results to those previously obtained; for example, to the DKK Hamiltonian (note that the latter is strictly for diamond but the Hamiltonian is the same if the asymmetry terms are neglected).

Let us recall two DKK matrix elements in JM_J basis using Elliott-LK notation (Table 3.6):

$$\left\langle \frac{3}{2}\frac{3}{2}|H|\frac{3}{2}\frac{3}{2}\right\rangle = \frac{1}{2}P = \frac{1}{2}(L+M)\left(k_x^2+k_y^2\right) + Mk_z^2,$$
$$\left\langle \frac{3}{2}\frac{3}{2}|H|\frac{3}{2}\frac{1}{2}\right\rangle = R = -N\frac{1}{\sqrt{3}}\left(k_x - \mathrm{i}k_y\right)k_z, \tag{5.152}$$

where the free-electron contribution to $\langle \frac{3}{2}\frac{3}{2}|H|\frac{3}{2}\frac{3}{2}\rangle$ has been discarded.

We now give the equivalent matrix elements following Cho's procedure. In the absence of external fields and strain, Tables 5.26 and 5.27 give

$$\left\langle \frac{3}{2}\frac{3}{2}|H_{\mathrm{S}}^{(v)} + H_{\mathrm{AS}}^{(v)}|\frac{3}{2}\frac{3}{2}\right\rangle = \left(a_1 + a_2 + 2\left(c_1 + \sqrt{2}c_2\right)\right)k_z^2$$
$$+\left(a_1 + a_2 - c_1 - \sqrt{2}c_2\right)\left(k_x^2 + k_y^2\right),$$
$$\left\langle \frac{3}{2}\frac{3}{2}|H_{\mathrm{S}}^{(v)} + H_{\mathrm{AS}}^{(v)}|\frac{3}{2}\frac{1}{2}\right\rangle = \frac{\sqrt{2}}{\sqrt{3}}(e_1 + 2e_2)(Y - \mathrm{i}X)$$
$$= \frac{\sqrt{2}}{\sqrt{3}}(e_1 + 2e_2)(k_x - \mathrm{i}k_y)k_z, \tag{5.153}$$

and the two results (DKK and Cho) agree symmetrywise. It is important to notice that time-reversal symmetry of the Hamiltonian ensures that there are no k_ik_j Γ_{15} contributions from $H_{\mathrm{AS}}^{(v)}$ to $\langle \frac{3}{2}\frac{3}{2}|H_{\mathrm{S}}^{(v)} + H_{\mathrm{AS}}^{(v)}|\frac{3}{2}\frac{1}{2}\rangle$. This is so since the $\bar{e}_1$ term in Eq. (5.146) involves three angular momenta and hence changes sign under time-reversal symmetry if $X = k_ik_j$ (k_ik_j is invariant under time-reversal symmetry). In other words, contributions from $\bar{c}_1$ term only arise when X changes sign under time-reversal.

5.10.2 Wurtzite

We proceed with a similar discussion for the WZ crystal system. In Tables 5.28 and 5.29, we list the various irreducible components and fields, respectively.

Note that **B** is an axial vector and hence transforms as the three-dimensional angular momentum matrices **I**. With the above tables, we can construct the Hamiltonian matrices in the two conduction states and the six valence states. This is done in Eqs. (5.154), (5.155), (5.156), and (5.157).

Table 5.27 $H_{\mathrm{AS}}^{(v)}$

$$
\frac{1}{3}\begin{pmatrix}
3\bar{a} & \sqrt{6}\bar{d}+\sqrt{3}\bar{h} & \sqrt{3}\bar{f}+\sqrt{6}\bar{r} & 3\bar{s} & \sqrt{3}\bar{d}-\sqrt{6}\bar{h} & \sqrt{6}\bar{f}-\sqrt{3}\bar{r} \\
 & 2\bar{b}-\bar{c}+\sqrt{2}\bar{g}_+ & 2\bar{l}+\bar{u}+\sqrt{8}\bar{e} & -\sqrt{6}\bar{r}-\sqrt{3}\bar{f} & \sqrt{2}(\bar{b}+\bar{c})+\bar{g}^*-2\bar{g} & \sqrt{2}(\bar{u}-\bar{l})+\bar{e} \\
 & & \bar{c}-2\bar{b}-\sqrt{2}\bar{g}_+ & \sqrt{3}\bar{h}+\sqrt{6}\bar{d} & \sqrt{2}(\bar{l}-\bar{u})^*-\bar{e}^* & \sqrt{2}(\bar{b}+\bar{c})+\bar{g}-2\bar{g}^* \\
 & & & -3\bar{a} & \sqrt{6}\bar{f}^*-\sqrt{3}\bar{r}^* & \sqrt{6}\bar{h}^*-\sqrt{3}\bar{d}^* \\
 & & & & \bar{b}-2\bar{c}-\sqrt{2}\bar{g}_+ & \sqrt{8}\bar{e}-(2\bar{u}+\bar{t}) \\
 & & & & & 2\bar{c}-\bar{b}+\sqrt{2}\bar{g}_+
\end{pmatrix},
$$

where

$$\bar{a} = (\sqrt{2}\bar{d}_1+\bar{d}_2+2\bar{d}_4)R,$$
$$\bar{b} = \bar{d}_2 R,$$
$$\bar{c} = (-\sqrt{2}\bar{d}_1+\bar{d}_2+2\bar{d}_4)R,$$
$$\bar{d} = (\bar{d}_1+\bar{d}_3)(P-\mathrm{i}Q)+\bar{e}_1(X+\mathrm{i}Y),$$
$$\bar{e} = (\bar{d}_1-\bar{d}_3)(P-\mathrm{i}Q)-\bar{e}_1(X+\mathrm{i}Y),$$
$$\bar{f} = -\mathrm{i}\bar{b}_1 T-\mathrm{i}\sqrt{2}\bar{c}_1 U+2\bar{e}_2 Z,$$
$$\bar{g}_+ = 4\bar{d}_3 R,$$
$$\bar{g} = -\mathrm{i}\sqrt{3}\bar{c}_1 V+2\bar{d}_3 R,$$
$$\bar{h} = (\bar{d}_2+\bar{d}_4)(P-\mathrm{i}Q)-\bar{e}_2(X+\mathrm{i}Y),$$
$$\bar{r} = -\mathrm{i}\sqrt{2}\bar{b}_1 T+\mathrm{i}\bar{c}_1 U-2\bar{e}_1 Z,$$
$$\bar{s} = (\bar{d}_4-\sqrt{2}\bar{d}_3)(P+\mathrm{i}Q)+(\sqrt{2}\bar{e}_1-\bar{e}_2)(X-\mathrm{i}Y),$$
$$\bar{t} = (\bar{d}_2+2\bar{d}_4)(P-\mathrm{i}Q)+2\bar{e}_2(X+\mathrm{i}Y),$$
$$\bar{u} = (\bar{d}_4+\sqrt{2}\bar{d}_3)(P-\mathrm{i}Q)-(\sqrt{2}\bar{e}_1+\bar{e}_2)(X+\mathrm{i}Y).$$

Table 5.28 Irreducible tensors for wurtzite

Γ_1	$S\ [1, z, l_z^2]$
Γ_2	$T\ [l_z]$
Γ_3	$U\ [(3x^2 - y^2)y]$
Γ_4	$V\ [(3y^2 - x^2)x]$
Γ_5	$X\ [x, -l_y]$
	$Y\ [y, l_x]$
Γ_6	$W\ [x^2 - y^2, l_x^2 - l_y^2]$
	$Z\ [2xy, 2l_x l_y]$

Table 5.29 Examples of fields and transformation properties for wurtzite. k_i, $\mathcal{E}_i$, B_i, and ε_{ij} denote wave vector, electric field, magnetic field, and strain field, respectively. A field transforming according to K_+ (K_-) is invariant (changes sign) under time reversal

		K_-	K_+
Γ_1	S	k_z	$k_x^2 + k_y^2, k_z^2, B_x^2 + B_y^2, B_z^2$
			$\mathcal{E}_z, \mathcal{E}_x^2 + \mathcal{E}_y^2, \mathcal{E}_z^2$
			$\varepsilon_{xx} + \varepsilon_{yy}, \varepsilon_{zz}$
Γ_2	T	B_z	
Γ_3	U		
Γ_4	V		
Γ_5	X	k_x, B_y	$\mathcal{E}_x, k_x k_z, \mathcal{E}_x \mathcal{E}_z, B_x B_z, \varepsilon_{xz}$
	Y	$k_y, -B_x$	$\mathcal{E}_y, k_y k_z, \mathcal{E}_y \mathcal{E}_z, B_y B_z, \varepsilon_{yz}$
Γ_6	W		$k_x^2 - k_y^2, \mathcal{E}_x^2 - \mathcal{E}_y^2, B_x^2 - B_y^2, \varepsilon_{xx} - \varepsilon_{yy}$
	Z		$2k_x k_y, 2\mathcal{E}_x \mathcal{E}_y, 2B_x B_y, 2\varepsilon_{xy}$

$$H_{\mathrm{S}}^{(c)} = \tau S, \tag{5.154}$$

$$H_{\mathrm{AS}}^{(c)} = 2\bar{\lambda}\sigma_{ez} + 2\bar{\mu}(Y\sigma_{ex} - X\sigma_{ey}), \tag{5.155}$$

$$\begin{aligned} H_{\mathrm{S}}^{(v)} = {} & \left[a_1 + a_2 l_z^2 + 2a_3 l_z \sigma_z + 2a_4(l_x\sigma_x + l_y\sigma_y)\right] S + \sqrt{2}b_1(l_x\sigma_y - l_y\sigma_x)T \\ & + 2\sqrt{2}e_1(Xl_x + Yl_y)\sigma_z + 2e_2(X\sigma_x + Y\sigma_y)l_z + 2\sqrt{2}e_3(X\{l_x l_z\} + Y\{l_y l_z\}) \\ & + f_1\left[W(l_x^2 - l_y^2) + 2Z\{l_x l_y\}\right] + \sqrt{2}f_2\left[(l_x\sigma_x - l_y\sigma_y)W + (l_x\sigma_y + l_y\sigma_x)Z\right], \end{aligned} \tag{5.156}$$

$$
\begin{aligned}
H_{\text{AS}}^{(v)} &= 2\sqrt{2}a_1\left(\{l_x l_z\}\sigma_y - \{l_y l_z\}\sigma_x\right) S \\
&+ \left[\overline{b}_1 l_z + 2\overline{b}_2\sigma_z + 2\overline{b}_3 l_z^2\sigma_z + 2\sqrt{2}b_4\left(\{l_x l_z\}\sigma_x + \{l_y l_z\}\sigma_y\right)\right] T \\
&+ \overline{c}_1\left[\left(l_x^2 - l_y^2\right)\sigma_x - 2\{l_x l_y\}\sigma_y\right] U + \overline{d}_1\left[\left(l_x^2 - l_y^2\right)\sigma_y + 2\{l_x l_y\}\sigma_x\right] V \\
&+ \sqrt{2}\overline{e}_1\left(l_x Y - l_y X\right) + 2\left(\overline{e}_2 + \overline{e}_4 l_z^2\right)(\sigma_x Y - \sigma_y X) + 4\sqrt{2}\overline{e}_3\left(\{l_x l_z\}Y - \{l_y l_z\}X\right)\sigma_z \\
&+ \overline{e}_5\left(\left[\left(l_x^2 - l_y^2\right)\sigma_x + 2\{l_x l_y\}\sigma_y\right] Y + \left[\left(l_x^2 - l_y^2\right)\sigma_y - 2\{l_x l_y\}\sigma_x\right] X\right) \\
&+ 2\overline{f}_1\left[(l_x^2 - l_y^2)\sigma_z Z - 2\{l_x l_y\}\sigma_z W\right] \\
&+ 2\sqrt{2}\overline{f}_2\left[\left(\{l_x l_z\}\sigma_y + \{l_y l_z\}\sigma_x\right) W - \left(\{l_x l_z\}\sigma_x - \{l_y l_z\}\sigma_y\right) Z\right].
\end{aligned}
\tag{5.157}
$$

The coefficients in Eqs. (5.154), (5.155), (5.156), and (5.157) agree with those in [82]. It is now possible to construct the representation of the conduction-band and valence-band Hamiltonians using a set of two and three basis states, respectively. This can be done, similar to the ZB case discussed above, accounting for mixing of S and Z states (as both belong to Γ_1) or using the quasi-cubic approximation for the WZ structure. The interested reader may consult Cho [82] for the details using the quasi-cubic approximation.

As an exercise, we will derive the final term in Eq. (5.157) using (1) Table 5.30 and (2) the tables in Koster et al. [33].

Table 5.30 Multiplication of irreducible components of representations for wurtzite

Γ_1	S	SS'	TT'	UU'	VV'	$XX' + YY'$	$WW' + ZZ'$
Γ_2	T	ST'		UV'		$XY' - YX'$	$WZ' - ZW'$
Γ_3	U	SU'	TV'			$XZ' + YW'$	
Γ_4	V	SV'	TU'			$YZ' - XW'$	
Γ_5	X	SX'	$-TY'$	UZ'	$-VW'$	$XZ' - YW'$	
	Y	SY'	TX'	UW'	VZ'	$YZ' + XW'$	
Γ_6	W	SW'	TZ'	$-UX'$	$-VX'$	$XX' - YY'$	$ZZ' - WW'$
	Z	SZ'	$-TW'$	UY'	VY'	$XY' + YX'$	$WZ' + ZW'$

Reprinted with permission from [82] following the coupling-constant tables in [33].

(1) We can form states $X' = -TY = -(l_z)(-l_x) = l_z l_x$ and $Y' = TX = (l_z)(l_y) = l_z l_y$ transforming as X and Y, respectively, as follows from the fourth column in the Γ_5 states (Table 5.30) and the Γ_5 states listed in Table 5.28. Further, σ_y and $-\sigma_x$ transform as X and Y, respectively, such that $(\sigma_y)(l_z l_x) - (-\sigma_x)(l_z l_y)$ and $(\sigma_y)(l_z l_y) + (-\sigma_x)(l_z l_x)$ transform as W' and Z', respectively, using the seventh column in the Γ_6 states (Table 5.30). Finally, we obtain from the eighth column in the Γ_1 states that $W'W + Z'Z = (\sigma_y l_z l_x + \sigma_x l_z l_y)W - (\sigma_x l_z l_x - \sigma_y l_z l_y)Z$. Symmetrization of products involving the same angular vector (e.g., $l_z l_x \to \{l_x l_z\}$) gives the final term in Eq. (5.157).

(2) Following the notation in [33] (pp. 17–18 (Sect. 4.3) and tables on pp. 68–70), we have either $[(\psi^5_{-1}, \psi^5_1) = (-\mathrm{i}(x + \mathrm{i}y), \mathrm{i}(x - \mathrm{i}y))$ or $(\psi^5_{-1}, \psi^5_1) = (-(S_x + \mathrm{i}S_y), (S_x - \mathrm{i}S_y))$ where, e.g., S_x denotes the x component of an axial vector.

We shall use three coupling-coefficient tables (pp. 68–70 in [33]) for this derivation reproduced below (Tables 5.31, 5.32, and 5.33).

Table 5.31 Coupling-constant tables for wurtzite [33]

	$u_2 v^5_{-1}$	$u_2 v^5_1$
ψ^5_{-1}	i	0
ψ^5_1	0	–i

Table 5.32 Coupling-constant tables for wurtzite [33]

	$u^5_{-1} v^5_{-1}$	$u^5_{-1} v^5_1$	$u^5_1 v^5_{-1}$	$u^5_1 v^5_1$
ψ_1	0	$\frac{1}{\sqrt{2}}$	$\frac{1}{\sqrt{2}}$	0
ψ_2	0	$\frac{\mathrm{i}}{\sqrt{2}}$	$-\frac{\mathrm{i}}{\sqrt{2}}$	0
ψ^6_{-1}	0	0	0	1
ψ^6_1	1	0	0	0

Table 5.33 Coupling-constant tables for wurtzite [33]

	$u^6_{-1} v^6_{-1}$	$u^6_{-1} v^6_1$	$u^6_1 v^6_{-1}$	$u^6_1 v^6_1$
ψ_1	0	$\frac{1}{\sqrt{2}}$	$\frac{1}{\sqrt{2}}$	0
ψ_2	0	$\frac{\mathrm{i}}{\sqrt{2}}$	$-\frac{\mathrm{i}}{\sqrt{2}}$	0
ψ^6_{-1}	0	0	0	1
ψ^6_1	1	0	0	0

u_1, ψ_1 and u_2, ψ_2 are states representing the one-dimensional representations Γ_1 and Γ_2, respectively. First, we shall construct states ψ^5_{-1} and ψ^5_1 using Table 5.31. These are: $(\psi^5_{-1}, \psi^5_1) = (\mathrm{i}u_2 v^5_{-1}, -\mathrm{i}u_2 v^5_1) = (-\mathrm{i}S_z(S_x + \mathrm{i}S_y), -\mathrm{i}S_z(S_x - \mathrm{i}S_y))$ with $u_2 = S_z$ and $(v^5_{-1}, v^5_1) = (-(S_x + \mathrm{i}S_y), S_x - \mathrm{i}S_y)$.

Combining the latter states with the ψ'^5 states: $(\sigma_x - \mathrm{i}\sigma_y, -(\sigma_x + \mathrm{i}\sigma_y))$ using Table 5.32 allows us to construct ψ^6_{-1} and ψ^6_1 states. We find: $(\psi^6_{-1}, \psi^6_1) = (\psi^5_1 \psi'^5_1, \psi^5_{-1}\psi'^5_{-1}) = (-\mathrm{i}(S_x - \mathrm{i}S_y)S_z(S'_x - \mathrm{i}S'_y), \mathrm{i}(S_x + \mathrm{i}S_y)S_z(S'_x + \mathrm{i}S'_y))$.

An invariant (transforming according to ψ_1) can be found by combining the latter ψ_6 states with a different set of Γ_6 states: $(\psi'^6_{-1}, \psi'^6_1) = (\psi'^5_1 \psi'^5_1, \psi'^5_{-1}\psi'^5_{-1}) = (\mathrm{i}(x - \mathrm{i}y)\mathrm{i}(x - \mathrm{i}y), -\mathrm{i}(x + \mathrm{i}y) - \mathrm{i}(x + \mathrm{i}y)] = [-W + \mathrm{i}Z, -W - \mathrm{i}Z)$ employing

Table 5.33. We obtain:

$$\psi_1 = \frac{1}{\sqrt{2}}\left(\psi^{6}_{-1}\psi'^{6}_{-1} + \psi^{6}_{1}\psi'^{6}_{1}\right) = \sqrt{2}\left[(S_z S_x S'_y + S_z S_y S'_x)W - (S_z S_x S'_x - S_z S_y S'_y)Z\right]. \tag{5.158}$$

Again, we note that the product of two identical angular vectors such as $S_z S_x$ is replaced by the symmetrized combination $\{S_z S_x\} = \frac{1}{2}(S_z S_x + S_x S_z)$. Further, using $S = l$, $S' = \sigma$ leads to the final term in Eq. (5.157).

5.11 Summary

The current chapter introduced the method of invariants, which allows one to list all the independent invariant contributions to the Hamiltonian based upon symmetry considerations. The parameters of the model can then be fitted to experiments or can be compared to a perturbative formulation of the Hamiltonian. The latter step requires that one expresses the invariant Hamiltonian in an explicit representation. The theory was illustrated first using the classic work of Luttinger on the valence band of diamond-type semiconductors in the presence of a magnetic field. This allowed us to introduce the Luttinger parameters. The theory was then generalized to Hamiltonians containing multiple representations, culminating in an exposition of a 14-band model for zincblende. The theory was also applied to the wurtzite structure. Finally, the method was revisited using the more general formulation given by K. Cho.

Chapter 6
Spin Splitting

6.1 Overview

The combined effect of inversion symmetry and time-reversal symmetry (Kramers degeneracy) lead to spin degeneracy in semiconductors. Hence, if these two symmetries are present, all levels are at least two-fold degenerate, a familiar result from atomic physics. The proof for crystalline solids is straightforward. Time-reversal symmetry implies that both the wave vector and the spin components of a state change sign:

$$E_{+}(\mathbf{k}) = E_{-}(-\mathbf{k}), \tag{6.1}$$

whereas inversion symmetry changes the sign of the wave vector but leaves the spin state unaffected:

$$E_{\pm}(\mathbf{k}) = E_{\pm}(-\mathbf{k}), \tag{6.2}$$

where upper (lower) subscripts match together. Thus, combining the two operations give spin degeneracy:

$$E_{+}(\mathbf{k}) = E_{-}(\mathbf{k}). \tag{6.3}$$

This means that, for diamond structures where inversion symmetry is present, spin-splitting terms with odd powers in $\mathbf{k}$ cannot appear in the energy-dispersion relations near the Γ point.

In zincblende and wurtzite structures, being inversion asymmetric, the spin-degeneracy at a given $\mathbf{k}$ point is lifted, even in the absence of an external magnetic field. In other words, if inversion is not a symmetry, both even and odd powers in $\mathbf{k}$ are generally allowed in the energy expansion around the Γ point. This effect occurs in bulk structures and is known as the Dresselhaus effect [32] or bulk inversion asymmetry (BIA).

Besides the Dresselhaus spin-splitting effect, inversion asymmetry in semiconductors can result due to the presence of external fields and/or material macroscopic inhomogenities (such as a heterostructure). Spin splittings arising from the

L.C. Lew Yan Voon, M. Willatzen, *The* k · p *Method*,
DOI 10.1007/978-3-540-92872-0_6,

latter type of inversion asymmetry are known as Rashba spin splittings or structure inversion asymmetry (SIA). The Rashba spin-splitting effects will be described in Chap. 13.

The $k \cdot p$ theory of spin splitting has been extensively studied for bulk [32, 48, 104] using both perturbation theory and the method of invariant. A detailed presentation of the effect was recently given by Winkler [21].

6.2 Dresselhaus Effect in Zincblende

One can study the spin splitting either within isolated conduction and valence bands or within a coupled-band model. Both approaches will be presented.

6.2.1 Conduction State

The conduction electrons (Γ_1 states) do not display spin splittings linear in k, whether due to the $\sigma \cdot (\mathbf{k} \times \nabla V)$ spin-orbit coupling term (∇V transforms as Γ_{15}) in first-order perturbation theory, or due to the $\sigma \cdot (\mathbf{p} \times \nabla V)$ term ($\mathbf{p} \times \nabla V$ transforms as Γ_{25}) coupled with the $k \cdot p$ term ($\mathbf{p}$ transforms as Γ_{15}) in second-order perturbation theory. This result (absence of linear-in-k in zincblende) was also obtained using the method of invariant (refer to Tables 5.23 and 5.24 and the expression for the conduction-band Hamiltonian in the same section). Hence, the lowest-order spin-splitting contribution in $\mathbf{k}$ is third-order in nature. Using the theory of invariants, it is possible to write down the general expression of third-order spin splittings in zincblende structures. Note that the appropriate expressions depend upon the $k \cdot p$ model being used.

Thus, using a two-band model for the conduction electrons, the result is [21, 107]

$$
\begin{aligned}
H_k^{6c6c} &= \gamma_c \left(\left\{k_x(k_y^2 - k_z^2)\right\}\sigma_x + \text{c.p.}\right) \\
&= \gamma_c \begin{pmatrix} \frac{1}{2}\left\{(k_+^2 + k_-^2)k_z\right\} & \frac{1}{4}\left\{(k_+^2 - k_-^2)k_-\right\} - \left\{k_z^2 k_+\right\} \\ \frac{1}{4}\left\{(k_-^2 - k_+^2)k_+\right\} - \left\{k_z^2 k_+\right\} & -\frac{1}{2}\left\{(k_+^2 + k_-^2)k_z\right\} \end{pmatrix},
\end{aligned} \tag{6.4}
$$

where $k_\pm = k_x \pm \mathrm{i}k_y$, $\{AB\} = \frac{1}{2}(AB + BA)$, and γ_c is a constant. Since the conduction states along the [100, 111] directions transform according to the two-dimensional Γ_{15} and Γ_{25} double-group representations of C_{2v} and C_{3v}, respectively, conduction-band spin splittings vanish along these directions. This fact is also immediately observable from the matrix in Eq. (6.4).

6.2.2 Valence States

For the Γ_{7v} and Γ_{8v} states, both linear and cubic spin splittings exist in zincblende. The first-order coupling $\sigma \cdot (\mathbf{k} \times \nabla V)$ and the (more important) second-order coupling through $\boldsymbol{\sigma} \cdot (\mathbf{p} \times \nabla V)$ and $k \cdot p$ to the remote Γ_{12}-like states [104] contribute to linear spin splittings in $\mathbf{k}$.

Again, spin splittings vanish along the [100] direction (though the dispersion curve can be linear in k) as the wave-vector group C_{2v} only has two-dimensional double-group irreducible representations. Along the [111] direction the heavy-hole states split as they transform according to the one-dimensional double-group representation Γ_{15} of C_{3v}. However, the light-hole states transform according to the two-dimensional double-group representation Γ_{25} of C_{3v} along [111] and are thus not split along this direction.

Dresselhaus derived a linear-in-k Hamiltonian for the Γ_8 states [32]:

$$H_k = C_k \begin{pmatrix} 0 & \frac{\mathrm{i}}{2}k_+ & -k_z & -\frac{\mathrm{i}\sqrt{3}}{2}k_- \\ -\frac{\mathrm{i}}{2}k_+ & 0 & -\frac{\mathrm{i}\sqrt{3}}{2}k_+ & k_z \\ -k_z & \frac{\mathrm{i}\sqrt{3}}{2}k_- & 0 & \frac{\mathrm{i}}{2}k_+ \\ \frac{\mathrm{i}\sqrt{3}}{2}k_+ & k_z & -\frac{\mathrm{i}}{2}k_- & 0 \end{pmatrix}, \tag{6.5}$$

with

$$C_k = -\frac{1}{2\sqrt{3}}\frac{\hbar^2}{m_0^2c^2}\left\langle x\left|\frac{\partial V}{\partial y}\right|z\right\rangle. \tag{6.6}$$

Solving for $E(\mathbf{k})$ gives four roots (keeping terms to first order in k only),

$$E(\mathbf{k}) = \pm C_k \left\{k^2 \pm [3(k_x^2k_y^2 + k_y^2k_z^2 + k_x^2k_z^2)]^{1/2}\right\}^{1/2}, \tag{6.7}$$

For $\mathbf{k} \parallel [100]$ the energy dispersion becomes (doubly degenerate)

$$E(\mathbf{k})_{[100]} = \pm C_k k, \tag{6.8}$$

while, for $\mathbf{k} \parallel [111]$,

$$\begin{aligned} E(\mathbf{k}/\sqrt{3})_{[111]} &= 0 \quad \text{(double)}, \\ E(\mathbf{k}/\sqrt{3})_{[111]} &= \sqrt{2}C_k k, \\ E(\mathbf{k}/\sqrt{3})_{[111]} &= -\sqrt{2}C_k k. \end{aligned} \tag{6.9}$$

However, one can show that Dresselhaus' C_k is nearly zero. Another more important contribution involves coupling to second-order in the $H_{k \cdot p}$ and H_{so} Hamiltonians with the Γ_{12} as intermediate states. Cardona et al. [104] estimates this second-order contribution to the linear spin-splitting coefficient to be

$$-A\frac{\Delta_{d,c}}{E(\Gamma_v^8) - E_{d,c}} + B\frac{\Delta_{d,a}}{E(\Gamma_v^8) - E_{d,a}}, \tag{6.10}$$

where $\Delta_{d,c}$ ($\Delta_{d,c}$) and $E_{d,c}$ ($E_{d,a}$) are the spin-orbit splittings of the outermost cation (anion) d levels and their energies, respectively. The coefficients A, B are constants for III-V materials ($A = 285$ meV Å and $B = 155$ meV Å).

Using a six-band model, invariants for the Γ_{7v} and Γ_{8v} states are [21, 107]

$$
\begin{aligned}
H_k^{8v8v} &= \frac{2}{\sqrt{3}} C_k \left[k_x\{J_x, J_y^2 - J_z^2\} + k_y\{J_y, J_z^2 - J_x^2\} + k_z\{J_z, J_x^2 - J_y^2\}\right] \\
&\quad + \gamma_{8v,1}\left[\{k_x, k_y^2 - k_z^2\}J_x + \{k_y, k_z^2 - k_x^2\}J_y + \{k_z, k_x^2 - k_y^2\}J_z\right] \\
&\quad + \gamma_{8v,2}\left[\{k_x, k_y^2 - k_z^2\}J_x^3 + \{k_y, k_z^2 - k_x^2\}J_y^3 + \{k_z, k_x^2 - k_y^2\}J_z^3\right] \\
&\quad + \gamma_{8v,3}\left[\{k_x, k_y^2 + k_z^2\}\{J_x, J_y^2 - J_z^2\} + \{k_y, k_z^2 + k_x^2\}\{J_y, J_z^2 - J_x^2\}\right. \\
&\quad \left. + \{k_z, k_x^2 + k_y^2\}\{J_z, J_x^2 - J_y^2\}\right] \\
&\quad + \gamma_{8v,4}\left[k_x^3\{J_x, J_y^2 - J_z^2\} + k_y^3\{J_y, J_z^2 - J_x^2\} + k_z^3\{J_z, J_x^2 - J_y^2\}\right], \qquad (6.11)\\
H_k^{7v7v} &= \gamma_{7v}\left[\{k_x, k_y^2 - k_z^2\}\sigma_x + \{k_y, k_z^2 - k_x^2\}\sigma_y + \{k_z, k_x^2 - k_y^2\}\sigma_z\right], \qquad (6.12)\\
H_k^{8v7v} &= -\mathrm{i}\sqrt{3} C_k \left[k_x U_{yz} + k_y U_{zx} + k_z U_{xy}\right] \\
&\quad + \gamma_{8v7v,1}\left(\{k_x, k_y^2 - k_z^2\}U_x + \{k_y, k_z^2 - k_x^2\}U_y + \{k_z, k_x^2 + k_y^2\}U_z\right) \\
&\quad + \gamma_{8v7v,2}\left(\{k_x, k_y^2 + k_z^2\}U_{yz} + \{k_y, k_z^2 + k_x^2\}U_{zx} + \{k_z, k_x^2 + k_y^2\}U_{xy}\right) \\
&\quad + \gamma_{8v7v,3}\left(k_x^3 U_{yz} + k_y^3 U_{zx} + k_z^3 U_{xy}\right). \qquad (6.13)
\end{aligned}
$$

In the expressions above, the J_i and U_{ij} matrices are given in Eqs. (5.56) and (5.108).

6.2.3 Extended Kane Model

If one, instead, uses the extended 14×14 Kane model, invariants for the Γ_{6c} and Γ_{7v}, Γ_{8v} states are (from Table 5.13)

$$
\begin{aligned}
H_k^{6c8v} &= \mathrm{i}B_{8v}^{+}\Big[T_x\{k_y k_z\} + T_y\{k_z k_x\} + T_z\{k_x k_y\}\Big] \\
&\quad + \frac{1}{2}B_{8v}^{-}\Big[(T_{xx} - T_{yy})\left(k_z^2 - \frac{1}{3}k^2\right) - T_{zz}\left(k_x^2 - k_y^2\right)\Big], \qquad (6.14)\\
H_k^{6c7v} &= -\frac{\mathrm{i}}{\sqrt{3}}B_{7v}\Big[\rho_x\{k_y k_z\} + \rho_y\{k_z k_x\} + \rho_z\{k_x k_y\}\Big]. \qquad (6.15)
\end{aligned}
$$

The T_{ij} matrices are given in Table 5.14.

Cardona et al. [104] were the first to derive $k \cdot p$ expressions for the cubic spin-splitting coefficients using fourth-order perturbation theory. The fourth-order perturbation employed in [104] involves three times the $k \cdot p$ interaction and once the spin-orbit interaction $(\nabla \times \mathbf{p}) \cdot \boldsymbol{\sigma}$. The corresponding diagrams are shown in Fig. 6.1. Evidently, the $k \cdot p$ interaction lines 1, 3, and 4 in Fig. 6.1a (and the $k \cdot p$ lines 1, 2, and 3 in Fig. 6.1b) contribute with factors proportional to P, Q, and P', respectively, such that the two diagrams (Fig.6.1a,b) lead to γ_c spin-splitting contributions proportional to $PP'Q$. Similarly, the $k \cdot p$ interaction lines 1, 3, and 4

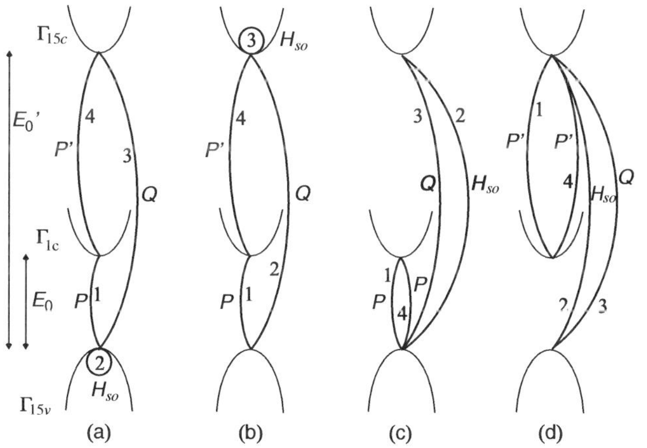

Fig. 6.1 Spin-splitting terms in perturbation theory for the lowest conduction band

in Fig. 6.1c correspond to factors P, Q, and P giving a γ_c contribution proportional to P^2Q. Finally, in Fig. 6.1d, two $k \cdot p$ interactions between Γ_1^c and Γ_{15}^c, one $k \cdot p$ interaction between Γ_{15}^c and Γ_{15}^v, and one spin-orbit interaction between Γ_{15}^c and Γ_{15}^v (Δ^-) lead to the fourth contribution proportional to P'^2Q. Adding the four terms gives [104, 108]

$$\gamma_c = A + B + C + D, \tag{6.16}$$

where

$$\begin{aligned}
A &= \frac{4}{3}PP'Q\frac{\Delta_0}{3E_0(E_0+\Delta_0')}\left[\frac{2}{E_0'-E_0+\Delta_0'}+\frac{1}{E_0'-E_0}\right],\\
B &= \frac{4}{3}PP'Q\frac{\Delta_0'}{3(E_0'-E_0)(E_0'-E_0+\Delta_0')}\left[\frac{2}{E_0}+\frac{1}{E_0+\Delta_0}\right],\\
C &= -\frac{4}{3}\frac{P^2Q\Delta^-}{\overline{E}_0^2(\overline{E}_0'-E_0)},\\
D &= -\frac{4}{3}\frac{P'^2Q\Delta^-}{\overline{E}_0(\overline{E}_0'-E_0)^2}.
\end{aligned} \tag{6.17}$$

Here, E_0 and E_0' refer to the $E_{\Gamma_6^c} - E_{\Gamma_8^v}$ and $E_{\Gamma_7^c} - E_{\Gamma_8^v}$ energies, Δ_0, Δ_0', and Δ^- are the $\Gamma_8^v - \Gamma_7^v$, $\Gamma_8^c - \Gamma_7^c$, and $3\langle(\frac{3}{2}\frac{3}{2})_v|H_{\mathrm{so}}|(\frac{3}{2}\frac{3}{2})_c\rangle$ spin-orbit energies. A bar above the E_0 and E_0' energies represents an average of the two spin-orbit split components with weight 2 for Γ_7 and 1 for Γ_8. We note, as already pointed out above, that all four terms A, B, C, D vanish in the absence of spin-orbit interaction.

We emphasize that the Δ_0 splitting term (Fig. 6.1a) was treated exactly in [104] since Δ_0 can be larger than the energies appearing in the denominators of the general perturbative expression (in InSb, $\Delta_0 = 0.8\,\text{eV}$ while $E_0 = 0.23\,\text{eV}$). Cardona, Christensen, and Fasol also derived, following a procedure similar to the one just outlined for γ_c, expressions for the γ_{lh}, γ_{hh}, γ_{sh}, γ_{le}, γ_{he}, and γ_{se} spin-splitting coefficients in [104]. In particular, it was shown perturbatively that $\gamma_{hh} = 0$. The cubic spin-splitting coefficients can be determined using third-order perturbation theory in the $k \cdot p$ coupling [104] and involves coupling between the Γ_{8v}, Γ_{7v} states, the Γ_{6c} states, and the Γ_{8c}, Γ_{7c} states (hence the product QPP' appears in all the cubic spin splittings). Such a calculation involves, for the conduction s band states,

$$\sum_\alpha \sum_\beta \frac{\langle s|H_{k\cdot p}|\alpha\rangle\langle\alpha|H_{k\cdot p}|\beta\rangle\langle\beta|H_{k\cdot p}|s\rangle}{(E - E_{\alpha\alpha})(E - E_{\beta\beta})}, \tag{6.18}$$

where α (β) denotes a Γ_{15}^c (Γ_{15}^v) state or vice versa. A similar result is obtained for the valence states. Carrying out this procedure gives

$$\gamma_c = -\frac{4\mathrm{i}}{3}PP'Q\left[\frac{1}{(E_0+\Delta_0)(E_0-E_0'-\Delta_0')} - \frac{1}{E_0(E_0-E_0')}\right], \tag{6.19}$$

$$\gamma_{8v,1} = \frac{\mathrm{i}}{6}PP'Q\frac{1}{E_0}\left[\frac{13}{E_0'} - \frac{5}{E_0'+\Delta_0'}\right], \tag{6.20}$$

$$\gamma_{8v,2} = \frac{2\mathrm{i}}{3}PP'Q\frac{1}{E_0}\left[\frac{1}{E_0'} - \frac{1}{E_0'+\Delta_0'}\right], \tag{6.21}$$

$$\gamma_{8v,3} = \frac{2\mathrm{i}}{9}PP'Q\frac{1}{E_0}\left[\frac{1}{E_0'+\Delta_0'} - \frac{1}{E_0'}\right], \tag{6.22}$$

$$\gamma_{8v,4} = \frac{4\mathrm{i}}{9}PP'Q\frac{1}{E_0}\left[\frac{1}{E_0'} - \frac{1}{E_0'+\Delta_0'}\right], \tag{6.23}$$

$$\gamma_{7v} = \frac{4\mathrm{i}}{3}PP'Q\frac{1}{(E_0+\Delta_0)(\Delta_0+E_0'+\Delta_0')}, \tag{6.24}$$

$$\gamma_{8v7v,1} = \mathrm{i}PP'Q\left[\frac{1}{E_0(E_0'+\Delta_0')} + \frac{1}{(E_0+\Delta_0)(\Delta_0+E_0'+\Delta_0')}\right], \tag{6.25}$$

$$\gamma_{8v7v,2} = -\frac{1}{3}PP'Q\left[\frac{1}{E_0E_0'} - \frac{1}{E_0(E_0'+\Delta_0')} + \frac{1}{(E_0+\Delta_0)(\Delta_0+E_0')}\right.$$

$$\left. - \frac{1}{(E_0+\Delta_0)(\Delta_0+E_0'+\Delta_0')}\right], \tag{6.26}$$

$$\gamma_{8v7v,3} = \gamma_{8v7v,2}. \tag{6.27}$$

"If Γ_7^c and Γ_8^c states are included among the A states (e.g., by changing from an 8×8 model to a 14×14 model), the Kane off-diagonal coefficients must be modified from B_{8v}^{+}, B_{8v}^{-}, B_{7v} to:

$$B_{8v,1} = B_{8v}^{+} - \frac{P'Q}{2\mathrm{i}}\left[\frac{1}{E_0 - E_0' - \Delta_0'} - \frac{1}{E_0' + \Delta_0'} + \frac{1}{E_0 - E_0'} - \frac{1}{E_0'}\right], \quad (6.28)$$

$$B_{8v,2} = B_{8v}^{-} - \frac{P'Q}{2\mathrm{i}}\left[\frac{1}{E_0 - E_0' - \Delta_0'} - \frac{1}{E_0' + \Delta_0'} - \frac{1}{E_0 - E_0'} + \frac{1}{E_0'}\right], \quad (6.29)$$

$$B_{7v}' = B_{7v} - \frac{P'Q}{\mathrm{i}}\left[\frac{1}{E_0 - E_0' - \Delta_0'} - \frac{1}{\Delta_0 + E_0' + \Delta_0'}\right]. \quad (6.30)$$

Some general comments are worthwhile to mention at this point. In inversion-symmetric structures, P' is always zero as Γ_{6c}, Γ_{7c}, Γ_{8c} are even-parity states while the momentum operator is a vector and hence changes under inversion. All coefficients except $\gamma_{8v,1}$, γ_{7v}, $\gamma_{8v7v,1}$ are zero if the spin-orbit interactions Δ_0, Δ_0' vanish. Even though the cubic $\mathbf{k}$ terms associated with the latter three are nonzero in the absence of spin-orbit interaction, these terms do not lead to spin splittings as they only couple states with the same spin orientation and do so similarly for the two spin orientations. In other words, spin splittings in bulk structures also require non-vanishing spin-orbit interactions Δ_0, Δ_0'. The cubic terms that remain in the absence of spin-orbit interactions are possible only due to degeneracy at the Γ point of the Γ_{7v} and Γ_{8v} states since in this case the presence of cubic terms does not conflict with time-reversal symmetry. It is important to mention that all cubic spin-splitting terms in Eqs. (6.19), (6.20), (6.21), (6.22), (6.23), (6.24), (6.25), (6.26), and (6.27) disappear in multiband Hamiltonian entries if the Γ_{6c} states are included among the 'interesting' states (the so-called A states in Löwdin perturbation theory) we are solving for. Hence, in 8×8 or 14×14 matrix calculations where Γ_{6c} belong to the set of A states, only the Kane B terms and the linear C_k spin-splitting terms are present. In Table 6.1, we list examples of spin-splitting coefficients for the zincblende materials GaAs, AlAs, InAs, and InSb [96, 104]. Values in Table 6.1 reveal that, near the Γ point and for the valence band the non-vanishing cubic terms in the absence of spin-orbit interaction: $\gamma_{8v,1}$, γ_{7v}, $\gamma_{8v7v,1}$ are the most important together with the linear spin-splitting coeffcient C_k. It should also be mentioned that according to LMTO calculations the heavy-hole linear and cubic coefficients of GaAs along the [110] direction are of opposite signs thus the spin-splitting coefficient for the heavy-hole band becomes zero at a (small) but finite $\mathbf{k} = \mathbf{k_c}$. Obviously then, for $\mathbf{k} < \mathbf{k_c}$ ($\mathbf{k} > \mathbf{k_c}$), the linear (cubic) spin-splitting is the most important and determines the sign of the spin splittings. In contrast, as mentioned above, from $\mathbf{k}\cdot\mathbf{p}$ perturbation theory, the cubic coefficient becomes zero for heavy holes along the [110] direction."

An example of spin splitting in the valence band can be observed in the 14-band model of GaAs in Fig. 5.3. It can be seen that the bands are spin split along the [011] direction but not along the [001] one. The magnitude of the spin splitting reaches 50 meV.

Table 6.1 Spin-splitting coefficients in some zincblende materials

	GaAs	AlAs	InAs	InSb
γ_c	27.58	18.53	27.18	760.1
$\gamma_{8v,1}$	−81.93	−33.51	−50.18	−934.8
$\gamma_{8v,2}$	.47	0.526	1.26	41.73
$\gamma_{8v,3}$	0.49	0.175	0.42	13.91
$\gamma_{8v,4}$	−0.98	−0.35	−0.84	−27.82
γ_{7v}	−58.71	−27.27	−22.31	−146.8
$\gamma_{8v7v,1}$	−101.9	−44.30	−51.29	709.5
$\gamma_{8v7v,2}$	−1.255i	−0.474i	−0.910i	−23.92i
C_k	−0.0034	0.0020	−0.0112	−0.0082
$B_{8v,1}$	−21.32	−34.81	−3.393	−32.20
$B_{8v,2}$	−0.5175	−1.468	−0.09511	−1.662
B_{7v}	−20.24	−32.84	−3.178	−27.77

6.2.4 Sign of Spin-Splitting Coefficients

Special attention is needed when evaluating the signs of spin-splitting coefficients (both linear and cubic terms). Evidently, the signs of P, P', Q, Δ^-, Δ_0, and Δ_0' determine the sign of the cubic spin-splitting coefficient γ_c (and likewise for the other cubic spin-splitting coefficients). Cardona et al. [104] provided a detailed discussion of the signs of matrix elements of the momentum operator $\mathbf{p}$ and the spin-orbit Hamiltonian H_{so} based on wave function phase conventions. They chose the anion to be located at the origin and the cation at $a_0(1, 1, 1)/4$ which, with their wavefunction phase convention, leads to positive values of P, P', and Q. The opposite choice (cation at the origin) leads to a sign reversal of P'. The same wave function phase convention also fixes the sign of Δ^- determined by renormalized atomic anion and cation spin-orbit splittings. We note, however, that the signs of Δ_0, and Δ_0' are independent of the wave function phase convention rather they are fixed by the (well-known) signs of the renormalized atomic anion and cation spin-orbit splittings only.

The linear spin-splitting coefficients C_k are determined by combining linear-muffin-tin-orbital (LMTO) calculations along the [111, 110] directions with $k \cdot p$ results for the cubic spin-splitting coefficients. For example, the splittings of the light-hole and heavy-hole states, ΔE_{lh} and ΔE_{hh}, respectively, for $\mathbf{k}$ along [110] are given by [104]

$$\begin{aligned} \Delta E_{lh} &= \frac{3\sqrt{3}}{2} C_k k, \\ \Delta E_{hh} &= \frac{\sqrt{3}}{2} C_k k, \end{aligned} \tag{6.31}$$

provided the difference in the quadratic-energy terms (effective-mass terms) is larger than ΔE_{lh} and ΔE_{hh}. The sign of C_k is determined by LMTO calculations of the $E - k$ dispersion: As the cubic spin-splittings interfere with the linear ones, the knowledge of the cubic spin-splitting coefficient values with sign from $k \cdot p$ perturbation theory expressions allows the sign of the linear spin-splitting coefficient to be determined.

A recent paper by Chantis et al. [109] uses the self-consistent GW approximation to perform first-principles calculations of the conduction-band spin splitting in GaAs under [110] strain. The spin-orbit interaction is perturbatively accounted for and added to the scalar relativistic Hamiltonian using a full-potential LMTO method [110, 111]. The GW scheme employed in [109] is expected to provide accurate spin-splitting values since band parameters are accurately reproduced by the GW approximation.

6.3 Linear Spin Splittings in Wurtzite

Two types of linear-in-k terms are possible in the band structure near the Γ point. One results from the splitting of spin degenerate doublets, while the other is a linear splitting between states which remain spin degenerate. We note that both are allowed for a zincblende crystal for example, at the top of the valence band, along the [111] and [100] directions. The second possibility requires a minimum four-fold degeneracy. While such degeneracy may exist in zincblende, it is reduced in wurtzite to two-fold by the hexagonal crystal field (all wurtzite double-group irreducible representations are two-dimensional). We also note that linear-in-k spin splitting in the S conduction states is possible in wurtzite but not in zincblende.

A detailed discussion of linear spin splittings for the lower conduction band and the upper valence band in wurtzite bulk semiconductors can be found in [48]. In this section, we give the main $k \cdot p$ results. In wurtzite, the upper valence band consists of two Γ_7 bands and a Γ_9 band while the lowest conduction band has Γ_7 symmetry. We point out that the normal ordering (in energy) of the valence band states typically is: Γ_9, Γ_7, and Γ_7 for, e.g., CdS and CdSe but Γ_7, Γ_9, and Γ_7 for ZnO. The reversal ordering of ZnO is due to a negative spin-orbit splitting of the valence bands.

In addition to linear spin-splitting terms, third-order terms in $\mathbf{k}$ appear like for zincblende structures [112, 113]. We note that linear spin-splitting terms also appear in the lower conduction-band dispersion around the Γ point even in the absence of strain in contrast to the result found in the zincblende case.

In Table B.9, we give the character table (C_{6v}) at the Γ point of wurtzite and associated basis functions. We note that C_{6v} is the group of the wave vector along the [0001] direction. Since all double-group representations $\Gamma_7, \Gamma_8, \Gamma_9$ are two-dimensional, spin splittings vanish along that direction. As before, linear spin splittings arise due to the term $\hbar^2(\nabla V \times \mathbf{k})\cdot\boldsymbol{\sigma}/(4m_0^2c^2)$ by use of first-order perturbation theory or the coupling between $\hbar\mathbf{k}\cdot\mathbf{p}/m_0$ and $\hbar(\nabla V \times \mathbf{p})\cdot\boldsymbol{\sigma}/(4m_0^2c^2)$ in second-order perturbation theory.

Before calculating spin-splitting results, we point out that the conduction and valence basis states for wurtzite, accounting properly for s, p, d state mixing, are:

$$
\begin{aligned}
|e\uparrow\rangle &= q_s|s\uparrow\rangle + q_z|z\uparrow\rangle,\\
|Z\uparrow\rangle &= -q_z|s\uparrow\rangle + q_s|z\uparrow\rangle,\\
|e\downarrow\rangle &= q_s|s\downarrow\rangle + q_z|z\downarrow\rangle,\\
|Z\downarrow\rangle &= -q_z|s\downarrow\rangle + q_s|z\downarrow\rangle,\\
|7\uparrow\rangle &= \sqrt{1-q_7^2}|u_5\downarrow\rangle - q_7|u_1\uparrow\rangle,\\
|9\uparrow\rangle &= \sqrt{1-q_9^2}|u_5\uparrow\rangle + q_9|u_6\downarrow\rangle,\\
|7\downarrow\rangle &= -\sqrt{1-q_7^2}|u_5^*\uparrow\rangle - q_7|u_1\downarrow\rangle,\\
|9\downarrow\rangle &= \sqrt{1-q_9^2}|u_5^*\downarrow\rangle - q_9|u_6\uparrow\rangle,\\
|7'\uparrow\rangle &= q_7|u_5\downarrow\rangle + \sqrt{1-q_7^2}|u_1\uparrow\rangle,\\
|9'\uparrow\rangle &= -q_9|u_5\uparrow\rangle + \sqrt{1-q_9^2}|u_6\downarrow\rangle,\\
|7'\downarrow\rangle &= -q_7|u_5^*\downarrow\rangle + \sqrt{1-q_7^2}|u_1\downarrow\rangle,\\
|9'\downarrow\rangle &= q_9|u_5^*\uparrow\rangle - \sqrt{1-q_9^2}|u_6\uparrow\rangle,
\end{aligned}
\tag{6.32}
$$

where

$$
\begin{aligned}
u_1 &= Z, u_3 = x\left(x^2-3y^2\right),\\
u_5 &= \frac{1}{\sqrt{2}}(x+\mathrm{i}y), u_6 = \frac{1}{2}(x+\mathrm{i}y)^2.
\end{aligned}
\tag{6.33}
$$

A popular approximation is to use the quasi-cubic model [49, 56] where the wurtzite crystal is assumed to be a zincblende one strained along the [111] direction. The quasi-cubic model is obtained from the above set of basis states by setting $q_s = 1, q_z = 0, q_7 = 0, q_9 = 0$ while the zincblende limit is obtained for $q_7^2 = 2/3$. We shall show below, however, that this approximation leads to significant errors in the discussion of spin splittings. In the following, in accordance with [48], the wurtzite valence A, B, C states are represented by the Γ_9, Γ_7, Γ_7' states, respectively.

Since (k_x, k_y) and (σ_x, σ_y) belong to the Γ_5 representation, the only possible spin-splitting linear-in-k Hamiltonian invariant is [104, 114]

$$
H_k \propto \left(k_x\sigma_y - k_y\sigma_x\right). \tag{6.34}
$$

Note that a linear-in-k_z spin-splitting Hamiltonian invariant does not exist in wurtzite.

Next, we combine each state of a given symmetry with its Kramers degenerate partner (at the Γ point) uniquely defined by

$$\mathcal{K}|u_i \uparrow\rangle = |u_i^* \downarrow\rangle,$$
$$\mathcal{K}|u_i \downarrow\rangle = -|u_i^* \uparrow\rangle, \quad (6.35)$$

with $\mathcal{K}$ defined in Eq. (3.54), and find the spin-split Hamiltonian of the Γ_m ($m = 9, 7, 7'$) states to be, using Eq. (6.34),

$$H_m(\mathbf{k}) = C_m \begin{pmatrix} 0 & \mathrm{i}(k_x - \mathrm{i}k_y) \\ -\mathrm{i}(k_x + \mathrm{i}k_y) & 0 \end{pmatrix}, \quad (6.36)$$

where C_m denotes the associated Γ_m spin-splitting coefficient. The spin-split energies E_m are found immediately from Eq. (6.36):

$$E_m = C_m \left(k_x^2 + k_y^2\right)^{1/2}. \quad (6.37)$$

Having obtained the general form of the linear-in-k spin-splitting Hamiltonian invariant, we can pursue with perturbation theory to determine the missing coefficient expressions.

6.3.1 Lower Conduction-Band e States

It is easy to show that the first-order perturbation contribution from $\hbar^2(\nabla V \times \mathbf{k}) \cdot \boldsymbol{\sigma}/(4m_0^2c^2)$ does not contribute to the spin splittings of the lower conduction-band states. Since $|S\rangle$ are eigenstates of the Hamiltonian at the Γ point in the absence of spin-orbit interaction, the following identity applies:

$$\langle S|\nabla V|S\rangle \propto \langle S|[H_0, \mathbf{p}]|S\rangle = (E_S - E_S)\langle S|\mathbf{p}|S\rangle = 0, \quad (6.38)$$

and matrix elements of $\hbar^2(\nabla V \times \mathbf{k}) \cdot \boldsymbol{\sigma}/(4m_0^2c^2)$ between S states vanish.

Next, coupling the electron Γ_7 states to the valence A, B, C states (denoted $\overline{L}$) using second-order perturbation theory yields a spin splitting

$$H_{m,2} = \frac{\hbar}{m_0} \sum_{\overline{L} \neq L} \frac{\langle L|H_{\mathrm{so}}|\overline{L}\rangle\langle\overline{L}|\mathbf{k}\cdot\mathbf{p}|L\rangle + \text{c.c.}}{E_L - E_{\overline{L}}}, \quad (6.39)$$

where c.c. denotes complex conjugation. This is a very good approximation since other (remote) bands are far away [48].

Hence, within the $eABC$ model (the notation e, A, B, C refers to the lowest conduction band and the topmost three valence bands in decreasing order of energy, respectively) we obtain linear spin-splitting contributions from Γ_7 and $\Gamma_{7'}$ (note that both Γ_7 and $\Gamma_{7'}$ states transform according to the Γ_7 representation) while Γ_9 does not contribute (the latter follows since $\langle S|H_{\mathrm{so}}|\Gamma_9\rangle \sim \Gamma_7 \otimes \Gamma_1 \otimes \Gamma_9 = 0$):

$$C_e = C_{e,7} + C_{e,7'}, \tag{6.40}$$

$$C_{e,7} = \frac{1 - q_7^2}{E_e - E_7} q_z q_s \Delta_{x,z} P_{x,s}, \tag{6.41}$$

$$C_{e,7'} = \frac{q_7^2}{E_e - E_{7'}} q_z q_s \Delta_{x,z} P_{x,s}, \tag{6.42}$$

where

$$\Delta_{x,z} = \frac{\hbar^2}{m_0^3 c^2} \langle x | \nabla_z V p_x - \nabla_x V p_z | z \rangle, \tag{6.43}$$

$$P_{x,s} = -\mathrm{i}\hbar \langle s | p_x | x \rangle. \tag{6.44}$$

Note that $C_{e,7}$ and $C_{e,7'}$ have the same sign and would be zero if there were no $s - p_z$ mixing.

6.3.2 *A, B, C* Valence States

For the valence-band states, we notice that first-order perturbation theory in $H_{\mathrm{so},k} = \hbar^2(\nabla V \times \mathbf{k}) \cdot \boldsymbol{\sigma}/(4m_0^2c^2)$ now leads to linear-in-k spin splittings. Using, Eq. (6.32), we find

$$C_{7,1} = -C_{7',1} = \frac{\sqrt{2}\hbar^2}{4m_0^2c^2} q_7 \sqrt{1 - q_7^2} \left\langle x \left| \frac{\partial V}{\partial x} \right| z \right\rangle. \tag{6.45}$$

Γ_9 linear-in-k spin-splittings are symmetry forbidden.

Considering next second-order perturbation theory contributions, we obtain for the spin-splitting coefficients of the valence Γ_7 band [48],

$$C_{7v} = C_{7,e} + C_{7,7'}, \tag{6.46}$$

with

$$C_{7,e} = -C_{e,7}, \tag{6.47}$$

and

$$C_{7,7'} = \frac{1}{\sqrt{2}} q_z \frac{1 - 2q_7^2}{E_7 - E_{7'}} P_{x,s} \left(q_7 \sqrt{1 - q_7^2} \Delta_{x,y} - \sqrt{2} q_s (1 - 2q_7^2) \Delta_{x,z} \right), \tag{6.48}$$

The expression for $\Delta_{x,y}$ is the same as for $\Delta_{x,z}$ in Eq. (6.43) with z replaced by y. Further, the $C_{7'v}$ coefficient becomes

$$C_{7'v} = C_{7',e} + C_{7',7} = -C_{7v}, \tag{6.49}$$

since

$$C_{7',e} = -C_{7,e}, \tag{6.50}$$

$$C_{7',7} = -C_{7,7'}. \tag{6.51}$$

In words, the $\Gamma_7 - \Gamma_{7'}$ interaction is equal and opposite for the Γ_7 and $\Gamma_{7'}$ states and arises due to the mixing of s-like states into the Γ_7 and $\Gamma_{7'}$ states (proportional to q_z).

6.3.3 Linear Spin Splitting

Compiling the various linear spin-splitting contributions within the $eABC$ model, we have

$$C_e \sim \left[\frac{1-q_7^2}{E_e - E_7} + \frac{q_7^2}{E_e - E_{7'}}\right]\frac{\mathcal{A}}{E_e - E_5}, \tag{6.52}$$

$$C_7 \sim -\frac{1-q_7^2}{E_e - E_7}\frac{\mathcal{A}}{E_e - E_5} + \frac{\mathcal{B}}{(E_7 - E_{7'})(E_e - E_5)} + \mathcal{C}, \tag{6.53}$$

$$C_{7'} \sim -\frac{q_7^2}{E_e - E_7}\frac{\mathcal{A}}{E_e - E_5} - \frac{\mathcal{B}}{(E_7 - E_{7'})(E_e - E_5)} - \mathcal{C}, \tag{6.54}$$

where $\mathcal{A}, \mathcal{B}, \mathcal{C}$ are constants approximately independent of the band-edge energies and E_5 is the Γ_5 (single-group) energy at $\mathbf{k} = \mathbf{0}$. The constant $\mathcal{A}$ is positive for semiconductors with normal ordering of the valence states while the term containing $\mathcal{B}$ is expected to be the largest one and that containing $\mathcal{C}$ the smallest in magnitude. A detailed discussion on the sign assignments of wurtzite linear spin-splitting coefficients is given in [48].

A combination of LMTO ab initio and $k \cdot p$ band structure calculations was carried out in [48]. The main results are listed in Table 6.2. CdS(u) and CdS(a) refer to $k \cdot p$ unadjusted and adjusted LMTO results, respectively. It has been shown (Table VIII in [48]) that $k \cdot p$ adjusted LMTO results are in good agreement with experimental results.

Table 6.2 Linear spin-splitting coefficients of bulk CdS and CdSe wurtzite

	CdS(u)	CdS(a)	CdSe(u)	CdSe(a)
$C_{7'}$(LMTO + $k \cdot p$)	+145	+70	+425	+133
$C_{7'}$(LMTO)	100	27	450	95

Adapted with permission from [48]. ©1996 by the American Physical Society

6.4 Summary

The theory of spin splitting at the zone-center of the Brillouin zone for both zincblende and wurtzite was given. Spin splitting results from a lack of inversion symmetry and from the spin-orbit coupling. For zincblende, the lowest (s-type) conduction band only has cubic splittings near the zone center, while the valence band has linear and cubic splittings. There is no spin splitting along Δ and a splitting along Λ for $\Lambda_{4,5}$ but not for Λ_6 [104]. For wurtzite, the lowest conduction band, if of Γ_7 type, does have linear splittings as well due to the admixture of p_z character into that state. Thus, models that do not use such extended basis states are unable to describe the linear spin splitting. Such linear spin splitting is not present for Γ_9 states. Finally, there is no spin splitting along the k_z direction.

Chapter 7
Strain

7.1 Overview

Experimentally, the application of an external stress on a bulk cubic semiconductor is known to lead to a modification of the band gap (with hydrostatic pressure) and/or a splitting of the heavy-hole–light-hole degeneracy (with uniaxial or biaxial stress). The $k \cdot p$ theory is known as a deformation-potential theory. Early works on the influence of homogeneous strain on electronic levels (critical points in diamond- and zincblende-type semiconductors) can be found in [1, 115–121]. In this chapter, we will discuss how the previously introduced $k \cdot p$ Hamiltonians need to be modified in order to incorporate strain effects. The resulting band structures will then be analyzed for a few cases.

7.2 Perturbation Theory

We first consider the effect of small homogeneous strains on the band structure by deriving the new Hamiltonian for a strained crystal.

7.2.1 Strain Hamiltonian

The following procedure was proposed by Bir and Pikus [1]. We need to first define the strain tensor. Let $\mathbf{x}'$ be the new coordinates in a deformed but still periodic crystal. In linear approximation, we define the strain tensor ε via

$$x_i' = x_i + \varepsilon_{ij} x_j. \tag{7.1}$$

It is known that the strain tensor is symmetric. Differentiating Eq. (7.1), we have

$$\frac{\partial x_i'}{\partial x_k'} = \delta_{ik} = (\delta_{ij} + \varepsilon_{ij}) \frac{\partial x_j}{\partial x_k'} = \left[(1+\varepsilon) \cdot (\mathbf{x}/\mathbf{x}')\right]_{ik},$$

or

L.C. Lew Yan Voon, M. Willatzen, *The* k · p *Method*,
DOI 10.1007/978-3-540-92872-0_7, © Springer-Verlag Berlin Heidelberg 2009

$$(\mathbf{x}/\mathbf{x}') = \frac{1}{(1+\varepsilon)}. \tag{7.2}$$

We now consider the transformation of the Schrödinger equation [Eq. (2.5)]

$$\left[\frac{p^2}{2m_0} + V(\mathbf{r}) + \frac{\hbar}{4m_0^2c^2}(\sigma \times \nabla V(\mathbf{r}))\cdot\mathbf{p}\right]\psi_{n\mathbf{k}}(\mathbf{r}) = E_n(\mathbf{k})\psi_{n\mathbf{k}}(\mathbf{r}),$$

under the strain. In the new coordinate system, we have

$$\left[\frac{p'^2}{2m_0} + V(\mathbf{r}') + \frac{\hbar}{4m_0^2c^2}\left(\sigma \times \nabla' V(\mathbf{r}')\right)\cdot\mathbf{p}'\right]\psi_{n\mathbf{k}}(\mathbf{r}') = E_n(\mathbf{k})\psi_{n\mathbf{k}}(\mathbf{r}'). \tag{7.3}$$

The goal is to re-express Eq. (7.3) in terms of unstrained coordinates and new terms linear in the strain tensor. In keeping with the unstrained problem, we first obtain the corresponding equation satisfied by the cellular function; it is clear that this is similar to the unstrained problem. Thus, if

$$\psi_{n\mathbf{k}}(\mathbf{r}') = \mathrm{e}^{\mathrm{i}\mathbf{k}\cdot\mathbf{r}'}u_{n\mathbf{k}}(\mathbf{r}'), \tag{7.4}$$

then

$$\left[\frac{p'^2}{2m_0} + V(\mathbf{r}') + \frac{\hbar}{4m_0^2c^2}\left(\sigma \times \nabla' V(\mathbf{r}')\right)\cdot\mathbf{p}' + \frac{\hbar}{m_0}\mathbf{k}\cdot\left(\mathbf{p}' + \frac{\hbar}{4m_0c^2}\left[\sigma \times \nabla' V(\mathbf{r}')\right]\right)\right]u_{n\mathbf{k}}(\mathbf{r}')$$
$$= \left[E_n(\mathbf{k}) - \frac{\hbar^2k^2}{2m_0^2}\right]u_{n\mathbf{k}}(\mathbf{r}'). \tag{7.5}$$

We now need a number of results:

$$\begin{aligned} V(\mathbf{r}') &= V(\mathbf{r}) + (\mathbf{r}' - \mathbf{r})\cdot\nabla' V(\mathbf{r}')|_{\mathbf{r}'=\mathbf{r}} + \cdots \\ &= V(\mathbf{r}) + \varepsilon_{ij}x_j\partial_i' V(\mathbf{r}')|_{\mathbf{r}'=\mathbf{r}} \equiv V(\mathbf{r}) + \varepsilon_{ij}V_{ij}(\mathbf{r}). \end{aligned} \tag{7.6}$$

Also, since

$$\frac{\partial}{\partial x_i'} = \frac{\partial x_j}{\partial x_i'}\frac{\partial}{\partial x_j} = \frac{\partial}{\partial x_i'}\left(x_j' - \varepsilon_{jk}x_k\right)\frac{\partial}{\partial x_j} \approx \frac{\partial}{\partial x_i} - \varepsilon_{ji}\frac{\partial}{\partial x_j} = \frac{\partial}{\partial x_i} - \varepsilon_{ij}\frac{\partial}{\partial x_j},$$

then

$$\begin{aligned} p_i' &\approx p_i - \varepsilon_{ij}p_j, \\ p'^2 &\approx p^2 - 2\varepsilon_{ij}p_ip_j. \end{aligned} \tag{7.7}$$

Thus

$$\left(\nabla' V(\mathbf{r}') \times \mathbf{p}'\right)_i = \varepsilon_{ijk}\frac{\partial V(\mathbf{r}')}{\partial x'_j}p'_k = \varepsilon_{ijk}\left[\frac{\partial V(\mathbf{r}')}{\partial x_j} - \varepsilon_{jp}\frac{\partial V(\mathbf{r}')}{\partial x_p}\right][p_k - \varepsilon_{kn}p_n]$$

$$\approx \varepsilon_{ijk}\left[\frac{\partial V(\mathbf{r})}{\partial x_j} + \varepsilon_{lm}\frac{\partial V_{lm}(\mathbf{r})}{\partial x_j} - \varepsilon_{jp}\frac{\partial V(\mathbf{r})}{\partial x_p}\right][p_k - \varepsilon_{kn}p_n]$$

$$\approx \varepsilon_{ijk}\left[\frac{\partial V}{\partial x_j}p_k + \varepsilon_{lm}\frac{\partial V_{lm}}{\partial x_j}p_k - \varepsilon_{jp}\frac{\partial V}{\partial x_p}p_k - \varepsilon_{kn}\frac{\partial V}{\partial x_j}p_n + \cdots\right]$$

$$= (\nabla V(\mathbf{r}) \times \mathbf{p})_i + \varepsilon_{lm}\varepsilon_{ijk}\frac{\partial V_{lm}}{\partial x_j}p_k - \varepsilon_{jp}\varepsilon_{ijk}\frac{\partial V}{\partial x_p}p_k - \varepsilon_{kn}\varepsilon_{ijk}\frac{\partial V}{\partial x_j}p_n.$$

The last term can be modified:

$$-\varepsilon_{kn}\varepsilon_{ijk}\frac{\partial V}{\partial x_j}p_n = +\varepsilon_{kp}\varepsilon_{ikj}\frac{\partial V}{\partial x_j}p_p = +\varepsilon_{jp}\varepsilon_{ijk}\frac{\partial V}{\partial x_k}p_p,$$

and

$$\left(\nabla' V(\mathbf{r}') \times \mathbf{p}'\right)_i$$

$$= (\nabla V(\mathbf{r}) \times \mathbf{p})_i + \varepsilon_{lm}\varepsilon_{ijk}\frac{\partial V_{lm}}{\partial x_j}p_k - \varepsilon_{jp}\varepsilon_{ijk}\frac{\partial V}{\partial x_p}p_k + \varepsilon_{jp}\varepsilon_{ijk}\frac{\partial V}{\partial x_k}p_p$$

$$= (\nabla V(\mathbf{r}) \times \mathbf{p})_i + \varepsilon_{lm}\left[\varepsilon_{ilk}\left(\frac{\partial V}{\partial x_k}p_m - \frac{\partial V}{\partial x_m}p_k\right) + \varepsilon_{ijk}\frac{\partial V_{lm}}{\partial x_j}p_k\right]$$

$$= (\nabla V(\mathbf{r}) \times \mathbf{p})_i + \varepsilon_{kl}\left[\varepsilon_{ijk}\left(\frac{\partial V}{\partial x_l}p_j - \frac{\partial V}{\partial x_j}p_l\right) + \varepsilon_{ijn}\frac{\partial V_{lk}}{\partial x_j}p_n\right]. \tag{7.8}$$

The operator on the LHS of Eq. (7.5) becomes

$$\frac{p^2}{2m_0} - \varepsilon_{ij}\frac{p_i p_j}{m_0} + V + \varepsilon_{ij}V_{ij} \quad + \frac{\hbar}{4m_0^2c^2}\boldsymbol{\sigma}\cdot(\nabla V \times \mathbf{p})$$

$$+\frac{\hbar}{4m_0^2c^2}\sigma_i\varepsilon_{kl}\left[\varepsilon_{ijk}\left(\frac{\partial V}{\partial x_l}p_j - \frac{\partial V}{\partial x_j}p_l\right) + \varepsilon_{ijn}\frac{\partial V_{lk}}{\partial x_j}p_n\right]$$

$$+\frac{\hbar}{m_0}\mathbf{k}\cdot\mathbf{p} - \frac{\hbar}{m_0}k_i\varepsilon_{ij}p_j + \frac{\hbar^2}{4m_0^2c^2}\mathbf{k}\cdot\left[\boldsymbol{\sigma}\times\nabla V + \boldsymbol{\sigma}\times\nabla V_{ij}\varepsilon_{ij}\right] - \frac{\hbar^2}{4m_0^2c^2}k_i\varepsilon_{ij}\boldsymbol{\sigma}\times\frac{\partial V}{\partial x_j}$$

$$= \frac{p^2}{2m_0} + V + \frac{\hbar}{4m_0^2c^2}\boldsymbol{\sigma}\cdot(\nabla V \times \mathbf{p}) + \frac{\hbar}{m_0}\mathbf{k}\cdot\left[\mathbf{p} + \frac{\hbar}{4m_0c^2}(\boldsymbol{\sigma}\times\nabla V)\right]$$

$$-\varepsilon_{ij}\frac{p_i p_j}{m_0} + \varepsilon_{ij}V_{ij} + \frac{\hbar}{4m_0^2c^2}\varepsilon_{kl}\sigma_i\left[\varepsilon_{ijk}\left(\frac{\partial V}{\partial x_l}p_j - \frac{\partial V}{\partial x_j}p_l\right) + \varepsilon_{ijn}\frac{\partial V_{lk}}{\partial x_j}p_n\right]$$

$$-\varepsilon_{ij}\frac{\hbar}{m_0}k_i p_j + \varepsilon_{ij}\frac{\hbar^2}{4m_0^2c^2}\mathbf{k}\cdot\left(\boldsymbol{\sigma}\times\nabla V_{ij}\right) - \frac{\hbar^2}{4m_0^2c^2}k_i\varepsilon_{ij}\boldsymbol{\sigma}\times\frac{\partial V}{\partial x_j}$$

$$= \frac{p^2}{2m_0} + V + \frac{\hbar}{4m_0^2c^2}\boldsymbol{\sigma}\cdot(\nabla V\times\mathbf{p}) + \frac{\hbar}{m_0}\mathbf{k}\cdot\left[\mathbf{p} + \frac{\hbar}{4m_0c^2}(\boldsymbol{\sigma}\times\nabla V)\right] + \varepsilon_{kl}D_{kl}, \tag{7.9}$$

where

$$D_{kl} = -\frac{p_k p_l}{m_0} + V_{kl} - \frac{\hbar}{m_0}k_k p_l - \frac{\hbar^2}{4m_0^2c^2}k_i\boldsymbol{\sigma}\times\frac{\partial V}{\partial x_j} \tag{7.10}$$

$$+\frac{\hbar}{4m_0^2c^2}\sigma_i\left[\varepsilon_{ijk}\left(\frac{\partial V}{\partial x_l}p_j - \frac{\partial V}{\partial x_j}p_l\right) + \varepsilon_{ijn}\frac{\partial V_{lk}}{\partial x_j}p_n\right] + \frac{\hbar^2}{4m_0^2c^2}\mathbf{k}\cdot(\boldsymbol{\sigma}\times\nabla V_{kl}).$$

Equation (7.9) gives the new Hamiltonian in the presence of linear, homogeneous strain. It reveals that strain allows for additional band coupling via both the orbital and spin-orbit terms. The strain Hamiltonian, Eq. (7.10), can be simplified if one neglects spin-orbit effects in the strain energy. Note, also, that the third term on the RHS is zero at an extremum. Then

$$D_{kl} \approx -\frac{p_k p_l}{m_0} + V_{kl}. \tag{7.11}$$

7.2.2 Löwdin Renormalization

One can now apply second-order perturbation theory in order to reveal the impact of the strain on the $k\cdot p$ Hamiltonian. To linear order in the strain, the renormalized matrix is given by [122]:

$$U_{mn} = H'_{mn} + \sum_{\alpha\in B}\frac{H'_{m\alpha}H'_{\alpha n}}{E - E_\alpha} + D_{mn} + \sum_{\alpha\in B}\frac{H'_{m\alpha}D'_{\alpha n} + D'_{m\alpha}H'_{\alpha n}}{E - E_\alpha}, \tag{7.12}$$

where the first two terms on the RHS are the conventional $k\cdot p$ terms and the last term couples states in class A and B via the strain; the latter effect is usually neglected.

There can, of course, be other contributions in higher orders of perturbation theory. For instance, one can develop terms that are linear in strain and quadratic in wave vector in third-order perturbation theory; they correspond to the strain dependence of the effective mass [123].

7.3 Valence Band of Diamond

Since the strain tensor is symmetric, its transformation properties are identical to $\{k_i k_j\}$. Thus, the extra strain Hamiltonian, H_ε, has the same form as the $k\cdot p$ one with the replacement $\{k_i k_j\} \to \varepsilon_{ij}$.

7.3.1 DKK Hamiltonian

Starting from the DKK Hamiltonian, Eq. (3.4), the strain Hamiltonian is [124]

$$H_\varepsilon = \begin{pmatrix} l\varepsilon_{xx} + m(\varepsilon_{yy} + \varepsilon_{zz}) & n\varepsilon_{xy} & n\varepsilon_{xz} \\ n\varepsilon_{xy} & l\varepsilon_{yy} + m(\varepsilon_{zz} + \varepsilon_{xx}) & n\varepsilon_{yz} \\ n\varepsilon_{xz} & n\varepsilon_{yz} & l\varepsilon_{zz} + m(\varepsilon_{xx} + \varepsilon_{yy}) \end{pmatrix}, \quad (7.13)$$

where l, m, n are the new strain parameters known as the valence-band deformation potentials.

7.3.2 Four-Band Bir–Pikus Hamiltonian

There are many forms of the $k \cdot p$ Hamiltonian with spin one can use. Following Bir and Pikus [1], we consider the four-band one but with our phase convention. This leads to the following $k \cdot p$ Hamiltonian (taken from Table 3.6):

$$H(\mathbf{k}) = \begin{pmatrix} |\frac{3}{2}\,\frac{3}{2}\rangle & |\frac{3}{2}\,\frac{1}{2}\rangle & |\frac{3}{2}\,-\frac{1}{2}\rangle & |\frac{3}{2}\,-\frac{3}{2}\rangle \\ F & H & I & 0 \\ H^* & G & 0 & -I \\ I^* & 0 & G & H \\ 0 & -I^* & H^* & F \end{pmatrix}, \quad (7.14)$$

where

$$\begin{aligned} F &= \frac{1}{2}(L+M)(k_x^2+k_y^2) + Mk_z^2 + \frac{\hbar^2 k^2}{2m_0}, \\ G &= \frac{1}{3}F + \frac{2}{3}[M(k_x^2+k_y^2) + Lk_z^2] + \frac{\hbar^2 k^2}{2m_0}, \\ H &= -\frac{N}{\sqrt{3}}(k_x - \mathrm{i}k_y)k_z, \\ I &= -\frac{1}{2\sqrt{3}}\left[(L-M)(k_x^2-k_y^2) - 2\mathrm{i}Nk_xk_y\right]. \end{aligned}$$

Then, the corresponding strain Hamiltonian is

$$H_\varepsilon = \begin{pmatrix} f & h & i & 0 \\ h^* & g & 0 & -i \\ i^* & 0 & g & h \\ 0 & -i^* & h^* & f \end{pmatrix}, \quad (7.15)$$

where

$$
\begin{aligned}
f &= \frac{1}{2}(l+m)(\varepsilon_{xx}+\varepsilon_{yy}) + m\varepsilon_{zz}, \\
g &= \frac{1}{3}f + \frac{2}{3}[m(\varepsilon_{xx}+\varepsilon_{yy}) + l\varepsilon_{zz}], \\
h &= -\frac{n}{\sqrt{3}}(\varepsilon_{xz} - \mathrm{i}\varepsilon_{yz}), \\
i &= -\frac{1}{2\sqrt{3}}\left[(l-m)(\varepsilon_{xx}-\varepsilon_{yy}) - 2\mathrm{i}n\varepsilon_{xy}\right].
\end{aligned}
\tag{7.16}
$$

Here we have the same three strain deformation potentials as in the DKK model.

7.3.3 Six-Band Hamiltonian

We can write the six-band Hamiltonian in the notation of the Burt–Foreman Hamiltonian to be derived in Sect. 12.4. We give it here for completeness. From Table 12.4, we have

$$
H_\varepsilon = \begin{pmatrix}
\left|\frac{3}{2}\ \frac{3}{2}\right\rangle & \left|\frac{3}{2}\ \frac{1}{2}\right\rangle & \left|\frac{3}{2}\ -\frac{1}{2}\right\rangle & \left|\frac{3}{2}\ -\frac{3}{2}\right\rangle & \left|\frac{1}{2}\ \frac{1}{2}\right\rangle & \left|\frac{1}{2}\ -\frac{1}{2}\right\rangle \\
p' & s & -r & 0 & -\frac{1}{\sqrt{2}}s & -\sqrt{2}r \\
s^* & p'' & 0 & r & -\sqrt{2}q & -\sqrt{\frac{3}{2}}s \\
-r^* & 0 & p'' & s & \sqrt{\frac{3}{2}}s^* & -\sqrt{2}q \\
0 & r^* & s^* & p' & \sqrt{2}r^* & -\frac{1}{\sqrt{2}}s^* \\
-\frac{1}{\sqrt{2}}s^* & -\sqrt{2}q^* & \sqrt{\frac{3}{2}}s & -\sqrt{2}r & p''' & 0 \\
\sqrt{2}r^* & -\sqrt{\frac{3}{2}}s^* & \sqrt{2}q^* & -\frac{1}{\sqrt{2}}s & 0 & p'''
\end{pmatrix},
\tag{7.17}
$$

where

$$
\begin{aligned}
p' &= \frac{1}{2}(l+m)(\varepsilon_{xx}+\varepsilon_{yy}) + m\varepsilon_{zz}, \\
p'' &= \frac{1}{6}(l+5m)(\varepsilon_{xx}+\varepsilon_{yy}) + \frac{1}{3}(2l+m)\varepsilon_{zz}, \\
p''' &= \frac{1}{3}(l+2m)(\varepsilon_{xx}+\varepsilon_{yy}+\varepsilon_{zz}), \\
q &= -\frac{1}{6}(l-m)(\varepsilon_{xx}+\varepsilon_{yy}+2\varepsilon_{zz}), \\
r &= \frac{1}{2\sqrt{3}}(l-m)(\varepsilon_{xx}-\varepsilon_{yy}) - \frac{\mathrm{i}}{\sqrt{3}}n\varepsilon_{xy}, \\
s &= -\frac{1}{\sqrt{3}}(f-g+h_1-h_2)(\varepsilon_{xz}-\mathrm{i}\varepsilon_{yz}).
\end{aligned}
\tag{7.18}
$$

For example,

$$
\begin{aligned}
p' &= \left\langle \frac{3}{2}\ \frac{3}{2} | H_\varepsilon | \frac{3}{2}\ \frac{3}{2} \right\rangle = \frac{1}{2} \langle (x+\mathrm{i}y) \uparrow | H_\varepsilon | (x+\mathrm{i}y) \uparrow \rangle \\
&= \frac{1}{2} (\langle x | H_\varepsilon | x \rangle + \langle y | H_\varepsilon | y \rangle) = \frac{1}{2} \left[(l+m)(\varepsilon_{xx} + \varepsilon_{yy}) + 2m\varepsilon_{zz} \right].
\end{aligned}
$$

We can also extend the four-band model to include all six bands of Table 3.6. In this case, we change the notation to be related to, e.g, Chuang's version [15]:

$$
H_\varepsilon = \begin{pmatrix}
\left|\frac{3}{2}\ \frac{3}{2}\right\rangle & \left|\frac{3}{2}\ \frac{1}{2}\right\rangle & \left|\frac{3}{2}\ -\frac{1}{2}\right\rangle & \left|\frac{3}{2}\ -\frac{3}{2}\right\rangle & \left|\frac{1}{2}\ \frac{1}{2}\right\rangle & \left|\frac{1}{2}\ -\frac{1}{2}\right\rangle \\
p+q & s & r & 0 & -\frac{1}{\sqrt{2}}s & \sqrt{2}r \\
s^* & p-q & 0 & -r & \sqrt{2}q & -\sqrt{\frac{3}{2}}s \\
r^* & 0 & p-q & s & \sqrt{\frac{3}{2}}s^* & \sqrt{2}q \\
0 & -r^* & s^* & p+q & -\sqrt{2}r^* & -\frac{1}{\sqrt{2}}s^* \\
-\frac{1}{\sqrt{2}}s^* & \sqrt{2}q^* & \sqrt{\frac{3}{2}}s & -\sqrt{2}r & p & 0 \\
\sqrt{2}r^* & -\sqrt{\frac{3}{2}}s^* & \sqrt{2}q^* & -\frac{1}{\sqrt{2}}s & 0 & p
\end{pmatrix}, \tag{7.19}
$$

where

$$
\begin{aligned}
p &= a\mathrm{Tr}\varepsilon, \\
q &= b(\varepsilon_{zz} - \frac{1}{2}\varepsilon_{xx} - \frac{1}{2}\varepsilon_{yy}), \\
r &= b\frac{\sqrt{3}}{2}(\varepsilon_{xx} - \varepsilon_{yy}) - d\mathrm{i}\varepsilon_{xy}, \\
s &= d(\varepsilon_{xz} - \mathrm{i}\varepsilon_{yz}).
\end{aligned} \tag{7.20}
$$

There are still three parameters, now written as a, b, and d. The three types of strain correspond to hydrostatic, uniaxial, and shear strain, respectively, as we will see in Sect. 7.4.

We note that Eq. (7.19) only differs marginally from Eq. (7.17) in the following replacements:

$$
p+q \leftrightarrow p',\ p-q \leftrightarrow p'',\ p \leftrightarrow p''',\ r \leftrightarrow -r,\ q \leftrightarrow -q.
$$

Thus, comparing the two forms of the six-band strain Hamiltonian allows us to relate the two sets of deformation potentials. For example

$$
\begin{aligned}
p' &= \frac{1}{2}(l+m)(\varepsilon_{xx}+\varepsilon_{yy}) + m\varepsilon_{zz} \\
&= p + q = a\mathrm{Tr}\varepsilon + b(\varepsilon_{zz} - \frac{1}{2}\varepsilon_{xx} - \frac{1}{2}\varepsilon_{yy}), \\
=> l &= a - 2b, m = a + b.
\end{aligned}
$$

We, therefore, get the correspondence given in Table 7.1.

Table 7.1 Comparison of the a, b, d deformation potentials to the l, m, n set

$l = a - 2b$
$m = a + b$
$n = -\sqrt{3}d$

7.3.4 Method of Invariants

We will now use the method of invariants to obtain the strain Hamiltonian. This follows from the fact that the strain tensor is symmetric; hence, its transformation properties are identical to $\{k_i k_j\}$. The extra terms can, therefore, be seen to mimick the $k \cdot p$ terms. We illustrate using the cubic case.

7.3.4.1 *LS* Basis

Following Eq. (5.6), the three new terms, in LS basis, can be written as

$$
\varepsilon_{kl} D_{kl} = a \sum_i \varepsilon_{ii} + 3b \sum_i \varepsilon_{ii}(I_i^2 - \frac{1}{3}I^2) + 2d\sqrt{3} \sum_{i<j} \varepsilon_{ij}\{I_i I_j\}, \tag{7.21}
$$

where

$$
\{I_i I_j\} = \frac{1}{2}(I_i I_j + I_j I_i).
$$

There is an overall sign difference compared to, e.g., Pollak [125]. Hence, the full Hamiltonian is

$$
\begin{aligned}
H(\mathbf{k}, \varepsilon) = E_0 + \frac{1}{3}\Delta_0 \mathbf{I} \cdot \boldsymbol{\sigma} + Ak^2 + a \sum_i \varepsilon_{ii} + 3 \sum_i (Bk_i^2 + b\varepsilon_{ii})(I_i^2 - \frac{1}{3}I^2) \\
+ 2\sqrt{3} \sum_{i<j} (Dk_i k_j + d\varepsilon_{ij})\{I_i I_j\}.
\end{aligned} \tag{7.22}
$$

In analogy to the relationships in Eq. (5.7), comparing Eq. (7.13) with Eq. (7.22) gives the same results as in Table 7.1.

The above has neglected strain-dependent spin-orbit terms. The latter can be written down just like Eq. (7.21) [125]:

$$H_\varepsilon^{(2)} = -a_2(\mathbf{I}\cdot\boldsymbol{\sigma})\sum_i \varepsilon_{ii} - 3b_2\sum_i \varepsilon_{ii}(I_i\sigma_i - \frac{1}{3}\mathbf{I}\cdot\boldsymbol{\sigma}) - 2d_2\sqrt{3}\sum_{i<j}\varepsilon_{ij}\{I_iI_j\}. \quad (7.23)$$

7.3.4.2 JM_J Basis

Similarly, in the JM_J basis, we have

$$H_{\frac{3}{2}}(\mathbf{k},\varepsilon) = Ak^2 + a\sum_i \varepsilon_{ii} + \sum_i (Bk_i^2 + b\varepsilon_{ii})(J_i^2 - \frac{1}{3}J^2) + \frac{2}{\sqrt{3}}\sum_{i<j}(Dk_ik_j + d\varepsilon_{ij})\{J_iJ_j\}, \quad (7.24)$$

and

$$H_{\mathrm{so}}(\mathbf{k},\varepsilon) = -\Delta_0 + Ak^2 + a\sum_i \varepsilon_{ii}. \quad (7.25)$$

In Yu and Cardona [16], the b and d terms have an extra factor of 3 in front. An explicit form of the strain matrix can be written down, using the matrices given in Chap. 5. Thus,

$$J_{x,y}^2 - \frac{1}{3}J^2 = \begin{pmatrix} -\frac{1}{2} & 0 & \pm\frac{\sqrt{3}}{2} & 0 \\ 0 & \frac{1}{2} & 0 & \pm\frac{\sqrt{3}}{2} \\ \pm\frac{\sqrt{3}}{2} & 0 & \frac{1}{2} & 0 \\ 0 & \pm\frac{\sqrt{3}}{2} & 0 & -\frac{1}{2} \end{pmatrix}, \quad J_z^2 - \frac{1}{3}J^2 = \begin{pmatrix} 1 & 0 & 0 & 0 \\ 0 & -1 & 0 & 0 \\ 0 & 0 & -1 & 0 \\ 0 & 0 & 0 & 1 \end{pmatrix}, \quad (7.26)$$

$$\sum_i \varepsilon_{ii}(J_i^2 - \frac{1}{3}J^2)$$
$$= \begin{pmatrix} (\varepsilon_{zz} - \frac{\varepsilon_{xx}}{2} - \frac{\varepsilon_{yy}}{2}) & 0 & \frac{\sqrt{3}}{2}(\varepsilon_{xx} - \varepsilon_{yy}) & 0 \\ 0 & (\frac{\varepsilon_{xx}}{2} + \frac{\varepsilon_{yy}}{2} - \varepsilon_{zz}) & 0 & \frac{\sqrt{3}}{2}(\varepsilon_{xx} - \varepsilon_{yy}) \\ \frac{\sqrt{3}}{2}(\varepsilon_{xx} - \varepsilon_{yy}) & 0 & (\frac{\varepsilon_{xx}}{2} + \frac{\varepsilon_{yy}}{2} - \varepsilon_{zz}) & 0 \\ 0 & \frac{\sqrt{3}}{2}(\varepsilon_{xx} - \varepsilon_{yy}) & 0 & (\varepsilon_{zz} - \frac{\varepsilon_{xx}}{2} - \frac{\varepsilon_{yy}}{2}) \end{pmatrix},$$

and, from Eq. (5.73),

$$\sum_{i<j}\varepsilon_{ij}\{J_iJ_j\} = \frac{\sqrt{3}}{2}\begin{pmatrix} 0 & (\varepsilon_{xz} - \mathrm{i}\varepsilon_{yz}) & -\mathrm{i}\varepsilon_{xy} & 0 \\ (\varepsilon_{xz} + \mathrm{i}\varepsilon_{yz}) & 0 & 0 & -\mathrm{i}\varepsilon_{xy} \\ \mathrm{i}\varepsilon_{xy} & 0 & 0 & -(\varepsilon_{xz} - \mathrm{i}\varepsilon_{yz}) \\ 0 & \mathrm{i}\varepsilon_{xy} & -(\varepsilon_{xz} + \mathrm{i}\varepsilon_{yz}) & 0 \end{pmatrix}.$$

Table 7.2 Four-band strain Hamiltonian

$$H_\varepsilon = \begin{pmatrix} a\mathrm{Tr}\varepsilon + b(\varepsilon_{zz} - \frac{1}{2}\varepsilon_{xx} - \frac{1}{2}\varepsilon_{yy}) & d(\varepsilon_{xz} - \mathrm{i}\varepsilon_{yz}) & b\frac{\sqrt{3}}{2}(\varepsilon_{xx} - \varepsilon_{yy}) - d\mathrm{i}\varepsilon_{xy} & 0 \\ d(\varepsilon_{xz} + \mathrm{i}\varepsilon_{yz}) & a\mathrm{Tr}\varepsilon - b(\varepsilon_{zz} - \frac{1}{2}\varepsilon_{xx} - \frac{1}{2}\varepsilon_{yy}) & 0 & b\frac{\sqrt{3}}{2}(\varepsilon_{xx} - \varepsilon_{yy}) - d\mathrm{i}\varepsilon_{xy} \\ b\frac{\sqrt{3}}{2}(\varepsilon_{xx} - \varepsilon_{yy}) + d\mathrm{i}\varepsilon_{xy} & 0 & a\mathrm{Tr}\varepsilon - b(\varepsilon_{zz} - \frac{1}{2}\varepsilon_{xx} - \frac{1}{2}\varepsilon_{yy}) & -d(\varepsilon_{xz} - \mathrm{i}\varepsilon_{yz}) \\ 0 & b\frac{\sqrt{3}}{2}(\varepsilon_{xx} - \varepsilon_{yy}) + d\mathrm{i}\varepsilon_{xy} & -d(\varepsilon_{xz} + \mathrm{i}\varepsilon_{yz}) & a\mathrm{Tr}\varepsilon + b(\varepsilon_{zz} - \frac{1}{2}\varepsilon_{xx} - \frac{1}{2}\varepsilon_{yy}) \end{pmatrix}.$$

Therefore, the four-band strain Hamiltonian is as given in Table 7.2. Note, however, that there are a couple of sign differences compared to the previously obtained form in Eq. (7.19). For example, they can be made the same by changing the sign of the $|\frac{3}{2} - \frac{3}{2}\rangle$ state.

7.4 Strained Energies

If one assumes the bands to be decoupled, then one can write, from Table 7.2, the band edges as

$$E_{hh}(\varepsilon) = E_{hh}(0) + a_v \mathrm{Tr}\varepsilon + b(\varepsilon_{zz} - \frac{1}{2}\varepsilon_{xx} - \frac{1}{2}\varepsilon_{yy}), \tag{7.27}$$

$$E_{lh}(\varepsilon) = E_{lh}(0) + a_v \mathrm{Tr}\varepsilon - b(\varepsilon_{zz} - \frac{1}{2}\varepsilon_{xx} - \frac{1}{2}\varepsilon_{yy}).$$

Otherwise, one can separately look at the solutions to the four-band and six-band models.

7.4.1 Four-Band Model

One can obtain the exact eigenvalues to the four-band problem in the presence of strain. One still has Eq. (5.52), with Eq. (5.53) modified to

$$\alpha_{ii} = Bk_i^2 + b\varepsilon_{ii}, \alpha_{ij} = Dk_i k_j + d\varepsilon_{ij} \; (i \neq j). \tag{7.28}$$

Then the solution is similar to previously obtained with

$$E_{1,2}(\mathbf{k}) = Ak^2 + a\sum_i \varepsilon_{ii} \pm (R_k + R_\varepsilon + R_{k\varepsilon})^{1/2}, \tag{7.29}$$

where

$$R_k = B^2k^4 + (D^2 - 3B^2)(k_y^2k_z^2 + k_z^2k_x^2 + k_x^2k_y^2), \tag{7.30}$$

$$R_\varepsilon = \frac{b^2}{2}\left[(\varepsilon_{yy} - \varepsilon_{zz})^2 + (\varepsilon_{zz} - \varepsilon_{xx})^2 + (\varepsilon_{xx} - \varepsilon_{yy})^2\right] + d^2(\varepsilon_{yz}^2 + \varepsilon_{zx}^2 + \varepsilon_{xy}^2), \tag{7.31}$$

$$R_{k\varepsilon} = -Bbk^2(\varepsilon_{xx} + \varepsilon_{yy} + \varepsilon_{zz}) + 3bB(k_x^2\varepsilon_{xx} + k_y^2\varepsilon_{yy} + k_z^2\varepsilon_{zz}) + 2Dd(k_yk_z\varepsilon_{yz} + k_zk_x\varepsilon_{zx} + k_xk_y\varepsilon_{xy}). \tag{7.32}$$

At $\mathbf{k} = \mathbf{0}$,

$$E_{1,2}(\Gamma) = a\sum_i \varepsilon_{ii} \pm \left\{ \frac{b^2}{2}\left[\left(\varepsilon_{yy}-\varepsilon_{zz}\right)^2 + (\varepsilon_{zz}-\varepsilon_{xx})^2 + \left(\varepsilon_{xx}-\varepsilon_{yy}\right)^2 \right] + d^2\left(\varepsilon_{yz}^2+\varepsilon_{zx}^2+\varepsilon_{xy}^2\right) \right\}^{1/2}. \tag{7.33}$$

Thus, exact solutions can be obtained at $\mathbf{k} = \mathbf{0}$ under various strains that can be realized experimentally.

7.4.1.1 Hydrostatic Pressure

The strain tensor is diagonal and isotropic. Writing

$$\sum_i \varepsilon_{ii} = \mathrm{Tr}\varepsilon,$$

we find that the valence-band top shifts under hydrostatic pressure by

$$\Delta E_v = a_v \mathrm{Tr}\varepsilon. \tag{7.34}$$

There is a corresponding shift of the conduction band given by

$$\Delta E_c = a_c \mathrm{Tr}\varepsilon. \tag{7.35}$$

7.4.1.2 [001] Strain

Again the strain tensor is diagonal; now, $\varepsilon_{xx} = \varepsilon_{yy} \neq \varepsilon_{zz}$. Then, Eq. (7.33) becomes

$$E_{1,2}(\Gamma) = a\sum_i \varepsilon_{ii} \pm b\varepsilon_{zz}, \tag{7.36}$$

and the hh and lh splits by

$$\Delta E = 2|b\varepsilon_{zz}|. \tag{7.37}$$

7.4.1.3 [111] Strain

One can write the strain tensor as

$$\varepsilon_{ij} = \frac{\varepsilon'_{111}}{3}(1-\delta_{ij}), \tag{7.38}$$

and the hh and lh splits by

$$\Delta E = \frac{2}{3}|d\varepsilon'_{111}|. \tag{7.39}$$

7.4.2 Six-Band Model

There are a few well-known results within the six-band model [125, 126]. Using Eq. (7.19), consider a uniaxial stress σ in the [001] direction. Then, $s = r = 0$ and the strain Hamiltonian block-diagonalizes. For example, one 3×3 strain block is

$$\begin{pmatrix} \left|\frac{3}{2}\ \frac{3}{2}\right\rangle & \left|\frac{3}{2}\ \frac{1}{2}\right\rangle & \left|\frac{1}{2}\ \frac{1}{2}\right\rangle \\ p+q & 0 & 0 \\ 0 & p-q & \sqrt{2}q \\ 0 & \sqrt{2}q & p \end{pmatrix}, \tag{7.40}$$

where

$$p = a\mathrm{Tr}\varepsilon, \tag{7.41}$$

$$q = b(\varepsilon_{zz} - \varepsilon_{xx}). \tag{7.42}$$

It can be seen that the hh state is decoupled.

7.4.3 Deformation Potentials

In general, one can define deformation potentials at each wave-vector point.

7.4.3.1 k = 0

Now that we understand the role of the deformation potentials, we study them in more detail. The standard values for the deformation potentials are such that: $a_c < 0, a_v > 0, b < 0, d < 0$. Examples are given in Table 7.3.

Table 7.3 Elasticity and strain parameters for a few bulk cubic semiconductors [127]

	C_{11} (kbar)	C_{12} (kbar)	C_{44} (kbar)	a_c (eV)	a_v (eV)	b (eV)	d (eV)
GaAs	1181	532	595	−7.17	1.16	−1.7	−4.55
GaSb	883.4	402.3	432.2	−6.85	0.8	−1.8	−4.6
InAs	832.9	452.6	395.9	−5.08	1.00	−1.8	−3.6

Sign of a_v agrees with Yu and Cardona [16] but opposite to that chosen by Marzin et al. [128].

A comparison of the various deformation potentials is given in Table 7.4.

For completeness, it is worth mentioning the very detailed and insightful study of deformation potentials by Blacha, Presting and Cardona [129], using pseudopotential and tight-binding theory.

Table 7.4 Notations for deformation potentials

	$\mathbf{k} = \mathbf{0}$			
	Bir-Pikus [1]	Kleiner-Roth [117]	Kane [119]	Trebin et al. [81]
a_c				C_1
a	$\frac{1}{3}(l+2m)$	D_d^v	$\frac{1}{\sqrt{3}}D_1^1$	D_d^v
b	$\frac{1}{3}(m-l)$	$-\frac{2}{3}D_u$	$\frac{1}{\sqrt{3}}D_3$	$-\frac{2}{3}D_u$
d	$-\frac{1}{\sqrt{3}}n$	$-\frac{2}{\sqrt{3}}D_{u'}$	$\frac{1}{\sqrt{2}}D_5$	$-\frac{2}{\sqrt{3}}D_{u'}$
				C_2
				C_4
				C_5'

	$\mathbf{k} \neq \mathbf{0}$	
Brooks [115]	Herring-Vogt [116]	Kane [119]
$\mathcal{E}_1$	$\Xi_d + \frac{1}{3}\Xi_u$	$\frac{1}{\sqrt{3}}D_1^1$
$\mathcal{E}_2$	Ξ_u	$\sqrt{\frac{2}{3}}D_1^3$ **k** along $\langle 001\rangle$
$\mathcal{E}_3$	$\frac{1}{2}\Xi_u$	$\frac{\sqrt{3}}{2}D_1^5$ **k** along $\langle 111\rangle$

7.4.3.2 $\mathbf{k} \neq \mathbf{0}$

A number of authors have also studied the deformation potentials at finite wave vector, particularly at band extrema and critical points [115, 116, 119]. In this case, a uniaxial stress can lead to either a reduction of the degeneracy of energy bands due to a lowering of the crystal symmetry or to a coupling between neighboring bands. For example, the energy change of a band extrema (shift and degeneracy splitting) is given by [125]

$$\Delta E = \hat{n} \cdot \left[\mathcal{E}_1 \mathrm{Tr}\varepsilon + \mathcal{E}_2 (\varepsilon - \frac{1}{3}\mathrm{Tr}\varepsilon \mathbb{1}) \right] \cdot \hat{n}, \tag{7.43}$$

where $\hat{n}$ is a unit vector in the direction of the band extremum, and $\mathcal{E}_1$ ($\mathcal{E}_2$) is the hydrostatic (shear) deformation potential. A comparison to other notations is given in Table 7.4. A more extensive discussion of the deformation potentials can be found in the review article by Pollak [125].

7.5 Eight-Band Model for Zincblende

There are some additional effects that occur for strain in an eight-band model. This problem has been studied by, e.g., Trebin et al. [81] using the method of invariant in a paper in 1979. A perturbative solution was given by Pollak in his review article [125]. We believe it is more instructive to present the perturbative treatment first, followed by the method of invariant.

7.5.1 Perturbation Theory

The Hamiltonian given by Pollak is in terms of a JM_J basis that differs in phase from ours but agrees with the one used by Kane (Table C.2). We give the complete Hamiltonian (including the unstrained terms) in Table 7.5.

The matrix elements are defined as follows:

$$
\begin{aligned}
H_{11} &= (1+D_c')\frac{\hbar^2 k^2}{2m_0} + a_c \mathrm{Tr}\varepsilon,\\
H_{13} &= \frac{\hbar}{2\sqrt{m_0}} \mathrm{i} P k_+,\\
H_{14} &= \frac{\hbar}{\sqrt{3m_0}} \mathrm{i} P k_z,\\
H_{33} &= \left[A'k^2 - \frac{B'}{2}\left(2k_z^2 - k_x^2 - k_y^2\right)\right]\frac{\hbar^2}{2m_0} - (a_1+a_2)\mathrm{Tr}\varepsilon - \frac{(b_1+2b_2)}{2}\left(2\varepsilon_{zz} - \varepsilon_{xx} - \varepsilon_{yy}\right),\\
H_{34} &= -\left[\frac{1}{\sqrt{3}} N' k_- k_z\right]\frac{\hbar^2}{2m_0} + (d_1+d_2)\left(\varepsilon_{xz} - \varepsilon_{yz}\right),\\
H_{35} &= -\left[\frac{\sqrt{3}}{2} B'\left(k_x^2 - k_y^2\right) + \frac{1}{\sqrt{3}} N' k_x k_z\right]\frac{\hbar^2}{2m_0} - \frac{\sqrt{3}(b_1+2b_2)}{2}\left(\varepsilon_{xx} - \varepsilon_{yy}\right)\\
&\quad + \mathrm{i}(d_1+2d_2)\varepsilon_{xy},\\
H_{37} &= \frac{1}{\sqrt{6}} N' k_- k_z \frac{\hbar^2}{2m_0} - (d_1-d_2)(\varepsilon_{xz} - i\varepsilon_{yz}),
\end{aligned}
\tag{7.44}
$$

Table 7.5 Eight-band zincblende strained Hamiltonian [125]

$$
\begin{pmatrix}
|S\uparrow\rangle & |S\downarrow\rangle & |\frac{3}{2}\,\frac{3}{2}\rangle & |\frac{3}{2}\,\frac{1}{2}\rangle & |\frac{3}{2}\,-\frac{1}{2}\rangle & |\frac{3}{2}\,-\frac{3}{2}\rangle & |\frac{1}{2}\,\frac{1}{2}\rangle & |\frac{1}{2}\,-\frac{1}{2}\rangle \\
E_0+H_{11} & 0 & H_{13} & H_{14} & \frac{1}{\sqrt{3}}H_{13}^* & 0 & \frac{1}{\sqrt{2}}H_{14}^* & -\sqrt{\frac{2}{3}}H_{13}^* \\
0 & E_0+H_{11} & 0 & \frac{1}{\sqrt{3}}H_{13} & H_{14} & H_{13}^* & \sqrt{\frac{2}{3}}H_{13} & \frac{1}{\sqrt{2}}H_{14} \\
H_{13}^* & 0 & H_{33} & H_{34} & H_{35} & 0 & H_{37} & H_{38} \\
H_{14}^* & \frac{1}{\sqrt{3}}H_{13}^* & H_{34}^* & H_{44} & 0 & H_{35} & H_{47} & H_{48} \\
\frac{1}{\sqrt{3}}H_{13} & H_{14}^* & H_{35}^* & 0 & H_{44} & H_{34} & H_{48} & -H_{47} \\
0 & H_{13} & 0 & H_{35}^* & H_{34}^* & H_{33} & -H_{38}^* & H_{37} \\
\frac{1}{\sqrt{2}}H_{14} & \sqrt{\frac{2}{3}}H_{13}^* & H_{37}^* & H_{47}^* & H_{48}^* & -H_{38} & -\Delta_0+H_{77} & 0 \\
-\sqrt{\frac{2}{3}}H_{13} & \frac{1}{\sqrt{2}}H_{14}^* & H_{38}^* & H_{48}^* & -H_{47}^* & H_{37}^* & 0 & -\Delta_0+H_{77}
\end{pmatrix}.
$$

$$H_{38} = \left[\frac{\sqrt{3}}{2}B'\left(k_x^2 - k_y^2\right) + \frac{1}{\sqrt{3}}N'k_xk_z\right]\frac{\hbar^2}{2m_0} + \sqrt{\frac{3}{2}}(b_1 - b_2)(\varepsilon_{xx} - \varepsilon_{yy})$$
$$+\mathrm{i}\sqrt{2}(d_1 - d_2)\varepsilon_{xy},$$
$$H_{44} = \left[A'k^2 + \frac{B'}{2}\left(2k_z^2 - k_x^2 - k_y^2\right)\frac{\hbar^2}{2m_0} - (a_1 + a_2)\mathrm{Tr}\varepsilon + \frac{(b_1 - b_2)}{2}(2\varepsilon_{zz} - \varepsilon_{xx} - \varepsilon_{yy}),\right.$$
$$H_{47} = \frac{B'}{\sqrt{2}}\left(2k_z^2 - k_x^2 - k_y^2\right)\frac{\hbar^2}{2m_0} - \frac{(b_1 - b_2)}{\sqrt{2}}(2\varepsilon_{zz} - \varepsilon_{xx} - \varepsilon_{yy}),$$
$$H_{48} = -\frac{1}{\sqrt{2}}N'k_-k_z\frac{\hbar^2}{2m_0} + \sqrt{\frac{3}{2}}(d_1 - d_2)(\varepsilon_{xz} - \mathrm{i}\varepsilon_{yz}),$$
$$H_{77} = A'_{\mathrm{so}}\frac{\hbar^2k^2}{2m_0} - (a_1 - 2a_2)\mathrm{Tr}\varepsilon,$$

with

$$P = -\mathrm{i}\sqrt{2}m_0\langle s|p_x|x\rangle,$$
$$D'_c = \frac{2}{m_0}\sum_j\frac{|\langle s|p_x|u_j\rangle|^2}{(E_c - E_j)}, \tag{7.45}$$
$$A'_{\mathrm{so}} = \frac{1}{3}(L'_{\mathrm{so}} + 2M'_{\mathrm{so}}) + 1.$$

7.5.2 Method of Invariants

We had already discussed the eight-band model of Trebin et al. in the absence of strain in Chap. 5. In Table 7.6, we give the additional linear-in-strain contributions.

Table 7.6 Strain contributions to the eight-band model for zincblende

$H_\varepsilon^{6c\,6c}$	$= C_1\mathrm{Tr}\varepsilon$
$H_\varepsilon^{8v\,8v}$	$= D_d\mathrm{Tr}\varepsilon + \frac{2}{3}D_u\left[\left(J_z^2 - \frac{1}{3}J^2\right)\varepsilon_{xx} + \mathrm{c.p.}\right] + \frac{2}{3}D'_u\left[2\{J_xJ_y\}\varepsilon_{xy} + \mathrm{c.p.}\right]$
$H_{\varepsilon k}^{8v\,8v}$	$= \left[C_4(\varepsilon_{yy} - \varepsilon_{zz})k_x + C'_5(\varepsilon_{xy}k_y - \varepsilon_{xz}k_z)\right]J_x + \mathrm{c.p.}$
$H_\varepsilon^{7v\,7v}$	$= D_d\mathrm{Tr}\varepsilon$
$H_\varepsilon^{6c\,8v}$	$= \sqrt{3}[\mathrm{i}C_2(T_x\varepsilon_{yz} + \mathrm{c.p.}) - 2P(T_x\sum_i\varepsilon_{xi}k_i + \mathrm{c.p.})]$
$H_\varepsilon^{6c\,7v}$	$= -\frac{1}{\sqrt{3}}\left[\mathrm{i}C_2(\rho_x\varepsilon_{yz} + \mathrm{c.p.}) - 2P(\sigma_x\sum_i\varepsilon_{xi}k_i + \mathrm{c.p.})\right]$
$H_\varepsilon^{8v\,7v}$	$= 2D_u(U_{xx}\varepsilon_{xx} + \mathrm{c.p.}) + 2D'_u(U_{xy}\varepsilon_{xy} + \mathrm{c.p.})$
$H_{\varepsilon k}^{8v\,7v}$	$= \frac{3}{2}\left[C_4(\varepsilon_{yy} - \varepsilon_{zz})k_x + C'_5(\varepsilon_{xy}k_y - \varepsilon_{xz}k_z)\right]U_x + \mathrm{c.p.}$

Adapted with permission from [21].

Note that this is actually the result quoted in the book by Winkler [21] as the latter has the linear-in-k coupling missing in Trebin et al.

Comparing Table 7.5 with Table 7.6, we find, e.g., that

$$a_c = C_1, \tag{7.46}$$

$$a_1 + a_2 = D_d. \tag{7.47}$$

The Pollak model does not have terms linear in k and ε. These are present both within the valence band, e.g., via the parameters C_4 and C_5', and they also couple the conduction band to the valence band.

7.6 Wurtzite

Historically, the method of invariants was used first to obtain the strain Hamiltonian but we will start with the expressions from perturbation theory.

7.6.1 Perturbation Theory

From Eq. (3.71), the strain Hamiltonian in X, Y, Z basis is given by

$$H_\varepsilon = \begin{pmatrix} |X\rangle & |Y\rangle & |Z\rangle \\ l_1\varepsilon_{xx} + m_1\varepsilon_{yy} + m_2\varepsilon_{zz} & n_1\varepsilon_{xy} & n_2\varepsilon_{xz} \\ n_1\varepsilon_{xy} & m_1\varepsilon_{xx} + l_1\varepsilon_{yy} + m_2\varepsilon_{zz} & n_2\varepsilon_{yz} \\ n_2\varepsilon_{xz} & n_2\varepsilon_{yz} & m_3(\varepsilon_{xx}+\varepsilon_{yy}) + l_2\varepsilon_{zz} \end{pmatrix}. \tag{7.48}$$

It is straightforward to convert to the u-basis (Table C.9). For example,

$$\begin{aligned} \langle u_1|H_\varepsilon|u_1\rangle &= \frac{1}{2}\langle X+\mathrm{i}Y|H_\varepsilon|X+\mathrm{i}Y\rangle \\ &= \frac{1}{2}\left(\langle X|H_\varepsilon|X\rangle + \langle Y|H_\varepsilon|Y\rangle + \mathrm{i}\langle X|H_\varepsilon|Y\rangle - \mathrm{i}\langle Y|H_\varepsilon|X\rangle\right) \\ &= \frac{1}{2}\left[(l_1+m_1)(\varepsilon_{xx}+\varepsilon_{yy}) + 2m_2\varepsilon_{zz}\right], \end{aligned} \tag{7.49}$$

$$\langle u_2|H_\varepsilon|u_2\rangle = \frac{1}{2}\langle X-\mathrm{i}Y|H_\varepsilon|X-\mathrm{i}Y\rangle = \langle u_1|H_\varepsilon|u_1\rangle, \tag{7.50}$$

$$\langle u_3|H_\varepsilon|u_3\rangle = \langle Z|H_\varepsilon|Z\rangle = m_3(\varepsilon_{xx}+\varepsilon_{yy}) + l_2\varepsilon_{zz}, \tag{7.51}$$

$$\begin{aligned} \langle u_1|H_\varepsilon|u_2\rangle &= -\frac{1}{2}\langle X+\mathrm{i}Y|H_\varepsilon|X-\mathrm{i}Y\rangle = -\frac{1}{2}\left(\langle X|H_\varepsilon|X\rangle - \langle Y|H_\varepsilon|Y\rangle - 2\mathrm{i}\langle X|H_\varepsilon|Y\rangle\right) \\ &= \frac{1}{2}(l_1-m_1)(\varepsilon_{yy}-\varepsilon_{xx}) + \mathrm{i}n_1\varepsilon_{xy}, \end{aligned} \tag{7.52}$$

Table 7.7 Strain Hamiltonian in u_i basis

$$H_\varepsilon = \begin{pmatrix} |u_1\rangle & |u_2\rangle & |u_3\rangle & |u_4\rangle & |u_5\rangle & |u_6\rangle \\ u_{11} & u_{12} & u_{13} & 0 & 0 & 0 \\ u_{12}^* & u_{11} & -u_{13}^* & 0 & 0 & 0 \\ u_{13}^* & -u_{13} & u_{33} & 0 & 0 & 0 \\ 0 & 0 & 0 & u_{11} & u_{12}^* & -u_{13}^* \\ 0 & 0 & 0 & u_{12} & u_{11} & u_{13} \\ 0 & 0 & 0 & -u_{13} & u_{13}^* & u_{33} \end{pmatrix},$$

with

$$u_{11} = \frac{1}{2}\left[(l_1 + m_1)(\varepsilon_{xx} + \varepsilon_{yy}) + 2m_2\varepsilon_{zz}\right],$$

$$u_{33} = m_3(\varepsilon_{xx} + \varepsilon_{yy}) + l_2\varepsilon_{zz},$$

$$u_{12} = \frac{1}{2}(l_1 - m_1)(\varepsilon_{yy} - \varepsilon_{xx}) + \mathrm{i}n_1\varepsilon_{xy},$$

$$u_{13} = -\frac{n_2}{\sqrt{2}}(\varepsilon_{xz} - \mathrm{i}\varepsilon_{yz}).$$

The result is given in Table 7.7.

7.6.2 Method of Invariants

Strain contributions to the Hamiltonians of Sect. 5.9 are now considered.

7.6.2.1 RSP Hamiltonian

Bir and Pikus [1] gave the following contributions:

$$H_{\mathrm{RSP}}(\varepsilon) = (D_1 + D_3 J_z^2)\varepsilon_{zz} + (D_2 + D_4 J_z^2)\varepsilon_\perp - D_5(J_+^2\varepsilon_- + J_-^2\varepsilon_+)$$
$$-2\mathrm{i}D_6(\{J_z J_+\}\varepsilon_{-z} - \{J_z J_-\}\varepsilon_{+z}),$$

where $\{AB\} = (AB + BA)/2$ and

$$\begin{aligned} \varepsilon_\perp &= \varepsilon_{xx} + \varepsilon_{yy}, \\ \varepsilon_{\pm z} &= \varepsilon_{xz} \pm \mathrm{i}\varepsilon_{yz}, \\ \varepsilon_\pm &= \varepsilon_{xx} + \varepsilon_{yy} \pm 2\mathrm{i}\varepsilon_{xy}, \end{aligned} \tag{7.53}$$

and Table 5.17 for the other quantities. There are, therefore, six deformation potentials for the valence band of wurtzite within this model (Table 7.8).

7.6.2.2 SJKLS Hamiltonian

Sirenko and coworkers [43] gave

$$H_{\mathrm{SJKLS}}(\varepsilon) = C_1\varepsilon_{zz} + C_3\varepsilon_{\perp} + C_2\varepsilon_{zz} + C_4\varepsilon_{\perp} \\ -C_5(J_+^2\varepsilon_- + J_-^2\varepsilon_+) - 2C_6(\{J_zJ_+\}\varepsilon_{-z} + \{J_zJ_-\}\varepsilon_{+z}). \quad (7.54)$$

7.6.2.3 CC Hamiltonian

Finally, the third form is the one by Chuang and Chang [46].

$$H_{\mathrm{CC}}(\varepsilon) = (D_1 + D_3J_z^2)\varepsilon_{zz} + (D_2 + D_4J_z^2)\varepsilon_{\perp} - D_5(J_+^2\varepsilon_- + J_-^2\varepsilon_+) \\ -2D_6(\{J_zJ_+\}\varepsilon_{-z} + \{J_zJ_-\}\varepsilon_{+z}). \quad (7.55)$$

Comparison of the different constants is given in Table 5.21.

Table 7.8 Comparison of deformation potentials for six-band Hamiltonian for wurtzite

CC96 [46]	BP75 [1]	SJKLS [43]
D_1	D_1	$-C_1$
D_2	D_2	$-C_3$
D_3	D_3	$-C_2$
D_4	D_4	$-C_4$
D_5	$-D_5$	C_5
D_6	$-D_6$	C_6

Table 7.9 Deformation potentials and elasticity constants for wurtzite

	GaN	AlN	ZnO	
	[57]	[57]	[53]	[130]
a_c (eV)				−6.05
D_1 (eV)	−3.0	−3.0	−2.66	−3.90
D_2 (eV)	3.6	3.6	2.82	−4.13
D_3 (eV)	8.82	9.6	−1.34	−1.15
D_4 (eV)	−4.41	−4.8	1.00	1.22
D_5 (eV)	−4.0	−4.0		1.53
D_6 (eV)	−5.1	−5.1		2.83
C_{11} (GPa)				209.7
C_{13} (GPa)				121.1
C_{13} (GPa)			105.1	105.1
C_{33} (GPa)			210.9	210.9

7.6.3 Examples

Example deformation potentials are given in Table 7.9 and band structures are shown in Fig. 7.1. The calculations were done using the model of Chuang and Chang. As can be seen, there is no further splitting of the valence band for WZ (as could have happened for ZB) but there is observable change in the energy gaps and dispersions. Since this model ignores the A_7 parameter, there is no spin splitting either.

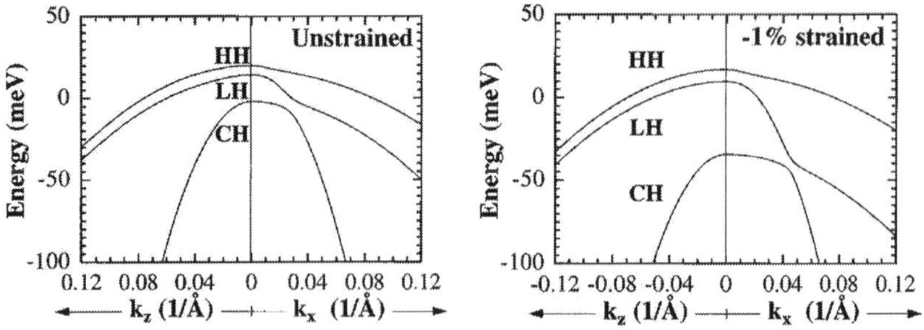

Fig. 7.1 GaN valence-band structure. Reprinted with permission from [46]. © 1996 by the American Physical society

7.7 Summary

The strain Hamiltonian has been written down for various cubic and hexagonal Hamiltonians. This follows again either from a perturbative treatment or an invariant one. Due to the formal similarity between the strain tensor and the symmetrized product $k_i k_j$, the strain Hamiltonian can be automatically written down once the quadratic $k \cdot p$ Hamiltonian is formed. For example, the strain Hamiltonian for the DKK Hamiltonian has three deformation potentials.

Part II
Nonperiodic Problem

Chapter 8
Shallow Impurity States

8.1 Overview

An electron in a periodic (crystalline) potential is described by a Bloch function When a nonperiodic potential is added to the problem (can be, e.g., an external electromagnetic field or an impurity atom or a heterojunction), translational symmetry is lost, and the Hamiltonian is no longer diagonal in a basis of Bloch functions. The question being posed is: how does one solve this new eigenvalue problem? This problem was probably first considered by Wannier [131] and Adams [132].

In theory, one does not need to appeal to quantum-mechanical perturbation theory. The Bloch functions form a complete set of orthonormal functions. Hence, one could expand any wave function as a linear combination of the Bloch functions. Schrödinger's equation then becomes an exact matrix equation which can, in principle, be diagonalized for the new solutions. In practice, the latter step is not feasible due to the band and wave-vector coupling, unless certain approximations are made.

One might wonder why use Bloch states as basis functions (so-called Bloch representation). It might seem reasonable if one is doing perturbation theory since the Bloch states are the unperturbed ones. However, there is already a simple obvious complication. The nonperiodic potential couples different **k** values (in addition to different bands). If the perturbation is such that large **k** Bloch functions are coupled to small **k** Bloch functions, this implies the exact solution is far from a Bloch basis. This is most clearly seen by Fourier transforming Schrödinger's equation. A sharply varying nonperiodic potential (e.g., a step function) will have large **k** components. These will couple a given **k** state to a much different **k**′ state. One can also understand the inadequacy of the Bloch function in real space. A lot of delocalized, i.e., Bloch, states will be needed to describe a sharply varying potential.

Such an expansion has been written down by, e.g., Kittel and Mitchell [133] for the impurity problem:

$$\psi(\mathbf{r}) = \sum_{n'} \int \mathrm{d}^3\mathbf{k}\, B_{n'}(\mathbf{k})\psi_{n'\mathbf{k}}(\mathbf{r}) = \sum_{n'} \int \mathrm{d}^3\mathbf{k}\, B_{n'}(\mathbf{k})\mathrm{e}^{\mathrm{i}\mathbf{k}\cdot\mathbf{r}} u_{n'\mathbf{k}}(\mathbf{r}), \tag{8.1}$$

L.C. Lew Yan Voon, M. Willatzen, *The* k · p *Method*,
DOI 10.1007/978-3-540-92872-0_8, © Springer-Verlag Berlin Heidelberg 2009

and operators would be evaluated via the matrix elements

$$O_{nn'}(\mathbf{k},\mathbf{k}') \equiv \langle n\mathbf{k}|\widehat{\mathcal{O}}|n'\mathbf{k}'\rangle = \int \mathrm{d}^3\mathbf{r}\,\psi^*_{n\mathbf{k}}(\mathbf{r})\mathcal{O}(\mathbf{r})\psi_{n'\mathbf{k}'}(\mathbf{r}). \tag{8.2}$$

Another example of the inadequacy of the Bloch representation is in the definition of the position operator. A number of authors have discussed that the latter is ill-defined in a Bloch representation at a degeneracy point [134–136].

Since Eq. (8.1) is general, one can expect to be able to get other representations by manipulating it. Indeed, the RHS can be shown to be exactly equivalent to [6]

$$\psi(\mathbf{r}) = \sum_{n'} \int \mathrm{d}^3\mathbf{k}\, A_{n'}(\mathbf{k})\mathrm{e}^{\mathrm{i}\mathbf{k}\cdot\mathbf{r}} u_{n'\mathbf{0}}(\mathbf{r}). \tag{8.3}$$

If we now view Eq. (8.3) as a special case of

$$\begin{aligned}\psi(\mathbf{r}) &= \sum_{n'} \int \mathrm{d}^3\mathbf{k}\, A_{n'}(\mathbf{k})\mathrm{e}^{\mathrm{i}\mathbf{k}\cdot\mathbf{r}}\mathrm{e}^{\mathrm{i}\mathbf{k}_0\cdot\mathbf{r}} u_{n'\mathbf{k}_0}(\mathbf{r})\\ &\equiv \sum_{n'} \int \mathrm{d}^3\mathbf{k}\, A_{n'}(\mathbf{k})\chi_{n'\mathbf{k}}(\mathbf{r}),\end{aligned} \tag{8.4}$$

and define matrix elements as

$$O_{nn'}(\mathbf{k},\mathbf{k}') \equiv (n\mathbf{k}|\widehat{\mathcal{O}}|n'\mathbf{k}') = \int \mathrm{d}^3\mathbf{r}\,\chi^*_{n\mathbf{k}}(\mathbf{r})\mathcal{O}(\mathbf{r})\chi_{n'\mathbf{k}'}(\mathbf{r}), \tag{8.5}$$

we have what is known as the Luttinger–Kohn (LK) representation. LK had shown that the $\chi_{n\mathbf{k}}$ form a complete set of orthonormal states. This is, of course, the basis used by LK in treating an impurity center and an external homogeneous magnetic field. A second basis set for representing operators is in terms of Wannier functions. This was used by Slater et al. [137].

8.2 Kittel–Mitchell Theory

Kittel and Mitchell [133] attempted to explain the data in Table 8.1. They considered what happens to the Wannier effective-mass Hamiltonian given the multiplicity of conduction-band energy minima and degeneracy of valence bands. Kittel and Mitchell (KM) compared the Fourier transform (i.e., plane-wave expansion) of the Wannier equation with the exact Hamiltonian in a Bloch function representation.

Table 8.1 Experimental ionization energies of substitutional impurities [133]

	Binding energies (eV)	
Host	Acceptor	Donor
Si	0.05	0.04
Ge	0.01	0.01

8.2.1 Exact Theory

The Schrödinger equation for the impurity problem is

$$(H_0 + U)\psi = E\psi, \tag{8.6}$$

where $U(\mathbf{r})$ is the additional potential due to the defect. One can expand the impurity wave function in terms of the Bloch functions of the perfect crystal:

$$\psi = \sum_{n\mathbf{k}} a_{n\mathbf{k}} \psi_{n\mathbf{k}}. \tag{8.7}$$

The energy of the impurity state is

$$E = \int \mathrm{d}^3\mathbf{r}\, \psi^*(H_0 + U(\mathbf{r}))\psi = \sum_{n\mathbf{k}} E_n(\mathbf{k})|a_{n\mathbf{k}}|^2 + \sum_{n\mathbf{k}n'\mathbf{k}'} a^*_{n\mathbf{k}} a_{n'\mathbf{k}'} \langle n\mathbf{k}|U(\mathbf{r})|n'\mathbf{k}'\rangle. \tag{8.8}$$

If we now expand the impurity potential in a Fourier series,

$$U(\mathbf{r}) = \sum_{\mathbf{K}} b_{\mathbf{K}} \mathrm{e}^{\mathrm{i}\mathbf{K}\cdot\mathbf{r}}, \tag{8.9}$$

then

$$\begin{aligned}\langle n\mathbf{k}|U|n'\mathbf{k}'\rangle &= \sum_{\mathbf{K}} b_{\mathbf{K}} \int \mathrm{d}^3\mathbf{r}\, \psi^*_{n\mathbf{k}}(\mathbf{r})\mathrm{e}^{\mathrm{i}\mathbf{K}\cdot\mathbf{r}}\psi_{n'\mathbf{k}'}(\mathbf{r}) \\ &= \sum_{\mathbf{K}} b_{\mathbf{K}} \int \mathrm{d}^3\mathbf{r}\, u^*_{n\mathbf{k}}(\mathbf{r})u_{n'\mathbf{k}'}(\mathbf{r})\mathrm{e}^{\mathrm{i}(\mathbf{k}'+\mathbf{K}-\mathbf{k})\cdot\mathbf{r}}.\end{aligned} \tag{8.10}$$

We will now prove via two separate methods that the integral is zero unless $\mathbf{k}' + \mathbf{K} = \mathbf{k}$. First, since the cellular functions are periodic,

$$u^*_{n\mathbf{k}}(\mathbf{r})u_{n'\mathbf{k}'}(\mathbf{r}) = \sum_{\mathbf{G}} c^{\mathbf{k},\mathbf{k}'}_{nn'}(\mathbf{G})\mathrm{e}^{\mathrm{i}\mathbf{G}\cdot\mathbf{r}}, \tag{8.11}$$

with $\mathbf{G}$ a reciprocal lattice vector and Eq. (8.10) becomes

$$\langle n\mathbf{k}|U(\mathbf{r})|n'\mathbf{k}'\rangle = \sum_{\mathbf{K},\mathbf{G}} b_{\mathbf{K}} c^{\mathbf{k},\mathbf{k}'}_{nn'}(\mathbf{G}) \int \mathrm{d}^3\mathbf{r}\, \mathrm{e}^{\mathrm{i}(\mathbf{k}'+\mathbf{K}-\mathbf{k}+\mathbf{G})\cdot\mathbf{r}}. \tag{8.12}$$

Using (with V the crystal volume)

$$\int \mathrm{d}^3\mathbf{r}\, \mathrm{e}^{\mathrm{i}(\mathbf{k}'+\mathbf{K}-\mathbf{k}+\mathbf{G})\cdot\mathbf{r}} = V\,\delta_{\mathbf{k}'+\mathbf{K}-\mathbf{k}+\mathbf{G}}, \tag{8.13}$$

$$\int \mathrm{d}^3\mathbf{r}\, u^*_{n\mathbf{k}}(\mathbf{r})u_{n'\mathbf{k}'}(\mathbf{r})\mathrm{e}^{-\mathrm{i}\mathbf{G}\cdot\mathbf{r}} = \sum_{\mathbf{G}'} c^{\mathbf{k},\mathbf{k}'}_{nn'}(\mathbf{G}') \int \mathrm{d}^3\mathbf{r}\, \mathrm{e}^{\mathrm{i}(\mathbf{G}'-\mathbf{G})\cdot\mathbf{r}} = V c^{\mathbf{k},\mathbf{k}'}_{nn'}(\mathbf{G}). \tag{8.14}$$

Equation (8.12) becomes

$$\langle n\mathbf{k}|U(\mathbf{r})|n'\mathbf{k}'\rangle = \frac{1}{V}\sum_{\mathbf{K}} b_{\mathbf{K}} c^{\mathbf{k},\mathbf{k}'}_{nn'}(\mathbf{k}'+\mathbf{K}-\mathbf{k}), \tag{8.15}$$

and

$$c^{\mathbf{k},\mathbf{k}'}_{nn'}(\mathbf{G}) = \frac{1}{V}\int \mathrm{d}^3\mathbf{r}\, u^*_{n\mathbf{k}}(\mathbf{r})u_{n'\mathbf{k}'}(\mathbf{r})\mathrm{e}^{-\mathrm{i}\mathbf{G}\cdot\mathbf{r}}. \tag{8.16}$$

Now, if $u^*_{n\mathbf{k}}(\mathbf{r})u_{n'\mathbf{k}'}(\mathbf{r})$ is smooth, the dominant term in its Fourier expansion has $\mathbf{G} = \mathbf{0}$. Therefore, $\mathbf{k} = \mathbf{k}' + \mathbf{K}$.

For the second method, consider

$$\begin{aligned}\int \mathrm{d}^3\mathbf{r}\, u^*_{n\mathbf{k}}(\mathbf{r})u_{n'\mathbf{k}'}(\mathbf{r})\mathrm{e}^{\mathrm{i}(\mathbf{k}'+\mathbf{K}-\mathbf{k})\cdot\mathbf{r}} &= \int \mathrm{d}^3\mathbf{r}\, u^*_{n\mathbf{k}}(\mathbf{r}-\mathbf{R})u_{n'\mathbf{k}'}(\mathbf{r}-\mathbf{R})\mathrm{e}^{\mathrm{i}(\mathbf{k}'+\mathbf{K}-\mathbf{k})\cdot\mathbf{r}}\\ &= \mathrm{e}^{\mathrm{i}(\mathbf{k}'+\mathbf{K}-\mathbf{k})\cdot\mathbf{R}}\int \mathrm{d}^3\mathbf{r}\, u^*_{n\mathbf{k}}(\mathbf{r})u_{n'\mathbf{k}'}(\mathbf{r})\mathrm{e}^{\mathrm{i}(\mathbf{k}'+\mathbf{K}-\mathbf{k})\cdot\mathbf{r}}.\end{aligned}$$

Since the latter must be valid $\forall\, \mathbf{R}$, we have $\mathbf{k} = \mathbf{k}' + \mathbf{K}$.

Hence, Eq. (8.10) is written as

$$\langle n\mathbf{k}|U(\mathbf{r})|n'\mathbf{k}'\rangle = \sum_{\mathbf{K}} b_{\mathbf{K}} \int \mathrm{d}^3\mathbf{r}\, u^*_{n\mathbf{k}}(\mathbf{r})u_{n'\mathbf{k}'}(\mathbf{r}) \equiv \sum_{\mathbf{K}} b_{\mathbf{K}} \Delta^{n'n}_{\mathbf{k}',\mathbf{k}'+\mathbf{K}}, \tag{8.17}$$

where

$$\Delta^{n'n}_{\mathbf{k}',\mathbf{k}'+\mathbf{K}} \equiv \int \mathrm{d}^3\mathbf{r}\, u^*_{n\mathbf{k}'+\mathbf{K}}(\mathbf{r})u_{n'\mathbf{k}'}(\mathbf{r}), \tag{8.18}$$

and Eq. (8.8) becomes

$$E = \sum_{n\mathbf{k}} E_n(\mathbf{k})|a_{n\mathbf{k}}|^2 + \sum_{n\mathbf{k}n'\mathbf{k}'}\sum_{\mathbf{K}} a^*_{n\mathbf{k}}a_{n'\mathbf{k}'}b_{\mathbf{K}}\Delta^{n'n}_{\mathbf{k}',\mathbf{k}'+\mathbf{K}}. \tag{8.19}$$

Equation (8.19) is exact. As for the bulk problem, one can now apply a variational principle to obtain a secular equation. The difficulty lies in the determination of $\Delta^{n'n}_{\mathbf{k}',\mathbf{k}'+\mathbf{K}}$. One can circumvent it for a smooth impurity potential. When that is the case, the Fourier expansion of the latter will only involve wave vectors $\mathbf{K} \approx \mathbf{0}$. With this approximation,

$$\Delta^{n'n}_{\mathbf{k}',\mathbf{k}'} \equiv \int \mathrm{d}^3\mathbf{r}\, u^*_{n\mathbf{k}'}(\mathbf{r}) u_{n'\mathbf{k}'}(\mathbf{r}) = \delta_{nn'},$$

and Eq. (8.19) becomes

$$E = \sum_n \left[\sum_{\mathbf{k}} E_n(\mathbf{k}) |a_{n\mathbf{k}}|^2 + \sum_{\mathbf{k}\mathbf{K}} a^*_{n\mathbf{k}+\mathbf{K}} a_{n\mathbf{k}} b_{\mathbf{K}} \right]. \tag{8.20}$$

Note that, in the latter equation, there is no coupling of different band states. Finally, we need to show that the latter is the Wannier equation.

8.2.2 Wannier Equation

The Wannier theorem states that, for a perturbed problem, one can replace the band electron by a free electron and the free Hamiltonian H_0 by $E(-\mathrm{i}\nabla)$. The Wannier equation for the wave function is then

$$[E_n(-\mathrm{i}\nabla) + U(\mathbf{r})]\, \phi_n = E\phi_n. \tag{8.21}$$

To relate Eq. (8.21) to Eq. (8.20), we expand ϕ_n in a plane-wave expansion:

$$\phi_n = \frac{1}{\sqrt{V}} \sum_{\mathbf{k}} a_{n\mathbf{k}} \mathrm{e}^{\mathrm{i}\mathbf{k}\cdot\mathbf{r}}. \tag{8.22}$$

Plugging into Eq. (8.21),

$$\sum_{\mathbf{k}} a_{n\mathbf{k}} \mathrm{e}^{\mathrm{i}\mathbf{k}\cdot\mathbf{r}} [E_n(\mathbf{k}) + U(\mathbf{r})] = E \sum_{\mathbf{k}} a_{n\mathbf{k}} \mathrm{e}^{\mathrm{i}\mathbf{k}\cdot\mathbf{r}},$$

or

$$\begin{aligned}
E &= \int \mathrm{d}^3\mathbf{r}\, \phi^*_n [E_n(-\mathrm{i}\nabla) + U(\mathbf{r})]\, \phi_n \\
&= \sum_{\mathbf{k},\mathbf{k}'} a^*_{n\mathbf{k}} a_{n\mathbf{k}'} E_n(\mathbf{k}') \int \frac{\mathrm{d}^3\mathbf{r}}{V} \mathrm{e}^{\mathrm{i}(\mathbf{k}-\mathbf{k}')\cdot\mathbf{r}} + \sum_{\mathbf{k},\mathbf{k}'} a^*_{n\mathbf{k}} a_{n\mathbf{k}'} \int \frac{\mathrm{d}^3\mathbf{r}}{V} U(\mathbf{r}) \mathrm{e}^{\mathrm{i}(\mathbf{k}-\mathbf{k}')\cdot\mathbf{r}} \\
&= \sum_{\mathbf{k}} |a_{n\mathbf{k}}|^2 E_n(\mathbf{k}) + \sum_{\mathbf{k}\mathbf{K}} a^*_{n\mathbf{k}+\mathbf{K}} a_{n\mathbf{k}} b_{\mathbf{K}},
\end{aligned} \tag{8.23}$$

since, from Eq. (8.9),

$$\int d^3\mathbf{r}\, U(\mathbf{r}) e^{-i\mathbf{G}\cdot\mathbf{r}} = \sum_{\mathbf{K}} b_{\mathbf{K}} e^{i(\mathbf{K}-\mathbf{G})\cdot\mathbf{r}} = V b_{\mathbf{G}}.$$

Therefore, Eq. (8.23) corresponds to a one-band version of Eq. (8.20). The theory is now applied to donor and acceptor states.

8.2.3 Donor States

We next examine the applicability of the Wannier equation to multiply degenerate valleys of conduction bands.

8.2.3.1 Single-Valley Approximation

If both $\mathbf{k}$ and $\mathbf{K}$ lie within the same valley, one can treat each spheroid separately. Given that, one still has to investigate the convergence of the sums

$$\sum_{\mathbf{kK}} a^*_{n\mathbf{k}+\mathbf{K}} a_{n\mathbf{k}} b_{\mathbf{K}}.$$

Following KM, we now provide an order of magnitude. Since the problem is hydrogenic, take

$$\phi = e^{-r/R_0} \tag{8.24}$$

as the ground-state function. Then,

$$\begin{aligned} a_{\mathbf{k}} &= \int d^3\mathbf{r}\, \phi e^{-i\mathbf{k}\cdot\mathbf{r}} = 2\pi \int_0^\infty dr\, r^2 e^{-r/R_0} \int_{-1}^{1} dx e^{-ikrx} = \frac{2\pi}{k} \int_0^\infty dr\, r e^{-r/R_0} \sin kr \\ &= \frac{2\pi}{ik} \int_0^\infty dr\, r \left[e^{r\left(-\frac{1}{R_0}+ik\right)} - e^{r\left(\frac{1}{R_0}-ik\right)} \right] = \frac{4\pi R_0^3}{(1+k^2R_0^2)^2}. \end{aligned} \tag{8.25}$$

For $U(\mathbf{r}) \sim 1/r$,

$$b_{\mathbf{K}} = \frac{1}{V} \int d^3\mathbf{r}\, U(\mathbf{r}) e^{-i\mathbf{K}\cdot\mathbf{r}} = \frac{1}{2\varepsilon_0 V} \int_0^\infty dr\, r \int_{-1}^{1} dx e^{-iKrx} = \frac{1}{\varepsilon_0 V K^2}, \tag{8.26}$$

$$\sum_{\mathbf{kK}} a^*_{n\mathbf{k}+\mathbf{K}} a_{n\mathbf{k}} b_{\mathbf{K}} \approx \sum_{\mathbf{k}} |a_{n\mathbf{k}}|^2 \int dK\, K^2 b_{\mathbf{K}} \approx \frac{1}{\left(1+k^2R_0^2\right)^4}. \tag{8.27}$$

Thus, the sum converges fast for $kR_0 > 1$. Example: $R_0 \sim 50a_0$, $k \sim 2\pi/a$, then $kR_0 \sim 250a_0/a \sim 25$. Hence, one can solve around one minimum.

8.2.3.2 Band-Decoupling Approximation

Now examine

$$\Delta^{n'n}_{\mathbf{k},\mathbf{k}+\mathbf{K}} \approx \delta_{n'n}.$$

Since $\mathbf{K}$ is assumed small, one can find $u_{\mathbf{k}+\mathbf{K}}$ by nondegenerate perturbation theory:

$$|i\rangle = |i\rangle_0 + \sum_k{}' \frac{|k\rangle_{00}\langle k|H'|i\rangle_0}{E_i^0 - E_k^0}.$$

Here

$$H' = \frac{\hbar}{m_0}\mathbf{K}\cdot\mathbf{p} = -\frac{\mathrm{i}\hbar^2}{m_0}\mathbf{K}\cdot\nabla.$$

Thus

$$u_{n\mathbf{k}+\mathbf{K}} = u_{n\mathbf{k}} - \frac{\mathrm{i}\hbar^2}{m_0}\sum_{n'} \frac{\mathbf{K}\cdot\int \mathrm{d}^3\mathbf{r}\, u^*_{n'\mathbf{k}}\nabla u_{n\mathbf{k}}}{E_n(\mathbf{k}) - E_{n'}(\mathbf{k})} u_{n'\mathbf{k}} + O(K^2). \tag{8.28}$$

As is well known, the state normalization is accurate to order $1 + O(K^2)$, i.e.,

$$\Delta^{n'n}_{\mathbf{k},\mathbf{k}+\mathbf{K}} = 1 + O(K^2).$$

Now

$$\begin{aligned}
\Delta^{n'n}_{\mathbf{k},\mathbf{k}+\mathbf{K}} &= \int \mathrm{d}^3\mathbf{r}\, u^*_{n'\mathbf{k}+\mathbf{K}} u_{n\mathbf{k}} \\
&= \delta_{n'n} + \frac{\mathrm{i}\hbar^2}{m_0}\sum_{n''} \frac{\mathbf{K}\cdot\int \mathrm{d}^3\mathbf{r}\, u^*_{n''\mathbf{k}}\nabla u_{n'\mathbf{k}}}{E_{n'}(\mathbf{k}) - E_{n''}(\mathbf{k})} \int \mathrm{d}^3\mathbf{r}'\, u^*_{n''\mathbf{k}}(\mathbf{r}') u_{n\mathbf{k}}(\mathbf{r}') + O(K^2) \\
&= \delta_{n'n} + \frac{\mathrm{i}\hbar^2}{m_0}\frac{\mathbf{K}\cdot\int \mathrm{d}^3\mathbf{r}\, u^*_{n\mathbf{k}}\nabla u_{n'\mathbf{k}}}{E_{n'}(\mathbf{k}) - E_n(\mathbf{k})} + O(K^2) \\
&= \delta_{n'n} - \frac{\hbar}{m_0}\frac{\mathbf{K}\cdot\langle u_{n\mathbf{k}}|\mathbf{p}|u_{n'\mathbf{k}}\rangle}{E_{n'}(\mathbf{k}) - E_n(\mathbf{k})} + O(K^2).
\end{aligned} \tag{8.29}$$

In order to estimate the latter, we will appeal to the effective-mass tensor to get rid of the momentum matrix elements. Thus,

$$\begin{aligned}
E_n(\mathbf{k}+\mathbf{K}) &= E_n(\mathbf{k}) + \frac{\hbar^2 K^2}{2m_0} + \left\langle u_{n\mathbf{k}}\left|\frac{\hbar}{m_0}\mathbf{K}\cdot\mathbf{p}\right| u_{n\mathbf{k}}\right\rangle + \sum_{n'} \frac{\left|\frac{\hbar}{m_0}\mathbf{K}\cdot\langle u_{n\mathbf{k}}|\mathbf{p}|u_{n'\mathbf{k}}\rangle\right|^2}{E_n(\mathbf{k}) - E_{n'}(\mathbf{k})} \\
&\equiv E_n(\mathbf{k}) + \frac{\hbar^2}{2}K_i\left(\frac{1}{m^*}\right)_{ij} K_j,
\end{aligned}$$

$$\Longrightarrow \quad \left(\frac{1}{m^*}\right)_{ij} = \frac{1}{m_0}\delta_{ij} + \frac{2}{m_0^2}\sum_{n'} \frac{\langle u_{n\mathbf{k}}|p_i|u_{n'\mathbf{k}}\rangle\langle u_{n'\mathbf{k}}|p_j|u_{n\mathbf{k}}\rangle}{E_n(\mathbf{k}) - E_{n'}(\mathbf{k})}.$$

Along the principal directions of the spheroid, the effective-mass tensor is diagonal. Choosing one of the two possible directions and also restricting to a two-band model, one finds

$$\frac{1}{m^*} \approx \frac{2\hbar^2 \left|\int \mathrm{d}^3 r\, u^*_{l'\mathbf{k}} \nabla u_{l\mathbf{k}}\right|^2}{m_0^2\, E_l(\mathbf{k}) - E_{l'}(\mathbf{k})} \approx \frac{2\hbar^2}{K^2}\left[E_l(\mathbf{k}) - E_{l'}(\mathbf{k})\right]\left|\Delta^{l'l}_{\mathbf{k}+\mathbf{K},\,\mathbf{k}}\right|^2$$

if $1/m^* >> 1/m_0$. Then,

$$\Delta^{n'n}_{\mathbf{k}+\mathbf{K},\mathbf{k}} \approx \left[\frac{\hbar^2 K^2/2m^*}{E_n(\mathbf{k}) - E_{n'}(\mathbf{k})}\right]^{1/2}. \tag{8.30}$$

Note that $\mathbf{K}$ follows from Fourier transforming the perturbation. Thus, the effective scale is $K \sim 1/R_0$, where R_0 is the donor radius. Therefore,

$$\Delta^{n'n}_{\mathbf{k}+\mathbf{K},\mathbf{k}} \approx \left[\frac{\text{impurity ionization energy}}{\text{band gap}}\right]^{1/2} \approx \left[\text{impurity ionization energy}\right]^{1/2}$$

if the band gap is around 1 eV.

8.2.3.3 Eigenvalue Problem

Finally, for a spheroidal valley, the perfect crystal has the following dispersion relation:

$$E(\Delta\,\mathbf{k}) = \hbar^2\left[\frac{(\Delta k_{\|})^2}{2m_1} + \frac{(\Delta k_{\perp})^2}{2m_2}\right], \tag{8.31}$$

where $\Delta\mathbf{k} = \mathbf{k} - \mathbf{k}_i$ and $\mathbf{k}_i$ is the location of one of the equivalent minima. One is, therefore, solving the eigenvalue equation

$$-\frac{\hbar^2}{2m_0}\left[\frac{1}{\alpha_1}\frac{\partial^2}{\partial x^2} + \frac{1}{\alpha_2}\left(\frac{\partial^2}{\partial y^2} + \frac{\partial^2}{\partial z^2}\right)\right]\phi + \frac{q}{4\pi\varepsilon r}\phi = E\phi, \tag{8.32}$$

where $m_i^* = \alpha_i m_0$ and ε is the static dielectric constant. This equation is not separable for $\alpha_1 \neq \alpha_2$. It is common to use the variational method to estimate the binding energies. KM [133] used

$$\phi = \left[\frac{ab^2}{\pi r_0^3}\right]^{1/2} \mathrm{e}^{-\left[a^2x^2+b^2(y^2+z^2)\right]^{1/2}/r_0}, \tag{8.33}$$

Table 8.2 Ionization energies of donors [133]

	Si	Ge
P	0.039	0.0120
As	0.049	0.0127
Sb	0.039	0.0096
Theory	0.030	0.0090

where $r_0 = 4\pi\varepsilon\hbar^2/me^2$. KM reported the data in Table 8.2.

8.2.4 Acceptor States

We treat here the case when the energy band minimum is degenerate. It will be found that the simple Wannier equation gets replaced by a set of coupled differential equations. Recall that the unperturbed Hamiltonian itself, e.g., the DKK one, is not diagonal; nor will the impurity perturbation be. Hence, the process is, in effect, one of simultaneously diagonalizing both terms.

Define n orthogonal states ϕ_0^i degenerate at $\mathbf{k} = \mathbf{0}$. To first order in perturbation theory, one obtains a new set of functions

$$\phi_{n\mathbf{k}} = \left[\phi_{n\mathbf{0}} - \frac{\mathrm{i}\hbar^2}{m_0}\sum_l \frac{\mathbf{k}\cdot\int \mathrm{d}^3\mathbf{r}\,\psi_{l\mathbf{0}}^*\nabla\phi_{n\mathbf{0}}}{E_n(\mathbf{0}) - E_l(\mathbf{0})}\psi_{l\mathbf{0}}\right]\mathrm{e}^{\mathrm{i}\mathbf{k}\cdot\mathbf{r}}. \tag{8.34}$$

The sum over l does not contain the degenerate states since the matrix elements among themselves are zero. $\phi_{\mathbf{k}}^j$ is an eigenstate of the translational operator but not of the unperturbed Hamiltonian. The ϕ's are then known as pseudo-Bloch functions. Nevertheless, they are appropriate for expanding the perturbed problem:

$$\psi = \sum_n a_{n\mathbf{k}}\phi_{n\mathbf{k}}. \tag{8.35}$$

As before, if

$$(H_0 + U(\mathbf{r}))\psi = E\psi,$$

then

$$\langle\phi_{n\mathbf{k}}|H_0 + U(\mathbf{r})|\psi\rangle = Ea_{n\mathbf{k}} = \sum_{n'\mathbf{k}'} a_{n'\mathbf{k}'}\langle\phi_{n\mathbf{k}}|H_0|\phi_{n'\mathbf{k}'}\rangle + \sum_{n'\mathbf{k}'} a_{n'\mathbf{k}'}\langle\phi_{n\mathbf{k}}|U(\mathbf{r})|\phi_{n'\mathbf{k}'}\rangle. \tag{8.36}$$

Also

$$U(\mathbf{r}) = \sum_{\mathbf{K}} b_{\mathbf{K}}\mathrm{e}^{\mathrm{i}\mathbf{K}\cdot\mathbf{r}},$$

and

$$\langle \phi_{\mathbf{k}}^{j} | U(\mathbf{r}) | \phi_{\mathbf{k}'}^{i} \rangle = b_{\mathbf{K}} \Delta_{\mathbf{k},\mathbf{k}-\mathbf{K}}^{ji}. \tag{8.37}$$

Using the same procedure as previously leads to the secular equation

$$\sum_{n'} a_{n'\mathbf{k}} H_{nn'}(\mathbf{k}) + a_{n\mathbf{k}-\mathbf{K}} b_{\mathbf{K}}, \tag{8.38}$$

where

$$H_{nn'}(\mathbf{k}) = \langle \phi_{n\mathbf{k}} | H_0 | \phi_{n'\mathbf{k}} \rangle. \tag{8.39}$$

Thus, the corresponding Wannier equation is

$$\sum_{n'} H_{nn'}(-\mathrm{i}\nabla) U_{n'}(\mathbf{r}) = E U_n(\mathbf{r}). \tag{8.40}$$

Kohn and Schechter [138] used a trial function of the form

$$a\mathrm{e}^{-r/r_1} \begin{pmatrix} 1 \\ 0 \\ 0 \\ 0 \end{pmatrix} + b\mathrm{e}^{r/r_2} \begin{pmatrix} z^2 - \frac{1}{2}(x^2 + y^2) \\ 0 \\ -\left(\frac{3}{2}\right)^{1/2}(x^2 - y^2) \\ 0 \end{pmatrix} + \mathrm{i}c\mathrm{e}^{-r/r_2} \begin{pmatrix} 0 \\ (x + \mathrm{i}y)z \\ xy \\ 0 \end{pmatrix}, \tag{8.41}$$

and the variational method for Ge and got 0.089 eV for the ionization energy. A very interesting recent calculation of acceptor levels in diamond resolved a long-standing problem with respect to the measurement of the spin-orbit energy and the difference between the spin-splitting of the band states and of the acceptor levels [139].

8.3 Luttinger–Kohn Theory

The problem discussed in the previous section is now treated using the canonical transformation technique of Luttinger and Kohn [6]. The goal of Luttinger and Kohn (LK) was also to generalize the Wannier equation for a single charge in a perturbed periodic field. They developed a more systematic theory than KM. They considered band structures with increasing level of complexity: simple band with minimum at $\mathbf{k} = \mathbf{0}$, simple band with minimum not at $\mathbf{k} = \mathbf{0}$, degenerate bands, and finally degenerate bands with spin-orbit coupling. We adopt the same presentation.

8.3.1 Simple Bands

The impurity envelope-function equation will be derived followed by a simple application.

8.3.1.1 Theory

The starting Schrödinger equation (without spin–orbit interaction) is again Eq. (8.6). The wave vector $\mathbf{k}$ is no longer a good quantum number since translational symmetry is lost. As in the KM theory, we expand the solution in terms of all Bloch functions:

$$\begin{aligned}\psi(\mathbf{r}) &= \sum_n \int d^3\mathbf{k} B_n(\mathbf{k})\psi_{n\mathbf{k}}(\mathbf{r}) = \sum_n \int d^3\mathbf{k} B_n(\mathbf{k}) e^{i\mathbf{k}\cdot\mathbf{r}} u_{n\mathbf{k}}(\mathbf{r}) \\ &= \sum_n \int d^3\mathbf{k} B_n(\mathbf{k}) e^{i\mathbf{k}\cdot\mathbf{r}} \sum_{n'} A_{nn'}(\mathbf{k}) u_{n'\mathbf{0}}(\mathbf{r}) \\ &= \sum_{n'} \int d^3\mathbf{k}\, e^{i\mathbf{k}\cdot\mathbf{r}} u_{n\mathbf{0}}(\mathbf{r}) \sum_n B_n(\mathbf{k}) A_{nn'}(\mathbf{k}) \\ &\equiv \sum_{n'} \int d^3\mathbf{k} \chi_{n'\mathbf{k}}(\mathbf{r}) A_{n'}(\mathbf{k}),\end{aligned} \tag{8.42}$$

where

$$\chi_{n\mathbf{k}} \equiv e^{i\mathbf{k}\cdot\mathbf{r}} u_{n\mathbf{0}}(\mathbf{r}), \tag{8.43}$$

$$A_{n'}(\mathbf{k}) \equiv \sum_n B_n(\mathbf{k}) A_{nn'}(\mathbf{k}). \tag{8.44}$$

The state in Eq. (8.43) is the Luttinger–Kohn basis state. Indeed, this establishes that the $\chi_{n\mathbf{k}}$'s form a complete set of states too. One would similarly expect them to be orthonormal. This is now demonstrated. The proof is based upon the orthonormality of the Bloch functions, Eq. (2.3). Similarly,

$$(n\mathbf{k}|n'\mathbf{k}') \equiv \int_V d^3\mathbf{r}\, \chi^*_{n\mathbf{k}} \chi_{n'\mathbf{k}'} = \int_V d^3\mathbf{r}\, e^{i(\mathbf{k}'-\mathbf{k})\cdot\mathbf{r}} u^*_{n\mathbf{0}}(\mathbf{r}) u_{n'\mathbf{0}}(\mathbf{r}).$$

From periodicity, one can expand the product of cellular functions in a Fourier series:

$$u^*_{n\mathbf{0}}(\mathbf{r}) u_{n'\mathbf{0}}(\mathbf{r}) = \sum_m B_m^{nn'} e^{-i\mathbf{K}_m\cdot\mathbf{r}}. \tag{8.45}$$

Then

$$(n\mathbf{k}|n'\mathbf{k}') = \sum_m B_m^{nn'} \int_V d^3\mathbf{r}\, e^{i(\mathbf{k}'-\mathbf{k}-\mathbf{K}_m)\cdot\mathbf{r}} = (2\pi)^3 \sum_m B_m^{nn'} \delta(\mathbf{k}'-\mathbf{k}-\mathbf{K}_m). \tag{8.46}$$

Since $\mathbf{k}, \mathbf{k}' \in$ FBZ, the delta function can only be satisfied if $\mathbf{K}_m = \mathbf{0}$ (write $m = 0$). Then, Eq. (8.46) becomes

$$(n\mathbf{k}|n'\mathbf{k}') = (2\pi)^3 B_0^{nn'} \delta(\mathbf{k} - \mathbf{k}'). \tag{8.47}$$

We now need to evaluate $B_0^{nn'}$. From Eq. (8.45),

$$B_m^{nn'} = \frac{1}{\Omega} \int_\Omega \mathrm{d}^3\mathbf{r}\, \mathrm{e}^{\mathrm{i}\mathbf{K}_m \cdot \mathbf{r}} u_{n\mathbf{0}}^*(\mathbf{r}) u_{n'\mathbf{0}}(\mathbf{r}),$$

$$\Longrightarrow \quad B_0^{nn'} = \frac{1}{\Omega} \int_\Omega \mathrm{d}^3\mathbf{r}\, u_{n\mathbf{0}}^*(\mathbf{r}) u_{n'\mathbf{0}}(\mathbf{r}). \tag{8.48}$$

Further,

$$\langle n\mathbf{k}|n'\mathbf{k}'\rangle \equiv \int_V \mathrm{d}^3\mathbf{r}\, \psi_{n\mathbf{k}}^* \psi_{n'\mathbf{k}'} = \int_V \mathrm{d}^3\mathbf{r}\, \mathrm{e}^{\mathrm{i}(\mathbf{k}'-\mathbf{k})\cdot\mathbf{r}} u_{n\mathbf{k}}^*(\mathbf{r}) u_{n'\mathbf{k}'}(\mathbf{r}).$$

Setting $\mathbf{k}' = \mathbf{k}$ and using Eq. (2.3),

$$\delta_{nn'}\delta(\mathbf{0}) = \int_V \mathrm{d}^3\mathbf{r}\, u_{n\mathbf{0}}^*(\mathbf{r}) u_{n'\mathbf{0}}(\mathbf{r}) = N \int_\Omega \mathrm{d}^3\mathbf{r}\, u_{n\mathbf{0}}^*(\mathbf{r}) u_{n'\mathbf{0}}(\mathbf{r}),$$

where N is the number of unit cells. Integrating over $\mathbf{k}$,

$$\delta_{nn'} \int \mathrm{d}^3\mathbf{k}\, \delta(\mathbf{0}) = N \int \mathrm{d}^3\mathbf{k} \int_\Omega \mathrm{d}^3\mathbf{r}\, u_{n\mathbf{0}}^*(\mathbf{r}) u_{n'\mathbf{0}}(\mathbf{r}),$$

or

$$\delta_{nn'} = N \frac{(2\pi)^3}{V} \langle u_{n\mathbf{0}}|u_{n'\mathbf{0}}\rangle_\Omega = \frac{(2\pi)^3}{\Omega} \langle u_{n\mathbf{0}}|u_{n'\mathbf{0}}\rangle_\Omega. \tag{8.49}$$

This, together with Eq. (8.48), gives

$$B_0^{nn'} = \frac{1}{(2\pi)^3} \delta_{nn'}.$$

Therefore, Eq. (8.47) becomes

$$(n\mathbf{k}|n'\mathbf{k}') = \delta_{nn'}\delta(\mathbf{k} - \mathbf{k}'). \tag{8.50}$$

Equation (8.49) can be written down in a more useful form:

$$\langle u_{n\mathbf{0}}|u_{n'\mathbf{0}}\rangle_\Omega \equiv \int_\Omega \mathrm{d}^3\mathbf{r}\, u_{n\mathbf{0}}^*(\mathbf{r}) u_{n'\mathbf{0}}(\mathbf{r}) = \frac{\Omega}{(2\pi)^3} \delta_{nn'}. \tag{8.51}$$

Also, from Eqs. (8.47) and (8.48),

$$(n\mathbf{k}|n'\mathbf{k}') \equiv \int_V \mathrm{d}^3\mathbf{r}\, \mathrm{e}^{\mathrm{i}(\mathbf{k}'-\mathbf{k})\cdot\mathbf{r}} u_{n\mathbf{0}}^*(\mathbf{r}) u_{n'\mathbf{0}}(\mathbf{r}) = \frac{(2\pi)^3}{\Omega} \int_\Omega \mathrm{d}^3\mathbf{r}\, u_{n\mathbf{0}}^*(\mathbf{r}) u_{n'\mathbf{0}}(\mathbf{r}) \delta(\mathbf{k}-\mathbf{k}'). \tag{8.52}$$

Indeed, the latter is trivially generalized to replacing the product of u's by any periodic function.

We are now ready to proceed with the derivation of the effective-mass equation. Putting Eq. (8.42) into Eq. (8.6):

$$\sum_{n'} \int \mathrm{d}^3\mathbf{k}' \left[\frac{p^2}{2m_0} + V(\mathbf{r}) + U(\mathbf{r})\right] \chi_{n'\mathbf{k}'} = E \sum_{n'} \int \mathrm{d}^3\mathbf{k}'\, \chi_{n'\mathbf{k}'} A_{n'}(\mathbf{k}'),$$

and multiplying by $\int_V \mathrm{d}^3\mathbf{r}\, \chi_{n\mathbf{k}}^*$ gives

$$\begin{aligned}
&\textstyle\sum_{n'} \int \mathrm{d}^3\mathbf{k}'\, A_{n'}(\mathbf{k}') \int_V \mathrm{d}^3\mathbf{r}\, \mathrm{e}^{\mathrm{i}(\mathbf{k}'-\mathbf{k})\cdot\mathbf{r}} u_{n\mathbf{0}}^* \left[\frac{(p+\hbar\mathbf{k}')^2}{2m_0} + V(\mathbf{r}) + U(\mathbf{r})\right] u_{n'\mathbf{0}} \\
&\quad = E \textstyle\sum_{n'} \int \mathrm{d}^3\mathbf{k}'\, A_{n'}(\mathbf{k}') \int_V \mathrm{d}^3\mathbf{r}\, \mathrm{e}^{\mathrm{i}(\mathbf{k}'-\mathbf{k})\cdot\mathbf{r}} u_{n\mathbf{0}}^* u_{n'\mathbf{0}} = E A_n(\mathbf{k}).
\end{aligned} \tag{8.53}$$

Next

$$\sum_{n'} \int \mathrm{d}^3\mathbf{k}'\, A_{n'}(\mathbf{k}') \int \mathrm{d}^3\mathbf{r}\, \mathrm{e}^{\mathrm{i}(\mathbf{k}'-\mathbf{k})\cdot\mathbf{r}} u_{n\mathbf{0}}^* V(\mathbf{r}) u_{n'\mathbf{0}} = \sum_{n'} A_{n'}(\mathbf{k}) \frac{(2\pi)^3}{\Omega} \int_\Omega \mathrm{d}^3\mathbf{r}\, u_{n\mathbf{0}}^* V(\mathbf{r}) u_{n'\mathbf{0}},$$

$$\begin{aligned}
&\sum_{n'} \int \mathrm{d}^3\mathbf{k}'\, A_{n'}(\mathbf{k}') \int \mathrm{d}^3\mathbf{r}\, \mathrm{e}^{\mathrm{i}(\mathbf{k}'-\mathbf{k})\cdot\mathbf{r}} u_{n\mathbf{0}}^* (p+\hbar\mathbf{k}')^2 u_{n'\mathbf{0}} \\
&\quad = \sum_{n'} A_{n'}(\mathbf{k}) \frac{(2\pi)^3}{\Omega} \int_\Omega \mathrm{d}^3\mathbf{r}\, u_{n\mathbf{0}}^* (p+\hbar\mathbf{k})^2 u_{n'\mathbf{0}} \\
&\quad = A_n(\mathbf{k})\hbar^2\mathbf{k}^2 + \sum_{n'} A_{n'}(\mathbf{k}) \frac{(2\pi)^3}{\Omega} \int_\Omega \mathrm{d}^3\mathbf{r}\, u_{n\mathbf{0}}^* (p^2 + 2\hbar\mathbf{k}\cdot\mathbf{p}) u_{n'\mathbf{0}},
\end{aligned}$$

giving

$$\begin{aligned}
&A_n(\mathbf{k}) \frac{\hbar^2\mathbf{k}^2}{2m_0} + \sum_{n'} A_{n'}(\mathbf{k}) \frac{(2\pi)^3}{\Omega} \int_\Omega \mathrm{d}^3\mathbf{r}\, u_{n\mathbf{0}}^* \left[H_0 + \mathbf{k}\cdot\mathbf{p}\right] u_{n'\mathbf{0}} \\
&\quad + \sum_{n'} \int \mathrm{d}^3\mathbf{k}'\, A_{n'}(\mathbf{k}') \int_V \mathrm{d}^3\mathbf{r}\, \mathrm{e}^{\mathrm{i}(\mathbf{k}'-\mathbf{k})\cdot\mathbf{r}} u_{n\mathbf{0}}^* U(\mathbf{r}) u_{n'\mathbf{0}} = E A_n(\mathbf{k}),
\end{aligned}$$

or

$$\left[E_n(\mathbf{0})+\frac{\hbar^2k^2}{2m_0}\right]A_n(\mathbf{k})+\sum_{n'}\frac{\hbar}{m_0}\mathbf{k}\cdot\mathbf{p}_{nn'}A_{n'}(\mathbf{k})+\sum_{n'}\int \mathrm{d}^3\mathbf{k}'\,(n\mathbf{k}|U(\mathbf{r})|n'\mathbf{k}')A_{n'}(\mathbf{k}')$$
$$= EA_n(\mathbf{k}), \tag{8.54}$$

where $\mathbf{p}_{nn'}$ was defined in Eq. (2.16) and

$$(n\mathbf{k}|U(\mathbf{r})|n'\mathbf{k}') \equiv \int_V \mathrm{d}^3\mathbf{r}\,\mathrm{e}^{\mathrm{i}(\mathbf{k}'-\mathbf{k})\cdot\mathbf{r}}u_{n\mathbf{0}}^*U(\mathbf{r})u_{n'\mathbf{0}}. \tag{8.55}$$

Note that this equation is exact. We have introduced the momentum matrix elements at $\mathbf{k}=\mathbf{0}$. Two useful properties are:

$$p_{nn}^i = 0;\; p_{nn'}^i = p_{n'n}^i = (p_{nn'}^i)^*. \tag{8.56}$$

The former is because $\mathbf{k}=\mathbf{0}$ is a band minimum:

$$\mathbf{p} = m_0\mathbf{v} = m_0\dot{\mathbf{r}} = \mathrm{i}\hbar m_0[H_0,\mathbf{r}] = m_0\frac{\partial H_0}{\partial \mathbf{p}} = \frac{m_0}{\hbar}\frac{\partial H_0}{\partial \mathbf{k}},$$
$$\Longrightarrow \quad p_{nn}^i = \frac{m_0}{\hbar}\frac{\partial}{\partial k_i}\int_\Omega \mathrm{d}^3\mathbf{r}\,u_{n\mathbf{0}}^*H_0u_{n\mathbf{0}} = \frac{m_0}{\hbar}\frac{\partial E_n(\mathbf{k})}{\partial k_i}|_{\mathbf{k}=\mathbf{0}} = 0.$$

The lack of translational symmetry of U mixes different wave vectors $\mathbf{k}$; U also mixes bands n. To uncouple the bands, one would need to remove the interband coupling to order k in both V and U; the resulting effective mass would depend upon U—an undesirable situation. One way out is to find out when $\langle n\mathbf{k}|U(\mathbf{r})|n'\mathbf{k}'\rangle = 0$ for $n \neq n'$. In analyzing the latter, we use Eq. (8.45). Then,

$$(n\mathbf{k}|U(\mathbf{r})|n'\mathbf{k}') = \int_V \mathrm{d}^3\mathbf{r}\,\mathrm{e}^{\mathrm{i}(\mathbf{k}'-\mathbf{k})\cdot\mathbf{r}}u_{n\mathbf{0}}^*U(\mathbf{r})u_{n'\mathbf{0}} = \sum_m B_m^{nn'}\int_V \mathrm{d}^3\mathbf{r}\,\mathrm{e}^{\mathrm{i}(\mathbf{k}'-\mathbf{k}+\mathbf{K}_m)\cdot\mathbf{r}}U(\mathbf{r})$$
$$\equiv (2\pi)^3\sum_m B_m^{nn'}\mathcal{U}(\mathbf{k}'-\mathbf{k}+\mathbf{K}_m), \tag{8.57}$$

where

$$\mathcal{U}(\mathbf{k}) = \frac{1}{(2\pi)^3}\int_V \mathrm{d}^3\mathbf{r}\,\mathrm{e}^{-\mathrm{i}\mathbf{k}\cdot\mathbf{r}}U(\mathbf{r}).$$

The orthogonality of the Bloch functions leads to a vanishing $B_m^{nn'}$ for $n \neq n'$ only if $\mathbf{K}_m = \mathbf{0}$. This implies that $\langle n\mathbf{k}|U|n'\mathbf{k}'\rangle \approx 0$ for $n \neq n'$ if all terms except $\mathbf{K}_m = \mathbf{0}$ can be neglected; i.e., $U(\mathbf{r})$ *is smooth over a lattice unit cell*. Then, $\mathcal{U}(\mathbf{k})$ is only nonzero for $\mathbf{k} << \mathbf{K}_m \neq \mathbf{0}$. Equation (8.54) becomes

$$\left[E_n(\mathbf{0})+\frac{\hbar^2k^2}{2m_0}\right]A_n(\mathbf{k})+\sum_{n'}\frac{\hbar}{m_0}\mathbf{k}\cdot\mathbf{p}_{nn'}A_{n'}(\mathbf{k})+\int \mathrm{d}^3\mathbf{k}'\,\mathcal{U}(\mathbf{k}-\mathbf{k}')A_n(\mathbf{k}')\approx EA_n(\mathbf{k}). \tag{8.58}$$

Equation (8.58) has a band-coupling term linear-in-k. An effective-mass equation should have no band coupling term and is only accurate to order k^2. This suggests that it is sufficient to remove the linear-in-k term to order k^2. This can be achieved via a canonical transformation:

$$A=\mathrm{e}^S B,$$

The theory was developed in the absence of the inhomogeneity in Sect. 2.4. Here,

$$H(\mathbf{k})=H_0+H_1+U, \tag{8.59}$$

with

$$\begin{aligned}(n\mathbf{k}|H_0|n'\mathbf{k}')&=\left(E_n(\mathbf{0})+\frac{\hbar^2k^2}{2m_0}\right)\delta_{nn'}\delta(\mathbf{k}-\mathbf{k}'),\\ (n\mathbf{k}|H_1|n'\mathbf{k}')&=\frac{\hbar}{m_0}\mathbf{k}\cdot\mathbf{p}_{nn'}\delta(\mathbf{k}-\mathbf{k}'),\\ (n\mathbf{k}|U|n'\mathbf{k}')&=\mathcal{U}(\mathbf{k}-\mathbf{k}')\delta_{nn'},\end{aligned} \tag{8.60}$$

and Eq. (2.21) becomes

$$\overline{H}=H_0+H_1+U+[H_0,S]+[H_1,S]+[U,S]+\frac{1}{2}[[H_0,S],S]+\frac{1}{2}[[U,S],S]+\ldots \tag{8.61}$$

As before, to remove H_1, we choose S such that

$$H_1+[H_0,S]=0. \tag{8.62}$$

In terms of the χ states,

$$(n\mathbf{k}|(H_0S-SH_0)|n'\mathbf{k}')=-(n\mathbf{k}|H_1|n'\mathbf{k}')=-\frac{\hbar}{m_0}\mathbf{k}\cdot\mathbf{p}_{nn'}\delta(\mathbf{k}-\mathbf{k}').$$

The LHS is

$$\begin{aligned}&=\sum_{n''}\int \mathrm{d}^3\mathbf{k}''\left[(n\mathbf{k}|H_0|n''\mathbf{k}'')(n''\mathbf{k}''|S|n'\mathbf{k}')-(n\mathbf{k}|S|n''\mathbf{k}'')(n''\mathbf{k}''|H_0|n'\mathbf{k}')\right]\\ &=\left(E_n(\mathbf{0})+\frac{\hbar^2k^2}{2m_0}\right)(n\mathbf{k}|S|n'\mathbf{k}')-\left(E_{n'}(\mathbf{0})+\frac{\hbar^2k'^2}{2m_0}\right)(n\mathbf{k}|S|n'\mathbf{k}')\\ &=\left[E_n(\mathbf{0})-E_{n'}(\mathbf{0})+\frac{\hbar^2k^2}{2m_0}-\frac{\hbar^2k'^2}{2m_0}\right](n\mathbf{k}|S|n'\mathbf{k}').\end{aligned}$$

Hence,

$$
\begin{aligned}
(n\mathbf{k}|S|n'\mathbf{k}') &= -\frac{\hbar}{m_0}\frac{\mathbf{k}\cdot\mathbf{p}_{nn'}}{[E_n(\mathbf{0})-E_{n'}(\mathbf{0})]+\frac{\hbar^2k^2}{2m_0}-\frac{\hbar^2k'^2}{2m_0}}\delta(\mathbf{k}-\mathbf{k}') \\
&= -\frac{\hbar}{m_0}\frac{\mathbf{k}\cdot\mathbf{p}_{nn'}}{[E_n(\mathbf{0})-E_{n'}(\mathbf{0})]}\delta(\mathbf{k}-\mathbf{k}') \equiv -\frac{\mathbf{k}\cdot\mathbf{p}_{nn'}}{m_0\omega_{nn'}}\delta(\mathbf{k}-\mathbf{k}'), \quad (8.63)
\end{aligned}
$$

where $\hbar\omega_{nn'} = E_n(\mathbf{0}) - E_{n'}(\mathbf{0})$. Set

$$(n\mathbf{k}|S|n\mathbf{k}') = 0.$$

Note that S is linear in k. We now look at the correction terms in the transformed Hamiltonian [Eq. (8.61)]. Corrections to H_0 and H_1 are both seen to be second order in k:

$$[H_1, S] + \frac{1}{2}[[H_0, S], S] = [H_1, S] - \frac{1}{2}[H_1, S] = \frac{1}{2}[H_1, S],$$

and

$$
\begin{aligned}
&\frac{1}{2}(n\mathbf{k}|[H_1, S]|n'\mathbf{k}') \\
&= \frac{1}{2}\sum_{n''}\int \mathrm{d}^3\mathbf{k}''\left[(n\mathbf{k}|H_1|n''\mathbf{k}'')(n''\mathbf{k}''|S|n'\mathbf{k}') - (n\mathbf{k}|S|n''\mathbf{k}'')(n''\mathbf{k}''|H_1|n'\mathbf{k}')\right] \\
&= -\frac{\hbar}{2m_0^2}\sum_{n''}\int \mathrm{d}^3\mathbf{k}''\left[\mathbf{k}\cdot\mathbf{p}_{nn''}\delta(\mathbf{k}-\mathbf{k}'')\frac{\mathbf{k}'\cdot\mathbf{p}_{n'n''}}{\omega_{n''n'}}\delta(\mathbf{k}'-\mathbf{k}'')[1-\delta_{n'n''}][1-\delta_{n''n}]\right. \\
&\quad\left.-\frac{\mathbf{k}\cdot\mathbf{p}_{nn''}}{\omega_{nn''}}\delta(\mathbf{k}-\mathbf{k}'')[1-\delta_{nn''}][1-\delta_{n'n''}]\mathbf{k}'\cdot\mathbf{p}_{n'n''}\delta(\mathbf{k}'-\mathbf{k}'')\right] \\
&= \frac{\hbar}{2m_0^2}\sum_{n''}[1-\delta_{n''n}][1-\delta_{n''n'}]\delta(\mathbf{k}-\mathbf{k}')\mathbf{k}\cdot\mathbf{p}_{nn''}\mathbf{k}'\cdot\mathbf{p}_{n''n'}\left(\frac{1}{\omega_{nn''}}+\frac{1}{\omega_{n'n''}}\right). \quad (8.64)
\end{aligned}
$$

Corrections to U are:

$$
\begin{aligned}
&(n\mathbf{k}|[U, S]|n'\mathbf{k}') \\
&= \sum_{n''}\int \mathrm{d}^3\mathbf{k}''\left[(n\mathbf{k}|U|n''\mathbf{k}'')(n''\mathbf{k}''|S|n'\mathbf{k}') - (n\mathbf{k}|S|n''\mathbf{k}'')(n''\mathbf{k}''|U|n'\mathbf{k}')\right] \\
&= -\frac{1}{m_0}\int \mathrm{d}^3\mathbf{k}''\left[\mathcal{U}(\mathbf{k}-\mathbf{k}'')\frac{\mathbf{k}'\cdot\mathbf{p}_{nn'}}{\omega_{nn'}}\delta(\mathbf{k}'-\mathbf{k}'') - \frac{\mathbf{k}\cdot\mathbf{p}_{nn'}}{\omega_{nn''}}\mathcal{U}(\mathbf{k}'-\mathbf{k}'')\delta(\mathbf{k}-\mathbf{k}'')\right] \\
&= \frac{(\mathbf{k}-\mathbf{k}')\cdot\mathbf{p}_{nn'}}{m_0\omega_{nn'}}\mathcal{U}(\mathbf{k}-\mathbf{k}')[1-\delta_{nn'}].
\end{aligned}
$$

This is seen to be linear in k but has magnitude $\sim \mathcal{U}(\overline{\mathbf{k}})/\overline{\omega}$. Since, the Fourier component of the perturbation determines the binding energy and for a shallow impurity, the band gap is much larger, one can safely drop this term. Similarly, one can also neglect the next-order correction $[[U, S], S]$.

The aim is next to obtain the new equation for the B's. We now present two different approaches. The basic difference in the two derivations involves treating the impurity potential separately and not. Since the latter is more straightforward, we give it first. It is based upon the realization that if

$$HA = EA,$$

then

$$\overline{H}B = EB.$$

This was shown in Sect. 2.4. The new transformed Hamiltonian is

$$\overline{H} = H_0 + U + \frac{1}{2}[H_1, S] + O(k^3). \tag{8.65}$$

Note that, in Eq. (8.64), interband terms ($n \neq n'$) will lead to band-diagonal terms at a higher order in k (i.e., $O(k^3)$). Since these terms are dropped, we will also set $n = n'$ in Eq. (8.64). Thus, we have

$$\left(E_n(\mathbf{0}) + \frac{\hbar^2 k^2}{2m_0} + \frac{\hbar}{m_0^2} k_i k_j \sum_{n''}{}' \frac{p^i_{nn''} p^j_{n''n}}{\omega_{nn''}}\right) B_n(\mathbf{k}) + \int \mathrm{d}^3\mathbf{k}' \, \mathcal{U}(\mathbf{k}-\mathbf{k}') B_n(\mathbf{k}') = E B_n(\mathbf{k}). \tag{8.66}$$

One can simplify this further by noting that the coefficient of $B_n(\mathbf{k})$ on the LHS is the expansion of $E_n(\mathbf{k})$ to second order. Using second-order nondegenerate perturbation theory about $\mathbf{k} = \mathbf{0}$ (first-order term vanishes due to symmetry),

$$\begin{aligned} E_n(\mathbf{k}) &= E_n(\mathbf{0}) + \frac{\hbar^2 k^2}{2m_0} + \sum_{n''}{}' \frac{\langle u_{n\mathbf{0}}|H_1|u_{n''\mathbf{0}}\rangle \langle u_{n''\mathbf{0}}|H_1|u_{n\mathbf{0}}\rangle}{E_n(\mathbf{0}) - E_{n''}(\mathbf{0})} \\ &= E_n(\mathbf{0}) + \frac{\hbar^2 k^2}{2m_0} + \frac{\hbar}{m_0^2} k_i k_j \sum_{n''}{}' \frac{p^i_{nn''} p^j_{n''n}}{\omega_{nn''}}. \end{aligned}$$

Incidentally, the last equation gives us the f-sum rule:

$$\frac{\partial^2 E_n(\mathbf{k})}{\partial k_i \partial k_j} = \frac{\hbar^2}{m_0}\delta_{ij} + 2\frac{\hbar}{m_0^2} \sum_{n''}{}' \frac{p^i_{nn''}\mathbf{k} p^j_{n''n}\mathbf{k}}{\omega_{nn''}}. \tag{8.67}$$

Finally, we get the effective-mass equation in "momentum space":

$$E_n(\mathbf{k}) B_n(\mathbf{k}) + \int \mathrm{d}^3\mathbf{k}' \, \mathcal{U}(\mathbf{k}-\mathbf{k}') B_n(\mathbf{k}') = E B_n(\mathbf{k}). \tag{8.68}$$

We now provide another derivation of the effective-mass equation. Rewrite Eq. (8.58) as

$$\int \mathrm{d}^3\mathbf{k}' \left[E_n(\mathbf{0}) + \frac{\hbar^2 k^2}{2m_0} \right] A_n(\mathbf{k}')\delta(\mathbf{k}-\mathbf{k}') + \int \mathrm{d}^3\mathbf{k}' \sum_{n'} \frac{\hbar}{m_0}\mathbf{k}'\cdot\mathbf{p}_{nn'}A_{n'}(\mathbf{k}')\delta(\mathbf{k}-\mathbf{k}')$$
$$+ \int \mathrm{d}^3\mathbf{k}'\,\mathcal{U}(\mathbf{k}-\mathbf{k}')A_n(\mathbf{k}') = \int \mathrm{d}^3\mathbf{k}'\,E(\mathbf{k}')A_n(\mathbf{k}')\delta(\mathbf{k}-\mathbf{k}'). \quad (8.69)$$

One can simplify the notation to

$$\sum_{n'} \int \mathrm{d}^3\mathbf{k}' \left[H_0^{nn'}(\mathbf{k},\mathbf{k}') + H_1^{nn'}(\mathbf{k},\mathbf{k}') \right] A_{n'}(\mathbf{k}')$$
$$= \int \mathrm{d}^3\mathbf{k}' \left[-\mathcal{U}(\mathbf{k}-\mathbf{k}') + E(\mathbf{k}')\delta(\mathbf{k}-\mathbf{k}') \right] A_n(\mathbf{k}'),$$

where

$$H_0^{nn'}(\mathbf{k},\mathbf{k}') = \left[E_n(\mathbf{0}) + \frac{\hbar^2 k^2}{2m_0} \right] \delta(\mathbf{k}-\mathbf{k}')\delta_{nn'} = (n\mathbf{k}|H_0|n'\mathbf{k}'),$$

$$H_1^{nn'}(\mathbf{k},\mathbf{k}') = \frac{\hbar}{m_0}\mathbf{k}'\cdot\mathbf{p}_{nn'}\delta(\mathbf{k}-\mathbf{k}') = (n\mathbf{k}|H_1|n'\mathbf{k}').$$

Formally, one can also write

$$H_{01}\cdot A = (E-\mathcal{U})\cdot A.$$

This simplifies finding the transformed equation since we now have

$$H_{01}\mathrm{e}^S B = (E-\mathcal{U})\mathrm{e}^S B,$$
$$\overline{H}_{01}B = (e^{-S}He^S)B = \mathrm{e}^{-S}(E-\mathcal{U})\mathrm{e}^S B \approx EB - \mathcal{U}B + S\mathcal{U}B - \mathcal{U}SB. \quad (8.70)$$

Simplifying each side in turn, the RHS contains $(SUB)_{n\mathbf{k}} - (USB)_{n\mathbf{k}} \equiv$ RHS

$$= \left[(n\mathbf{k}|S|n''\mathbf{k}'')(n''\mathbf{k}''|U|n'\mathbf{k}') - (n\mathbf{k}|U|n''\mathbf{k}'')(n''\mathbf{k}''|S|n'\mathbf{k}') \right] (n'\mathbf{k}'|B\rangle$$
$$= \frac{\hbar}{m_0}\left[-\frac{\mathbf{k}\cdot\mathbf{p}_{nn''}}{[E_n(\mathbf{0})-E_{n''}(\mathbf{0})]}(n''\mathbf{k}|U|n'\mathbf{k}') + \frac{\mathbf{k}'\cdot\mathbf{p}_{n''n'}}{[E_{n''}(\mathbf{0})-E_{n'}(\mathbf{0})]}(n\mathbf{k}|U|n'\mathbf{k}') \right] (n'\mathbf{k}'|B\rangle.$$

But

$$(n''\mathbf{k}|U|n'\mathbf{k}') = \delta_{n'n''}\mathcal{U}(\mathbf{k}-\mathbf{k}'),$$

then

$$\mathrm{RHS} = \frac{\hbar}{m_0}\left[-\frac{\mathbf{k}\cdot\mathbf{p}_{nn'}}{[E_n(\mathbf{0})-E_{n'}(\mathbf{0})]}\mathcal{U}(\mathbf{k}-\mathbf{k}') + \frac{\mathbf{k}'\cdot\mathbf{p}_{nn'}}{[E_n(\mathbf{0})-E_{n'}(\mathbf{0})]}\mathcal{U}(\mathbf{k}-\mathbf{k}') \right] (n'\mathbf{k}'|B\rangle$$

$$= \frac{\hbar}{m_0} \frac{(\mathbf{k}' - \mathbf{k}) \cdot \mathbf{p}_{nn'}}{[E_n(\mathbf{0}) - E_{n'}(\mathbf{0})]} \mathcal{U}(\mathbf{k} - \mathbf{k}')(n'\mathbf{k}'|B\rangle.$$

As for the LHS, the analysis follows exactly that for the perfect crystal. Thus,

$$\overline{H}_{01} = H_0 + \frac{1}{2}[H_1, S] + O(k^3),$$

and

$$\overline{H}_{01}^{nn'}(\mathbf{k}, \mathbf{k}') - H_0^{nn'}(\mathbf{k}, \mathbf{k}') + \frac{1}{2}(n\mathbf{k}|[H_1, S]|n'\mathbf{k}') - H_0^{nn'}(\mathbf{k}, \mathbf{k}') \\ + \frac{1}{2}\sum_{n''}\Big[(n\mathbf{k}|H_1|n''\mathbf{k}'')(n''\mathbf{k}''|S|n'\mathbf{k}') - (n\mathbf{k}|S|n''\mathbf{k}'')(n''\mathbf{k}''|H_1|n'\mathbf{k}')\Big],$$

giving for the LHS of Eq. (8.70)

$$(n\mathbf{k}|\overline{H}_{01}B|n'\mathbf{k}') \\ = \left(E_n(\mathbf{0}) + \frac{\hbar^2 k^2}{2m_0}\right) B_n(\mathbf{k}) + \frac{\hbar^2}{m_0^2} {\sum_{n''}}' \left[\frac{\mathbf{k} \cdot \mathbf{p}_{nn''}\, \mathbf{k} \cdot \mathbf{p}_{n''n'}}{[E_{n'}(\mathbf{0}) - E_{n''}(\mathbf{0})]} + \frac{\mathbf{k} \cdot \mathbf{p}_{nn''}\, \mathbf{k} \cdot \mathbf{p}_{n''n'}}{[E_n(\mathbf{0}) - E_{n''}(\mathbf{0})]}\right] B_{n'}(\mathbf{k}).$$

Hence, we have

$$\left(E_n(\mathbf{0}) + \frac{\hbar^2 k^2}{2m_0}\right) B_n(\mathbf{k}) + \frac{\hbar^2}{m_0^2} {\sum_{n''}}' \left[\frac{\mathbf{k} \cdot \mathbf{p}_{nn''}\, \mathbf{k} \cdot \mathbf{p}_{n''n'}}{[E_{n'}(\mathbf{0}) - E_{n''}(\mathbf{0})]} + \frac{\mathbf{k} \cdot \mathbf{p}_{nn''}\, \mathbf{k} \cdot \mathbf{p}_{n''n'}}{[E_n(\mathbf{0}) - E_{n''}(\mathbf{0})]}\right] B_{n'}(\mathbf{k}) \\ + \int \mathrm{d}^3\mathbf{k}'\, \mathcal{U}(\mathbf{k} - \mathbf{k}') B_n(\mathbf{k}') + \frac{\hbar}{m_0} \frac{(\mathbf{k}' - \mathbf{k}) \cdot \mathbf{p}_{nn'}}{[E_n(\mathbf{0}) - E_{n'}(\mathbf{0})]} \mathcal{U}(\mathbf{k} - \mathbf{k}') B_{n'}(\mathbf{k}') \\ = E(\mathbf{k}) B_n(\mathbf{k}). \tag{8.71}$$

It is worthwhile to summarize the essense of Eq. (8.71). There were two band-coupling terms in the original Hamiltonian: V and U. We have attempted to remove these interband terms to first order in k via a canonical transformation. For the crystal potential, this was successful, just as for the perfect-crystal case, and one can now absorb the interband terms to second order into an effective mass. However, this time, there is a residual term linear in k (and in the impurity potential). Such a term is, nevertheless, smaller compared to the original linear-in-k term (the $k \cdot p$ term) by a factor of $\mathcal{U}/\Delta E$. This term is, therefore, indeed smaller if the impurity is shallow, i.e., the binding energy is smaller than the band gap. This, of course, relies on the conventional definition of a shallow impurity and also assume the only relevant band coupling is at the fundamental gap. Furthermore, we note that this impurity term differs from the other impurity term in the LHS by a factor of $k \cdot p/\Delta E \sim S$ (indeed, arising from $[U, S]$). Also, being an interband term, it actually differs by a factor of S^2. Since we have assumed S small in the theory, the neglect of the linear-in-k term in Eq. (8.71) is consistent. Finally, it should be reminded that the equation is also only valid for small k. The gist of the argument is that, the effective-mass equation valid to order k^2 is

$$\left[E_n(\mathbf{0})+\frac{\hbar^2k^2}{2m_n}\right]B_n(\mathbf{k})+\int \mathrm{d}^3\mathbf{k}'\,\mathcal{U}(\mathbf{k}-\mathbf{k}')B_n(\mathbf{k}')=E(\mathbf{k})B_n(\mathbf{k}), \tag{8.72}$$

which is the same as Eq. (8.68). Equation (8.72) is similar to Schrödinger's equation for a potential U in momentum space. An important distinction, however, is that the wave vector is here restricted to the FBZ instead of the whole space. *Were the wave vector not restricted to the FBZ, this equation would then describe a particle of mass m_n moving in a potential $U(\mathbf{r})$.*

Nevertheless, our goal is to see how close to this description one can get. Thus, we define a real-space function $F(\mathbf{r})$ corresponding to the Fourier coefficients $B_n(\mathbf{k})$:

$$F_n(\mathbf{r})\equiv\int_{\mathrm{FBZ}}\mathrm{d}^3\mathbf{k}\,\mathrm{e}^{\mathrm{i}\mathbf{k}\cdot\mathbf{r}}B_n(\mathbf{k}). \tag{8.73}$$

Then, Eq. (8.72) becomes

$$\int_{\mathrm{FBZ}}\mathrm{d}^3\mathbf{k}\left[E_n(\mathbf{0})-\frac{\hbar^2\nabla^2}{2m_n}\right]\mathrm{e}^{\mathrm{i}\mathbf{k}\cdot\mathbf{r}}B_n(\mathbf{k})+\int \mathrm{d}^3\mathbf{k}\int \mathrm{d}^3\mathbf{k}'\,\mathcal{U}(\mathbf{k}-\mathbf{k}')\mathrm{e}^{\mathrm{i}\mathbf{k}\cdot\mathbf{r}}B_n(\mathbf{k}')$$
$$=E\int \mathrm{d}^3\mathbf{k}\,\mathrm{e}^{\mathrm{i}\mathbf{k}\cdot\mathbf{r}}B_n(\mathbf{k})=E\,F_n(\mathbf{r}).$$

On the RHS, the energy E is outside the integral since it is a parameter to be found by solving the equation. Recall that

$$\mathcal{U}(\mathbf{k}-\mathbf{k}')=\frac{1}{(2\pi)^3}\int_{V'}\mathrm{d}V'\,\mathrm{e}^{\mathrm{i}(\mathbf{k}-\mathbf{k}')\cdot\mathbf{r}'}U(\mathbf{r}'),$$

and the impurity term on the LHS of Eq. (8.72) becomes

$$\frac{1}{(2\pi)^3}\int \mathrm{d}^3\mathbf{k}\int \mathrm{d}^3\mathbf{k}'\int_{V'}\mathrm{d}V'\,U(\mathbf{r}')\mathrm{e}^{\mathrm{i}(\mathbf{k}-\mathbf{k}')\cdot\mathbf{r}'}B_n(\mathbf{k}')\mathrm{e}^{\mathrm{i}\mathbf{k}\cdot\mathbf{r}}$$
$$=\int_{V'}\mathrm{d}V\,U(\mathbf{r}')\frac{1}{(2\pi)^3}\int \mathrm{d}^3\mathbf{k}\,\mathrm{e}^{\mathrm{i}\mathbf{k}\cdot(\mathbf{r}-\mathbf{r}')}\int \mathrm{d}^3\mathbf{k}'\,\mathrm{e}^{\mathrm{i}\mathbf{k}'\cdot\mathbf{r}'}B_n(\mathbf{k}')$$
$$=\int_{V'}\mathrm{d}V'\,U(\mathbf{r}')\Delta(\mathbf{r}-\mathbf{r}')F_n(\mathbf{r}').$$

The function $\Delta(\mathbf{r})$ is approximately a delta function if multiplied by a slowly-varying function since $\Delta(\mathbf{r})$ has a width of a; it would be a delta function exactly if k is in all space. Note that $\int_{\text{all space}}\mathrm{d}V\,\Delta(\mathbf{r})=1$. Hence, *if we assume $F_n(\mathbf{r})$ to be fairly smooth*, one has

$$-\frac{\hbar^2}{2m_n}\nabla^2 F_n(\mathbf{r}) + U(\mathbf{r})F_n(\mathbf{r}) = [E - E_n(\mathbf{0})]F_n(\mathbf{r}), \tag{8.74}$$

which is Schrödinger's equation for a particle of mass m_n moving in a potential $U(\mathbf{r})$. It can also be rewritten down as

$$[E_n(-\mathrm{i}\nabla) + U(\mathbf{r})]\, F_n(\mathbf{r}) = E F_n(\mathbf{r}). \tag{8.75}$$

The smoothness of $F_n(\mathbf{r})$ remains to be studied and of its connection to the real wave function ψ. A smooth $F_n(\mathbf{r})$ implies that its Fourier coefficients [see Eq. (8.73)] will only be appreciable for small $\mathbf{k}$; i.e., for wave vectors smaller than the size of the FBZ ($\mathbf{k} << \mathbf{K}_m \forall\, m \neq 0$). Small k implies that S is small and, therefore,

$$A_n(\mathbf{k}) = \mathrm{e}^S B_n(\mathbf{k}) \approx B_n(\mathbf{k}).$$

Finally, the wave function becomes [starting with Eq. (8.42)]

$$\psi(\mathbf{r}) \approx \sum_n \int \mathrm{d}^3\mathbf{k}\, \mathrm{e}^{\mathrm{i}\mathbf{k}\cdot\mathbf{r}} u_{n\mathbf{0}}(\mathbf{r}) B_n(\mathbf{k}) = \sum_n F_n(\mathbf{r}) u_{n\mathbf{0}}(\mathbf{r}). \tag{8.76}$$

Note also that our effective-mass equation has no interband coupling. Hence, it would be consistent to write

$$\psi(\mathbf{r}) = F_n(\mathbf{r}) u_{n\mathbf{0}}(\mathbf{r}). \tag{8.77}$$

It is thus seen that the envelope function has replaced the plane-wave factor for the Bloch solution near the zone center.

8.3.1.2 Band with Minimum not at Center

For a band minimum at $\mathbf{k} = \mathbf{k}_0$, one uses as basis a state with a wave vector $\mathbf{k}$ away from the Bloch state at $\mathbf{k}_0$:

$$\varphi_{n\mathbf{k}} = \mathrm{e}^{\mathrm{i}\mathbf{k}\cdot\mathbf{r}}(\mathrm{e}^{\mathrm{i}\mathbf{k}_0\cdot\mathbf{r}} u_{n\mathbf{k}_0}). \tag{8.78}$$

One can show that the $\varphi_{n\mathbf{k}}$'s satisfy the same properties as the $\chi_{n\mathbf{k}}$'s. For example,

$$\langle \varphi_{n\mathbf{k}} | \varphi_{n'\mathbf{k}'} \rangle = \delta_{nn'}\delta(\mathbf{k} - \mathbf{k}'),$$

and

$$p^i_{nn}(\mathbf{k}_0) = 0.$$

Hence, one can repeat the above derivations and get essentially the same results. Namely, the wave function can be written down as

$$\psi(\mathbf{r}) = F(\mathbf{r})\psi_{n\mathbf{k}_0}(\mathbf{r}), \tag{8.79}$$

where F is the solution of the effective-mass equation

$$\left[E\left(\mathbf{k}_0 + \frac{1}{\mathrm{i}}\nabla\right) + U(\mathbf{r})\right] F(\mathbf{r}) = EF(\mathbf{r}), \tag{8.80}$$

where E is expanded to second order in $\frac{1}{\mathrm{i}}\nabla$ near $\mathbf{k}_0$.

Generally, the bulk band structure for $\mathbf{k}_0 \neq \mathbf{0}$ has a many-valley structure. For an arbitrary impurity potential, the energy degeneracy of the valleys is broken. Then, the various solutions are independent solutions of the Schrödinger equation. These solutions will also be approximately orthogonal since, if the envelope function is fairly constant within a cell (as is required by the effective-mass theory), then the orthogonality follows from Eq. (8.79). If, however, the impurity potential is either spherically symmetric or has the point-group symmetry, then the valleys stay degenerate. Then, a linear combination of the multivalley states is required and this will lead to a multiplet splitting of the energy states.

8.3.1.3 Solution

The solution of Eq. (8.74) for a Coulombic potential is well known. One can rewrite the equation explicitly as

$$\left[-\frac{\hbar^2}{2m^*}\nabla^2 - \frac{e^2}{4\pi\varepsilon_r\varepsilon_0 r}\right] F_n(\mathbf{r}) = [E - E_n(\mathbf{0})]F_n(\mathbf{r}). \tag{8.81}$$

This is a hydrogenic problem with an effective mass m^* and a relative permittivity ε_r. The ground bound state is the $1S$ state

$$F_{1S}(\mathbf{r}) = \frac{1}{(\pi a_0^3)^{1/2}} \mathrm{e}^{-r/a_0}, \tag{8.82}$$

where a_0 is the effective Bohr radius:

$$a_0 = \frac{4\pi\varepsilon_r\varepsilon_0\hbar^2}{m^* e^2} = 0.53 \frac{\varepsilon_r}{m^*/m_0} \text{ Å}. \tag{8.83}$$

For example, GaAs has $\varepsilon_r = 12.85$ and $m^* = 0.067m_0$. These give $a_0 \sim 100$ Å and $E - E_n(\mathbf{0}) \sim 5.5$ meV.

8.3.2 Degenerate Bands

Without loss of generality, we will assume $\mathbf{k}_0 = \mathbf{0}$. To simplify the discussion, we will also neglect spin for now; the latter will be considered in the next subsection.

Consider a set of states ϕ_s degenerate at $\mathbf{k} = \mathbf{0}$; other states will be denoted by ϕ_r in analogy to the remote set of Löwdin. The idea is to again establish a complete set of orthonormal functions with properties similar to those of $\chi_{n\mathbf{k}}$. Once this is done, the earlier theory follows through almost unchanged, particularly if there are no coupling terms from the perturbation to couple the degenerate states.

First we note that, barring accidental degeneracy, the ϕ_s must belong to an irreducible representation of the crystal point group. For DM crystals, inversion ensures that the ϕ_s's have definite parity (indeed, even for the valence-band states). Thus, the matrix elements of the momentum operator between those states vanish:

$$p^i_{ss'}(\mathbf{0}) = 0. \tag{8.84}$$

Similar to the nondegenerate case, the basis states for the perturbed problem will be taken to be

$$\phi_{n\mathbf{k}} = \mathrm{e}^{\mathrm{i}\mathbf{k}\cdot\mathbf{r}}\phi_n, \tag{8.85}$$

where n include both the degenerate and distant states. The impurity wave function is then written as

$$\psi(\mathbf{r}) = \sum_n \int \mathrm{d}^3\mathbf{k}\, A_n(\mathbf{k})\phi_{n\mathbf{k}}(\mathbf{r}). \tag{8.86}$$

The Schrödinger equation becomes

$$\sum_n \int \mathrm{d}^3\mathbf{k}'\,(n\mathbf{k}|H_0 + U|n'\mathbf{k}')A_{n'}(\mathbf{k}') = EA_n(\mathbf{k}). \tag{8.87}$$

Recall that, for a simple band, we had

$$\left[E_n(\mathbf{0}) + \frac{\hbar^2k^2}{2m_0}\right]A_n(\mathbf{k}) + \sum_{n'}\frac{\hbar}{m_0}\mathbf{k}\cdot\mathbf{p}_{nn'}A_{n'}(\mathbf{k}) + \int \mathrm{d}^3\mathbf{k}'\,\mathcal{U}(\mathbf{k}-\mathbf{k}')A_n(\mathbf{k}') = EA_n(\mathbf{k}).$$

For $n = s$, since $p^i_{ss'} = 0$, we have

$$p^i_{sn'} = p^i_{sr}\delta_{n'r}, \tag{8.88}$$

and

$$\left[E_n(\mathbf{0}) + \frac{\hbar^2k^2}{2m_0}\right]A_s(\mathbf{k}) + \sum_r\frac{\hbar}{m_0}\mathbf{k}\cdot\mathbf{p}_{sr}A_r(\mathbf{k}) + \int \mathrm{d}^3\mathbf{k}'\,\mathcal{U}(\mathbf{k}-\mathbf{k}')A_s(\mathbf{k}') = EA_s(\mathbf{k}). \tag{8.89}$$

We again transform away the linear-in-k interband term using

$$(n\mathbf{k}|S|n'\mathbf{k}') = \begin{cases} -\frac{\mathbf{k}\cdot\mathbf{p}_{nn'}}{m_0\omega_{nn'}}\delta(\mathbf{k}-\mathbf{k}'), & n, n' \neq s, \\ 0, & n \text{ or } n' = s. \end{cases} \tag{8.90}$$

In zeroth-order effective-mass theory,

$$A_s(\mathbf{k}) = \sum_n \int \mathrm{d}^3\mathbf{k}'\,(s\mathbf{k}|T|n\mathbf{k}')B_n(\mathbf{k}') \approx \sum_n \int \mathrm{d}^3\mathbf{k}'\,(s\mathbf{k}|1+S|n\mathbf{k}')B_n(\mathbf{k}') \approx B_s(\mathbf{k}).$$

The relevant simple-band results were

$$\frac{1}{2}(n\mathbf{k}|[H_1, S]|n'\mathbf{k}') = \frac{\hbar}{2m_0^2}\sum_{n''\neq n,n'} \mathbf{k}\cdot\mathbf{p}_{nn''}\mathbf{k}\cdot\mathbf{p}_{n''n'}\left(\frac{1}{\omega_{nn''}} + \frac{1}{\omega_{n'n''}}\right)\delta(\mathbf{k}-\mathbf{k}'),$$

$$\sum_n \int \mathrm{d}^3\mathbf{k}\,(n\mathbf{k}|\overline{H}|n'\mathbf{k}')B_{n'}(\mathbf{k}') = EB_n(\mathbf{k}),$$

$$\overline{H} = H_0 + U + \frac{1}{2}[H_1, S] + O(k^3).$$

Since we are interested in the degenerate bands s, we will drop all other bands. Hence, Eq. (8.89) becomes, dropping the $n = r$ matrix element,

$$\left[E_n(\mathbf{0}) + \frac{\hbar^2k^2}{2m_0}\right]B_s(\mathbf{k}) + \int \mathrm{d}^3\mathbf{k}'\,\mathcal{U}(\mathbf{k}-\mathbf{k}')B_s(\mathbf{k}')$$
$$+\sum_{s'}\int \mathrm{d}^3\mathbf{k}'\left(s\mathbf{k}\left|\frac{1}{2}[H_1, S]\right|s'\mathbf{k}'\right)B_{s'}(\mathbf{k}') = EB_s(\mathbf{k}).$$

Now

$$\sum_{s'}\int \mathrm{d}^3\mathbf{k}'\left(s\mathbf{k}\left|\frac{1}{2}[H_1, S]\right|s'\mathbf{k}'\right)B_{s'}(\mathbf{k}') = \sum_{s'}\frac{\hbar}{m_0^2}\sum_{n''\neq s,s'}\frac{\mathbf{k}\cdot\mathbf{p}_{sn''}\mathbf{k}\cdot\mathbf{p}_{n''s'}}{\omega_{sn''}}B_{s'}(\mathbf{k}).$$

Thus,

$$\sum_{s'}\left[E_s(\mathbf{0})\delta_{ss'} + \frac{\hbar^2k^2}{2m_0}\delta_{ss'} + \frac{\hbar}{m_0^2}k_ik_j\sum_{n''\neq s,s'}\frac{p^i_{sn''}p^j_{n''s'}}{\omega_{sn''}}\right]B_{s'}(\mathbf{k})$$
$$+\int \mathrm{d}^3\mathbf{k}'\,\mathcal{U}(\mathbf{k}-\mathbf{k}')B_s(\mathbf{k}') = EB_s(\mathbf{k}). \tag{8.91}$$

In the absence of the perturbation, Eq. (8.91) is the secular equation for the B_s, with the matrix elements being given by the momentum matrix elements at $\mathbf{k} = \mathbf{0}$. It is now conventional to either shift the valence-band energy $E_n(\mathbf{0})$ to the RHS of the equation or to set it to zero. Then, one can rewrite Eq. (8.91) as

$$\sum_{s'}(D_{ss'}^{ij}k_ik_j)B_{s'}(\mathbf{k}) + \int \mathrm{d}^3\mathbf{k}'\,\mathcal{U}(\mathbf{k}-\mathbf{k}')B_s(\mathbf{k}') = EB_s(\mathbf{k}), \tag{8.92}$$

where the matrix D is the same as previously given [Eq. (3.6)]; it plays the role of the inverse effective mass in a one-band model. Note that Eq. (8.92) is still in momentum space. We can transform to coordinate space as before by introducing the envelope function

$$F_s(\mathbf{r}) \equiv \int_{\mathrm{FBZ}} \mathrm{d}^3\mathbf{k}\,\mathrm{e}^{\mathrm{i}\mathbf{k}\cdot\mathbf{r}} B_s(\mathbf{k}). \tag{8.93}$$

As we have seen already, this is implemented by replacing k_α by $\frac{1}{\mathrm{i}}\nabla_\alpha$. Thus, Eq. (8.92) becomes

$$\sum_{s'}\left[D_{ss'}^{ij}(\frac{1}{\mathrm{i}}\nabla_i)(\frac{1}{\mathrm{i}}\nabla_j) + U(\mathbf{r})\delta_{ss'}\right]F_{s'}(\mathbf{r}) = EF_s(\mathbf{r}). \tag{8.94}$$

The latter is the effective-mass equation for impurity states in a degenerate-band semiconductor. It is a set of coupled differential equations and this result is identical to that of Kittel–Mitchell [133]. The impurity wave function is seen to be

$$\begin{aligned}\psi(\mathbf{r}) &= \sum_n \int \mathrm{d}^3\mathbf{k} A_n(\mathbf{k})\phi_{n\mathbf{k}}(\mathbf{r}) = \sum_n \int \mathrm{d}^3\mathbf{k} A_n(\mathbf{k})\mathrm{e}^{\mathrm{i}\mathbf{k}\cdot\mathbf{r}}\phi_n(\mathbf{r}) \\ &= \sum_s \int \mathrm{d}^3\mathbf{k} A_s(\mathbf{k})\mathrm{e}^{\mathrm{i}\mathbf{k}\cdot\mathbf{r}}\phi_s(\mathbf{r}) \approx \sum_s \int \mathrm{d}^3\mathbf{k} B_s(\mathbf{k})\mathrm{e}^{\mathrm{i}\mathbf{k}\cdot\mathbf{r}}\phi_s(\mathbf{r}) \\ &\equiv \sum_s F_s(\mathbf{r})\phi_s(\mathbf{r}). \end{aligned} \tag{8.95}$$

8.3.3 Spin-Orbit Coupling

We now add the spin-orbit coupling to the above discussion. The Hamiltonian remains periodic and the unperturbed Hamiltonian

$$\overline{H}_0 = H_0 + H_{\mathrm{so}}. \tag{8.96}$$

still has Bloch functions as solutions. Let the solutions be given by

$$\overline{H}_0\overline{\psi}_{n\mathbf{k}} = \overline{E}_n(\mathbf{k})\overline{\psi}_{n\mathbf{k}}, \tag{8.97}$$

with

$$\overline{\psi}_{n\mathbf{k}} = \mathrm{e}^{\mathrm{i}\mathbf{k}\cdot\mathbf{r}}\overline{u}_{n\mathbf{k}}. \tag{8.98}$$

The $\overline{u}_{n\mathbf{k}}$'s differ from the $u_{n\mathbf{k}}$'s via the addition of a spinor function. In discussing the perturbed problem, it is now necessary to distinguish between the nondegenerate and degenerate cases.

8.3.3.1 Nondegenerate Case

Introduce a complete set of functions

$$\overline{\chi}_{n\mathbf{k}} = \mathrm{e}^{\mathrm{i}\mathbf{k}\cdot\mathbf{r}}\overline{u}_{n0}, \tag{8.99}$$

and let the perturbed wave function be

$$\psi = \sum_n \int \mathrm{d}^3\mathbf{k} A_n(\mathbf{k})\overline{\chi}_{n\mathbf{k}}. \tag{8.100}$$

The earlier theory follows through unchanged leading to a similarly looking Schrödinger equation for $A_n(\mathbf{k})$ except for the replacement

$$\mathbf{p}_{nn'} \to \boldsymbol{\pi}_{nn'}. \tag{8.101}$$

The π operator has the same properties as $\mathbf{p}$: $\boldsymbol{\pi}_{nn}$ is still zero at a band minimum.

8.3.3.2 Degenerate Case

While there are twice as many states as without spin, some of the degeneracy might be lifted by the spin-orbit Hamiltonian. We will treat specifically the case of a diamond crystal for which the three-fold valence band without spin, becomes six-fold with spin, and then split into a four-fold and a two-fold subspace in the presence of the spin-orbit Hamiltonian. If the latter is treated as a perturbation, then the zeroth-order wave functions are the $|JM_J\rangle$ states (Table 3.4). We note that LK used a different set of $|JM_J\rangle$ states from us (Table C.2).

In the JM_J basis, one now needs to generalize the D matrix. LK used the DKK Hamiltonian with spin. For conciseness, we only write down the case of a four-band model. It was already obtained starting with the DKK Hamiltonian in Chap. 3 (Table 3.6). Thus, in the presence of a shallow impurity, the problem to solve is the equation in Table 8.3.

8.4 Baldereschi–Lipari Model

Baldereschi and Lipari have reformulated the multiband impurity problem by separating the Hamiltonian into a spherical and a cubic term [30, 140]. We give a brief introduction to this theory.

Table 8.3 Four-band impurity equation

$$\begin{pmatrix} \frac{1}{2}P(-\mathrm{i}\nabla)+U(\mathbf{r})-E & R(-\mathrm{i}\nabla) & S(-\mathrm{i}\nabla) & 0 \\ R^*(-\mathrm{i}\nabla) & \frac{1}{6}P(-\mathrm{i}\nabla)+\frac{2}{3}Q(-\mathrm{i}\nabla)+U(\mathbf{r})-E & 0 & -S(-\mathrm{i}\nabla) \\ S^*(-\mathrm{i}\nabla) & 0 & \frac{1}{6}P(-\mathrm{i}\nabla)+\frac{2}{3}Q(-\mathrm{i}\nabla)+U(\mathbf{r})-E & R(-\mathrm{i}\nabla) \\ 0 & -S^*(-\mathrm{i}\nabla) & R^*(-\mathrm{i}\nabla) & \frac{1}{2}P(-\mathrm{i}\nabla)+U(\mathbf{r})-E \end{pmatrix} \begin{pmatrix} F_1 \\ F_2 \\ F_3 \\ F_4 \end{pmatrix} = 0.$$

8.4.1 Hamiltonian

The starting point of their model is the acceptor Hamiltonian in Luttinger form:

$$H = \left(\gamma_1 + \frac{5}{2}\gamma_2\right)\frac{p^2}{2m_0} - \frac{\gamma_2}{2m_0}\left(p_x^2 J_x^2 + p_y^2 J_y^2 + p_z^2 J_z^2\right)$$
$$-\frac{2\gamma_3}{2m_0}\left(\{p_x p_y\}\{J_x J_y\} + \{p_y p_z\}\{J_y J_z\} + \{p_z p_x\}\{J_z J_x\}\right) - \frac{e^2}{4\pi\varepsilon r}. \quad (8.102)$$

Here **p** is the hole linear momentum operator. The first term in Eq. (8.102) looks like a kinetic-energy one while the last one is the Coulomb potential; the other terms can be viewed as effective spin-orbit terms. This analogy can be strengthened by introducing spherical tensors.

First, one can define second-rank symmetric Cartersian tensors with zero trace:

$$P_{ij} = 3p_i p_j - \delta_{ij} p^2, \quad (8.103)$$
$$J_{ij} = 3\{J_i J_j\} - \delta_{ij} J^2. \quad (8.104)$$

Then Eq. (8.102) becomes

$$H = \frac{\gamma_1}{2m_0}p^2 - \frac{e^2}{4\pi\varepsilon r} - \frac{1}{9m_0}[\gamma_3 - (\gamma_3 - \gamma_2)\delta_{ij}]P_{ij}J_{ij}. \quad (8.105)$$

The cartesian tensors are now expressed in terms of irreducible spherical ones of rank $l = 0$, 1, and 2. The $l = 1$ component is absent due to the symmetry of the tensors and the $l = 0$ component due to being traceless. Then Eq. (8.105) becomes

$$H = \left(\frac{\gamma_1}{2m_0}p^2 - \frac{e^2}{4\pi\varepsilon r}\right) - \frac{3\gamma_3 + 2\gamma_2}{45m_0}(P^{(2)} \cdot J^{(2)}) \quad (8.106)$$
$$+\frac{(\gamma_3 - \gamma_2)}{18m_0}\left(\left[P^{(2)} \times J^{(2)}\right]_{-4}^{(4)} + \frac{1}{5}\sqrt{70}\left[P^{(2)} \times J^{(2)}\right]_0^{(4)} + \left[P^{(2)} \times J^{(2)}\right]_4^{(4)}\right),$$

where

$$\left[U^{(k_1)} \times V^{(k_2)}\right]_k^{(q)} = (-1)^{k_1-k_2+q}(2k+1)^{1/2}\sum_{q_1,q_2}\begin{pmatrix} k_1 & k_2 & k \\ q_1 & q_2 & -q \end{pmatrix} U_{q_1}^{(k_1)} V_{q_2}^{(k_2)}. \quad (8.107)$$

In Eq. (8.106), the first term is spherical, the third cubic, and the second one again represents an effective spin-orbit coupling. Defining new parameters and units,

$$\mu = \frac{6\gamma_3 + 4\gamma_2}{5\gamma_1},$$

$$\delta = \frac{\gamma_3 - \gamma_2}{\gamma_1}, \tag{8.108}$$

$$R_0 = \frac{e^4 m_0}{16\pi^2\varepsilon^2\hbar^2\gamma_1},$$

$$a_0 = \frac{4\pi\varepsilon\hbar^2\gamma_1}{e^2 m_0},$$

then Eq. (8.106) becomes

$$H = \frac{1}{\hbar^2}p^2 - \frac{2}{r} - \frac{\mu}{9\hbar^2}\left(P^{(2)} \cdot J^{(2)}\right) + \frac{\delta}{9\hbar^2}\left(\left[P^{(2)} \times J^{(2)}\right]^{(4)}_{-4} + \frac{1}{5}\sqrt{70}\left[P^{(2)} \times J^{(2)}\right]^{(4)}_{0} + \left[P^{(2)} \times J^{(2)}\right]^{(4)}_{4}\right). \tag{8.109}$$

It is clear that band parameters only appear in μ and δ; representative values are given in Table 8.4.

Table 8.4 Representative valence-band parameters [30, 140]

	μ	δ
Si	0.483	0.249
Ge	0.766	0.108
GaAs	0.767	0.114
GaP	0.701	0.178

8.4.2 Solution

Since $\delta << \mu$, one can treat the problem at two levels. The first is to retain the μ term in a spherical model as a perturbation on the hydrogenic problem. One can then add the cubic term as an additional perturbation.

As examples of the validity of the spherical model, the valence-band dispersions for a few semiconductors obtained by BL are given in Fig. 8.1.

Next, the reader is referred to the paper by BL for a detailed treatment of the μ term. This was done separately for large μ (large "spin-orbit coupling") and small μ. For large μ, BL found an exact solution of the spherical model for $P_{F=\frac{1}{2}}$, where $\mathbf{F} = \mathbf{L} + \mathbf{S}$:

$$E(nP_{\frac{1}{2}}) = \frac{R_0}{(1+\mu)n^2}, \quad (n = 2, 3, \ldots). \tag{8.110}$$

Binding energies within both the spherical and cubic models are shown in Fig. 8.2.

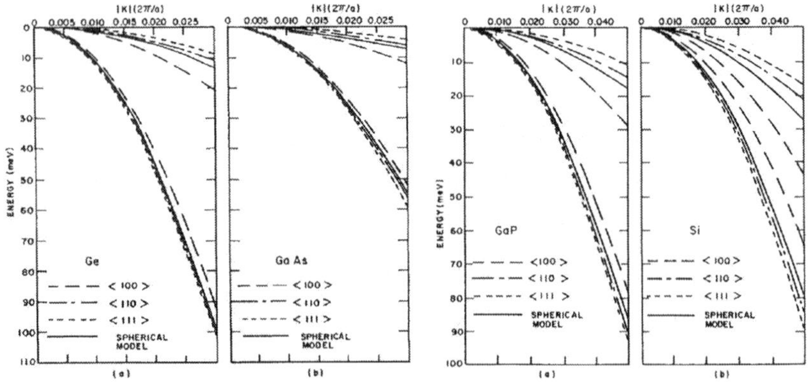

Fig. 8.1 Valence-band dispersion for Ge, GaAs, GaP and Si along different directions and within the spherical model. Reprinted with permission from [140]. ©1974 by the American Physical Society

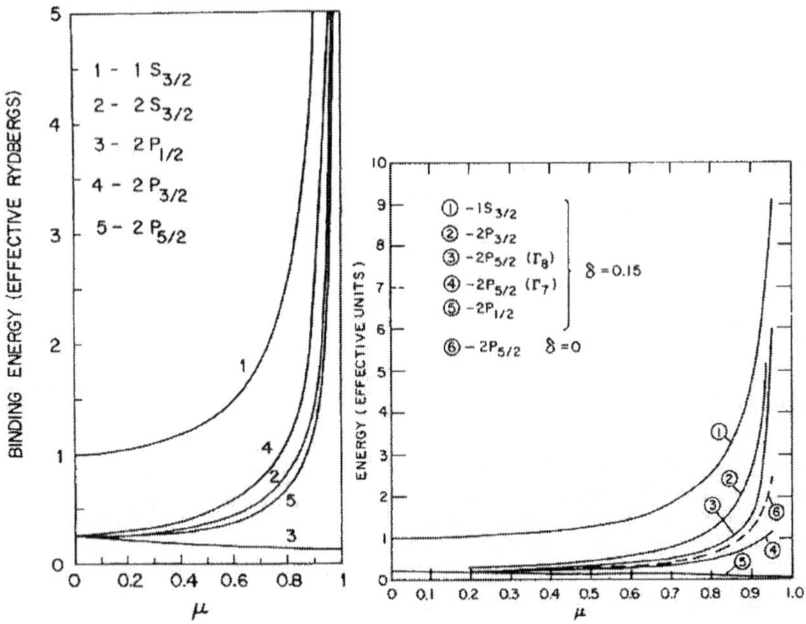

Fig. 8.2 Acceptor binding energy within the spherical and cubic models. Reprinted with permission from [30, 140]. ©1973, 1974 by the American Physical Society

8.5 Summary

Two theories (Kittel–Mitchell based upon a Fourier expansion and Luttinger–Kohn based upon a canonical transformation) were introduced in order to derive the $k \cdot p$ Hamiltonian in the presence of a shallow impurity. The basic result is a Wannier-like effective equation for an envelope function. The one-band model can be solved exactly as a hydrogenic problem in which the binding energies are independent of the impurity atom and depend upon the host via its band parameters and dielectric function. The multiband problem has no exact solution and is typically solved using a variational technique. We discussed briefly the approach of Baldereschi and Lipari which involved separating the problem into a spherical and cubic part.

Chapter 9
Magnetic Effects

9.1 Overview

One of the earliest $k \cdot p$ papers, the DKK one [2] already treated magnetic effects via the cyclotron-resonance effect. A classical analysis was presented. For a single parabolic band, the band is characterized solely by the effective mass and there is one cyclotron frequency; in the quantum picture, this corresponds to equidistant magnetic Landau levels (Fig. 9.1).

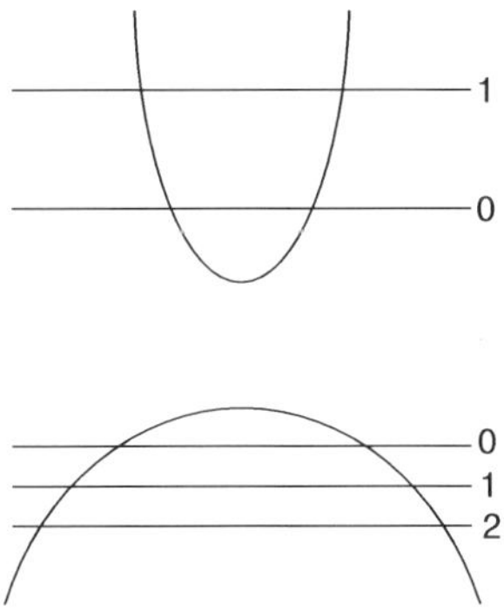

Fig. 9.1 Magnetic levels for two simple parabolic bands

The quantum theory of cyclotron resonance was developed by Luttinger in 1956 [4] using a four-band model for diamond semiconductors, and extended to an eight-band model by Pidgeon and Brown [86]. Inclusion of strain was considered by Suzuki and Hensel [80] but within a six-band model for germanium, and by Trebin et al. [81, 141] for eight-band zincblende. The band coupling and nonparabolicity results in a nonlinear behaviour of the Landau levels with magnetic field. This effect is typically observed in a magneto-optical experiment.

The first derivation of the g-factor equation was performed by Roth et al. [25]. They also presented the first theory of magnetoabsorption in diamond-type semiconductors. Weiler, Aggrawal and Lax [98] studied magnetoabsorption in zincblende-type semiconductors.

L.C. Lew Yan Voon, M. Willatzen, *The* k · p *Method*,
DOI 10.1007/978-3-540-92872-0_9, © Springer-Verlag Berlin Heidelberg 2009

In the following, we will first present the analysis of how to include an external magnetic field using the canonical transformation of Luttinger and Kohn [6]. We have included a discussion on calculating the g-factor. The application of the theory of invariants to obtaining the magnetic Hamiltonian was already given in Chap. 5. An excellent exposition of the theory of band electrons in a magnetic field can be found in the book by Zeiger and Pratt [12].

9.2 Canonical Transformation

The procedure here follows closely the derivation given for the Wannier equation of an impurity in Chap. 8. We will leave out the electron spin for now and introduce it in Sect. 9.5.

9.2.1 One-Band Model

We first look at the theory of an electron in a simple band in an external homogeneous magnetic field. We will choose the field to be in the z direction with the crystal at an arbitrary orientation to it. The magnetic field is then

$$\mathbf{B} = (0, 0, B).$$

As before, we choose the Landau gauge ($\nabla \cdot \mathbf{A} = 0$) for the vector potential:

$$\mathbf{A} = (-By, 0, 0).$$

The magnetic field is introduced into the Hamiltonian via minimal coupling ($\mathbf{p} \rightarrow \mathbf{p} + e\mathbf{A}$ with $e > 0$):

$$\begin{aligned} H &= \frac{1}{2m_0}(\mathbf{p} + e\mathbf{A})^2 + V(\mathbf{r}) = \frac{1}{2m_0}(p^2 + 2e\mathbf{A}\cdot\mathbf{p} + e^2A^2) + V(\mathbf{r}) \\ &= H_0 + \frac{1}{2m_0}(e^2B^2y^2 - 2eByp_x) \\ &\equiv H_0 - \frac{\hbar s}{m_0}yp_x + \frac{\hbar s^2}{2m_0}y^2, \end{aligned} \tag{9.1}$$

where $s \equiv eB/\hbar$ is the inverse square of the magnetic length. We will be expressing the Hamiltonian in the LK $\chi_{n\mathbf{k}}$ basis. Matrix elements of yp_x and y^2 are now needed (those of H_0 were obtained in Sect. 8.3.1):

$$
\begin{aligned}
(n\mathbf{k}|yp_x|n'\mathbf{k}') &= \int_V \mathrm{d}^3\mathbf{r}\,\mathrm{e}^{\mathrm{i}(\mathbf{k}'-\mathbf{k})\cdot\mathbf{r}} u^*_{n\mathbf{0}}(\mathbf{r})y[\hbar k'_x - \mathrm{i}\hbar\nabla_x]u_{n'\mathbf{0}}(\mathbf{r}) \\
&= -\frac{1}{\mathrm{i}}\frac{\partial}{\partial k_y}\int_V \mathrm{d}^3\mathbf{r}\,\mathrm{e}^{\mathrm{i}(\mathbf{k}'-\mathbf{k})\cdot\mathbf{r}} u^*_{n\mathbf{0}}(\mathbf{r})[\hbar k'_x - \mathrm{i}\hbar\nabla_x]u_{n'\mathbf{0}}(\mathbf{r}) \\
&= -\frac{1}{\mathrm{i}}\frac{\partial}{\partial k_y}\left[\hbar k'_x\int_V \mathrm{d}^3\mathbf{r}\,\mathrm{e}^{\mathrm{i}(\mathbf{k}'-\mathbf{k})\cdot\mathbf{r}} u^*_{n\mathbf{0}}u_{n'\mathbf{0}} + \int_V \mathrm{d}^3\mathbf{r}\,\mathrm{e}^{\mathrm{i}(\mathbf{k}'-\mathbf{k})\cdot\mathbf{r}} u^*_{n\mathbf{0}}p_x u_{n'\mathbf{0}}\right] \\
&= -\frac{1}{\mathrm{i}}\frac{\partial}{\partial k_y}\left[\hbar k'_x\delta_{nn'}\delta(\mathbf{k}-\mathbf{k}') + p^x_{nn'}\delta(\mathbf{k}-\mathbf{k}')\right] \\
&= \left[\hbar k_x\delta_{nn'} + p^x_{nn'}\right]\frac{1}{\mathrm{i}}\frac{\partial}{\partial k'_y}\delta(\mathbf{k}-\mathbf{k}'), \qquad (9.2)
\end{aligned}
$$

$$
\begin{aligned}
(n\mathbf{k}|y^2|n'\mathbf{k}') &= \int_V \mathrm{d}^3\mathbf{r}\,\mathrm{e}^{\mathrm{i}(\mathbf{k}'-\mathbf{k})\cdot\mathbf{r}} u^*_{n\mathbf{0}}(\mathbf{r})y^2u_{n'\mathbf{0}}(\mathbf{r}) \\
&= \left(\frac{1}{\mathrm{i}}\frac{\partial}{\partial k'_y}\right)^2\int_V \mathrm{d}^3\mathbf{r}\,\mathrm{e}^{\mathrm{i}(\mathbf{k}'-\mathbf{k})\cdot\mathbf{r}} u^*_{n\mathbf{0}}(\mathbf{r})u_{n'\mathbf{0}}(\mathbf{r}) \\
&= -\delta_{nn'}\frac{\partial^2}{\partial k'^2_y}\delta(\mathbf{k}-\mathbf{k}'). \qquad (9.3)
\end{aligned}
$$

The Hamiltonian is then, using Eq. (8.60),

$$
\begin{aligned}
(n\mathbf{k}|H|n'\mathbf{k}') &= (n\mathbf{k}|H_0|n'\mathbf{k}') - \frac{s\hbar}{m_0}(n\mathbf{k}|yp_x|n'\mathbf{k}') + \frac{s^2\hbar^2}{2m_0}(n\mathbf{k}|y^2|n'\mathbf{k}') \\
&= \left[\left(E_n(\mathbf{0}) + \frac{\hbar^2k^2}{2m_0}\right)\delta_{nn'} + \frac{\hbar}{m_0}\mathbf{k}\cdot\mathbf{p}_{nn'}\right]\delta(\mathbf{k}-\mathbf{k}') \\
&\quad -\frac{s\hbar}{m_0}\left[\hbar k_x\delta_{nn'} + p^x_{nn'}\right]\frac{1}{\mathrm{i}}\frac{\partial}{\partial k'_y}\delta(\mathbf{k}-\mathbf{k}') + \frac{s^2\hbar^2}{2m_0}\delta_{nn'}\frac{\partial^2}{\partial k'^2_y}\delta(\mathbf{k}-\mathbf{k}') \\
&= \left[\left(E_n(\mathbf{0}) + \frac{\hbar^2k^2}{2m_0}\right)\delta(\mathbf{k}-\mathbf{k}') - \frac{s\hbar^2k_x}{m_0}\frac{1}{\mathrm{i}}\frac{\partial}{\partial k'_y}\delta(\mathbf{k}-\mathbf{k}') - \frac{s^2\hbar^2}{2m_0}\frac{\partial^2}{\partial k'^2_y}\delta(\mathbf{k}-\mathbf{k}')\right]\delta_{nn'} \\
&\quad +\frac{1}{m_0}\left[\hbar\mathbf{k}\cdot\mathbf{p}_{nn'}\delta(\mathbf{k}-\mathbf{k}') - \frac{s\hbar}{\mathrm{i}}p^x_{nn'}\frac{\partial}{\partial k'_y}\delta(\mathbf{k}-\mathbf{k}')\right]. \qquad (9.4)
\end{aligned}
$$

There are thus intraband and interband terms. The interband term is small. The term linear in k is small in the sense that $k\cdot p$ perturbation applies. The one proportional to s is small for weak magnetic fields. We will again make a canonical transformation $T' = \exp(S')$ in order to remove the interband term to first order. Write

$$
\begin{aligned}
(n\mathbf{k}|H_a|n'\mathbf{k}') &= -\frac{s\hbar^2 k_x}{m_0}\frac{1}{\mathrm{i}}\frac{\partial}{\partial k'_y}\delta(\mathbf{k}-\mathbf{k}')\delta_{nn'}, \\
(n\mathbf{k}|H_b|n'\mathbf{k}') &= -\frac{s^2\hbar^2}{2m_0}\frac{\partial^2}{\partial k'^2_y}\delta(\mathbf{k}-\mathbf{k}')\delta_{nn'}, \\
(n\mathbf{k}|H_1|n'\mathbf{k}') &= \frac{1}{m_0}\left[\hbar\mathbf{k}\cdot\mathbf{p}_{nn'}\delta(\mathbf{k}-\mathbf{k}') - \frac{s\hbar}{\mathrm{i}}p^x_{nn'}\frac{\partial}{\partial k'_y}\delta(\mathbf{k}-\mathbf{k}')\right].
\end{aligned}
\tag{9.5}
$$

Then

$$
\begin{aligned}
H &= H_0 + H_a + H_b + H_1, \\
\overline{H} &= \mathrm{e}^{-S'} H \mathrm{e}^{S'} \\
&= H_0 + H_a + H_b + H_1 + [H_0, S'] + [H_a, S'] + [H_b, S'] + [H_1, S'] + \frac{1}{2}[[H_0, S'], S'] + \dots
\end{aligned}
\tag{9.6}
$$

As before, one removes the interband term to first order via

$$
H_1 + [H_0, S'] = 0, \tag{9.7}
$$

which then leads to

$$
\overline{H} = H_0 + H_a + H_b + [H_a, S'] + \frac{1}{2}[H_1, S'] + \dots \tag{9.8}
$$

S' is similarly obtained to S [Eq. (8.63)]. Have

$$
(n\mathbf{k}|S'|n'\mathbf{k}') = \begin{cases} -\frac{(n\mathbf{k}|H_1|n'\mathbf{k}')}{\hbar\omega_{nn'}} & n \neq n', \\ 0 & n = n', \end{cases} \tag{9.9}
$$

where $\hbar\omega_{nn'} = E_n(\mathbf{0}) - E_{n'}(\mathbf{0})$. We can now obtain the matrix elements of $\overline{H}$; we only need the intraband elements. Since $(n\mathbf{k}|H_a|n'\mathbf{k}') \propto \delta_{nn'}$,

$$
(n\mathbf{k}|[H_a, S']|n\mathbf{k}') = 0, \tag{9.10}
$$

because $S'_{nn} = 0$. Also,

$$
\begin{aligned}
&(n\mathbf{k}|[H_1, S']|n'\mathbf{k}') \\
&= \sum_{n''}\int \mathrm{d}^3\mathbf{k}'' \left[(n\mathbf{k}|H_1|n''\mathbf{k}'')(n''\mathbf{k}''|S'|n'\mathbf{k}') - (n\mathbf{k}|S'|n''\mathbf{k}'')(n''\mathbf{k}''|H_1|n'\mathbf{k}')\right] \\
&= -\sum_{n''}\int \mathrm{d}^3\mathbf{k}'' \left[\frac{(n\mathbf{k}|H_1|n''\mathbf{k}'')(n''\mathbf{k}''|H_1|n'\mathbf{k}')}{\omega_{n''n'}} - \frac{(n\mathbf{k}|H_1|n''\mathbf{k}'')(n''\mathbf{k}''|H_1|n'\mathbf{k}')}{\omega_{nn''}}\right] \\
&= \frac{1}{\hbar}\sum_{n''}\int \mathrm{d}^3\mathbf{k}''\,(n\mathbf{k}|H_1|n''\mathbf{k}'')(n''\mathbf{k}''|H_1|n'\mathbf{k}')\left[\frac{1}{\omega_{nn''}} - \frac{1}{\omega_{n''n'}}\right]
\end{aligned}
$$

$$= \frac{\hbar}{m_0^2}\sum_{n''}\int \mathrm{d}^3\mathbf{k}''\left[\mathbf{k}\cdot\mathbf{p}^i_{nn''}\delta(\mathbf{k}-\mathbf{k}'')+\frac{s}{\mathrm{i}}p^x_{nn''}\frac{\partial}{\partial k''_y}\delta(\mathbf{k}''-\mathbf{k})\right]$$
$$\times\left[\mathbf{k}''\cdot\mathbf{p}_{n''n'}\delta(\mathbf{k}''-\mathbf{k}')-\frac{s}{\mathrm{i}}p^x_{n''n'}\frac{\partial}{\partial k'_y}\delta(\mathbf{k}'-\mathbf{k}'')\right]\left[\frac{1}{\omega_{nn''}}-\frac{1}{\omega_{n''n'}}\right], \tag{9.11}$$

and

$$(n\mathbf{k}|[H_1,S']|n\mathbf{k}')$$
$$= \frac{\hbar}{m_0^2}\sum_{n'}\int \mathrm{d}^3\mathbf{k}''\left[\mathbf{k}\cdot\mathbf{p}_{nn'}\delta(\mathbf{k}-\mathbf{k}'')-\frac{s}{\mathrm{i}\hbar}p^x_{nn'}\frac{\partial}{\partial k''_y}\delta(\mathbf{k}''-\mathbf{k})\right]$$
$$\times\left[\mathbf{k}''\cdot\mathbf{p}_{n'n}\delta(\mathbf{k}''-\mathbf{k}')-\frac{s}{\mathrm{i}}p^x_{n'n}\frac{\partial}{\partial k'_y}\delta(\mathbf{k}'-\mathbf{k}'')\right]\left[\frac{1}{\omega_{nn'}}-\frac{1}{\omega_{n'n}}\right]$$
$$= \frac{2\hbar}{m_0^2}\sum_{n'}{}'\left[k_ik_jp^i_{nn'}p^j_{n'n}\delta(\mathbf{k}-\mathbf{k}')-s(k_i+k'_i)p^i_{nn'}p^x_{n'n}\frac{1}{\mathrm{i}}\frac{\partial}{\partial k'_y}\delta(\mathbf{k}'-\mathbf{k})\right.$$
$$\left.-s^2p^x_{nn'}p^x_{n'n}\frac{\partial^2}{\partial k'^2_y}\delta(\mathbf{k}'-\mathbf{k})\right]\frac{1}{\omega_{nn'}}. \tag{9.12}$$

We can simplify using the f-sum rule, [Eq. (8.67)], giving

$$(n\mathbf{k}|[H_1,S']|n\mathbf{k}') - -\frac{1}{m_0}\left\{k_ik_j\left[\hbar^2\delta_{ij}-m_0\frac{\partial^2E_n(\mathbf{k})}{\partial k_i\partial k_j}\right]_0\delta(\mathbf{k}-\mathbf{k}')\right.$$
$$-s\left[\hbar^2(k_x+k'_x)-m_0(k_i+k'_i)\left(\frac{\partial^2E_n(\mathbf{k})}{\partial k_i\partial k_x}\right)_0\right]\frac{1}{\mathrm{i}}\frac{\partial}{\partial k'_y}\delta(\mathbf{k}'-\mathbf{k})$$
$$\left.-s^2\left[\hbar^2-m_0\left(\frac{\partial^2E_n(\mathbf{k})}{\partial k_x^2}\right)_0\right]\frac{\partial^2}{\partial k'^2_y}\delta(\mathbf{k}'-\mathbf{k})\right\}. \tag{9.13}$$

Hence,

$$(n\mathbf{k}|\overline{H}|n\mathbf{k}') = (n\mathbf{k}|H_0|n\mathbf{k}')+(n\mathbf{k}|H_a|n\mathbf{k}')+(n\mathbf{k}|H_b|n\mathbf{k}')+\frac{1}{2}(n\mathbf{k}|[H_1,S']|n\mathbf{k}')$$
$$=\left[E_n(\mathbf{0})+\frac{\hbar^2k^2}{2m_0}\right]\delta(\mathbf{k}-\mathbf{k}')-\frac{s\hbar^2k_x}{m_0}\frac{1}{\mathrm{i}}\frac{\partial}{\partial k'_y}\delta(\mathbf{k}-\mathbf{k}')-\frac{s^2\hbar^2}{2m_0}\frac{\partial^2}{\partial k'^2_y}\delta(\mathbf{k}-\mathbf{k}')$$
$$-\frac{1}{2m_0}\left\{k_ik_j\left[\hbar^2\delta_{ij}-m_0\frac{\partial^2E_n(\mathbf{k})}{\partial k_i\partial k_j}\right]_0\delta(\mathbf{k}-\mathbf{k}')\right.$$
$$-s\left[\hbar^2(k_x+k'_x)-m_0(k_i+k'_i)\left(\frac{\partial^2E_n(\mathbf{k})}{\partial k_i\partial k_x}\right)_0\right]\frac{1}{\mathrm{i}}\frac{\partial}{\partial k'_y}\delta(\mathbf{k}'-\mathbf{k})$$
$$\left.-s^2\left[\hbar^2-m_0\left(\frac{\partial^2E_n(\mathbf{k})}{\partial k_x^2}\right)_0\right]\frac{\partial^2}{\partial k'^2_y}\delta(\mathbf{k}'-\mathbf{k})\right\}$$

$$
\begin{aligned}
&= E_n(\mathbf{0})\delta(\mathbf{k}-\mathbf{k}') - \frac{1}{2m_0}\Bigg\{ - k_i k_j m_0 \left(\frac{\partial^2 E_n(\mathbf{k})}{\partial k_i \partial k_j}\right)_0 \delta(\mathbf{k}-\mathbf{k}') \\
&\quad + s\left[\hbar^2(k_x - k_x') + m_0(k_i + k_i')\left(\frac{\partial^2 E_n(\mathbf{k})}{\partial k_i \partial k_x}\right)_0\right]\frac{1}{\mathrm{i}}\frac{\partial}{\partial k_y'}\delta(\mathbf{k}'-\mathbf{k}) \\
&\quad + s^2 m_0 \left(\frac{\partial^2 E_n(\mathbf{k})}{\partial k_x^2}\right)_0 \frac{\partial^2}{\partial k_y'^2}\delta(\mathbf{k}'-\mathbf{k})\Bigg\}.
\end{aligned}
\tag{9.14}
$$

Now

$$
(k_i - k_i')\frac{\partial}{\partial k_y'}\delta(\mathbf{k}'-\mathbf{k}) = \delta_{iy}\delta(\mathbf{k}'-\mathbf{k}). \tag{9.15}
$$

One can prove, e.g., by making the change $\mathbf{k} \to -\mathbf{k}$, $\mathbf{k}' \to -\mathbf{k}'$, or by using

$$
\int_{-\infty}^{\infty} \mathrm{d}x\, f(x)\delta(x - x') = -f'(x).
$$

Then

$$
\begin{aligned}
(n\mathbf{k}|\overline{H}|n\mathbf{k}') &= \left[E_n(\mathbf{0}) + \frac{k_i k_j}{2}\left(\frac{\partial^2 E_n(\mathbf{k})}{\partial k_i \partial k_j}\right)_0\right]\delta(\mathbf{k}-\mathbf{k}') \\
&\quad -\frac{1}{2}s(k_i + k_i')\left(\frac{\partial^2 E_n(\mathbf{k})}{\partial k_i \partial k_x}\right)_0 \frac{1}{\mathrm{i}}\frac{\partial}{\partial k_y'}\delta(\mathbf{k}'-\mathbf{k}) - \frac{s^2}{2}\left(\frac{\partial^2 E_n(\mathbf{k})}{\partial k_x^2}\right)_0 \frac{\partial^2}{\partial k_y'^2}\delta(\mathbf{k}'-\mathbf{k}).
\end{aligned}
$$

Next

$$
E_n(\mathbf{0}) + \frac{k_i k_j}{2}\left(\frac{\partial^2 E_n(\mathbf{k})}{\partial k_i \partial k_j}\right)_0 \equiv E_n(\mathbf{k}) \tag{9.16}
$$

to second order and writing

$$
k_i + k_i' = k_i' - k_i + 2k_i,
$$

we have

$$
\begin{aligned}
(n\mathbf{k}|\overline{H}|n\mathbf{k}') &= E_n(\mathbf{k})\delta(\mathbf{k}-\mathbf{k}') - s k_i \left(\frac{\partial^2 E_n(\mathbf{k})}{\partial k_i \partial k_x}\right)_0 \frac{1}{\mathrm{i}}\frac{\partial}{\partial k_y'}\delta(\mathbf{k}'-\mathbf{k}) \\
&\quad -\frac{\mathrm{i}s}{2}(k_i - k_i')\left(\frac{\partial^2 E_n(\mathbf{k})}{\partial k_i \partial k_x}\right)_0 \frac{\partial}{\partial k_y'}\delta(\mathbf{k}'-\mathbf{k}) - \frac{s^2}{2}\left(\frac{\partial^2 E_n(\mathbf{k})}{\partial k_x^2}\right)_0 \frac{\partial^2}{\partial k_y'^2}\delta(\mathbf{k}'-\mathbf{k}).
\end{aligned}
$$

Finally,

$$(n\mathbf{k}|\overline{H}|n\mathbf{k}') = E_n(\mathbf{k})\delta(\mathbf{k}-\mathbf{k}') - sk_i\left(\frac{\partial^2 E_n(\mathbf{k})}{\partial k_i \partial k_x}\right)_0 \frac{1}{\mathrm{i}}\frac{\partial}{\partial k'_y}\delta(\mathbf{k}'-\mathbf{k})$$
$$-\frac{\mathrm{i}s}{2}\left(\frac{\partial^2 E_n(\mathbf{k})}{\partial k_y \partial k_x}\right)_0 \delta(\mathbf{k}'-\mathbf{k}) - \frac{s^2}{2}\left(\frac{\partial^2 E_n(\mathbf{k})}{\partial k_x^2}\right)_0 \frac{\partial^2}{\partial k_y'^2}\delta(\mathbf{k}'-\mathbf{k}). \quad (9.17)$$

Equation (9.17) is the effective-mass Hamiltonian in momentum space. As for the impurity problem, the Schrödinger equation for the transformed Hamiltonian can be written succintly as

$$\overline{H}B = EB,$$

i.e.,

$$E_n(\mathbf{k})B_n(\mathbf{k}) + \int \mathrm{d}^3\mathbf{k}' \left[-sk_i\left(\frac{\partial^2 E_n(\mathbf{k})}{\partial k_i \partial k_x}\right)_0 \frac{1}{\mathrm{i}}\frac{\partial}{\partial k'_y}\delta(\mathbf{k}'-\mathbf{k})B_n(\mathbf{k}')\right.$$
$$\left.-\frac{\mathrm{i}s}{2}\left(\frac{\partial^2 E_n(\mathbf{k})}{\partial k_y \partial k_x}\right)_0 \delta(\mathbf{k}'-\mathbf{k})B_n(\mathbf{k}') - \frac{s^2}{2}\left(\frac{\partial^2 E_n(\mathbf{k})}{\partial k_x^2}\right)_0 \frac{\partial^2}{\partial k_y'^2}\delta(\mathbf{k}'-\mathbf{k})B_n(\mathbf{k}')\right]$$
$$= EB_n(\mathbf{k}),$$

or

$$E_n(\mathbf{k})B_n(\mathbf{k}) \; -s\left[k_i\left(\frac{\partial^2 E_n(\mathbf{k})}{\partial k_i \partial k_x}\right)_0 \mathrm{i}\frac{\partial B_n(\mathbf{k})}{\partial k_y} + \frac{\mathrm{i}}{2}\left(\frac{\partial^2 E_n(\mathbf{k})}{\partial k_x \partial k_y}\right)_0 B_n(\mathbf{k})\right]$$
$$+\frac{s^2}{2}\left(\frac{\partial^2 E_n(\mathbf{k})}{\partial k_x^2}\right)_0 \mathrm{i}^2\frac{\partial^2 B_n(\mathbf{k})}{\partial k_y^2} = EB_n(\mathbf{k}). \quad (9.18)$$

To convert to real space, we again introduce an envelope function via Eq. (8.73). Thus, we multiply Eq. (9.18) by $\mathrm{e}^{\mathrm{i}\mathbf{k}\cdot\mathbf{r}}$ and integrate over the FBZ:

$$\int \mathrm{d}^3\mathbf{k}\, E_n(\mathbf{k})\mathrm{e}^{\mathrm{i}\mathbf{k}\cdot\mathbf{r}}B_n(\mathbf{k}) \; -s\int \mathrm{d}^3\mathbf{k}\left[k_i\left(\frac{\partial^2 E_n(\mathbf{k})}{\partial k_i \partial k_x}\right)_0 \mathrm{i}\frac{\partial B_n(\mathbf{k})}{\partial k_y} + \frac{\mathrm{i}}{2}\left(\frac{\partial^2 E_n(\mathbf{k})}{\partial k_x \partial k_y}\right)_0 B_n(\mathbf{k})\right]\mathrm{e}^{\mathrm{i}\mathbf{k}\cdot\mathbf{r}}$$
$$+\frac{s^2}{2}\int \mathrm{d}^3\mathbf{k}\left(\frac{\partial^2 E_n(\mathbf{k})}{\partial k_x^2}\right)_0 \mathrm{i}^2\mathrm{e}^{\mathrm{i}\mathbf{k}\cdot\mathbf{r}}\frac{\partial^2 B_n(\mathbf{k})}{\partial k_y^2} = E\int \mathrm{d}^3\mathbf{k}\,\mathrm{e}^{\mathrm{i}\mathbf{k}\cdot\mathbf{r}}B_n(\mathbf{k}).$$

Consider the $O(s)$ term:

$$\left(\frac{\partial^2 E_n(\mathbf{k})}{\partial k_i \partial k_x}\right)_0 \int \mathrm{d}^3\mathbf{k}\left[\mathrm{i}k_i\frac{\partial B_n(\mathbf{k})}{\partial k_y}\mathrm{e}^{\mathrm{i}\mathbf{k}\cdot\mathbf{r}}\right] + \left(\frac{\partial^2 E_n(\mathbf{k})}{\partial k_x \partial k_y}\right)_0 \frac{\mathrm{i}}{2}\int \mathrm{d}^3\mathbf{k}\, B_n(\mathbf{k})\mathrm{e}^{\mathrm{i}\mathbf{k}\cdot\mathbf{r}}$$
$$= \left(\frac{\partial^2 E_n(\mathbf{k})}{\partial k_i \partial k_x}\right)_0 \int \mathrm{d}^3\mathbf{k}\left[\mathrm{i}k_i\frac{\partial B_n(\mathbf{k})}{\partial k_y}\mathrm{e}^{\mathrm{i}\mathbf{k}\cdot\mathbf{r}}\right] + \frac{\mathrm{i}}{2}\left(\frac{\partial^2 E_n(\mathbf{k})}{\partial k_x \partial k_y}\right)_0 F_n(\mathbf{r})$$
$$= \left(\frac{\partial^2 E_n(\mathbf{k})}{\partial k_i \partial k_x}\right)_0 \int \mathrm{d}^3\mathbf{k}\left[\mathrm{i}k_i\frac{\partial B_n(\mathbf{k})}{\partial k_y}\mathrm{e}^{\mathrm{i}\mathbf{k}\cdot\mathbf{r}}\right] + \frac{\mathrm{i}}{2}\left(\frac{\partial^2 E_n(\mathbf{k})}{\partial k_x \partial k_i}\right)_0 (\nabla_i y)F_n(\mathbf{r})$$

$$= \frac{1}{2}\left(\frac{\partial^2 E_n(\mathbf{k})}{\partial k_i \partial k_x}\right)_0 \left[\int \mathrm{d}^3\mathbf{k}\, 2\mathrm{i}k_i \frac{\partial B_n(\mathbf{k})}{\partial k_y} \mathrm{e}^{\mathrm{i}\mathbf{k}\cdot\mathbf{r}} + \mathrm{i}\nabla_i(yF_n) + \frac{1}{\mathrm{i}} y\nabla_i F_n\right].$$

But

$$2\mathrm{i}k_i \frac{\partial B_n(\mathbf{k})}{\partial k_y} \mathrm{e}^{\mathrm{i}\mathbf{k}\cdot\mathbf{r}} = 2\frac{\partial B_n(\mathbf{k})}{\partial k_y} \nabla_i \mathrm{e}^{\mathrm{i}\mathbf{k}\cdot\mathbf{r}} = 2\nabla_i \left(\frac{\partial B_n(\mathbf{k})}{\partial k_y} \mathrm{e}^{\mathrm{i}\mathbf{k}\cdot\mathbf{r}}\right)$$
$$= 2\nabla_i \left[\frac{\partial}{\partial k_y}(B_n(\mathbf{k})\mathrm{e}^{\mathrm{i}\mathbf{k}\cdot\mathbf{r}}) - \mathrm{i}yB_n(\mathbf{k})\mathrm{e}^{\mathrm{i}\mathbf{k}\cdot\mathbf{r}}\right],$$

and the integral of the first term gives zero; hence,

$$\int \mathrm{d}^3\mathbf{k}\, 2\mathrm{i}k_i \frac{\partial B_n(\mathbf{k})}{\partial k_y} \mathrm{e}^{\mathrm{i}\mathbf{k}\cdot\mathbf{r}} = \frac{2}{\mathrm{i}} y\nabla_i(Fy).$$

Combining with the other terms, we finally get

$$\left[E_n(-\mathrm{i}\nabla) - \frac{s}{2}\left(\frac{\partial^2 E_n(\mathbf{k})}{\partial k_i \partial k_x}\right)_0 (y\frac{1}{\mathrm{i}}\nabla_i + \frac{1}{\mathrm{i}}\nabla_i y) + \frac{s^2}{2}\left(\frac{\partial^2 E_n(\mathbf{k})}{\partial k_x^2}\right)_0 y^2\right] F_n(\mathbf{r})$$
$$= EF_n(\mathbf{r}). \tag{9.19}$$

Equation (9.19) is the differential equation satisfied by the envelope function F_n in the presence of a uniform magnetic field. Using Eq. (9.16), we can interpret it as a Schrödinger equation with the Hamiltonian being given only by the kinetic energy operator which is the expansion of the band structure to second order, and replacing $\mathbf{k}$ by $-\mathrm{i}\nabla + e\mathbf{A}/\hbar$. In the expansion of $E_n(\mathbf{k})$, any product of noncommuting factors should be symmetrized [as can be seen, e.g., in the $O(s)$ term of Eq. (9.19)].

Away from the zone center, F satisfies

$$E_n\left(\mathbf{k}_0 + \frac{1}{\mathrm{i}}\nabla + \frac{e}{\hbar}\mathbf{A}\right) F_n(\mathbf{r}) = EF_n(\mathbf{r}). \tag{9.20}$$

9.2.1.1 Exact Solution

For a one-band parabolic model in a static magnetic field, there is an exact solution: equally-spaced Landau levels as in standard quantum mechanics. Again choosing the magnetic field along the z direction, one has

$$H = \frac{p_x^2}{2m_e} + \frac{p_z^2}{2m_e} + \frac{1}{2m_e}\left(p_y + eBx\right)^2. \tag{9.21}$$

Eigenfunctions can be separated in the y and z parts:

$$f(\mathbf{r}) = \frac{1}{\sqrt{L_y L_z}} \mathrm{e}^{\mathrm{i}(k_y y + k_z z)} \phi(x), \tag{9.22}$$

where ϕ satisfies the differential equation

$$\left(\frac{p_x^2}{2m_e} + \frac{1}{2}m_e\omega_c^2(x + x_0)^2 + \frac{\hbar^2 k_z^2}{2m_e}\right)\phi(x) = E\phi(x), \tag{9.23}$$

with

$$x_0 = \frac{\hbar k_y}{eB}. \tag{9.24}$$

Solutions to this problem (the harmonic oscillator) can be written in terms of Hermite polynomials:

$$\phi_n(x) = \exp\left[-eB\frac{(x - x_0)^2}{2\hbar}\right] H_n(\sqrt{s}(x - x_0)), \tag{9.25}$$

$$H_n(x) = (-1)^n \mathrm{e}^{x^2} \frac{\mathrm{d}^n}{\mathrm{d}x^n}[\mathrm{e}^{-x^2}]. \tag{9.26}$$

The eigenvalues of Eq. (9.23) (Landau levels) are independent of the center point x_0:

$$E(n, k_y, k_z) = \left(n + \frac{1}{2}\right)\hbar\omega_c + \frac{\hbar^2 k_z^2}{2m_e}, \tag{9.27}$$

where $\hbar\omega_c = eB\hbar/m_e$. Clearly, for a given spin direction, E vs. k_z are a set of one-dimensional parabolas evenly spaced by $\hbar\omega_c$. To get a feeling about the magnitude of the Landau-level spacing, we note that for an effective mass of approximately $0.1\,m_0$ and a magnetic-field strength of $B = 1$ T, the level spacing is approximately 1 meV.

9.2.1.2 Magnetooptics

Roth, Lax and Zwerdling [25] studied optical magneto-absorption using a simple model of one conduction and one valence isotropic and parabolic band. The matrix element governing the transition is proportional to

$$\begin{aligned} M &\propto \mathcal{E}_0 \int \mathrm{d}V\, f_c^* u_c^* \mathbf{p} \cdot \varepsilon f_v u_v \propto \mathcal{E}_0 \int_\Omega \mathrm{d}V\, u_c^* \mathbf{p} \cdot \varepsilon u_v \int \mathrm{d}V\, f_c^* f_v \\ &= \mathcal{E}_0(\mathbf{p}_{cv} \cdot \varepsilon) \int \mathrm{d}V\, f_c^* f_v, \end{aligned} \tag{9.28}$$

where $\mathcal{E}_0$ is the amplitude of the electric field and the absorption coefficient is given by

$$\alpha(B,E) \propto |M|^2 \left(\frac{2m_c m_h}{m_c + m_h}\right)^{1/2} B \sum_n \frac{1}{(E - E_n)^{-1/2}}, \tag{9.29}$$

where m_i are the two effective masses, n is the Landau-level index, and

$$E_n = E_0 + (n + \frac{1}{2})\hbar(\omega_{cc} + \omega_{cv}), \tag{9.30}$$

where ω_{cc} (ω_{cc}) stands for the cyclotron frequency in the conduction (valence) band. Thus, the selection rule for interband magnetoabsorption (for direct transitions) is $\Delta n = 0$, the absorption curve is related to the density of states of a one-dimensional band, and the oscillations in the curve are separated by an amount equal to multiples of the sum of the two cyclotron frequencies. Of course, a more complicated band structure (degeneracies, warping, inversion asymmetry, ...) will lead to more complex spectra [142, 143]. Finally, note that magnetooptics involves both a magnetic and an electric field. A study of cyclotron resonance in crossed electric and magnetic fields for narrow-gap semiconductors was performed by Zawadzki and coworkers [144].

9.2.2 Degenerate Bands

The derivation is not substantially different from the simple band case. For the latter case, we had Eq. (9.4). Recall that this is an exact equation; it still applies to the degenerate subspace. In particular, for centrosymmetric crystals,

$$(r\mathbf{k}|H|r'\mathbf{k}') = 0, \tag{9.31}$$

where r and r belong to the degenerate subspace. There is thus no coupling within the latter. Equations (9.8) and (9.9) are unchanged. Note that the term $H_0 + H_a + H_b$ is diagonal, while $[H_1, S']$ is off-diagonal [Eq. (9.13)]. Then, the effective Hamiltonian, Eq. (9.8), in the degenerate subspace is

$$\begin{aligned}(r\mathbf{k}|\overline{H}|r'\mathbf{k}') = &\left[\left(E_r(\mathbf{0}) + \frac{\hbar^2 k^2}{2m_0}\right)\delta(\mathbf{k}-\mathbf{k}') - \frac{s\hbar^2 k_x}{m_0}\frac{1}{\mathrm{i}}\frac{\partial\delta(\mathbf{k}-\mathbf{k}')}{\partial k'_y} - \frac{s^2\hbar^2}{2m_0}\frac{\partial^2\delta(\mathbf{k}-\mathbf{k}')}{\partial k'^2_y}\right]\delta_{rr'} \\ &+\frac{\hbar}{m_0^2}\sum_v\left[k_i k_j p^i_{rv} p^j_{vr'}\delta(\mathbf{k}-\mathbf{k}') - s(k_i + k'_i)p^i_{rv}p^x_{vr'}\frac{1}{\mathrm{i}}\frac{\partial\delta(\mathbf{k}-\mathbf{k}')}{\partial k'_y}\right. \\ &\left.- s^2 p^x_{rv}p^x_{vr'}\frac{\partial^2\delta(\mathbf{k}-\mathbf{k}')}{\partial k'^2_y}\right]\frac{1}{\omega_{rv}},\end{aligned} \tag{9.32}$$

with

$$\sum_{r'}\int \mathrm{d}^3\mathbf{k}'\,(r\mathbf{k}|\overline{H}|r'\mathbf{k}')B_{r'}(\mathbf{k}') = EB_r(\mathbf{k}), \tag{9.33}$$

and ν is a state outside the degenerate subspace (in the so-called Löwdin B set). We rewrite Eq. (9.32) as (with $E_r(\mathbf{0}) = 0$)

$$
\begin{aligned}
(r\mathbf{k}|\overline{H}|r'\mathbf{k}') = & \left[\frac{\hbar^2 k^2}{2m_0}\delta_{rr'} + \frac{\hbar}{m_0^2}\sum_{\nu} k_i k_j \frac{p^i_{r\nu} p^j_{\nu r'}}{\omega_{r\nu}} \right] \delta(\mathbf{k}-\mathbf{k}') \\
& - s^2\hbar^2 \left[\frac{1}{2m_0}\frac{\partial^2 \delta(\mathbf{k}-\mathbf{k}')}{\partial k_y'^2}\delta_{rr'} + \frac{1}{m_0^2}\sum_{\nu}\frac{p^x_{r\nu}p^x_{\nu r'}}{\hbar\omega_{r\nu}}\frac{\partial^2\delta(\mathbf{k}-\mathbf{k}')}{\partial k_y'^2} \right] \\
& - s\hbar \left[\frac{\hbar k_x}{m_0}\frac{1}{\mathrm{i}}\frac{\partial\delta(\mathbf{k}-\mathbf{k}')}{\partial k_y'}\delta_{rr'} + (k_i + k_i')\frac{1}{m_0^2}\sum_{\nu}\frac{p^i_{r\nu}p^x_{\nu r'}}{\omega_{r\nu}}\frac{1}{\mathrm{i}}\frac{\partial\delta(\mathbf{k}-\mathbf{k}')}{\partial k_y'} \right].
\end{aligned}
\tag{9.34}
$$

We can now separate the terms in $\overline{H}$ in terms of powers of s. Using the definition of the D matrix given in Eq. (3.6),

$$
O(1) = D^{ij}_{rr'}k_i k_j \delta(\mathbf{k}-\mathbf{k}'), \tag{9.35}
$$

$$
O(s^2) = -s^2 D^{xx}_{rr'}\delta(\mathbf{k}-\mathbf{k}')\frac{\partial^2}{\partial k_y^2}. \tag{9.36}
$$

To simplify the $O(s)$ term, we use the following result obtained earlier [and Eq. (9.15)]:

$$
\begin{aligned}
(k_i + k_i')\frac{\partial\delta(\mathbf{k}-\mathbf{k}')}{\partial k_y'} &= 2k_i\frac{\partial\delta(\mathbf{k}-\mathbf{k}')}{\partial k_y'} + (k_i' - k_i)\frac{\partial\delta(\mathbf{k}-\mathbf{k}')}{\partial k_y'} \\
&= 2k_i\frac{\partial\delta(\mathbf{k}-\mathbf{k}')}{\partial k_y'} + \delta_{iy}(k_y' - k_y)\frac{\partial\delta(\mathbf{k}-\mathbf{k}')}{\partial k_y'} \\
&= 2k_i\frac{\partial\delta(\mathbf{k}-\mathbf{k}')}{\partial k_y'} - \delta_{iy}\delta(\mathbf{k}-\mathbf{k}').
\end{aligned}
$$

Then the $O(s)$ term becomes

$$
\begin{aligned}
&= -s\hbar\left\{ \frac{\hbar k_x}{m_0}\frac{1}{\mathrm{i}}\frac{\partial\delta(\mathbf{k}-\mathbf{k}')}{\partial k_y'}\delta_{rr'} + \left[2k_i\frac{\partial\delta(\mathbf{k}-\mathbf{k}')}{\partial k_y'} - \delta_{iy}\delta(\mathbf{k}-\mathbf{k}')\right]\frac{1}{m_0^2}\sum_{\nu}\frac{p^i_{r\nu}p^x_{\nu r'}}{\omega_{r\nu}}\frac{1}{\mathrm{i}} \right\} \\
&= s\hbar\left\{ \frac{\hbar k_x}{m_0}\frac{1}{\mathrm{i}}\delta(\mathbf{k}-\mathbf{k}')\delta_{rr'} + \frac{2k_i}{m_0^2}\sum_{\nu}\frac{p^i_{r\nu}p^x_{\nu r'}}{\omega_{r\nu}}\frac{1}{\mathrm{i}}\delta(\mathbf{k}-\mathbf{k}') \right\}\frac{\partial}{\partial k_y} \\
&\quad + s\delta_{iy}\delta(\mathbf{k}-\mathbf{k}')\frac{\hbar^2}{m_0^2}\sum_{\nu}\frac{p^i_{r\nu}p^x_{\nu r'}}{\omega_{r\nu}}\frac{1}{\mathrm{i}} \\
&= s\hbar\left\{ \frac{2\hbar k_i}{\mathrm{i}}D^{ix}_{rr'}\frac{\partial}{\partial k_y} + \frac{1}{\mathrm{i}}D^{xy}_{rr'} \right\}\delta(\mathbf{k}-\mathbf{k}').
\end{aligned}
\tag{9.37}
$$

Thus, Eqs. (9.32) and (9.33) become

$$\sum_{r'}\left[D_{rr'}^{ij}k_i k_j B_{r'}(\mathbf{k}) - \mathrm{i}s\left(2k_i D_{rr'}^{ix}\frac{\partial B_{r'}(\mathbf{k})}{\partial k_y} + D_{rr'}^{xy}B_{r'}(\mathbf{k})\right) - s^2 D_{rr'}^{xx}\frac{\partial^2 B_{r'}(\mathbf{k})}{\partial k_y^2}\right] = E B_r(\mathbf{k}). \tag{9.38}$$

The latter is the effective-mass equation in a degenerate-band semiconductor in momentum space. Using Eq. (8.93),

$$\int \mathrm{d}^3\mathbf{k}\, D_{rr'}^{ij}k_i k_j B_{r'}(\mathbf{k})\mathrm{e}^{\mathrm{i}\mathbf{k}\cdot\mathbf{r}} = \int \mathrm{d}^3\mathbf{k}\, D_{rr'}^{ij}\left(\frac{1}{\mathrm{i}}\nabla_i\right)\left(\frac{1}{\mathrm{i}}\nabla_j\right) B_{r'}(\mathbf{k})\mathrm{e}^{\mathrm{i}\mathbf{k}\cdot\mathbf{r}},$$

$$\begin{aligned}\int \mathrm{d}^3\mathbf{k}\,\mathrm{i}k_i\frac{\partial B_{r'}(\mathbf{k})}{\partial k_y}\mathrm{e}^{\mathrm{i}\mathbf{k}\cdot\mathbf{r}} &= \int \mathrm{d}^3\mathbf{k}\,\nabla_i\frac{\partial B_{r'}(\mathbf{k})}{\partial k_y}\mathrm{e}^{\mathrm{i}\mathbf{k}\cdot\mathbf{r}}\\ &= \nabla_i\int \mathrm{d}^3\mathbf{k}\left[\frac{\partial}{\partial k_y}(B_{r'}(\mathbf{k})\mathrm{e}^{\mathrm{i}\mathbf{k}\cdot\mathbf{r}}) - \mathrm{i}yB_{r'}(\mathbf{k})\right]\\ &= \frac{1}{\mathrm{i}}\nabla_i(y)\int \mathrm{d}^3\mathbf{k}\,B_{r'}(\mathbf{k})\mathrm{e}^{\mathrm{i}\mathbf{k}\cdot\mathbf{r}},\end{aligned}$$

$$-\int \mathrm{d}^3\mathbf{k}\,\frac{\partial^2 B_{r'}(\mathbf{k})}{\partial k_y^2}\mathrm{e}^{\mathrm{i}\mathbf{k}\cdot\mathbf{r}} = \mathrm{i}y\int \mathrm{d}^3\mathbf{k}\,\frac{\partial B_{r'}(\mathbf{k})}{\partial k_y}\mathrm{e}^{\mathrm{i}\mathbf{k}\cdot\mathbf{r}} = y^2\int \mathrm{d}^3\mathbf{k}\,B_{r'}(\mathbf{k})\mathrm{e}^{\mathrm{i}\mathbf{k}\cdot\mathbf{r}} = y^2 F_{r'}(\mathbf{r}),$$

we have

$$\sum_{r'}\left[D_{rr'}^{ij}\left(\frac{1}{\mathrm{i}}\nabla_i\right)\left(\frac{1}{\mathrm{i}}\nabla_j\right) - 2sD_{rr'}^{ix}\frac{1}{\mathrm{i}}\nabla_i(y) + \mathrm{i}sD_{rr'}^{xy} + s^2y^2D_{rr'}^{xx}\right]F_{r'}(\mathbf{r}) = EF_r(\mathbf{r}). \tag{9.39}$$

But

$$\begin{aligned}&(\frac{1}{\mathrm{i}}\nabla_i - sy\delta_{ix})(\frac{1}{\mathrm{i}}\nabla_j - sy\delta_{jx})\\ &= \left(\frac{1}{\mathrm{i}}\nabla_i\right)\left(\frac{1}{\mathrm{i}}\nabla_j\right) + s^2y^2\delta_{ix}\delta_{jx} - s\frac{1}{\mathrm{i}}\nabla_i y\delta_{jx} - sy\delta_{ix}\frac{1}{\mathrm{i}}\nabla_j\\ &= \left(\frac{1}{\mathrm{i}}\nabla_i\right)\left(\frac{1}{\mathrm{i}}\nabla_j\right) + s^2y^2\delta_{ix}\delta_{jx} - 2s\frac{1}{\mathrm{i}}\delta_{jx}(\nabla_i y) + \frac{s}{\mathrm{i}}\delta_{ix}\delta_{jx}.\end{aligned}$$

Therefore,

$$\sum_{r'}\left[D_{rr'}^{ij}\left(\frac{1}{\mathrm{i}}\nabla_i - sy\delta_{ix}\right)\left(\frac{1}{\mathrm{i}}\nabla_j - sy\delta_{jx}\right)\right]F_{r'}(\mathbf{r}) = EF_r(\mathbf{r}). \tag{9.40}$$

9.2.3 Spin-Orbit Coupling

In cyclotron-resonance experiments, the cyclotron energies are typically smaller than the spin-orbit energy [4]. Then, the four-band model is adequate. We derive it now. Since the vector potential is only a function of y in our choice of gauge, periodicity is maintained in the other two directions. Thus, the envelope function can be written as

$$F_j(\mathbf{r}) = \mathrm{e}^{\frac{\mathrm{i}}{\hbar}(p_x x + p_z z)} f_j(y). \tag{9.41}$$

We again make the substitution $k_i \to \frac{1}{\mathrm{i}}\nabla_i - sy\delta_{ix}$ in the D matrix. For example, the (1, 1) matrix element from Table 3.6 becomes

$$\begin{aligned}
&\frac{1}{2}P(\frac{1}{\mathrm{i}}\nabla_i - sy\delta_{ix})F_j(\mathbf{r}) \\
&= \frac{1}{2}\left\{(L + M + \frac{\hbar^2}{m_0})\left[(-\mathrm{i}\nabla_x - sy)^2 + (-\mathrm{i}\nabla_y)^2\right] + 2(M + \frac{\hbar^2}{2m_0})(-\mathrm{i}\nabla_z)^2\right\} \\
&\quad \times \mathrm{e}^{\frac{\mathrm{i}}{\hbar}(p_x x + p_z z)} f_j(y) \\
&= \frac{\mathrm{e}^{\frac{\mathrm{i}}{\hbar}(p_x x + p_z z)}}{2}\left\{(L + M + \frac{\hbar^2}{m_0})\left[\left(\frac{p_x}{\hbar} - sy\right)^2 - \frac{\partial^2}{\partial y^2}\right] + \frac{2(M + \frac{\hbar^2}{2m_0})}{\hbar^2}p_z^2\right\} f_j(y) \\
&= \mathrm{e}^{\frac{\mathrm{i}}{\hbar}(p_x x + p_z z)}\left\{\frac{(L + M + \frac{\hbar^2}{m_0})}{2}\left[-\frac{\partial^2}{\partial y^2} + s^2y^2\right] + \frac{M + \frac{\hbar^2}{m_0}}{\hbar^2}p_z^2\right\} f_j(y),
\end{aligned}$$

if y is rescaled so that $p_x/\hbar - sy = 0$. Continuing,

$$\begin{aligned}
RF_j &= -\frac{N}{\sqrt{3}}\left[(-\mathrm{i}\nabla_x - sy - \nabla_y)(-\mathrm{i}\nabla_z)\right]\mathrm{e}^{\frac{\mathrm{i}}{\hbar}(p_x x + p_z z)} f_j(y) \\
&= \frac{Np_z}{\sqrt{3}\hbar}\mathrm{e}^{\frac{\mathrm{i}}{\hbar}(p_x x + p_z z)}\left(sy + \frac{\partial}{\partial y}\right) f_j(y), \\
SF_j &= -\frac{1}{2\sqrt{3}}\left\{(L - M)\left[(-\mathrm{i}\nabla_x - sy)^2 - (-\mathrm{i}\nabla_y)^2\right]\right. \\
&\quad \left. -2\mathrm{i}N(-\mathrm{i}\nabla_x - sy)(-\mathrm{i}\nabla_y)\right\}\mathrm{e}^{\frac{\mathrm{i}}{\hbar}(p_x x + p_z z)} f_j(y) \\
&= -\frac{\mathrm{e}^{\frac{\mathrm{i}}{\hbar}(p_x x + p_z z)}}{2\sqrt{3}}\left\{(L - M)\left(s^2y^2 + \frac{\partial^2}{\partial y^2}\right) + Ns\left(y\frac{\partial}{\partial y} + \frac{\partial}{\partial y}y\right)\right\} f_j(y), \\
\left[\frac{1}{6}P + \frac{2}{3}Q\right]F_j &= \frac{\mathrm{e}^{\frac{\mathrm{i}}{\hbar}(p_x x + p_z z)}}{3}\left\{\frac{(L + M)}{2}\left[-\frac{\partial^2}{\partial y^2} + s^2y^2\right] + \frac{M}{\hbar^2}p_z^2\right\} f_j(y) \\
&\quad + \frac{2}{3}\left\{M\left[(-\mathrm{i}\nabla_x - sy)^2 - (-\mathrm{i}\nabla_y)^2\right] + L(-\mathrm{i}\nabla_z)^2\right\}\mathrm{e}^{\frac{\mathrm{i}}{\hbar}(p_x x + p_z z)} f_j(y) \\
&= \frac{\mathrm{e}^{\frac{\mathrm{i}}{\hbar}(p_x x + p_z z)}}{3}\left\{\frac{(L + M)}{2}\left[-\frac{\partial^2}{\partial y^2} + s^2y^2\right] + \frac{M}{\hbar^2}p_z^2\right\} f_j(y) \\
&\quad + \mathrm{e}^{\frac{\mathrm{i}}{\hbar}(p_x x + p_z z)}\left[\frac{2M}{3}\left(s^2y^2 - \frac{\partial^2}{\partial y^2}\right) + \frac{2L}{3}\frac{p_z^2}{\hbar^2}\right] f_j(y) \\
&= \mathrm{e}^{\frac{\mathrm{i}}{\hbar}(p_x x + p_z z)}\left\{\frac{(L + 5M)}{6}\left[-\frac{\partial^2}{\partial y^2} + s^2y^2\right] + \frac{(2L + M)}{3}\frac{p_z^2}{\hbar^2}\right\} f_j(y).
\end{aligned}$$

Hence, we obtain the coupled differential equations given in Table 9.1.

Table 9.1 Envelope-function equation for four-band LK Hamiltonian with magnetic field

$$
\begin{bmatrix}
\left|\frac{3}{2}\,\frac{3}{2}\right\rangle & \left|\frac{3}{2}\,\frac{1}{2}\right\rangle & \left|\frac{3}{2}\,-\frac{1}{2}\right\rangle & \left|\frac{3}{2}\,-\frac{3}{2}\right\rangle \\
\frac{\left(L+M+\frac{\hbar^2}{m_0}\right)}{2}\left(-\frac{\partial^2}{\partial y^2}+s^2y^2\right) + \frac{M+\frac{\hbar^2}{m_0}}{\hbar^2}p_z^2-\lambda & \frac{Np_z}{\sqrt{3}\hbar}\left(sy+\frac{\partial}{\partial y}\right) & -\frac{1}{2\sqrt{3}}\left[(L-M)\left(s^2y^2+\frac{\partial^2}{\partial y^2}\right) - Ns\left(y\frac{\partial}{\partial y}+\frac{\partial}{\partial y}y\right)\right] & 0 \\
-\frac{Np_z}{\sqrt{3}\hbar}\left(-sy+\frac{\partial}{\partial y}\right) & \frac{(L+5M)+\frac{6\hbar^2}{2m_0}}{6}\left(-\frac{\partial^2}{\partial y^2}+s^2y^2\right) + \frac{(2L+M)+\frac{3\hbar^2}{2m_0}}{3}\frac{p_z^2}{\hbar^2}-\lambda & 0 & \frac{1}{2\sqrt{3}}\left[(L-M)\left(s^2y^2+\frac{\partial^2}{\partial y^2}\right) + Ns\left(y\frac{\partial}{\partial y}+\frac{\partial}{\partial y}y\right)\right] \\
-\frac{1}{2\sqrt{3}}\left[(L-M)\left(s^2y^2+\frac{\partial^2}{\partial y^2}\right)\right] + Ns\left(y\frac{\partial}{\partial y}+\frac{\partial}{\partial y}y\right)\Big] & 0 & \frac{(L+5M)+\frac{6\hbar^2}{2m_0}}{6}\left(-\frac{\partial^2}{\partial y^2}+s^2y^2\right) + \frac{(2L+M)+\frac{3\hbar^2}{2m_0}}{3}\frac{p_z^2}{\hbar^2}-\lambda & +\frac{Np_z}{\sqrt{3}\hbar}\left(sy+\frac{\partial}{\partial y}\right) \\
0 & \frac{1}{2\sqrt{3}}\left[(L-M)\left(s^2y^2+\frac{\partial^2}{\partial y^2}\right) + Ns\left(y\frac{\partial}{\partial y}+\frac{\partial}{\partial y}y\right)\right] & +\frac{Np_z}{\sqrt{3}\hbar}\left(sy+\frac{\partial}{\partial y}\right) & \frac{(L+M)+\frac{\hbar^2}{m_0}}{2}\left(-\frac{\partial^2}{\partial y^2}+s^2y^2\right) + \frac{M+\frac{\hbar^2}{m_0}}{\hbar^2}p_z^2-\lambda
\end{bmatrix}
\begin{bmatrix} f_1 \\ f_2 \\ f_3 \\ f_4 \end{bmatrix} = 0.
$$

9.3 Valence-Band Landau Levels

The $k \cdot p$ models in the presence of a magnetic field can be solved to obtain the Landau levels. The procedure is well described in the paper by Hensel and Suzuki [145].

9.3.1 Exact Solution

A simplified solution was given by LK. They chose $p_z = 0$ and ignored the electron spin. The Hamiltonian becomes the one given in Table 9.2. One can now introduce ladder operators:

$$y = \sqrt{\frac{1}{2s}}(a + a^\dagger),\ p_y = -\mathrm{i}\hbar\sqrt{\frac{s}{2}}(a - a^\dagger), \tag{9.42}$$

giving

$$[a, a^\dagger] = 1. \tag{9.43}$$

Then

$$\begin{aligned} p_y^2 + s^2\hbar^2 y^2 &= \hbar^2 s(aa^\dagger + a^\dagger a) = \hbar^2 s(2N+1),\ N = a^\dagger a, \\ s^2y^2 - p_y^2 &= \hbar^2 s(a^2 + (a^\dagger)^2), \\ s(yp_y + p_y y) &= \frac{\hbar s}{\mathrm{i}}(a^2 - (a^\dagger)^2). \end{aligned} \tag{9.44}$$

LK now made the spherical approximation: $L - M = N$. We will also write

$$\begin{aligned} \alpha &= \frac{(L + M + \frac{\hbar^2}{m_0})}{2}, \\ \beta &= \frac{(L + 5M) + \frac{6\hbar^2}{2m_0}}{6}, \\ \alpha - \beta &= \frac{L - M}{3}, \\ \gamma &= \sqrt{3}(\alpha - \beta). \end{aligned} \tag{9.45}$$

This finally gives the Hamiltonian as

$$\begin{pmatrix} \alpha(2N+1) & -\gamma(a^\dagger)^2 & 0 & 0 \\ -\gamma a^2 & \beta(2N+1) & 0 & 0 \\ 0 & 0 & \beta(2N+1) & \gamma(a^\dagger)^2 \\ 0 & 0 & \gamma a^2 & \alpha(2N+1) \end{pmatrix}, \tag{9.46}$$

Table 9.2 Four-band LK Hamiltonian with magnetic field with $p_z = 0$

$$
\begin{pmatrix}
\left|\frac{3}{2}\ \frac{3}{2}\right\rangle & \left|\frac{3}{2}\ -\frac{1}{2}\right\rangle & \left|\frac{3}{2}\ \frac{1}{2}\right\rangle & \left|\frac{3}{2}\ -\frac{3}{2}\right\rangle \\
\frac{\left(L+M+\frac{\hbar^2}{m_0}\right)}{2\hbar^2}\left(p_y^2+s^2\hbar^2y^2\right) & -\frac{1}{2\sqrt{3}}\left[\frac{(L-M)}{\hbar^2}\left(s^2\hbar^2y^2-p_y^2\right) + \mathrm{i}N\frac{s}{\hbar}\left(yp_y+p_yy\right)\right] & 0 & 0 \\
-\frac{1}{2\sqrt{3}}\left[\frac{(L-M)}{\hbar^2}\left(s^2y^2-p_y^2\right) + \mathrm{i}N\frac{s}{\hbar}\left(yp_y+p_yy\right)\right] & \frac{(L+5M)+\frac{6\hbar^2}{2m_0}}{6\hbar^2}\left(p_y^2+s^2y^2\right) & 0 & 0 \\
0 & 0 & \frac{(L+5M)+\frac{6\hbar^2}{2m_0}}{6\hbar^2}\left(p_y^2+s^2\hbar^2y^2\right) & \frac{1}{2\sqrt{3}}\left[\frac{(L-M)}{\hbar^2}\left(s^2\hbar^2y^2-p_y^2\right) + \mathrm{i}N\frac{s}{\hbar}\left(yp_y+p_yy\right)\right] \\
0 & 0 & \frac{1}{2\sqrt{3}}\left[\frac{(L-M)}{\hbar^2}\left(s^2\hbar^2y^2-p_y^2\right) + \mathrm{i}N\frac{s}{\hbar}\left(yp_y+p_yy\right)\right] & \frac{(L+M)+\frac{\hbar^2}{m_0}}{2\hbar^2}\left(p_y^2+s^2\hbar^2y^2\right)
\end{pmatrix}
$$

where an overall factor of s has been left out. Due to the presence of the ladder operators, one can use as basis the oscillator eigenstates:

$$f_j = \sum_n f_j(n)\phi_n. \tag{9.47}$$

We know that

$$a^\dagger \phi_n = (n+1)^{1/2}\phi_{n+1},\ a\phi_n = n^{1/2}\phi_{n-1}. \tag{9.48}$$

Thus, for the top 2×2 block,

$$\begin{aligned}[\alpha(2n+1) - \lambda]f_1(n) - \gamma n^{1/2}(n-1)^{1/2} f_2(n-2) &= 0,\\ -\gamma(n+1)^{1/2}(n+2)^{1/2} f_1(n+2) + [\beta(2n+1) - \lambda]f_2(n) &= 0.\end{aligned} \tag{9.49}$$

This system of equations can be solved by making use of the following *Ansatz*:

$$f_1(n) = C_1\delta_{n,n_0},\ f_2(n) = C_2\delta_{n+2,n_0}. \tag{9.50}$$

Then

$$\begin{aligned}[\alpha(2n_0+1) - \lambda]C_1 - \gamma n_0^{1/2}(n_0-1)^{1/2}C_2 &= 0,\ n_0 = 0,1,2,\ldots\\ -\gamma n_0^{1/2}(n_0-1)^{1/2}C_1 + [\beta(2n_0-3) - \lambda]C_2 &= 0,\ n_0 = 2,3,\ldots\end{aligned}$$

and for a nontrivial solution,

$$\begin{vmatrix} \alpha(2n_0+1) - \lambda & -\gamma n_0^{1/2}(n_0-1)^{1/2} \\ -\gamma(n_0)^{1/2}(n_0-1)^{1/2} & \beta(2n_0-3) - \lambda \end{vmatrix} = 0,$$

or

$$[\alpha(2n_0+1) - \lambda][\beta(2n_0-3) - \lambda] = \gamma^2 n_0(n_0-1).$$

If $n_0 = 0, 1$,

$$\lambda = \alpha(2n_0+1).$$

If $n_0 > 1$,

$$\begin{aligned}\lambda = \frac{1}{2}\Big\{&(\alpha+\beta)(2n_0+1) - 4\beta\\ &\pm\left[[(\alpha+\beta)(2n_0+1) - 4\beta]^2 - 4\alpha\beta(2n_0+1)(2n_0-3) + 4\gamma^2 n_0(n_0-1)\right]^{1/2}\Big\}.\end{aligned}$$

The latter can be simplified by noting

$$\begin{aligned}&[(\alpha+\beta)(2n_0+1) - 4\beta]^2 - 4\alpha\beta(2n_0+1)^2 + 16\alpha\beta(2n_0+1)\\ &= [(\alpha-\beta)(2n_0+1) + 4\beta]^2.\end{aligned}$$

Hence, if we similarly work out the eigenvalues of the lower block of Eq. (9.46), all the energies are (putting back the factor of s)

$$\lambda = \alpha s(2n_0 + 1), \tag{9.51}$$

$$\lambda = \beta s(2n_0 + 1), \tag{9.52}$$

for $n_0 = 0$, 1, and

$$\lambda_\pm = \frac{s}{2}\left\{(\alpha+\beta)(2n_0+1) - 4\beta \pm \left[[(\alpha-\beta)(2n_0+1)+4\beta]^2 + 4\gamma^2 n_0(n_0-1)\right]^{1/2}\right\},$$

$$\lambda'_\pm = \frac{s}{2}\left\{(\alpha+\beta)(2n_0+1) - 4\alpha \pm \left[[(\beta-\alpha)(2n_0+1)+4\alpha]^2 + 4\gamma^2 n_0(n_0-1)\right]^{1/2}\right\},$$

for $n_0 > 1$. The levels are no longer degenerate, due to the breaking of time-reversal symmetry by the magnetic field. Note that the Landau levels are again equally spaced.

The result was obtained by introducing the magnetic field right from the start into the Hamiltonian. An alternative approach might have been to obtain the dispersion relation first, then apply the magnetic field; LK called the latter the "classical" expression. The four-band problem has an exact solution as we saw in Chap. 3; within the spherical approximation,

$$\begin{aligned} E(\mathbf{k}) &= \frac{1}{3}(L+2M)k^2 + \frac{\hbar^2 k^2}{2m_0} \pm \frac{1}{3}(L-M)k^2 \\ &= \frac{1}{3}(2L+M)k^2 + \frac{\hbar^2 k^2}{2m_0},\ Mk^2 + \frac{\hbar^2 k^2}{2m_0} \\ &= \frac{1}{2}(3\alpha-\beta)k^2,\ \frac{1}{2}(3\beta-\alpha)k^2. \end{aligned} \tag{9.53}$$

Note that each level remains doubly degenerate. Each also behaves like a free electron. Application of a magnetic field is known to lead to Landau levels. One can also apply the effective-mass theory. The energy levels become

$$E(n_0) = \begin{cases} \frac{1}{2}(3\alpha-\beta)(2n_0+1), \\ \frac{1}{2}(3\beta-\alpha)(2n_0+1), \end{cases} \tag{9.54}$$

for $n_0 = 0, 1, 2, \ldots$ The differences compared to the earlier results are obvious. Nevertheless, the two are related at high quantum numbers [4].

9.3.2 General Solution

In general, the magnetic Hamiltonian is infinite dimensional and the eigenvector is a linear combination of harmonic oscillator $|\phi_n\rangle$ and Bloch states $|M_J\rangle$:

$$\psi = \sum_{n,M_J} a_n(M_J)|\phi_n\rangle|M_J\rangle. \tag{9.55}$$

In practice, the matrix is truncated. For zincblende, Bell and Rodgers solved the full magnetic problem for a magnetic field along $\langle 100\rangle$ [146]. An example of the solution for $k_B = 0$ for Ge is shown in Fig. 9.2. Following Hensel and Suzuki [145], the Landau levels are labeled by (N_n, K^π). Here, N is the Landau quantum number for the envelope function and n labels the various M_J states for each N (for example, there are four n values for the Luttinger model). The quantum number $K = 0, 1, \ldots, \nu - 1$ labels the rotational symmetry about B for the latter along a crystal axis of ν-fold rotational symmetry, and π is the parity of the envelope function. One has [145]

$$N = M_J + \frac{3}{2} + n,\ K \equiv N \ (\mathrm{mod}\ \nu). \tag{9.56}$$

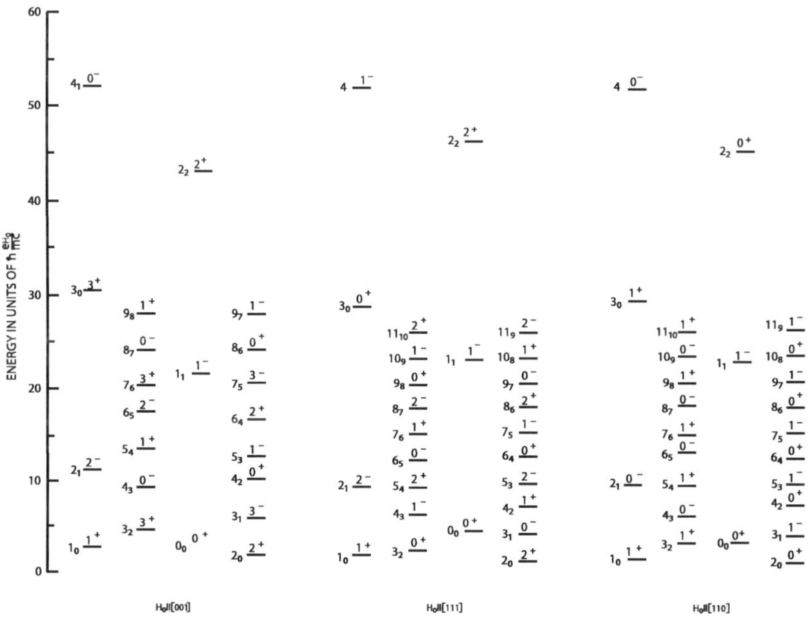

Fig. 9.2 Valence-band Landau levels for B along different directions and $k_B = 0$ for Ge. Reprinted with permission from [145]. ©1974 by the American Physical Society

9.4 Extended Kane Model

The invariant form of the magnetic interaction within a 14-band model for zincblende is given in Table 9.3. The valence parameters (κ', q') are clearly the Luttinger ones with the conduction band subtracted out. Similarly, there is a g-factor term in the conduction band (see Sect. 9.5 for details). However, in addition, there is a conduction–valence band-coupling term that is linear in the magnetic field and is parametrized by D'.

Table 9.3 Invariant expansion of magnetic interactions of the Hamiltonian for a 14-band model of zincblende. $\{AB\} = \frac{1}{2}(AB + BA)$, c.p. stands for cyclic permutation. Adapted with permission from [21, 81]. ©1979 by the American Physical Society

$$\begin{aligned}
H_B^{6c6c} &= \frac{1}{2}g'\mu_B(\sigma \cdot \mathbf{B}),\\
H_B^{8v8v} &= -2\mu_B\left[\kappa'\mathbf{J}\cdot\mathbf{B} + q'(J_x^3 B_x + \text{c.p.})\right],\\
H_B^{7v7v} &= -2\mu_B\kappa'\boldsymbol{\sigma}\cdot\mathbf{B},\\
H_B^{6c8v} &= \frac{\mathrm{i}}{\sqrt{3}}\mu_B D'(T_{yz}B_x + \text{c.p.}),\\
H_B^{8v7v} &= -3\mu_B\kappa'\mathbf{U}\cdot\mathbf{B}.
\end{aligned}$$

9.5 Landé g-Factor

We now consider the more complete problem and rewrite the one-particle Hamiltonian for an electron in a static magnetic field with the electron spin term:

$$H = \frac{(\mathbf{p} + e\mathbf{A})^2}{2m_0} + V(\mathbf{r}) + \frac{\hbar}{4m_0^2c^2}(\sigma \times \nabla V)\cdot\mathbf{p} + g_0\mu_B\boldsymbol{\sigma}\cdot\mathbf{B}, \tag{9.57}$$

where g_0 is the free-electron Landé g-factor ($g_0 \approx 2$). Thus, the focus is on the conduction band.

The conduction-band wave function ψ near the Γ point is, in the one-band model, written as:

$$\psi = f_e(\mathbf{r})|S\rangle, \tag{9.58}$$

where $|S\rangle$ is the Bloch-periodic part and f_e is the envelope function. As we have seen, to zeroth order in $\mathbf{A}$, the term $(\mathbf{p} + e\mathbf{A})^2/(2m_0)$ leads to the effective mass in the one-band Hamiltonian for the envelope function f_e using second-order Löwdin perturbation theory through interactions with remote states (including the valence-band $|P\rangle$ states among remote states). To first order in $\mathbf{A}$, the same term leads to the electron effective g-factor in the one-band Hamiltonian for the envelope function f_e using second-order perturbation theory.

Specifically, we had from Sect. 5.4.2.1 that

$$D_{nm} = D^{ij}_{nm'}k_ik_j = \frac{1}{2}[D^{ij}_{nm'}, D^{ji}_{nm}][k_i, k_j] + \{D^{ij}_{nm'}, D^{ji}_{nm}\}\{k_i, k_j\}.$$

The second, symmetric terms leads to the effective mass. For the spin-up S state of a cubic semiconductor, the Löwdin perturbation term with the antisymmetric term is

$$\begin{aligned}\Delta E - \frac{\hbar^2}{m_0^2}\sum_l \Bigg\{&\left[\frac{\langle S\uparrow|p_x|l\rangle\langle l|p_y|S\uparrow\rangle - \langle S\uparrow|p_y|l\rangle\langle l|p_x|S\uparrow\rangle}{E_s - E_l}\right][k_x, k_y] \\ &+\left[\frac{\langle S\uparrow|p_z|l\rangle\langle l|p_x|S\uparrow\rangle - \langle S\uparrow|p_x|l\rangle\langle l|p_z|S\uparrow\rangle}{E_s - E_l}\right][k_z, k_x] \\ &+\left[\frac{\langle S\uparrow|p_y|l\rangle\langle l|p_z|S\uparrow\rangle - \langle S\uparrow|p_z|l\rangle\langle l|p_y|S\uparrow\rangle}{E_s - E_l}\right][k_y, k_z]\Bigg\},\end{aligned}$$

where l denote remote states.

We have seen that the commutator of the wave vector is proportional to the magnetic field. Assuming a magnetic field along the z-direction, we then have

$$\begin{aligned}\Delta E &= -\mathrm{i}e\frac{\hbar}{m_0^2}B_z\sum_l \frac{\langle S\uparrow|p_x|l\rangle\langle l|p_y|s\uparrow\rangle - \langle s\uparrow|p_y|l\rangle\langle l|p_x|s\uparrow\rangle}{E_s - E_l} \\ &= \Delta g\mu_B\sigma_z B_z.\end{aligned} \tag{9.59}$$

Hence, using second-order perturbation theory, there is an additional g-factor contribution to the one-band electron effective-mass Hamiltonian:

$$\Delta g = \frac{g_0}{\mathrm{i}m_0}\sum_l \frac{\langle S\uparrow|p_x|l\rangle\langle l|p_y|S\uparrow\rangle - \langle S\uparrow|p_y|l\rangle\langle l|p_x|S\uparrow\rangle}{E_S - E_l}, \tag{9.60}$$

For valence states, the effective g factor arises from the κ (isotropic) and q terms in the Luttinger Hamiltonian [87].

9.5.1 Zincblende

We now give the explicit derivation of the g-factor for the conduction electron in a zincblende semiconductor. In this case, the conduction state will be written as $|\mathrm{i}S\rangle$ and we rewrite Eq. (9.60) as

$$\Delta g = \frac{g_0}{\mathrm{i}m_0}\sum_l \frac{\langle \mathrm{i}S\uparrow|p_x|l\rangle\langle l|p_y|\mathrm{i}S\uparrow\rangle - \langle \mathrm{i}S\uparrow|p_y|l\rangle\langle l|p_x|\mathrm{i}S\uparrow\rangle}{E_S - E_l}. \tag{9.61}$$

Table 9.4 Matrix elements for computing the g factor for zincblende

l	$\langle iS\uparrow \vert p_x \vert l\rangle$	$\langle l \vert p_x \vert iS\uparrow\rangle$	$\langle iS\uparrow \vert p_y \vert l\rangle$	$\langle l \vert p_y \vert iS\uparrow\rangle$
$hh\uparrow$	$\frac{P}{\sqrt{2}}$	$\frac{P}{\sqrt{2}}$	$\frac{iP}{\sqrt{2}}$	$-\frac{iP}{\sqrt{2}}$
$hh\downarrow$	0	0	0	0
$lh\uparrow$	0	0	0	0
$lh\downarrow$	$-\frac{P}{\sqrt{6}}$	$-\frac{P}{\sqrt{6}}$	$\frac{iP}{\sqrt{6}}$	$-\frac{iP}{\sqrt{6}}$
$sh\uparrow$	0	0	0	0
$sh\downarrow$	$-\frac{P}{\sqrt{3}}$	$-\frac{P}{\sqrt{3}}$	$\frac{iP}{\sqrt{3}}$	$-\frac{iP}{\sqrt{3}}$

First, we will assume the only relevant states in the summation are from the valence-band hh, lh (Γ_{8v}), and sh (Γ_{7v}) bands. We can use the Hamiltonian in Table 4.1 in order to evaluate the matrix elements in Eq. (9.61). All the needed matrix elements are given in Table 9.4.

Then

$$\begin{aligned}\frac{\Delta g}{g_0} &= \frac{1}{im_0}\left\{\frac{\left(\frac{P}{\sqrt{2}}\right)\left(-\frac{iP}{\sqrt{2}}\right)-\left(\frac{iP}{\sqrt{2}}\right)\left(\frac{P}{\sqrt{2}}\right)}{E_0}+\frac{\left(-\frac{P}{\sqrt{6}}\right)\left(-\frac{iP}{\sqrt{6}}\right)-\left(\frac{iP}{\sqrt{6}}\right)\left(-\frac{P}{\sqrt{6}}\right)}{E_0}\right.\\ &\quad\left.+\frac{\left(-\frac{P}{\sqrt{3}}\right)\left(-\frac{iP}{\sqrt{3}}\right)-\left(\frac{iP}{\sqrt{3}}\right)\left(-\frac{P}{\sqrt{3}}\right)}{E_0+\Delta_0}\right\}\\ &= -\frac{2P^2}{3m_0}\frac{\Delta_0}{E_0\left(E_0+\Delta_0\right)},\end{aligned} \tag{9.62}$$

and the total (effective) g-factor g^* becomes:

$$g^* = g_0 + \Delta g = g_0\left(1-\frac{2\,P^2}{3\,m_0}\frac{\Delta_0}{E_0\left(E_0+\Delta_0\right)}\right). \tag{9.63}$$

Thus, the term $g_0\mu_B\sigma\cdot\mathbf{B}$ must be replaced by $g^*\mu_B\sigma\cdot\mathbf{B}$ in the effective-mass one-band electron Hamiltonian similar to the replacement of the free-electron mass by the effective mass. The Landé spin-splitting effect is small; at $B = 1$ T, it amounts to approximately 0.1 meV for an effective g factor equal to 2. Hermann and Weisbuch [147] include the upper conduction bands into the summation explicitly:

$$\begin{aligned}\frac{\Delta g}{g_0} &= -\frac{2P^2}{3m_0}\frac{\Delta_0}{E_0\left(E_0+\Delta_0\right)}+\frac{2P'^2}{3m_0}\left(\frac{1}{E_{\Gamma_{6c}}-E_{\Gamma_{8c}}}-\frac{1}{E_{\Gamma_{6c}}-E_{\Gamma_{7c}}}\right)+C'\\ &= -\frac{2P^2}{3m_0}\frac{\Delta_0}{E_0\left(E_0+\Delta_0\right)}+\frac{2P'^2}{3m_0}\frac{\Delta_1}{E_1\left(E_1+\Delta_1\right)}+C',\end{aligned} \tag{9.64}$$

where C' accounts for more remote bands.

9.5.2 Wurtzite

Hermann and Weisbuch [147] also evaluated the g-factor of wurtzite semiconductors using the Gutsche–Jahne basis (Table C.9). In this case, the valence band consists of the A (Γ_{9v}), B (Γ_{7v}), and C (Γ_{7v}) states and the g-factor is anisotropic. The necessary matrix elements for evaluating $g_{||}$ are given in Table 9.5.

Table 9.5 Matrix elements for computing the $g_{||}$ factor for wurtzite

l	$\langle \mathrm{i}S\uparrow \vert p_x\vert l\rangle$	$\langle l\vert p_x\vert \mathrm{i}S\uparrow\rangle$	$\langle \mathrm{i}S\uparrow \vert p_y\vert l\rangle$	$\langle l\vert p_y\vert \mathrm{i}S\uparrow\rangle$
$A\uparrow$	$\frac{P}{\sqrt{2}}$	$\frac{P}{\sqrt{2}}$	$\frac{\mathrm{i}P}{\sqrt{2}}$	$-\frac{\mathrm{i}P}{\sqrt{2}}$
$A\downarrow$	0	0	0	0
$B\uparrow$	0	0	0	0
$B\downarrow$	$\frac{(1-q_7^2)^{1/2}}{\sqrt{2}}P$	$\frac{(1-q_7^2)^{1/2}}{\sqrt{2}}P$	$-\mathrm{i}\frac{(1-q_7^2)^{1/2}}{\sqrt{2}}P$	$\mathrm{i}\frac{(1-q_7^2)^{1/2}}{\sqrt{2}}P$
$C\uparrow$	0	0	0	0
$C\downarrow$	$-\frac{q_z}{\sqrt{2}}P$	$-\frac{q_z}{\sqrt{2}}P$	$\mathrm{i}\frac{q_z}{\sqrt{2}}P$	$-\mathrm{i}\frac{q_z}{\sqrt{2}}P$

Then

$$\begin{aligned}\frac{\Delta g_{||}}{g_0} &= \frac{1}{\mathrm{i}m_0}\left\{\frac{\left(\frac{P}{\sqrt{2}}\right)\left(-\frac{\mathrm{i}P}{\sqrt{2}}\right)-\left(\frac{\mathrm{i}P}{\sqrt{2}}\right)\left(\frac{P}{\sqrt{2}}\right)}{E_e-E_A}\right.\\ &\quad+\frac{\left(\frac{(1-q_7^2)^{1/2}P}{\sqrt{2}}\right)\left(\frac{\mathrm{i}(1-q_7^2)^{1/2}P}{\sqrt{2}}\right)-\left(\frac{-\mathrm{i}(1-q_7^2)^{1/2}P}{\sqrt{2}}\right)\left(\frac{(1-q_7^2)^{1/2}P}{\sqrt{2}}\right)}{E_e-E_B}\\ &\quad\left.+\frac{\left(-\frac{q_zP}{\sqrt{2}}\right)\left(-\frac{\mathrm{i}q_zP}{\sqrt{2}}\right)-\left(\frac{\mathrm{i}q_zP}{\sqrt{2}}\right)\left(-\frac{q_zP}{\sqrt{2}}\right)}{E_e-E_C}\right\}\\ &= \frac{P^2}{m_0}\left(\frac{1}{E_A}-\frac{(1-q_7^2)}{E_B}-\frac{q_z^2}{E_C}\right),\end{aligned} \tag{9.65}$$

where E_e was set to zero. Similarly, one obtains

$$\frac{\Delta g_\perp}{g_0} = -\frac{P^2}{m_0}\left(\frac{1}{E_C}-\frac{1}{E_B}\right)[2q_z(1-q_7^2)]^{1/2}. \tag{9.66}$$

The effective g-factor in a semiconductor was first discussed by Luttinger [4]. Other studies on the electron g-factor can be found in [25, 147]. Typical experimental values as compiled by Hermann and Weisbuch [147] are given in Table 9.6.

Table 9.6 Experimental g^*-factor values. Adapted with permission from [147]. ©1977 by the American Physical Society

	g^*
InSb	−51.3
InAs	−14.8
InP	1.26
GaSb	−9.25
GaAs	−0.44

9.6 Summary

It was shown how to include an external static magnetic field into the $k \cdot p$ Hamiltonian using the method of canonical transformations of Luttinger and Kohn. The answer is the minimal-coupling procedure $\mathbf{k} \rightarrow \mathbf{k} + e\mathbf{A}/\hbar$, where $\mathbf{A}$ is the vector potential, applied to the effective Hamiltonian. An exact solution for the degenerate valence-band problem discovered by Luttinger was then presented. The theory of the g factor for a simple conduction band for both zincblende and wurtzite was also given.

Chapter 10
Electric Field

10.1 Overview

The formalism for studying a static electric field in an infinite solid is fundamentally an ill-defined problem [148]. The reason is because the corresponding potential energy is given by

$$V(\mathbf{r}) = -\mathcal{E} \cdot \mathbf{r}, \tag{10.1}$$

which is a nonperiodic and unbounded function. A representation in terms of extended Bloch states is, therefore, infinite [134]. Furthermore, being unbounded from below means that there is no stable ground state. This causes difficulty with variational formulations of the energy functional [149, 150]. For the most part, the study of this problem has evolved independently within the ab initio and empirical communities. The problem was extensively discussed within $k \cdot p$ theory by Blount in 1962 [134]. It was shown that there is no difficulty unless there is a degeneracy in the energy spectrum. A solution for a degenerate band structure was recently proposed by Foreman [135, 136, 151].

10.2 One-Band Model of Stark Effect

A general solution to the one-band effective-mass Hamiltonian,

$$H = \frac{p_x^2}{2m^*} + e\mathcal{E}x + \frac{p_y^2}{2m^*} + \frac{p_z^2}{2m^*}, \tag{10.2}$$

with $\mathcal{E}$ the electric field applied along the x direction is:

$$\psi(\mathbf{r}) = \phi(x)\mathrm{e}^{\mathrm{i}k_y y}\mathrm{e}^{\mathrm{i}k_z z}, \tag{10.3}$$

where $\phi(x)$ is the solution to the differential equation

L.C. Lew Yan Voon, M. Willatzen, *The* k · p *Method*,
DOI 10.1007/978-3-540-92872-0_10, © Springer-Verlag Berlin Heidelberg 2009

$$\left(\frac{p_x^2}{2m^*} + e\mathcal{E}x\right)\phi(x) = E_x\phi(x), \tag{10.4}$$

with the boundary condition:

$$\lim_{x\to\pm\infty} |\phi(x)| < \infty. \tag{10.5}$$

The energy E_x is related to the total energy E by:

$$E = E_x + \frac{\hbar^2 k_y^2}{2m^*} + \frac{\hbar^2 k_z^2}{2m^*}. \tag{10.6}$$

Solutions ϕ are given in terms of Bessel J and K functions of fractional order. Since the potential term $e\mathcal{E}x$ is arbitrarily negative as $x \to -\infty$, the spectrum is continuous. In particular, bounded solutions with any $E_x < 0$ are allowed whenever $\mathcal{E} \neq 0$. Similarly, for holes, a continuous spectrum exists above the valence-band edge whenever $\mathcal{E} \neq 0$.

10.3 Multiband Stark Problem

Adding an unbounded operator such as the Stark term to the Hamiltonian and performing a multiband analysis involving degenerate states, require a careful analysis [134, 151]. We shall follow Foreman's discussion [151].

10.3.1 Basis Functions

Using the Bloch functions $\psi_{n\mathbf{k}}$, the cell-periodic parts $u_{n\mathbf{k}}$, and the Schrödinger Hamiltonian H as defined in Eqs. (2.1), (2.2), (2.3), (2.4), (2.5), (2.6), (2.7), (2.8), (2.9), (2.10), we obtain

$$\begin{aligned}\mathbf{p}_{nn'}(\mathbf{k}) &= \frac{(2\pi)^3}{\Omega}\int_\Omega \mathrm{d}^3\mathbf{r}\, \psi^\dagger_{n\mathbf{k}}(\mathbf{r})\mathbf{p}\psi_{n'\mathbf{k}}(\mathbf{r})\\ &= \frac{(2\pi)^3}{\Omega}\int_\Omega \mathrm{d}^3\mathbf{r}\, u^*_{n\mathbf{k}}(\mathbf{r})\mathbf{p}u_{n'\mathbf{k}}(\mathbf{r}) + \hbar\mathbf{k}\delta_{nn'}, \end{aligned} \tag{10.7}$$

$$\boldsymbol{\pi}_{nn'}(\mathbf{k}) = \frac{(2\pi)^3}{\Omega}\int_\Omega \mathrm{d}^3\mathbf{r}\, u^*_{n\mathbf{k}}(\mathbf{r})\boldsymbol{\pi} u_{n'\mathbf{k}}(\mathbf{r}) + \hbar\mathbf{k}\delta_{nn'}, \tag{10.8}$$

$$\mathbf{r}_{nn'}(\mathbf{k}) = \frac{(2\pi)^3}{\Omega}\int_\Omega \mathrm{d}^3\mathbf{r}\, u^*_{n\mathbf{k}}(\mathbf{r})\mathbf{r}u_{n'\mathbf{k}}(\mathbf{r}) + \hbar\mathbf{k}\delta_{nn'}. \tag{10.9}$$

It is known that the Bloch functions $\psi_{n\mathbf{k}}$ and the cellular part $u_{n\mathbf{k}}$ above are *not* analytical functions in the wave vector $\mathbf{k}$. In fact, they are discontinuous in $\mathbf{k}$ at a degeneracy (say, at $\mathbf{k}_0$) [152], and depend critically on the direction $\mathbf{k}-\mathbf{k}_0$ for small $\mathbf{k}-\mathbf{k}_0$ [1]. This non-analytical behavior in $\mathbf{k}$ of the Bloch functions, as we shall see, leads to severe problems in the calculation of Stark matrix elements.

For that reason, one should use the LK basis states, Eq. (8.43), which are entire functions of $\mathbf{k}$ [6]:

$$\chi_{n\mathbf{k}}(\mathbf{r}) = \mathrm{e}^{\mathrm{i}\mathbf{k}\cdot\mathbf{r}}\psi_{n\mathbf{k}_0}(\mathbf{r}) = \mathrm{e}^{\mathrm{i}(\mathbf{k}+\mathbf{k}_0)\cdot\mathbf{r}}U_n(\mathbf{r}), \tag{10.10}$$

where

$$U_n(\mathbf{r}) = u_{n\mathbf{k}_0}(\mathbf{r}), \tag{10.11}$$

and $\mathbf{k}_0$ is some fixed point in $\mathbf{k}$ space. The LK basis states $\chi_{n\mathbf{k}}(\mathbf{r})$ are entire functions in $\mathbf{k}$ as the $\mathbf{k}$ dependence is through the entire function $\exp(\mathrm{i}\mathbf{k}\cdot\mathbf{r})$ only.

The periodicity of $U_n(\mathbf{r})$ implies it can be written as a Fourier series:

$$U_n(\mathbf{r}) = \sum_m U_{nm}\mathrm{e}^{\mathrm{i}\mathbf{K}_m\cdot\mathbf{r}}, \tag{10.12}$$

$$U_{nm} = \frac{1}{\Omega}\int_\Omega \mathrm{d}^3\mathbf{r}\, U_n(\mathbf{r})\mathrm{e}^{-\mathrm{i}\mathbf{K}_m\cdot\mathbf{r}}, \tag{10.13}$$

where $\mathbf{K}_m$ is a reciprocal lattice vector. The set $\{U_n\}$ is orthogonal and complete with respect to cell-periodic functions, hence [153]:

$$\frac{1}{\Omega}\int_\Omega \mathrm{d}^3\mathbf{r}\, U_n^*(\mathbf{r})U_{n'}(\mathbf{r}) = \sum_m U_{nm}^* U_{n'm} = \delta_{nn'}, \tag{10.14}$$

$$\frac{1}{\Omega}\sum_n U_n(\mathbf{r})U_n^*(\mathbf{r}') = I\sum_{\mathbf{R}}\delta(\mathbf{r}-\mathbf{r}'-\mathbf{R}), \tag{10.15}$$

where $\mathbf{R}$ is a Bravais lattice vector and I is the 2×2 identity matrix.

We will write the LK basis functions in Dirac notation:

$$\chi_{n\mathbf{k}}^s(\mathbf{r}) = \langle s\mathbf{r}|n\mathbf{k}), \tag{10.16}$$

where s is a spin index and $|s\mathbf{r}\rangle$ an eigenket of spin and position. Additional properties of the LK basis states can be found in Sect. 8.3.

From the definition of the LK basis states [Eq. (10.10)], we can write down an expression for the coordinate matrix elements as follows:

$$(n\mathbf{k}|\mathbf{r}|n'\mathbf{k}') = \mathrm{i}\nabla_{\mathbf{k}}(n\mathbf{k}|n'\mathbf{k}'). \tag{10.17}$$

Further, using Eq. (8.46) yields:

$$(n\mathbf{k}|\mathbf{r}|n'\mathbf{k}') = \mathrm{i}\sum_m B_m^{nn'}\nabla_{\mathbf{k}}\delta(\mathbf{k}-\mathbf{k}'+\mathbf{K}_m). \tag{10.18}$$

An expression of integrals involving the product of a general function $A(\mathbf{r})$ and a cell-periodic function $B(\mathbf{r})$ is useful for the further analysis. We have

$$A(\mathbf{r}) = \frac{1}{(2\pi)^{3/2}} \int \mathrm{d}\mathbf{k}\, A(\mathbf{k}) \mathrm{e}^{\mathrm{i}\mathbf{k}\cdot\mathbf{r}}, \tag{10.19}$$

$$A(\mathbf{k}) = \frac{1}{(2\pi)^{3/2}} \int \mathrm{d}^3\mathbf{r}\, A(\mathbf{r}) \mathrm{e}^{-\mathrm{i}\mathbf{k}\cdot\mathbf{r}}, \tag{10.20}$$

$$B(\mathbf{r}) = \sum_m B_m \mathrm{e}^{\mathrm{i}\mathbf{K}_m\cdot\mathbf{r}}, \tag{10.21}$$

$$B_m = \frac{1}{\Omega} \int_\Omega \mathrm{d}^3\mathbf{r}\, B(\mathbf{r}) \mathrm{e}^{-\mathrm{i}\mathbf{K}_m\cdot\mathbf{r}}. \tag{10.22}$$

Combining these expressions allows us to write a modified form of Parseval's theorem:

$$\int \mathrm{d}^3\mathbf{r}\, A(\mathbf{r})B(\mathbf{r}) = \sum_m \int \mathrm{d}^3\mathbf{r}\, A(\mathbf{r}) B_m \mathrm{e}^{\mathrm{i}\mathbf{K}_m\cdot\mathbf{r}} = (2\pi)^{3/2} \sum_m A_{-m} B_m. \tag{10.23}$$

10.3.2 Matrix Elements of the Coordinate Operator

We seek to determine an expression for the coordinate matrix element evaluated between Bloch states to emphasize the problem in using a basis set of Bloch states in computing the Stark effect multiband problem. First, notice that Parseval's theorem [Eq. (10.23)], the orthogonality relation [Eq. (2.4)], and the periodicity condition: $\psi_{n,\mathbf{k}}(\mathbf{r}) = \psi_{n,\mathbf{k}+\mathbf{G}}(\mathbf{r})$ lead to:

$$\int_\Omega \mathrm{d}^3\mathbf{r}\, \psi^\dagger_{n,\mathbf{k}}(\mathbf{r}) \psi_{n',\mathbf{k}'} = \sum_{\mathbf{G}} \delta\left(\mathbf{k} - \mathbf{k}' + \mathbf{G}\right) \delta_{nn'}. \tag{10.24}$$

Taking the gradient with respect to $\mathbf{k}$ of this equation gives [134]

$$\int_\Omega \mathrm{d}^3\mathbf{r}\, \psi^\dagger_{n,\mathbf{k}}(\mathbf{r}) \mathbf{r} \psi_{n',\mathbf{k}'} = \sum_{\mathbf{G}} \left[\mathrm{i}\nabla_{\mathbf{k}}\delta\left(\mathbf{k} - \mathbf{k}' + \mathbf{G}\right)\delta_{nn'} + \xi_{nn'}(\mathbf{k})\delta\left(\mathbf{k} - \mathbf{k}' + \mathbf{G}\right)\right], \tag{10.25}$$

where

$$\xi_{nn'}(\mathbf{k}) = \frac{1}{\Omega} \int_\Omega \mathrm{d}^3\mathbf{r}\, u^*_{n\mathbf{k}}(\mathbf{r}) [\mathrm{i}\nabla_{\mathbf{k}} u_{n'\mathbf{k}}(\mathbf{r})]. \tag{10.26}$$

Evidently, matrix elements of the coordinate operator are ill-defined as $\nabla_{\mathbf{k}} u_{n'\mathbf{k}}(\mathbf{r})$ is singular at a point of degeneracy [1, 152] (refer to the discussion above). Hence, we have demonstrated the problem in using a basis set of Bloch states to solve the multiband Stark effect problem. The solution to the Stark problem is, as mentioned above, to define matrix elements in terms of a basis set of entire functions such as the LK basis set. In the following section, we shall elaborate further on this and derive a Hamiltonian convenient for obtaining the electronic band structure of the upper valence-band states and the lower conduction-band states (possibly including the p conduction states). This Hamiltonian includes effects to terms of order k^3, $k\mathcal{E}$, and $\mathcal{E}^2$ in an appropriately chosen quasi-degenerate set of eigenstates using two subsequent unitary transformations of the LK basis states. This combination of unitary transformations leads to a Hamiltonian which does not contain, in the non-relativistic approximation, dipole-like terms.

10.3.3 Multiband Hamiltonian

Consider the case where the potential V in the Pauli equation is composed of the crystal potential and the Stark term:

$$U(\mathbf{r}) = e\mathcal{E}\cdot\mathbf{r}. \tag{10.27}$$

Hence, the Pauli equation reads

$$H = H_0 + U(\mathbf{r}) + \frac{e\hbar}{4m_0^2c^2}\mathcal{E}\cdot(\mathbf{p}\times\boldsymbol{\sigma}), \tag{10.28}$$

with H_0 as given by Eq. (2.5).

Employing Eqs. (2.1), (2.2), (2.3), (2.4), (2.5), (2.6) we obtain

$$\left(n\mathbf{k}|H|n'\mathbf{k}'\right) = \left[\left(E_n(\mathbf{k}_0) + \frac{\hbar^2k^2}{2m_0}\right)\delta_{nn'} + \frac{\hbar}{m_0}\mathbf{k}\cdot\boldsymbol{\pi}_{nn'}(\mathbf{k}_0)\right]\delta(\mathbf{k}-\mathbf{k}_0), \tag{10.29}$$

$$\left(n\mathbf{k}|U|n'\mathbf{k}'\right) = \mathrm{i}e\mathcal{E}\cdot\sum_m B_m^{nn'}\nabla_{\mathbf{k}}\delta(\mathbf{k}-\mathbf{k}'+\mathbf{K}_m), \tag{10.30}$$

$$\left(n\mathbf{k}|W|n'\mathbf{k}'\right) = \frac{e\hbar}{4m_0^2c^2}[\mathcal{E}\cdot(\mathbf{p}\times\boldsymbol{\sigma})_{nn'} + (\mathcal{E}\times\hbar\mathbf{k})\cdot\sigma_{nn'}]\delta(\mathbf{k}-\mathbf{k}_0), \tag{10.31}$$

where the matrix elements $(\mathbf{p}\times\boldsymbol{\sigma})_{nn'}$ and $\sigma_{nn'}$ are evaluated at $\mathbf{k}_0$ in analogy with Eqs. (10.7), (10.8), (10.9).

In the following, we shall restrict our analysis to effects far away from the zone boundary and neglect terms in the crystal potential involving reciprocal lattice vectors $\mathbf{K}_m \neq \mathbf{0}$. Hence, all matrix elements of operators $\mathcal{O}$ appearing in the subsequent analysis can be written as

$$\left(n\mathbf{k}|\mathcal{O}|n'\mathbf{k}'\right) = \left(n|\mathcal{O}|n'\right)\delta\left(\mathbf{k}-\mathbf{k}'\right), \tag{10.32}$$

where $(n|\mathcal{O}|n')$ is a $\mathbf{k}$-dependent expression [refer to Eqs. (10.29), (10.30), (10.31)].

Next, we split up the contribution from H in three terms:

$$H = H^A + H^B + H^C, \tag{10.33}$$

where

$$\left(n|H^A|n'\right) = \left(E_n(\mathbf{k}_0) + \frac{\hbar^2 k^2}{2m_0}\right)\delta_{nn'}, \tag{10.34}$$

$$\left(n|H^B|n'\right) = \frac{\hbar}{m_0}\sum_{i=1}^{3} k_i \pi^i_{nn'}(1 - \Delta_{nn'}), \tag{10.35}$$

$$\left(n|H^C|n'\right) = \frac{\hbar}{m_0}\sum_{i=1}^{3} k_i \pi^i_{nn'}\Delta_{nn'}, \tag{10.36}$$

where $\pi^i_{nn'}$ is the i'th component of $\pi_{nn'}$ ($i = 1, 2, 3$ equivalent to x, y, z components, respectively). Here, $\Delta_{nn'}$ is the function

$$\begin{aligned} \Delta_{nn'} &= 1, & &\text{if } E_n \approx E_{n'}, \\ \Delta_{nn'} &= 0, & &\text{otherwise.} \end{aligned} \tag{10.37}$$

Similarly, we split contributions from W in two terms, $W = W^A + W^B$, and retaining only the $\mathbf{K}_m = \mathbf{0}$ term in the U matrix elements [Eq. (10.30)] as just argued:

$$\left(n|U|n'\right) = \mathrm{i}e\mathcal{E}\cdot\delta_{nn'}\nabla_{\mathbf{k}}, \tag{10.38}$$

$$\left(n|W^A|n'\right) = \frac{e\hbar}{4m_0^2c^2}\mathcal{E}\cdot(\mathbf{p}\times\sigma)_{nn'}, \tag{10.39}$$

$$\left(n|W^B|n'\right) = \frac{e\hbar^2}{4m_0^2c^2}(\mathcal{E}\times\mathbf{k})\cdot\sigma_{nn'}. \tag{10.40}$$

With these results, we are equipped to solve the multiband problem. First, we wish to decouple the quasi-degenerate set, we are solving for, from all other states to a certain specified order in $\mathbf{k}$ and $\mathcal{E}$ so as to effectively reduce the Hamiltonian problem to block-diagonal form. We will make use of the technique of canonical transformation again.

Let us therefore recast our problem in a new set of states $|\overline{n\mathbf{k}})$ related to the old set $|n\mathbf{k})$ by a unitary matrix $\exp(S)$:

$$\left|\overline{n\mathbf{k}}\right) = \mathrm{e}^S\left|n\mathbf{k}\right), \tag{10.41}$$

The original Hamiltonian in the new basis set relates to the new Hamiltonian in the old states as follows [6]:

$$\left(\overline{n'\mathbf{k}'}|H|\overline{n\mathbf{k}}\right) = \left(n'\mathbf{k}'|\overline{H}|n\mathbf{k}\right), \tag{10.42}$$

$$\begin{aligned} \overline{H} &= \mathrm{e}^{-S}H\mathrm{e}^{S} = H + [H,S] \\ &\quad + \frac{1}{2!}[[H,S],S] + \frac{1}{3!}[[[H,S],S],S] + \ldots. \end{aligned} \tag{10.43}$$

The link between the two Hamiltonians becomes

$$\overline{H} = H^A + H^B + H^C + U + W^A + W^B + \left[H^A, S\right] + \left[H^B, S\right] + \left[H^C, S\right]$$
$$+[U, S] + \frac{1}{2!}\left[\left[H^A, S\right], S\right] + \frac{1}{2!}\left[\left[H^B, S\right], S\right] + \frac{1}{2!}\left[\left[H^C, S\right], S\right]$$
$$+\frac{1}{2!}[[U, S], S] + \frac{1}{3!}\left[\left[\left[H^A, S\right], S\right], S\right], \tag{10.44}$$

which includes all terms of order k^3, $k\mathcal{E}$, and $\mathcal{E}^2$ except those stemming from $[W^A + W^V, S]$. To obtain a block-diagonal form we seek to eliminate the coupling H^B by choosing appropriately S. This is achieved if

$$H^B + \left[H^A, S\right] = 0, \tag{10.45}$$

giving [6]

$$\left(n|S|n'\right) = -\sum_{i=1}^{3} \frac{\hbar k_i \pi_{nn'}^i}{m_0(E_n - E_{n'})}(1 - \Delta_{nn'}) = -\mathrm{i}\sum_{i=1}^{3} k_i \xi_{nn'}^i (1 - \Delta_{nn'}), \tag{10.46}$$

with $\xi_{nn'}$ as defined in Eq. (10.26). In obtaining the last equality, we have used the commutator relation

$$\boldsymbol{\pi} \equiv \tfrac{m_0}{\mathrm{i}\hbar}[\mathbf{r}, H_0] = \mathbf{p} + \tfrac{\hbar}{4m_0c^2}(\boldsymbol{\sigma} \times \nabla V). \tag{10.47}$$

to write

$$\xi_{nn'} = -\frac{\mathrm{i}\hbar}{m_0}\frac{\pi_{nn'}}{E_n - E_{n'}}. \tag{10.48}$$

With this choice of S the Hamiltonian $\overline{H}$ reads

$$\overline{H} = H^A + H^C + U + W^A + W^B + \frac{1}{2}\left[H^B, S\right]$$
$$+[U, S] + \frac{1}{3}\left[\left[H^B, S\right], S\right] + \frac{1}{2}\left[\left[H^C, S\right], S\right] + \frac{1}{2}[[U, S], S], \tag{10.49}$$

discarding the term $[H^C, S]$ appearing in Eq. (10.44) because its contributions are all off-diagonal of order k^2.

Let us next write down expressions for the remaining matrix elements. Using Eqs. (10.34), (10.35), (10.36) and (10.46), (10.47), (10.48), we find

$$\frac{1}{2}\left(n|[H^B,S]|n'\right)=\sum_{i,j=1}^{3}\frac{\hbar^2k_ik_j}{2m_0^2}\sum_{\alpha}{}'\pi^i_{n\alpha}\pi^j_{\alpha n'}\left(\frac{1}{E_n-E_\alpha}+\frac{1}{E_{n'}-E_\alpha}\right),\tag{10.50}$$

$$\frac{1}{3}\left(n|[[H^B,S],S]|n'\right)=-\frac{\hbar}{3m_0}\sum_{l,m,n=1}^{3}k_lk_mk_n$$

$$\times\sum_{\alpha,\beta}{}'\left(\pi^l_{n\alpha}\xi^m_{\alpha\beta}\xi^n_{\beta n'}-2\xi^l_{n\alpha}\pi^m_{\alpha\beta}\xi^n_{\beta n'}+\xi^l_{n\alpha}\xi^m_{\alpha\beta}\pi^n_{\beta n'}\right),\tag{10.51}$$

$$\left(n|[U,S]|n'\right)=e\mathcal{E}\cdot\xi_{nn'}(1-\Delta_{nn'}),\tag{10.52}$$

$$\frac{1}{2}\left(n|[[U,S],S]|n'\right)=\frac{\mathrm{i}e}{2m_0^2}\sum_{i,j=1}^{3}\left(\mathcal{E}_ik_j-\mathcal{E}_jk_i\right)\sum_{\alpha}{}'\frac{\hbar^2\pi^i_{n\alpha}\pi^j_{\alpha n'}}{(E_n-E_\alpha)(E_\alpha-E_{n'})}.\tag{10.53}$$

It is important to observe that the dipole-like matrix elements: $(n|[U,S]|n')$ vanish between quasi-degenerate states due to the presence of the term: $1-\Delta_{nn'}$. Thus, the problem with singularities in the energy denominator expression of $\xi_{nn'}$ is removed by formulating the Hamiltonian problem in the LK transformed basis states: $|\overline{n\mathbf{k}})$ which are entire functions of $\mathbf{k}$.

Performing yet another transformation to eliminate the dipole term $[U,S]$ in Eq. (10.52), which is linear in $\mathcal{E}$ and generated by the first transformation, is next done. The necessary unitary transformation: e^R superimposed the first unitary transformation e^S implies that the Hamiltonian problem is solved using a basis set $|\widehat{n\mathbf{k}})$ connected to the original basis set by

$$|\widehat{n\mathbf{k}})=\mathrm{e}^R\mathrm{e}^S|n\mathbf{k}),\tag{10.54}$$

with an associated Hamiltonian $\widehat{H}$ given as

$$\widehat{H}=\mathrm{e}^{-R}\mathrm{e}^{-S}H\mathrm{e}^S\mathrm{e}^R,\tag{10.55}$$

The operator R is chosen such that

$$[U,S]+\left[H^A,R\right]=0,\tag{10.56}$$

yielding

$$\left(n|R|n'\right)=\sum_{i=1}^{3}\frac{\mathrm{i}e\hbar\mathcal{E}_i\pi^i_{nn'}}{m_0(E_n-E_{n'})^2}(1-\Delta_{nn'}).\tag{10.57}$$

Then, we obtain

$$\widehat{H} = H^A + H^C + U + W^A + W^B + \frac{1}{2}\left[H^B, S\right] + \frac{1}{3}\left[\left[H^B, S\right], S\right] + \frac{1}{2}\left[\left[H^C, S\right], S\right] + \frac{1}{2}[[U, S], R], \quad (10.58)$$

where the relation $[U, R] = 0$ has been used and the (generated) matrix element $[H^C, R]$ neglected since this term contributes off-block-diagonally only to the order $k\mathcal{E}$. The remaining matrix element, that we keep, generated by this last transformation is

$$\frac{1}{2}\left(n|[[U, S], R]|n'\right) = -\sum_{i,j=1}^{3} \frac{e^2\mathcal{E}_i\mathcal{E}_j}{2m_0^2} \sum_{i,j=1}^{3} \sum_{\alpha}{}' \frac{\hbar^2\pi^i_{n\alpha}\pi^j_{\alpha n'}}{(E_n - E_\alpha)(E_\alpha - E_{n'})} \times \left(\frac{1}{E_n - E_\alpha} + \frac{1}{E_{n'} - E_\alpha}\right). \quad (10.59)$$

In conclusion, we have in Eq. (10.58) obtained a LK-like Hamiltonian for degenerate bands valid to order k^3, $k\mathcal{E}$, and $\mathcal{E}^2$. We anticipate that linear-in-$\mathcal{E}$ terms being independent of k are still present in this Hamiltonian due to the small relativistic term W^A [refer to Eq. (10.39)]. This term W^A is similar to a dipole term since $(\mathbf{p} \times \boldsymbol{\sigma})_{nn'}$ transforms as the vector $\xi_{nn'}$ under time-reversal symmetry and point-group operations. The third-order in k term: $\frac{1}{2}[[H^C, S], S]$ is zero in the absence of spin-orbit interaction and exists only in the absence of inversion symmetry [104] (the term is proportional to P' in the 14-band model of Cardona and Pollak [5]).

10.3.4 Explicit Form of Hamiltonian Matrix Contributions

Above, a general form was derived for the Hamiltonian of a set of quasi-degenerate states accounting for Stark effects. We shall now give explicit expressions of the matrix elements in the framework of Cardona, Christensen, and Fasol's three-band model [104]. First, we give the Dresselhaus, Kip, and Kittel version of the Hamiltonian H_{DKK} in the Γ^v_{15} states (X, Y, Z) due to the $k \cdot p$ interaction [2] and parameters within the three-band model:

$$H_{\mathrm{DKK}}(\mathbf{k}) = \begin{pmatrix} Lk_x^2 + M\left(k_y^2 + k_z^2\right) & Nk_xk_y & Nk_xk_z \\ Nk_xk_y & Lk_y^2 + M\left(k_x^2 + k_z^2\right) & Nk_yk_z \\ Nk_xk_z & Nk_yk_z & Lk_z^2 + M\left(k_x^2 + k_y^2\right) \end{pmatrix} + \begin{pmatrix} 0 & \mathrm{i}G\left(k_x^2 - k_y^2\right)k_z & \mathrm{i}G\left(k_x^2 - k_z^2\right)k_y \\ \mathrm{i}G\left(k_y^2 - k_x^2\right)k_z & 0 & \mathrm{i}G\left(k_y^2 - k_z^2\right)k_x \\ \mathrm{i}G\left(k_z^2 - k_x^2\right)k_z & \mathrm{i}G\left(k_z^2 - k_y^2\right)k_x & 0 \end{pmatrix}, \quad (10.60)$$

where

$$
\begin{aligned}
L &= \frac{\hbar^2}{2m_0} - \frac{P^2}{E_0}, \\
M &= \frac{\hbar^2}{2m_0} - \frac{Q^2}{E_0'}, \\
N &= -\left(\frac{P^2}{E_0} + \frac{Q^2}{E_0'}\right), \\
G &= 2\frac{PP'Q}{E_0E_0'},
\end{aligned}
\tag{10.61}
$$

and matrix elements P, Q and energy gaps E_0, E_0' have their usual meaning.

The contribution $\frac{1}{2}[H^B, S]$ to the total Hamiltonian $\widehat{H}$ is the first matrix appearing in Eq. (10.60) while, in the three-band model, the contribution $\frac{1}{3}[[H^B, S], S]$ is the second matrix appearing in Eq. (10.60).

In the three-band model, the term $\frac{1}{2}[[U, S], S]$ takes the following form:

$$
\begin{pmatrix}
0 & in\left(\mathcal{E}_x k_y - k_x \mathcal{E}_y\right) & in\left(\mathcal{E}_x k_z - k_x \mathcal{E}_z\right) \\
in\left(\mathcal{E}_y k_x - k_y \mathcal{E}_x\right) & 0 & in\left(\mathcal{E}_y k_z - k_z \mathcal{E}_y\right) \\
in\left(\mathcal{E}_z k_x - k_z \mathcal{E}_z\right) & in\left(\mathcal{E}_z k_y - k_z \mathcal{E}_y\right) & 0
\end{pmatrix},
\tag{10.62}
$$

where

$$
n = -\frac{e}{2}\left(\frac{P^2}{E_0^2} - \frac{Q^2}{E_0'^2}\right).
\tag{10.63}
$$

Finally, it is found that the term $\frac{1}{2}[[U, S], R]$ in the three-band model reads

$$
\begin{pmatrix}
\lambda\mathcal{E}_x^2 + \mu\left(\mathcal{E}_y^2 + \mathcal{E}_z^2\right) & \nu\mathcal{E}_x\mathcal{E}_y & \nu\mathcal{E}_x\mathcal{E}_z \\
\nu\mathcal{E}_x\mathcal{E}_y & \lambda\mathcal{E}_y^2 + \mu\left(\mathcal{E}_x^2 + \mathcal{E}_z^2\right) & \nu\mathcal{E}_y\mathcal{E}_z \\
\nu\mathcal{E}_x\mathcal{E}_z & \nu\mathcal{E}_y\mathcal{E}_z & \lambda\mathcal{E}_z^2 + \mu\left(\mathcal{E}_x^2 + \mathcal{E}_y^2\right)
\end{pmatrix},
\tag{10.64}
$$

where

$$
\begin{aligned}
\lambda &= -e^2\frac{P^2}{E_0^3}, \\
\mu &= -e^2\frac{Q^2}{E_0'^3}, \\
\nu &= -e^2\left(\frac{P^2}{E_0^3} + \frac{Q^2}{E_0'^3}\right).
\end{aligned}
\tag{10.65}
$$

This completes the Hamiltonian derivation in the three-band model. We note, as a final comment, that the form of the matrices above is in agreement with the method of invariants for zincblende. In particular, we observe that the form of the matrix $\frac{1}{2}[[U, S], R]$ in the electric-field-squared components is the same as the form of the first matrix in Eq. (10.60) in the wave-vector-squared components. Since both $\mathbf{k}$ and $\mathcal{E}$ transform as vectors this is indeed consistent with the method of invariants.

10.4 Summary

The addition of an external static electric field to the $k \cdot p$ Hamiltonian is problematic and still under dispute. The most recent model of Foreman was discussed in detail. A static electric field is also responsible for a number of interesting effects such as Wannier-Stark localization [154], Stark effect [155], Fano resonance [156], and Pockels effect [135, 136, 151].

Chapter 11
Excitons

11.1 Overview

An exciton, in its simplest form, is a composite particle of an electron and a hole in a semiconductor. They are typically formed by optical excitation. Since they are observed as a modification of the single-particle optical spectrum primarily via the formation of discrete features, their properties are relevant to a band-structure discussion. In effect, the problem statement we are interested in is how is the single-particle band structure modified in the presence of the many-body interactions (Coulomb and exchange).

An exciton has many degrees of freedom due to its translational motion (characterized by a wave vector), the relative motion between the electron and the hole (atomic-like quantum numbers), and internal spin degrees of freedom (due to the spin and orbital degrees of freedom of the particles). One will find that, in the simplest case of excitons formed from nondegenerate (except for spin) parabolic band states, there is an exact solution equivalent to a hydrogen-like problem. However, when the valence band is degenerate, as for cubic semiconductors, the above degrees of freedom are mixed, including the translational and relative motions, and there is no exact solution; a perturbative approach is then conventional. One can also introduce external perturbations such as electric, magnetic and strain fields in order to remove any degeneracy in the level structure as that would allow a clearer experimental analysis of the states. Theoretically, this can be viewed as a symmetry-breaking mechanism inducing further level mixing and splitting. Thus, we will also describe the development of an effective exciton Hamiltonian in the presence of external fields.

The first observation of excitons in semiconductors were both direct and indirect excitons in Si and Ge by MacFarlane and coworkers in 1957 [157]. However, the theory goes back farther. Excitons in simple band structures were studied within the effective-mass picture by Wannier in 1937 [131] (excitons confined to atomic sites, so-called Frenkel excitons, will not be discussed in this book, nor will more complicated composite particles such as biexcitons, ...). The theory of direct excitons in degenerate bands was first done by G. Dresselhaus in 1956 [158], in which he showed the similarity to the impurity problem. Baldereschi

L.C. Lew Yan Voon, M. Willatzen, *The* k · p *Method*,

DOI 10.1007/978-3-540-92872-0_11,

and Lipari proposed perturbative solutions for the ground and excited states [159]. The effect of strain on excitons has also been investigated [106, 160]. Altarelli and Lipari [161, 162] extended the Baldereschi and Lipari work with the inclusion of a magnetic field along the ⟨001⟩ direction. Cho et al. [163] considered the problem starting from the method of invariants in order to account for arbitrary magnetic fields and then compared to the perturbative solution; they also added the exchange interaction. Cho [82] subsequently generalized the above formalism to include various symmetry-breaking perturbations and for wurtzite semiconductors. Dresselhaus [158], Kane [88] and Altarelli and Lipari [164] studied the problem of exciton dispersion.

11.2 Excitonic Hamiltonian

The first step in a theory of exciton is to derive the Schrödinger equation satisfied by the excitonic wave function. The proper approach is to start with the many-electron Schrödinger equation and consider a state with one electron and one hole. We outline here the treatment given by Haken [165]. The Hamiltonian in second-quantized notation is (neglecting electron–electron and hole–hole interactions and the ground-state energy)

$$H = H_e + H_h + H_{eh}, \tag{11.1}$$

$$H_e = \sum_{\mathbf{k}} E_c(\mathbf{k}) a^\dagger_{\mathbf{k}} a_{\mathbf{k}}, \tag{11.2}$$

$$H_h = -\sum_{\mathbf{k}} E_v(\mathbf{k}) d^\dagger_{\mathbf{k}} d_{\mathbf{k}}, \tag{11.3}$$

$$H_{eh} = -\sum_{\mathbf{k}_1 \dots \mathbf{k}_4} a^\dagger_{\mathbf{k}_1} a_{\mathbf{k}_4} d^\dagger_{\mathbf{k}_3} d_{\mathbf{k}_2} \left[W\left(\begin{smallmatrix} \mathbf{k}_1 \mathbf{k}_4 \\ c\, v \end{smallmatrix} \mid \begin{smallmatrix} \mathbf{k}_2 \mathbf{k}_3 \\ v\, c \end{smallmatrix} \right) - W\left(\begin{smallmatrix} \mathbf{k}_4 \mathbf{k}_1 \\ v\, c \end{smallmatrix} \mid \begin{smallmatrix} \mathbf{k}_2 \mathbf{k}_3 \\ v\, c \end{smallmatrix} \right) \right], \tag{11.4}$$

where $a^\dagger$ ($d^\dagger$) creates an electron (a hole) from the vacuum state $|V\rangle$ (completely filled valence bands and empty conduction bands), $E_{C(V)}(\mathbf{k})$ are the band energies, and W is the Coulomb matrix element,

$$W\left(\begin{smallmatrix} \mathbf{k}_1 \mathbf{k}_4 \\ n\, m \end{smallmatrix} \mid \begin{smallmatrix} \mathbf{k}_2 \mathbf{k}_3 \\ m\, n \end{smallmatrix} \right) = \int \mathrm{d}^3\mathbf{r}\, \mathrm{d}^3\mathbf{r}'\, \psi^*_{n\mathbf{k}_1}(\mathbf{r}) \psi^*_{m\mathbf{k}_4}(\mathbf{r}') \frac{e^2}{4\pi\varepsilon|\mathbf{r}-\mathbf{r}'|} \psi_{m\mathbf{k}_2}(\mathbf{r}') \psi_{n\mathbf{k}_3}(\mathbf{r}), \tag{11.5}$$

and $\psi_{n\mathbf{k}}$ is a Bloch function. Using Eq. (11.1) with a state of one electron and one hole,

$$\Phi = \sum_{\mathbf{k}_1,\mathbf{k}_2} c_{\mathbf{k}_1,\mathbf{k}_2} a^\dagger_{\mathbf{k}_1} d^\dagger_{\mathbf{k}_2} |V\rangle\,, \tag{11.6}$$

then one obtains

$$[E_c(\mathbf{k}_1) - E_v(\mathbf{k}_2)]c_{\mathbf{k}_1,\mathbf{k}_2} - \sum_{\mathbf{k}_3,\mathbf{k}_4}\left[W\left(\begin{smallmatrix}\mathbf{k}_1\mathbf{k}_4\\ c\,v\end{smallmatrix}\;\middle|\;\begin{smallmatrix}\mathbf{k}_2\mathbf{k}_3\\ v\,c\end{smallmatrix}\right) - W\left(\begin{smallmatrix}\mathbf{k}_4\mathbf{k}_1\\ v\,c\end{smallmatrix}\;\middle|\;\begin{smallmatrix}\mathbf{k}_2\mathbf{k}_3\\ v\,c\end{smallmatrix}\right)\right]c_{\mathbf{k}_3,\mathbf{k}_4}$$
$$= Ec_{\mathbf{k}_1,\mathbf{k}_2}\,. \tag{11.7}$$

One can now show that the coefficient is the exciton envelope function. This is typically done by neglecting the exchange interaction and considering only Bloch waves near $\mathbf{k} = \mathbf{0}$ (the usual effective-mass excitons) [165]. Then

$$\psi(\mathbf{r}_1, \mathbf{r}_2) = \frac{1}{V}\sum_{\mathbf{k}_1,\mathbf{k}_2} c_{\mathbf{k}_1,\mathbf{k}_2} \mathrm{e}^{\mathrm{i}(\mathbf{k}_1\cdot\mathbf{r}_1 - \mathbf{k}_2\cdot\mathbf{r}_2)} \tag{11.8}$$

satisfies the two-particle Schrödinger equation

$$\left[E_c(-\mathrm{i}\nabla_1) - E_v(-\mathrm{i}\nabla_2) - \frac{e^2}{4\pi\epsilon|\mathbf{r}_1 - \mathbf{r}_2|}\right]\psi(\mathbf{r}_1, \mathbf{r}_2) = E\psi(\mathbf{r}_1, \mathbf{r}_2). \tag{11.9}$$

The above result can be expressed for the multiband problem. A discussion in terms of the Luttinger Hamiltonian was also carried out by Dimmock [166] and Cho et al. [163]. This is achieved by realizing that, within the effective-mass model, $E_c(-\mathrm{i}\nabla_1)$ $[E_v(-\mathrm{i}\nabla_2)]$ corresponds to the conduction-band (valence-band) Hamiltonian $H^c_{ii'}(\mathbf{k}_e)$ $[H^v_{j'j}(-\mathbf{k}_h)]$. The subtlety concerns the hole Hamiltonian: the band states are transposed and the wave vector changes sign compared to the valence-band picture. This can be seen by noting that a hole state is the time-reversed state of an electron state [1, 163, 166]:

$$\psi^h_{n'} = \mathcal{K}\psi^e_n = \psi^e_{Kn}, \tag{11.10}$$

where $\mathcal{K}$ is defined in Eq. (3.52). Thus, the appropriate exciton envelope equation, in the presence of a magnetic field, is [163]

$$\sum_{i'j'}\left[\delta_{jj'}H^c_{ii'}(\mathbf{p}_e + e\mathbf{A}) - \delta_{ii'}H^v_{j'j}(-\mathbf{p}_h + e\mathbf{A}) - \delta_{ii'}\delta_{jj'}\frac{e^2}{4\pi\epsilon|\mathbf{r}_e - \mathbf{r}_h|}\right]\overline{B}_{i'j'}(\mathbf{r}_e, \mathbf{r}_h)$$
$$= E\overline{B}_{ij}(\mathbf{r}_e, \mathbf{r}_h). \tag{11.11}$$

In Eq. (11.11), the basis matrices $\mathbf{J}$ for the valence-band Hamiltonian has its sign reversed compared to the multiband $k \cdot p$ Hamiltonian.

11.3 One-Band Model of Excitons

Let us first consider an idealized exciton where the electron and the hole are treated using one-band model descriptions (later, we shall take into account multiband $k \cdot p$ effects for the exciton system). Since the average distance between the electron and the hole forming the exciton is at least an order of magnitude larger than the lattice constant (this will be shown below), the Coulomb interaction is described using the static permittivity ε of the bulk semiconductor and we may write for the two-particle effective-mass Hamiltonian,

$$\left[E_0 + \frac{p_e^2}{2m_e} + \frac{p_h^2}{2m_h} - \frac{e^2}{4\pi\varepsilon|\mathbf{r}_e - \mathbf{r}_h|}\right]\psi(\mathbf{r}_e, \mathbf{r}_h) = E\psi(\mathbf{r}_e, \mathbf{r}_h), \tag{11.12}$$

where $\mathbf{r}_e$, $\mathbf{r}_h$, p_e, p_h, m_e, m_h, E, E_0, and $\psi(\mathbf{r}_e, \mathbf{r}_h)$ are the electron coordinate, hole coordinate, electron momentum, hole momentum, electron effective mass, hole effective mass, exciton energy, energy gap, and the exciton wave function, respectively. Note that this idealized model comes about simply by adding the Coulomb interaction to the sum of the one-band Hamiltonians of the one-particle electrons and holes.

The Coulomb term is the only term combining electron and hole coordinates and since this term is relative in the particle coordinates, it is convenient to introduce the coordinates

$$\mathbf{r} = \mathbf{r}_e - \mathbf{r}_h, \tag{11.13}$$

$$\mathbf{R} = \frac{m_e\mathbf{r}_e + m_h\mathbf{r}_h}{m_e + m_h}, \tag{11.14}$$

where $\mathbf{R}$ is the center-of-mass coordinate and $\mathbf{r}$ is the relative coordinate. Further, writing

$$\begin{aligned} \mathbf{P} &= -\mathrm{i}\hbar\frac{\partial}{\partial\mathbf{R}}, \\ \mathbf{p} &= -\mathrm{i}\hbar\frac{\partial}{\partial\mathbf{r}}, \\ \mu &= \frac{m_e m_h}{m_e + m_h}, \end{aligned} \tag{11.15}$$

Eq. (11.12) can be recast as

$$\left[\frac{P^2}{2(m_e + m_h)} + \frac{p^2}{2\mu} - \frac{e^2}{4\pi\epsilon r}\right]\psi(\mathbf{r}, \mathbf{R}) = (E - E_0)\psi(\mathbf{r}, \mathbf{R}). \tag{11.16}$$

Since the Hamiltonian does not depend on $\mathbf{R}$, $\mathbf{P}$ is a good quantum number and solutions can be found in the form

$$\psi(\mathbf{r}, \mathbf{R}) = \mathrm{e}^{\mathrm{i}\mathbf{K}\cdot\mathbf{R}}\phi(\mathbf{r}). \tag{11.17}$$

Equation (11.16) now reads

$$\left(\frac{p^2}{2\mu} - \frac{e^2}{4\pi\varepsilon r}\right)\phi(\mathbf{r}) = \left[E - E_0 - \frac{\hbar^2 K^2}{2(m_e + m_h)}\right]\phi(\mathbf{r}). \qquad (11.18)$$

The latter equation is the same as the donor impurity problem except that the reduced mass μ replaces the electron mass m_e. Evidently, bound two-particle systems (excitons) correspond to the case where $E - E_0 - \hbar^2 K^2/(2(m_e + m_h))$ is negative. The latter problem is well-known in quantum mechanics and the ground-state wave function is known to be the $1S$ hydrogenic wave function:

$$\phi(\mathbf{r}) = \exp(-\frac{r}{a^{\mathrm{ex}}}), \qquad (11.19)$$

where a^{ex} is the exciton Bohr radius,

$$a^{\mathrm{ex}} = \frac{4\pi\varepsilon\hbar^2}{\mu e^2}. \qquad (11.20)$$

The binding energy is the exciton Rydberg energy:

$$R^{\mathrm{ex}} = \frac{\mu e^4}{2(4\pi\varepsilon)^2\hbar^2}. \qquad (11.21)$$

The value of the Bohr radius is larger than the corresponding one for the donor impurity problem since the reduced mass μ is smaller than the electron effective mass m_e. For the same reason, the exciton binding energy is smaller as compared to the electron-donor binding energy. In bulk semiconductors, the Bohr radius and the Rydberg energy are typically several hundreds Ångström and a few meV, respectively (for bulk InAs, the values are 370 Å and 1.3 meV, respectively).

11.4 Multiband Theory of Excitons

In realistic semiconductors, band-mixing effects must be accounted for. We will be discussing the cases of excitons formed from s and p states in cubic and hexagonal crystals.

11.4.1 Formalism

Consider first the diamond case. The contributions from the electron kinetic energy and the Coulomb interaction between electron and holes are included by adding a diagonal matrix to the LK 6×6 effective-mass matrix problem [158] in the valence-band hole states – following Baldereschi and Lipari [159]:

$$H_{\text{ex}}(\mathbf{p}) = H^{\text{LK}}_{6\times 6}(\mathbf{p}) + \left[E_0 + \frac{p^2}{2m_e} - \frac{e^2}{4\pi \varepsilon r} \right] \mathbf{1}, \tag{11.22}$$

where $\mathbf{1}$ denotes the 6×6 unit matrix. This Hamiltonian is defined in relative coordinates $\mathbf{r}$, $\mathbf{p}$ neglecting the contribution from the exciton center-of-mass motion.

The full 6×6 exciton Hamiltonian is next recast as

$$H_{\text{ex}}(\mathbf{p}) = H_s + H_d, \tag{11.23}$$

where

$$H_s = \begin{pmatrix} P & 0 & 0 & 0 & 0 & 0 \\ 0 & P & 0 & 0 & 0 & 0 \\ 0 & 0 & P & 0 & 0 & 0 \\ 0 & 0 & 0 & P & 0 & 0 \\ 0 & 0 & 0 & 0 & P+\Delta_0 & 0 \\ 0 & 0 & 0 & 0 & 0 & P+\Delta_0 \end{pmatrix}, \tag{11.24}$$

and

$$H_d = \begin{pmatrix} Q & L & M & 0 & \mathrm{i}/\sqrt{2}L & -\mathrm{i}\sqrt{2}M \\ L^* & -Q & 0 & M & -\mathrm{i}\sqrt{2}Q & \mathrm{i}\sqrt{\frac{3}{2}}L \\ M^* & 0 & -Q & -L & -\mathrm{i}\sqrt{\frac{3}{2}}L^* & -\mathrm{i}\sqrt{2}Q \\ 0 & M^* & -L^* & Q & -\mathrm{i}\sqrt{2}M^* & -\mathrm{i}\sqrt{2}L^* \\ -\mathrm{i}\sqrt{2}L^* & \mathrm{i}\sqrt{2}Q & \mathrm{i}\sqrt{\frac{3}{2}}L & \mathrm{i}\sqrt{2}M & 0 & 0 \\ \mathrm{i}\sqrt{2}M^* & -\mathrm{i}\sqrt{\frac{3}{2}}L^* & \mathrm{i}\sqrt{2}Q & \mathrm{i}\sqrt{2}L^* & 0 & 0 \end{pmatrix}. \tag{11.25}$$

The coefficients appearing in the two matrices read

$$P = \frac{p^2}{2\mu_0} - \frac{e^2}{4\pi\varepsilon r}, \tag{11.26}$$

$$Q = \frac{p_x^2 + p_y^2 - 2p_z^2}{2\mu_1}, \tag{11.27}$$

$$L = -\mathrm{i}\frac{(p_x - \mathrm{i}p_y)p_z}{2\mu_2}, \tag{11.28}$$

$$M = \sqrt{3}\frac{p_x^2 - p_y^2}{2\mu_1} - \mathrm{i}\frac{p_x p_y}{2\mu_2}, \tag{11.29}$$

where the masses are

$$\frac{1}{\mu_0} = \frac{1}{m_e} + \frac{\gamma_1}{m_0}, \tag{11.30}$$

$$\frac{1}{\mu_1} = \frac{\gamma_2}{m_0}, \tag{11.31}$$

$$\frac{1}{\mu_2} = 2\sqrt{3}\frac{\gamma_3}{m_0}. \tag{11.32}$$

This splitting of the full Hamiltonian is convenient as it renders a perturbation analysis possible. Since the electron effective mass is much smaller than the free-electron mass and the Luttinger parameters γ_i usually are considerably smaller than m_0/m_e, it is meaningful to consider H_d a perturbation to H_s. In actual fact, the solutions to the unperturbed problem are hydrogen-like functions corresponding to a reduced mass μ_0 moving in a central Coulomb potential characterized by a permittivity ϵ. The solutions can be classified as

$$|nlm, i\rangle = |nlm\rangle|i\rangle, \tag{11.33}$$

for the discrete solutions (bound solutions), and

$$|klm, i\rangle = |klm\rangle|i\rangle, \tag{11.34}$$

for continuous solutions (unbound solutions). The parameters n, l, m denote the (atomic-like) main quantum number, the orbital angular quantum, and the magnetic quantum number, respectively, and k is a continuous parameter characterizing the unbound solutions. The state $|i\rangle$ is a 6-spinor of the form:

$$|i\rangle = \begin{pmatrix} \delta_{i1} \\ \delta_{i2} \\ \delta_{i3} \\ \delta_{i4} \\ \delta_{i5} \\ \delta_{i6} \end{pmatrix}, \tag{11.35}$$

where δ_{ij} is the Kronecker delta function. The corresponding energies in units of the Rydberg energy R_0^{ex} [which differs from R^{ex} in Eq. (11.21) by replacing μ by μ_0] are

$$E_{n,i} = -\frac{1}{n^2}, \qquad i = 1, 2, 3, 4, \tag{11.36}$$

$$E_{n,i} = \overline{\Delta} - \frac{1}{n^2}, \quad i = 5, 6, \tag{11.37}$$

with $\overline{\Delta} = \Delta / R_0^{\text{ex}}$. The energies for the unbound states are

$$E_{k,i} = k^2, \qquad i = 1, 2, 3, 4, \tag{11.38}$$

$$E_{k,i} = \overline{\Delta} + k^2, \quad i = 5, 6. \tag{11.39}$$

Now consider the lowest exciton states of S nature, namely $1S$ and $2S$ for the main series ($i = 1, 2, 3, 4$) and the split-off series $i = 5, 6$. Both states ($1S$ and $2S$) do not couple to other unperturbed degenerate states in first-order perturbation theory (with H_d the perturbation), as H_d matrix entries have d symmetry while unperturbed degenerate states of $1S$ and $2S$ are s and s, p-like, respectively. Thus, it suffices to use second-order non-degenerate perturbation theory so as to find the energy shifts due the H_d,

$$\Delta E_d(1S) = \sum_{i=1}^{6}\left(\sum_{n,l,m}\frac{|\langle nlm, i|H_d|100, 1\rangle|^2}{-1 - E_{n,i}} + \int_0^{\infty}\mathrm{d}k\frac{|\langle klm, i|H_d|100, 1\rangle|^2}{-1 - E_{k,i}}\right),$$
$$\Delta E_d(2S) = \sum_{i=1}^{6}\left(\sum_{n,l,m}\frac{|\langle nlm, i|H_d|200, 1\rangle|^2}{-\frac{1}{4} - E_{n,i}} + \int_0^{\infty}\mathrm{d}k\frac{|\langle klm, i|H_d|200, 1\rangle|^2}{-\frac{1}{4} - E_{k,i}}\right), \tag{11.40}$$

for the main series, while for the split-off series,

$$\Delta E_d^{\mathrm{so}}(1S) = \sum_{i=1}^{4}\left(\sum_{n,l,m}\frac{|\langle nlm, i|H_d|100, 5\rangle|^2}{\overline{\Delta} - 1 - E_{n,i}} + \int_0^{\infty}\mathrm{d}k\frac{|\langle klm, i|H_d|100, 5\rangle|^2}{\overline{\Delta} - 1 - E_{k,i}}\right),$$
$$\Delta E_d^{\mathrm{so}}(2S) = \sum_{i=1}^{4}\left(\sum_{n,l,m}\frac{|\langle nlm, i|H_d|200, 5\rangle|^2}{\overline{\Delta} - \frac{1}{4} - E_{n,i}} + \int_0^{\infty}\mathrm{d}k\frac{|\langle klm, i|H_d|200, 5\rangle|^2}{\overline{\Delta} - \frac{1}{4} - E_{k,i}}\right). \tag{11.41}$$

Note that the sum for the split-off series extends to 4 only as there are no couplings in H_d between states 5 and 6. Further, the summations over n, l, m, i in Eqs. (11.4.1) and (11.4.1) encompass states with different energy as compared to the initial state. The integral and summation expressions obtained can be written as

$$\begin{aligned}
\Delta E_d(1S) &= -\frac{4}{5}\Phi(\mu_0, \mu_1, \mu_2)\left[S_1(0) + S_1(\overline{\Delta})\right],\\
\Delta E_d(2S) &= -\frac{1}{10}\Phi(\mu_0, \mu_1, \mu_2)\left[S_2(0) + S_2(\overline{\Delta})\right],\\
\Delta E_d^{\mathrm{so}}(1S) &= -\frac{6}{5}\Phi(\mu_0, \mu_1, \mu_2)\left[T_1(\overline{\Delta})\right],\\
\Delta E_d^{\mathrm{so}}(2S) &= -\frac{1}{5}\Phi(\mu_0, \mu_1, \mu_2)\left[T_1(\overline{\Delta})\right],
\end{aligned} \tag{11.42}$$

where

$$\Phi(\mu_0, \mu_1, \mu_2) = 8\left(\frac{\mu_0}{\mu_1}\right)^2 + \left(\frac{\mu_0}{\mu_2}\right)^2,$$
$$S_1(x) = \sum_{n=3}^{\infty}\frac{|I_n|^2}{x + 1 - 1/(n^2)} + \int_0^{\infty}\frac{|I_k|^2}{x + 1 + k^2}\mathrm{d}k,$$

$$
\begin{aligned}
S_2(x) &= \sum_{n=3}^{\infty} \frac{|J_n|^2}{x+\frac{1}{4}-1/(n^2)} + \int_0^{\infty} \frac{|J_k|^2}{x+\frac{1}{4}+k^2}\mathrm{d}k, \qquad (11.43)\\
T_1(x) &= S_1(-x),\\
T_2(x) &= S_2(-x).
\end{aligned}
$$

The functions I_n and J_n are integral expressions:

$$
\begin{aligned}
I_n &= \int_0^{\infty} \mathrm{d}r\, R_{n2}(r)\left(r+r^2\right)\mathrm{e}^{-r},\\
J_n &= \int_0^{\infty} \mathrm{d}r\, R_{n2}(r)\left(r+\frac{1}{2}r^2-\frac{1}{8}r^3\right)\mathrm{e}^{-\frac{r}{2}},
\end{aligned}
$$

where R_{nl} are the normalized hydrogen-like radial functions given in Landau and Lifshitz [167]. Analogous expressions define I_k and J_k (with k replacing n in the subscripts to the radial functions).

We give the details for obtaining the expression for $\Delta E_d(1S)$. Equation (11.4.1) can, by use of Eq. (11.25), be written as

$$
\begin{aligned}
\Delta E_d(1S) = &\sum_{n,l,m} \frac{|\langle nlm|\overline{Q}|100\rangle|^2}{1/(n^2)-1} + \int_0^{\infty}\mathrm{d}k\,\frac{|\langle klm|\overline{Q}|100\rangle|^2}{-k^2-1}\\
&+\sum_{n,l,m} \frac{|\langle nlm|\overline{L}^*|100\rangle|^2}{1/(n^2)-1} + \int_0^{\infty}\mathrm{d}k\,\frac{|\langle klm|\overline{L}^*|100\rangle|^2}{-k^2-1}\\
&+\sum_{n,l,m} \frac{|\langle nlm|\overline{M}^*|100\rangle|^2}{1/(n^2)-1} \int_0^{\infty}\mathrm{d}k\,\frac{|\langle klm|\overline{M}^*|100\rangle|^2}{-k^2-1}\\
&+\sum_{n,l,m} \frac{|\langle nlm|-\mathrm{i}/(\sqrt{2})\overline{L}^*|100\rangle|^2}{1/(n^2)-1-\overline{\Delta}} + \int_0^{\infty}\mathrm{d}k\,\frac{|\langle klm|-\mathrm{i}/(\sqrt{2})\overline{L}^*|100\rangle|^2}{-k^2-1-\overline{\Delta}}\\
&+\sum_{n,l,m} \frac{|\langle nlm|\mathrm{i}\sqrt{2}\overline{M}^*|100\rangle|^2}{1/(n^2)-1-\overline{\Delta}} + \int_0^{\infty}\mathrm{d}k\,\frac{|\langle klm|\mathrm{i}\sqrt{2}\overline{M}^*|100\rangle|^2}{-k^2-1-\overline{\Delta}}, \qquad (11.44)
\end{aligned}
$$

where the bars over Q, L, M denote their values in exciton Rydberg units. Further, inserting in Eq. (11.44) the relations

$$
\begin{aligned}
\overline{Q}|100\rangle &= -\frac{4}{\sqrt{5}}\frac{\mu_0}{\mu_1}\left[\left(1+\frac{1}{r}\right)\mathrm{e}^{-r}\right]Y_2^0,\\
\overline{L}|100\rangle &= -\mathrm{i}\frac{2\sqrt{2}}{\sqrt{15}}\frac{\mu_0}{\mu_2}\left[\left(1+\frac{1}{r}\right)\mathrm{e}^{-r}\right]Y_2^{-1}, \qquad (11.45)
\end{aligned}
$$

$$\overline{M}|100\rangle = \left[\left(\frac{2\sqrt{2}}{\sqrt{5}}\frac{\mu_0}{\mu_1} - \frac{\sqrt{2}}{\sqrt{15}}\frac{\mu_0}{\mu_2}\right)Y_2^2 + \left(\frac{2\sqrt{2}}{\sqrt{5}}\frac{\mu_0}{\mu_1} + \frac{\sqrt{2}}{\sqrt{15}}\frac{\mu_0}{\mu_2}\right)Y_2^{-2}\right]$$
$$\left[\left(1 + \frac{1}{r}\right)e^{-r}\right],$$

where Y_l^m are the spherical harmonics [167], performing the summation over l, m, and using the orthogonal relations in the spherical harmonics leads to

$$\Delta E_d(1S) = -\frac{4}{5}\Phi(\mu_0, \mu_1, \mu_2)\left[\sum_{n=3}^{\infty}\frac{|I_n|^2}{1 - 1/(n^2)} + \int_0^{\infty} \mathrm{d}k\, \frac{|I_k|^2}{1 + k^2}\right.$$
$$\left. + \sum_{n=3}^{\infty}\frac{|I_n|^2}{\overline{\Delta} + 1 - 1/(n^2)} + \int_0^{\infty}\mathrm{d}k\, \frac{|I_k|^2}{\overline{\Delta} + 1 + k^2}\mathrm{d}k\right]. \quad (11.46)$$

11.4.2 Results and Discussions

With the above analytical results for the energy changes due to H_d, one can compute the exciton binding energies. In Table 11.1, we give the parameter values used in the calculations.

In Table 11.2, the various binding energies along with the individual contributions to the $1S$ binding energy for the Γ_8 (main) series computed by Baldereschi and Lipari [159] are given. Note that there is good agreement between theory and experiment for GaAs, InAs, GaSb, and InSb but not so for GaP (probably due to uncertainties in some of the parameters used in the calculation). Further, observe that the binding energies are small for all material systems (less than 10 meV). It is also evident that the H_s contribution is significantly larger than the H_d contribution hence ascerting the rationale in treating the latter as a perturbation.

In Fig. 11.1, a plot of the exciton $1S$ and $2S$ binding energies is shown as a function of the spin-orbit splitting $\overline{\Delta}$ for an ideal semiconductor with $\Phi(\mu_0, \mu_1, \mu_2) = 0.5$.

Table 11.1 Material parameters. Adapted with permission from [159]. The vacuum permittivity is ε_0. ©1971 by the American Physical Society

	$\varepsilon/\varepsilon_0$	m_e/m_0	μ_0/m_0	μ_1/m_0	μ_2/m_0	E_0 (eV)	Δ_0 (eV)
GaAs	12.5	0.066	0.048	0.823	0.148	1.52	0.34
InAs	11.8	0.024	0.018	0.157	0.042	0.41	0.38
GaSb	15.2	0.047	0.035	0.444	0.100	0.81	0.80
InSb	16.8	0.015	0.012	0.117	0.032	0.24	0.81
GaP	11.1	0.13	0.075	−18.182	0.271	2.74	0.09

Table 11.2 Exciton binding energies computed using the Hamiltonian in Eq. (11.22). The parameters E_s, E_d^{intra}, E_d^{inter}, $E_{\text{ex}}(1S)$, $E_{\text{ex}}^{\text{exp}}(1S)$, $E_{\text{ex}}(2S)$, and $E_{\text{ex,so}}(1S)$ denote the contribution from H_s (Γ_8), the intraband series contribution from H_d (Γ_8), the interband series contribution from H_d (Γ_8), the total theoretical exciton binding energy for the $1S$ state (Γ_8), the total exciton binding energy for the $1S$ state (experimental), the total exciton binding energy for the $2S$ state (theory), and the total exciton binding energy for the $1S$ split-off series, respectively. All energies are in meV. Adapted with permission from [159]. ©1971 by the American Physical Society

	E_s	E_d^{intra}	E_d^{inter}	$E_{\text{ex}}(1S)$	$E_{\text{ex}}^{\text{exp}}(1S)$	$E_{\text{ex}}(2S)$	$E_{\text{ex,so}}(1S)$
GaAs	4.16	0.10	0.00	4.26	4.4	1.08	4.14
InAs	1.74	0.09	0.00	1.83	–	0.47	1.74
GaSb	2.05	0.06	0.00	2.11	–	0.54	2.05
InSb	0.56	0.02	0.00	0.58	≈ 0.4	0.15	0.56
GaP	8.30	0.11	0.03	8.44	3.5	2.13	8.23

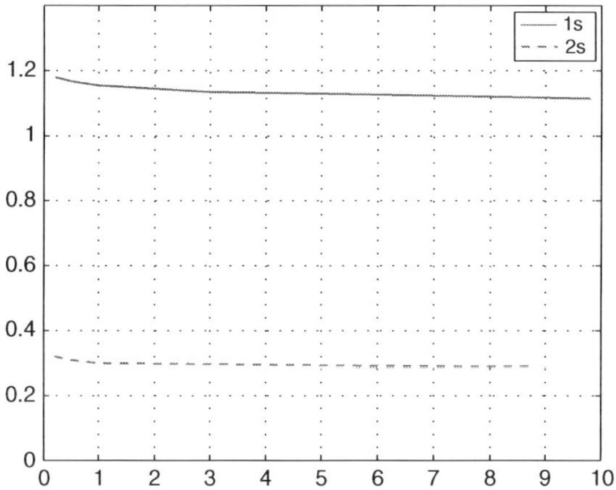

Fig. 11.1 Exciton $1S$ and $2S$ binding energies (in units of effective Rydberg) as a function of the spin-orbit splitting $\overline{\Delta}$ for an ideal semiconductor with $\Phi(\mu_0, \mu_1, \mu_2) = 0.5$. Adapted with permission from [159]. ©1971 by the American Physical Society

11.4.3 Zincblende

For T_d, one can simply add a linear-in-p Hamiltonian to the O_h one and, if still expected to be a small contribution, then treat it perturbatively as well. For the ground state of the main series, Baldereschi and Lipari [159] argues that the spin-split-off band can be dropped and they only considered a 4×4 perturbation:

$$H_p(\mathbf{p}) = \begin{pmatrix} 0 & U & -V & \sqrt{3}U^* \\ U^* & 0 & -\sqrt{3}U & V \\ -V & -\sqrt{3}U^* & 0 & U \\ \sqrt{3}U & V & U^* & 0 \end{pmatrix}, \tag{11.47}$$

where

$$U = \frac{\sqrt{3}}{2}\frac{\mu_0}{\mu_3}\left(\frac{\partial}{\partial x} + \mathrm{i}\frac{\partial}{\partial y}\right), \tag{11.48}$$

$$V = \mathrm{i}\sqrt{3}\frac{\mu_0}{\mu_3}\frac{\partial}{\partial z}, \tag{11.49}$$

and there is now a new inversion-asymmetry parameter

$$\mu_3 = \frac{\sqrt{3}}{2}\frac{\hbar^2}{a_0^{\mathrm{ex}}}\frac{1}{C_k}, \tag{11.50}$$

and C_k is the Dresselhaus spin-splitting parameter. Due to the paucity of data at the time, they were only able to consider the effect of H_p for InSb, for which they found $\mu_3 = 0.872m_0$, $E_b = 0.7$ meV, and $\Delta E_p(1S) = 0.0005$ meV. Thus, this correction term is likely not to be important for most materials.

11.5 Magnetoexciton

While the problem of an exciton, by itself, is an interesting one in its own right, experiments are often carried out in the presence of a magnetic field (for example, magnetophotoluminescence). As already mentioned, the magnetic field acts as a perturbation that breaks the original degeneracies of the exciton and introduces fine structures. These, in turn, help one analyze the parameters of the model. Given its importance, we will, therefore, provide an overview of the magnetoexciton theory.

The extension of the Baldereschi and Lipari work to include a magnetic field was done by Altarelli and Lipari [161, 162] and subsequently extended by Cho et al. [163]. We discuss the latter work for the zincblende case.

Cho et al. [163] used both the method of invariants and perturbation theory to write down the excitonic Hamiltonian to second order in the magnetic field (in order to account for the diamagnetic shift). We follow Cho et al. in combining Tables 5.11 and 5.12 in the absence of strain and for the exciton wave vector $\mathbf{K} = \mathbf{0}$, and rewriting as Table 11.3. One can now build a number of invariants from the quantities in Table 11.3. Since a number of them do not contribute to their perturbative expression, Cho et al. proposed the following effective exciton Hamiltonian:

Table 11.3 Symmetrized quantities for the zincblende exciton [163]

	Γ_1	Γ_2	Γ_{12}	Γ_{25}	Γ_{15}
σ	1			$(\sigma_x, \sigma_y, \sigma_z)$	
$\mathbf{J}$	1	$J_xJ_yJ_z + J_zJ_yJ_x$	(J_x^2, J_y^2)	(J_x, J_y, J_z)	(V_x, V_y, V_z)
				(J_x^3, J_y^3, J_z^3)	(U_x, U_y, U_z)
$\mathbf{B}$	B^2		(B_x^2, B_y^2)	(B_x, B_y, B_z)	(B_yB_z, B_zB_x, B_xB_y)

$$H = E_b + \widetilde{\Delta}_1 \mathbf{J}\cdot\boldsymbol{\sigma} + \widetilde{\Delta}_2(\sigma_x J_x^3 + \sigma_y J_y^3 + \sigma_z J_z^3) + \widetilde{g}_c \mu_B \boldsymbol{\sigma}\cdot\mathbf{B}$$
$$-2\mu_B\left[\widetilde{\kappa}\mathbf{J}\cdot\mathbf{B} + \widetilde{q}(B_x J_x^3 + B_y J_y^3 + B_z J_z^3)\right] \tag{11.51}$$
$$+\left(\frac{e a_0^{\text{ex}}}{2c}\right)^2 \frac{1}{\mu_0}\left[c_1 B^2 + c_2(\mathbf{J}\cdot\mathbf{B})^2 + c_3(B_x B_y\{J_x J_y\} B_y B_z\{J_y J_z\} + B_z B_x\{J_z J_x\})\right].$$

This model, therefore, contains 9 parameters: E_b, $\widetilde{\Delta}_1$, $\widetilde{\Delta}_2$, $\widetilde{g}_c$, $\widetilde{\kappa}$, $\widetilde{q}$, c_1, c_2, c_3. Note that these parameters are not (necessarily) the corresponding Luttinger parameters; this will be established by comparing to the perturbative expression.

In the effective-mass approach based upon Eq. (11.11), the model used by Cho et al. has

$$H^{(c)} = \frac{\hbar^2 k^2}{2m_e} + g_c \mu_B \boldsymbol{\sigma}\cdot\mathbf{B}, \tag{11.52}$$

$$H^{(v)} = -\frac{\hbar^2}{m_0}\Big\{(\gamma_1 + \frac{5}{2}\gamma_2)\frac{k^2}{2} - \gamma_2\left(k_x^2 J_x^2 + k_y^2 J_y^2 + k_z^2 J_z^2\right)$$
$$-2\gamma_3\left(\{k_x k_y\}\{J_x J_y\} + \{k_y k_z\}\{J_y J_z\} + \{k_z k_x\}\{J_z J_x\}\right) \tag{11.53}$$
$$-\frac{e}{\hbar}\kappa\left(B_x J_x + B_y J_y + B_z J_z\right) - \frac{e}{\hbar}q\left(B_x J_x^3 + B_y J_y^3 + B_z J_z^3\right)\Big\}$$
$$+K_l(k_x\{(J_y^2 - J_z^2)J_x\} + \text{c.p.}),$$

$$H_{\text{exch}} = \Delta_0 + \Delta_1 \mathbf{J}\cdot\boldsymbol{\sigma} + \Delta_2(\sigma_x J_x^3 + \sigma_y J_y^3 + \sigma_z J_z^3). \tag{11.54}$$

Using Eqs. (11.13) and (11.14), then

$$H = H^{(c)}(\mathbf{p} + e\mathbf{A} + \frac{m_e}{M}\hbar\mathbf{K}) - H^{(v)}(-\mathbf{p} + e\mathbf{A} + \frac{m_h}{M}\hbar\mathbf{K}) - \frac{e^2}{4\pi\epsilon r} + H_{\text{exch}}. \tag{11.55}$$

For $\mathbf{K} = \mathbf{0}$ excitonic states and in analogy to previous work, one can now write

$$H = H_s + H', \tag{11.56}$$

$$H_s = \frac{p^2}{2\mu_0} - \frac{e^2}{4\pi\epsilon r}\delta_{ii'}\delta_{jj'}, \tag{11.57}$$

$$H' = H_d + H_l + H_q + H_{k1} + H_{k2} + H_{\text{exch}}, \tag{11.58}$$

$$H_d = \frac{\gamma_2}{m_0}\left[\frac{5}{4}p^2 - (p_x^2 J_x^2 + p_y^2 J_y^2 + p_z^2 J_z^2)\right]$$
$$-\frac{2\gamma_3}{m_0}\left(\{k_x k_y\}\{J_x J_y\} + \{k_y k_z\}\{J_y J_z\} + \{k_z k_x\}\{J_z J_x\}\right), \tag{11.59}$$

$$H_l = \frac{eB}{2}\left(\frac{1}{m_e} - \frac{(\gamma_1 + \frac{5}{2}\gamma_2)}{m_0}\right)\left[\xi(y p_z - z p_y) + \text{c.p.}\right] + \frac{eB}{m_0}\gamma_2\left[(\overline{\eta} z - \overline{\xi} y)p_x J_x^2 + \text{c.p.}\right]$$

$$+\frac{eB}{m_0}\gamma_3\left(\left[\overline{\zeta}(xp_x-yp_y)-\overline{\xi}zp_x+\overline{\eta}zp_y\right]\{J_xJ_y\}+\text{c.p.}\right) \tag{11.60}$$

$$+\mu_B\left[-2\kappa\mathbf{J}\cdot\mathbf{B}-2qB(\overline{\xi}J_x^3+\overline{\eta}J_y^3+\overline{\zeta}J_z^3)+g_c\boldsymbol{\sigma}\cdot\mathbf{B}\right], \tag{11.61}$$

$$H_q=\left(\frac{eB}{2}\right)^2\left\{\left(\frac{1}{2\mu_0}+\frac{5}{4}\frac{\gamma_2}{m_0}\right)[(\overline{\eta}z-\overline{\zeta}y)^2+\text{c.p.}]-\frac{\gamma_2}{m_0}[(\overline{\eta}z-\overline{\zeta}y)^2J_x^2+\text{c.p.}]\right.$$
$$\left.-\frac{2\gamma_3}{m_0}[(\overline{\eta}z-\overline{\zeta}y)(\overline{\zeta}x-\overline{\xi}z)\{J_xJ_y\}+\text{c.p.}]\right\}, \tag{11.62}$$

$$H_{k1}=-K_l[p_x\{(J_y^2-J_z^2)J_x\}+\text{c.p.}], \tag{11.63}$$

$$H_{k2}=\frac{eB}{2}K_l[(\overline{\eta}z-\overline{\zeta}y)\{(J_y^2-J_z^2)J_x\}+\text{c.p.}], \tag{11.64}$$

where

$$\mathbf{A}=\frac{1}{2}(\mathbf{B}\times\mathbf{r}),$$
$$\mathbf{B}=B(\overline{\xi},\overline{\eta},\overline{\zeta}).$$

Comparing the Hamiltonian from the two methods allows one to relate the two parameter sets. Cho et al. [163] found:

$$\begin{aligned}
E_b &= -R_0^{\text{ex}}+\Delta_0-b_1-b_2,\\
\widetilde{\Delta}_1 &= \Delta_1,\ \widetilde{\Delta}_2=\Delta_2,\\
\widetilde{g}_c &= g_c,\\
\widetilde{\kappa} &= \kappa-d-\frac{13}{6}d(\tau-1)+\frac{7}{6}f,\\
\widetilde{q} &= q+\frac{2}{3}d(\tau-1)-\frac{2}{3}f,\\
c_1 &= 1-\nu-\frac{5}{4}\delta',\ c_2=\delta',\ c_3=2\delta'\left(\frac{1}{\tau}-1\right)+2\sigma',\\
d &= \frac{32}{5}\frac{\mu_0}{m_0}\gamma_3^2M,\ \tau=\frac{\gamma_2}{\gamma_3},\ f=4\frac{m_0}{\mu_0}G\left(\frac{\mu_0a_0^{\text{ex}}K_l}{\hbar}\right)^2,\\
\nu &= \frac{16}{15}\left(\frac{\mu_0\gamma_2}{m_0}\right)^2(2+3/\tau^2)(3N+W),\\
\delta' &= (1+16W/15)\mu_0\gamma_2/m_0,\ \sigma'=2F\left(\frac{\mu_0a_0^{\text{ex}}K_l}{\hbar}\right)^2,\\
M &= 0.281,\ G=0.375,\ N=0.469,\ W=0.719,\ F=0.884.
\end{aligned} \tag{11.65}$$

11.6 Summary

We have shown how a many-electron problem can be converted into an effective-mass theory for excitons. The translational motion of the exciton is then factored out and models for the internal motion presented. The spherical, nondegenerate band problem reduces to a hydrogen-like problem. The cubic, degenerate band problem can be formulated either using an effective Hamiltonian or via the effective-mass equations. In the hole formulation, the latter are not exactly the same as the band Hamiltonian. The Hamiltonian can be separated into a spherical and a cubic anisotropic part and the latter is usually treated using perturbation theory. One can similarly treat the problem of excitons in the presence of magnetic fields. A brief review of the theory for magnetoexcitons in zincblende semiconductors was presented. The wurtzite problem was also studied by Cho and coworkwers, and by Blattner et al. [168] who used it to analyze excitons and polaritons in CdS and ZnO up to 20 T. An application of the theory to wurtzite quantum dots was carried out by Bardoux et al. [169].

Chapter 12
Heterostructures: Basic Formalism

12.1 Overview

The starting point of any theory of electronic states in semiconductor nanostructures is to explain the origin of the discrete energy states, such as is evident in an optical spectrum or in the tunneling current through a resonant-tunneling device (Fig. 12.1).

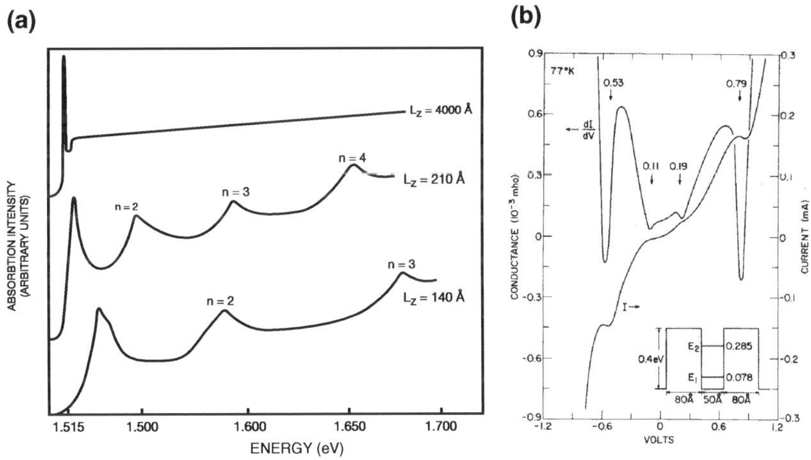

Fig. 12.1 **(a)** Absorption spectra at 2 K for GaAs quantum wells. Reprinted with permission from [170]. ©1974 by the American Physical Society. **(b)** Current and conductance characteristics at 77 K for a GaAs double-barrier structure. Reprinted with permission from [171]. ©1974, American Institute of Physics

The theory of Wannier–Luttinger–Kohn for the effective equation satisfied by a Bloch electron in the presence of a slowly-varying perturbation was applied early on to graded semiconductors [172] and to semiconductor inversion layers [173]. With the growth of atomically-sharp heterojunctions using molecular-beam epitaxy and metal-organo chemical vapor deposition techniques in the 1970s, the same theory was applied to the latter case even though the perturbation is no longer slowly

L.C. Lew Yan Voon, M. Willatzen, *The* k · p *Method*,
DOI 10.1007/978-3-540-92872-0_12, © Springer-Verlag Berlin Heidelberg 2009

varying [170, 174–176]. An equivalent problem would be to describe the electronic states in colloidal nanocrystals [177]. In all cases, linear dimensions could be as small as 1–10 nm. The fact that the Wannier–Luttinger–Kohn theory leads to an effective Schrödinger equation for an external potential indicates how this might work for a nanostructure. Thus, the external potential, due to the difference in band edges, would be a confining one leading to a problem similar to the particle-in-a-box problem in textbook quantum mechanics. There are, nevertheless, at least three differences. First, the mass is now the effective mass. Second, as we have seen before, a treatment of the valence band requires a multiband description. Third, contrary to the earlier cases, the material properties (e.g., effective masses and Kane parameters) are now position dependent and do not generally commute with spatial differential operators that appear in the Wannier-like equation. Hence, a proper theory should be able to account for all three effects right from the start.

Nevertheless, there have been three main approaches to the theory of the electronic structure of semiconductor nanostructures. The simplest approach is the one-band or particle-in-a-box model [170, 177, 178]. Another is the envelope-function theory of Bastard [175, 176, 179]. They typically require an ad hoc solution to the operator-ordering problem refered to above. Finally, the third approach is the more recent first-principles envelope-function theory of Burt and Foreman [34, 153, 180, 181]. The latter differs primarily from the previous two in attempting to derive the effective Hamiltonian from first principles. A number of review articles have appeared both on the early theories [95, 182–184] and on the more recent ones [185, 186].

The application of the theory has been to many different types of nanostructures; in particular, to quantum wells (QW's), superlattices, quantum wires and nanowires, and quantum dots. Quantum wells and superlattices are layered materials where one dimension is on the nanoscale. Quantum wires and nanowires display two-dimensional quantum confinement, the term nanowires being often associated with free-standing wires. Finally, quantum dots display three-dimensional quantum confinement with many analogies to atomic systems. Much of the explicit presentation of the theory in this book will be done for the case of QW's for simplicity and concreteness; the theory carries over trivially to the other cases with higher degree of quantum confinement. Excellent reviews are available specifically for the quantum-wire [187] and quantum-dot [188] problems.

12.2 Bastard's Theory

Bastard's theory was developed in two articles and expanded in his book [7, 175, 179].

12.2.1 Envelope-Function Approximation

The basic envelope-function model used was to assume that, in each layer of a heterostructure (here labeled A and B), the wave function can be expanded in terms of the periodic parts of the Bloch functions at a given bulk wave vector $\mathbf{k}_0$ (e.g., the Γ point),

$$\psi(\mathbf{r}) = \sum_n f_n^{(A,B)}(\mathbf{r}) u_{n\mathbf{k}_0}^{(A,B)}(\mathbf{r}), \tag{12.1}$$

and to then match the solutions across the boundaries. This can be seen to be an approximation to the theory expounded in Chap. 8, whereby distinct bulk wave vectors are assumed uncoupled. This is true if the perturbation is slowly varying, which would also result in a slowly-varying envelope function f_n. Models that do couple different bulk wave vectors (e.g., the Γ and X points [189, 190]) have been introduced but they will not be presented in this book.

Next, one assumes that the cellular functions in the layers are the same. This is a good approximation for chemically (e.g., a common ion) and structurally (e.g., lattice constant) similar materials; an example is the GaAs–AlAs system. Indeed, we have also seen that the momentum matrix elements do not change significantly among the III–V materials. Hence, as a first approximation, one might argue that the main properties that change from one layer to another are the energy gaps, effective masses and band offset. The latter is a new property of the heterointerface which reflects the different band alignments of different materials with respect to a reference level [127, 191, 192].

For a planar interface (e.g., QW), one can use Bloch periodicity in the plane to write

$$f_n^{(A,B)}(\mathbf{r}_{||}, z) = \frac{1}{\sqrt{S}} e^{i\mathbf{k}_{||}\cdot\mathbf{r}_{||}} \chi_n^{(A,B)}(z), \tag{12.2}$$

where $\mathbf{r}_{||}$ is a position vector in the QW plane and S is the surface area in the same plane. How good this theory is has been studied at length for QW's. Bastard [7] reports up to 0.3 eV or 10% of the first Brillouin zone for GaAs/AlAs structures. Gershoni et al. [193] have reported calculations up to $k = 0.15\,\text{Å}^{-1}$.

The goal is to now derive an effective equation for the envelope functions. The full Schrödinger equation can be written as

$$H\psi(\mathbf{r}) = E\psi(\mathbf{r}), \tag{12.3}$$

where

$$H = \frac{p^2}{2m_0} + V_A(\mathbf{r})\Theta_A + V_B(\mathbf{r})\Theta_B, \tag{12.4}$$

assuming there are two layers A and B and Θ is a step function. The effective equation can now be obtained using the method presented in Chap. 8. A simplified exposition was given by Bastard [7], whereby the key result was tonote that, for a slowly-varying envelope function, the integral over a cell-periodic function is such that

$$\int_V \mathrm{d}^3\mathbf{r}\, f(\mathbf{r})u(\mathbf{r}) \approx \frac{1}{V}\int_V \mathrm{d}^3\mathbf{r}\, f(\mathbf{r})\int_V \mathrm{d}^3\mathbf{r}\, u(\mathbf{r}). \tag{12.5}$$

Thus, inserting Eqs. (12.1) and (12.2) into Eq. (12.3), premultiplying by $\exp(-\mathrm{i}\mathbf{k}_{\|}\cdot\mathbf{r}_{\|})\chi_n^{(A,B)*}(z)$ and integrating gives [7]

$$D\left(z, -\mathrm{i}\hbar\frac{\partial}{\partial z}\right)\chi = E\chi, \tag{12.6}$$

where

$$\begin{aligned} D_{mn}^{(1)}\left(z, -\mathrm{i}\hbar\frac{\partial}{\partial z}\right) &= \left[E_m^{(A)}(\mathbf{0})\Theta_A + E_m^{(B)}(\mathbf{0})\Theta_B + \frac{\hbar^2 k_{\|}^2}{2m_0} - \frac{\hbar^2}{2m_0}\frac{\partial^2}{\partial z^2}\right]\delta_{mn} \\ &+ \frac{\hbar\mathbf{k}_{\|}}{m_0}\cdot\langle m|\mathbf{p}_{\|}|n\rangle - \frac{\mathrm{i}\hbar}{m_0}\langle m|p_z|n\rangle\frac{\partial}{\partial z}, \end{aligned} \tag{12.7}$$

if the first-order Kane model is used or

$$\begin{aligned} D_{mn}^{(2)}\left(z, -\mathrm{i}\hbar\frac{\partial}{\partial z}\right) &= \left[E_m^{(A)}(\mathbf{0})\Theta_A + E_m^{(B)}(\mathbf{0})\Theta_B + \frac{\hbar^2 k_{\|}^2}{2m_0} - \frac{\hbar^2}{2m_0}\frac{\partial^2}{\partial z^2}\right]\delta_{mn} \\ &+ \frac{\hbar\mathbf{k}_{\|}}{m_0}\cdot\langle m|\mathbf{p}_{\|}|n\rangle - \frac{\mathrm{i}\hbar}{m_0}\langle m|p_z|n\rangle\frac{\partial}{\partial z} - \frac{\hbar^2}{2}\frac{\partial}{\partial z}\frac{1}{M_{mn}^{zz}}\frac{\partial}{\partial z} \\ &- \frac{\mathrm{i}\hbar^2}{2}\sum_{i=x,y}\left[k_i\frac{1}{M_{mn}^{iz}}\frac{\partial}{\partial z} + \frac{\partial}{\partial z}\frac{1}{M_{mn}^{zi}}k_i\right] + \frac{\hbar^2}{2}\sum_{i,j=x,y} k_i\frac{1}{M_{mn}^{ij}}k_j, \end{aligned} \tag{12.8}$$

with

$$\frac{m_0}{M_{mn}^{ij}} = \frac{2}{m_0}\sum_\nu \langle m|p_i|\nu\rangle\frac{1}{E - E_\nu^{(A)}(\mathbf{0}) - V_\nu(z)}\langle\nu|p_j|n\rangle, \tag{12.9}$$

[and $V_\nu(z) = E_\nu^{(B)}(\mathbf{0}) - E_\nu^{(A)}(\mathbf{0})$] if the second-order Kane model is used. As for the impurity problem, the matrix D is related to the bulk $k\cdot p$ matrix with the replacement $\widehat{k}_z \to -\mathrm{i}\hbar\partial/\partial z$ if the periodicity along z is removed. Note that a symmetrization scheme has been imposed on Eq. (12.8).

12.2.2 Solution

First, we note that the heterostructure problem involves solving a coupled set of second-order differential equations, the number of equations being given by the size of the bulk multiband model. As is well-known from the simpler quantum mechanics problem, this requires matching the boundary conditions at the interfaces and specifying the asymptotic boundary conditions on the envelope functions. The latter is more straightforward. For a QW (with trivial extensions to higher degree of confinement), the envelope function must go to zero at large distances:

$$\lim_{|z|\to\infty} \chi(z) = 0. \tag{12.10}$$

Note, however, that in a numerical implementation, convergence is faster if a Neumann condition (rather than the Dirichlet one) is implemented [194]:

$$\lim_{|z|\to\infty} \frac{\partial\chi(z)}{\partial z} = 0. \tag{12.11}$$

It is, of course, very popular to use the particle-in-a-box model, whereby the barriers are assumed infinite. In that case, it is possible to obtain analytic solutions even for the four-band model [178, 195].

The interfacial boundary conditions are obtained by integrating the differential equations [196, 197]. These ensure current conservation. For the Hamiltonian of Eq. (12.8), one obtains the continuity of χ and $A\chi$, where [7]

$$A_{mn} = \left[\left(\delta_{mn} + \frac{m_0}{M_m^{zz}}\right)\frac{\partial}{\partial z} + \frac{2\mathrm{i}}{\hbar}\langle m|p_z|n\rangle + \mathrm{i}\sum_{i=x,y}\frac{m_0}{M_{mn}^{zi}}k_i\right], \tag{12.12}$$

and

$$M_{mn}^{zz} = M_m^{zz}\delta_{mn}. \tag{12.13}$$

Note that, for a one-band model,

$$A_m = 1 + \frac{2}{m_0}\sum_\nu \langle m|p_z|\nu\rangle \frac{1}{E_m - E_\nu^{(A)}(\mathbf{0}) - V_\nu(z)}\langle \nu|p_z|m\rangle,$$

and thus $\partial\chi/\partial z$ by itself is not continuous.

12.2.3 Example Models

Bastard has applied his theory to a couple of models. We would like to present these briefly for illustration of the theory and also of the technique of band folding.

12.2.3.1 Two-Band Model for Conduction Electron

The first example will illustrate downfolding from a first-order, eight-band Kane model to a two-band model. We carry this out using the Hamiltonian in Table 4.1. If one rewrites the matrix equation as a set of linear equations, then the approach is to eliminate all the other envelope-function components except for the two electron states of interest. Labeling the states as

$$u_1 = \mathrm{i}|s\uparrow\rangle,\ u_3 = \left|\frac{3}{2}\ \frac{1}{2}\right\rangle,\ u_5 = \left|\frac{3}{2}\ \frac{3}{2}\right\rangle,\ u_7 = \left|\frac{1}{2}\ \frac{1}{2}\right\rangle,$$
$$u_2 = \mathrm{i}|s\downarrow\rangle,\ u_4 = \left|\frac{3}{2}\ -\frac{1}{2}\right\rangle,\ u_6 = \left|\frac{3}{2}\ -\frac{3}{2}\right\rangle,\ u_8 = \left|\frac{1}{2}\ -\frac{1}{2}\right\rangle,$$

and neglecting the free-electron term, one can write

$$\begin{aligned}
\chi_3 &= \frac{1}{(E-E^{(A)}-V(z))}\left[-\sqrt{\frac{2}{3}}P\widehat{k}_z\chi_1+\sqrt{\frac{1}{6}}Pk_-\chi_2\right],\\
\chi_4 &= \frac{1}{(E-E^{(A)}-V(z))}\left[-\sqrt{\frac{1}{6}}Pk_+\chi_1-\sqrt{\frac{2}{3}}P\widehat{k}_z\chi_2\right],\\
\chi_5 &= \frac{1}{((E-E^{(A)}-V(z))}\sqrt{\frac{1}{2}}Pk_-\chi_1,\\
\chi_6 &= \frac{1}{(E-E^{(A)}-V(z))}\sqrt{\frac{1}{2}}Pk_+\chi_2,\\
\chi_7 &= \frac{1}{(E+\Delta_0-E^{(A)}-V(z))}\left[\sqrt{\frac{1}{3}}P\widehat{k}_z\chi_1+\sqrt{\frac{1}{3}}Pk_-\chi_2\right],\\
\chi_8 &= \frac{1}{(E+\Delta_0-E^{(A)}-V(z))}\left[-\sqrt{\frac{1}{3}}Pk_+\chi_1+\sqrt{\frac{1}{3}}P\widehat{k}_z\chi_2\right].
\end{aligned} \tag{12.14}$$

In the above equations, $\widehat{k}_z$ is an operator and the brackets do not all commute. Using these equations to eliminate all but χ_1 and χ_2, one gets the following 2×2 Hamiltonian for the conduction states:

$$\begin{aligned}
H_{11} = H_{22} &= E^{(A)} + V(z) + \frac{1}{3}P^2\widehat{k}_z\left(\frac{2}{(E-E^{(A)}-V(z))}\right.\\
&\left.+\frac{1}{(E+\Delta_0-E^{(A)}-V(z))}\right)\widehat{k}_z\\
&+\frac{1}{2}P^2k_+\left(\frac{1}{(E-E^{(A)}-V(z))}\right)k_-\\
&+\frac{1}{6}P^2k_-\left(\frac{1}{(E-E^{(A)}-V(z))}+\frac{2}{(E+\Delta_0-E^{(A)}-V(z))}\right)k_+,
\end{aligned} \tag{12.15}$$

$$H_{12} = -\frac{1}{3}P^2\widehat{k}_z\left(\frac{1}{(E-E^{(A)}-V(z))}-\frac{1}{(E+\Delta_0-E^{(A)}-V(z))}\right)k_-$$
$$+\frac{1}{3}P^2k_-\left(\frac{1}{(E-E^{(A)}-V(z))}-\frac{1}{(E+\Delta_0-E^{(A)}-V(z))}\right)\widehat{k}_z. \quad (12.16)$$

We have assumed a constant P. An interesting result is the nonvanishing off-diagonal H_{12} term which couples the two spins. It does vanish if $\Delta_0 = 0$ which means the coupling is a spin-orbit effect.

12.2.3.2 One-Band Model

One can also neglect the band coupling from the last model in order to obtain a one-band model for a heterostructure. For simplicity, we will also ignore the spin-hole bands and set $k_{||} = 0$. Reintroducing the free-electron term gives the one-band model as

$$\left[\widehat{k}_z\left(\frac{\hbar^2}{2m_0}+\frac{2P^2}{3(E-V(z))}\right)\widehat{k}_z + V(z)\right]\chi(z) = E\chi(z), \quad (12.17)$$

with $E^{(A)} = 0$. Integrating gives us the boundary condition that

$$\left(\frac{\hbar^2}{2m_0}+\frac{2P^2}{3(E-V(z))}\right)\frac{d\chi}{dz} \quad (12.18)$$

must be continuous across an interface. This is also equivalent to

$$\frac{1}{m^*(E,z)}\frac{d\chi}{dz}$$

being continuous. The energy dependence in the effective mass corresponds to non-parabolicity.

12.2.4 General Properties

The solutions to the Bastard model have a number of interesting properties. In no specific order, we present a few in this section.

Regarding the envelope functions, we recall that they are continuous but their first derivatives are not. In particular, for material systems whereby the effective mass changes sign (e.g., the HgTe–CdTe system), there are kinks in the envelope functions at the interfaces [7].

Within the eight-band Kane model, for $k_{||} = 0$, the hh and lh states are decoupled (see Table 4.1). In reality, it turns out that there is a coupling and those states anti-cross [198]. An example of such rules was obtained by Altarelli using a three-band (six with spin degeneracy) model [176]. His results are shown in Fig. 12.2.

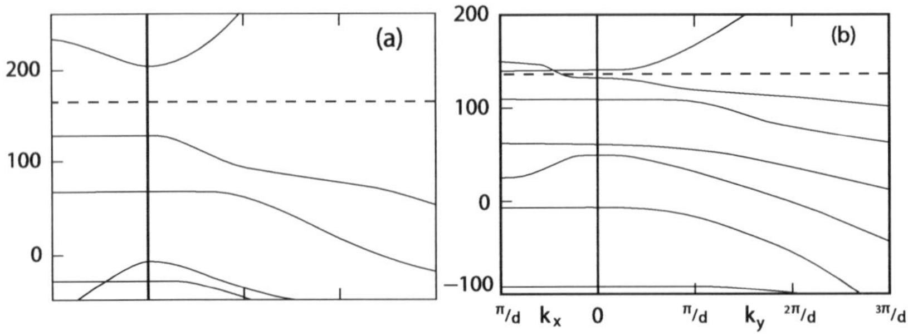

Fig. 12.2 Band structures of two InAs–GaSb superlattices. **(a)** 60–60 and **(b)** 90–90 Å. Here k_x is in the growth direction. Energies are in meV. Reprinted with permission from [176]. ©1983 by the American Physical Society

For a (symmetric) QW, there are useful symmetry properties of the envelope functions [199, 200]. Thus, the full wave function has definite parity. However, the effective Hamiltonian need not. For example, the Hamiltonian in Table 4.1 applied to a [001] QW has mixed behaviour as $\widehat{k}_z \to -\widehat{k}_z$. Since under reflection m_z

$$m_z|\uparrow\rangle = |\uparrow\rangle, m_z|\downarrow\rangle = -|\downarrow\rangle,$$

the cellular functions u_1, u_4, u_5, u_8 are even while u_2, u_3, u_6, u_7 are odd. The corresponding envelope-function components have the same behaviour. A symmetry analysis of the electron states in QW's and heterostructures within the eight-band model was also performed by Jorda and Rössler [201].

12.3 One-Band Models

We now give a more general discussion of one-band models.

12.3.1 Derivation

Equation (12.17), or indeed Eq. (12.8), can be rewritten, for a QW, as

$$\left[-\frac{\hbar^2}{2}\frac{\partial}{\partial z}\frac{1}{m(z)}\frac{\partial}{\partial z}+\frac{\hbar^2 k_{\|}^2}{2m(z)}+V_n(z)\right]\chi(z)=E_n\chi(z), \tag{12.19}$$

where $V_n(z)$ is the band-edge variation and $m(z) = m_A, m_B$ depending upon the layers. This is known as the Ben Daniel-Duke equation. A detailed discussion of the numerical solution to the Ben Daniel-Duke equation can be found in the book by Bastard [7].

Nevertheless, the above derivation was not entirely rigorous. In particular, it has the approximation of the same Kane parameter in both layers and had assumed the applicability of the envelope-function approximation even for a discontinuous potential profile. This has led others to propose different effective-mass equations.

The general one-band model is defined by the following equation,

$$H = \frac{1}{4}\left[m^{\alpha}\hat{p}m^{\beta}\hat{p}m^{\gamma} + m^{\gamma}\hat{p}m^{\beta}\hat{p}m^{\alpha}\right] + V(z), \tag{12.20}$$

where the solution is the envelope function. This is known as the von Roos Hamiltonian [202]. Numerous work on the correctness of the one-band model exists [203–206], most commonly as an approximate description of the conduction-band electron. Popular choices for the parameters are given in Table 12.1. The study of one-band models can be separated into the various types of mass functions used: piecewise constant, graded functions and infinite differentiable functions.

Table 12.1 Hermitian Hamiltonians for one-band models

α	β	γ	
0	0	-1	Gora-Williams [207], Bastard [175]
0	$-\frac{1}{2}$	$-\frac{1}{2}$	Li-Kuhn [208], Cavalcante et al. [209]
$-\frac{1}{2}$	0	$-\frac{1}{2}$	Ando and Mori [210], Zhu and Kroemer [211]
0	-1	0	Ben Daniel-Duke [212], White and Sham [213], Galbraith and Duggan [214], Einevoll et al. [215]

12.3.1.1 Piecewise Constant

- Morrow and Brownstein [216] studied the boundary conditions at a *single heterojunction* by integrating the differential equation. They obtained

$$\begin{aligned} \alpha &= \gamma, \\ m^{\alpha}\chi &= \text{continuous}, \\ m^{\alpha+\beta}\partial_z\chi &= \text{continuous}. \end{aligned} \tag{12.21}$$

These boundary conditions are consistent with the following general form (for two layers A and B) [20]

$$\chi_A = t_{11}\chi_B + t_{12}\frac{m_0}{m_B}a_0\partial_z\chi_B, \tag{12.22}$$

$$\chi_B = t_{21}\chi_B + t_{22}\frac{m_0}{m_B}a_0\partial_z\chi_B, \tag{12.23}$$

if t_{12} and t_{21} are small; such is apparently the case for the AlGaAs system [20]. Equations (12.23) give a conserved current

$$J_z = \frac{ie\hbar}{2}\left[\chi_i^* \partial_z \chi_i - (\partial_z \chi_i)^* \chi_i\right] \tag{12.24}$$

if

$$t_{11}t_{22} - t_{12}t_{21} = 1. \tag{12.25}$$

12.3.1.2 Infinitely Differentiable

- Li and Kuhn 1993 [208] showed that the von Roos Hamiltonian can be rewritten as the Ben Daniel-Duke one with an effective potential:

$$-\frac{\hbar^2}{2}\frac{\mathrm{d}}{\mathrm{d}z}\left(\frac{1}{m(z)}\frac{\mathrm{d}\psi(z)}{\mathrm{d}z}\right) + V_{\mathrm{eff}}(z,\alpha,\beta,\gamma)\chi(z) = E\chi(z), \tag{12.26}$$

$$V_{\mathrm{eff}}(z,\alpha,\beta,\gamma) = -\frac{\hbar^2}{2m^3(z)}\left[(1+\beta-\alpha\gamma)[m'(z)]^2 - \frac{1}{2}(1+\beta)m(z)m''(z)\right] + V(z).$$

  They argued that γ need not equal α for a non-abrupt interface.
- Balian et al. (1993) [217] showed that the von Roos Hamiltonian can be rewritten as a constant-mass one with an effective potential.
- Pistol (1999) [218], while he stated that he was solving for the piecewise continuous problem, really only solved for continuous coefficients. He applied the standard Sturm-Liouville theory to show that the homogeneous boundary condition necessary to define a hermitian operator is independent of α, β, and γ and given by a whole class of boundary conditions:

$$\left[\bar{u}\frac{1}{m(z)}v' - v'\frac{1}{m(z)}\bar{u}\right]_a^b = 0, \tag{12.27}$$

  where u and v are two solutions. This is not surprising since (1) Li and Kuhn had shown that all these Hamiltonians can be reduced to the Ben Daniel-Duke form, and (2) the boundary condition Pistol derived is the standard Surm-Liouville boundary condition. The latter is also consistent with current conservation. The omissions in Pistol's work were addressed in [219].

In general, the one-band model is fairly accurate for wide wells (e.g., 100 Å and above) for the nondegenerate electron states of DM, ZB and WZ semiconductor nanostructures. It can also be applied to the heavy-hole states if the latter are not mixed to the light-hole states (e.g., when $k_{||} = 0$).

12.4 Burt–Foreman Theory

A first-principles envelope-function theory was formulated by Michael Burt in the 1980s. This has the advantage of not imposing an ad hoc operator symmetrization but it will be seen that the commonly-used version makes a number of approximations.

12.4.1 Overview

Michael Burt has presented an exact envelope-function theory [153, 180, 181, 185, 186, 220–224]. Further development has been carried out by Brad Foreman [34, 225–231]. Burt's idea was a radical opposite to previous conceptualization of the heterostructure problem. Instead of proposing a heuristic differential equation and asking what kind of solutions are permitted, he imposed a constraint on the envelope function (that they be continuous and infinitely differentiable) and asked what kind of differential equation they satisfy. In a one-band formulation, he ended up with a derived Hamiltonian that basically looked like the heuristic one (for the Ben Daniel-Duke case) but had an extra nonlocal term. He showed the latter not to be important except within a few Ångström of a sharp interface [153].

Nonetheless, the theory was largely ignored until Brad Foreman derived an explicit six-band Hamiltonian from Burt's theory [34] and showed that it can lead to significant differences compared to the symmetrized Luttinger–Kohn Hamiltonian for the heavy-hole in an (001) QW structure [17]. He subsequently extended his model to an eight-band one but treating the extra conduction-valence coupling in a symmetrized fashion and neglecting linear-in-k terms. We will call this the Burt–Foreman (BF) model.

12.4.2 Envelope-Function Expansion

Burt first published his exact envelope-function theory in 1987, applicable to a one-dimensional unstrained multilayer microstructure with no spin-orbit interaction. The essence of the theory is that the wave function can be written as an envelope-function expansion:

$$\psi(z) = \sum_n F_n(z)\, U_n(z). \tag{12.28}$$

What differentiates the Burt formalism from others is to require that the envelope function $F_n(z)$ be a smooth continuous function and the $U_n(z)$ to be *a* complete set of orthogonal periodic functions over the whole structure, i.e., it is not in general the bulk solution in each layer.

The constraint that $F_n(z)$ be smooth implies that, in a plane-wave expansion, only Fourier components within the FBZ will be necessary. Substitution of Eq. (12.28) into the Schrödinger equation led to an exact equation for the envelope function which is a generalization of the standard envelope-function equation. The equation in [180] was corrected in an erratum published the same year [181]. Details of the derivation were given in 1988 [153]. We now go over the derivation.

We will first establish a number of properties and results of the envelope-function expansion given in Eq. (12.28). A key property is that, in a plane-wave representation of the envelope function $F_n(z)$, only wave vectors in the FBZ need be included. This will also be found to be related to the uniqueness of the expansion [153].

12.4.2.1 Envelope Function $F_n(z)$

One starts by expanding the wave function in terms of plane waves. Our lattice structure is a multilayer (e.g., the two layers in the case of a bilayer superlattice). We will assume that the lattice constant is the same for all the N unit cells making up the two layers. However, the chemical composition and electronic potential energy need not be (if they are, the system reduces to a bulk crystal). Take the lattice constant of each unit cell to be a. The crystal length is $L = Na$. As is traditional, we now impose cyclic boundary conditions. Then, the plane-wave expansion of ψ is given by

$$\psi(z) = \sum_{kG} \widetilde{\psi}_G(k) \mathrm{e}^{\mathrm{i}(k+G)z}. \tag{12.29}$$

Note that the plane-wave expansion of ψ is over the whole wave-number space. However, this can be split to account for the two inherent length scales. The underlying unit-cell periodicity allows the definition of reciprocal-lattice vectors. Furthermore, the cyclic boundary condition allows the introduction of wave numbers within the FBZ. One can obtain the expansion coefficients via the standard technique:

$$\int_L \frac{\mathrm{d}z}{L} \psi(z) \mathrm{e}^{-\mathrm{i}(k+G)z} = \sum_{k'G'} \widetilde{\psi}_{G'}(k') \int_L \frac{\mathrm{d}z}{L} \mathrm{e}^{\mathrm{i}(k'+G'-k-G)z} = \sum_{k'G'} \widetilde{\psi}_{G'}(k') \delta_{k'+G'-k-G,0},$$

giving

$$\widetilde{\psi}_G(k) = \int_L \frac{\mathrm{d}z}{L} \psi(z) \mathrm{e}^{-\mathrm{i}(k+G)z}. \tag{12.30}$$

We now appeal to the fact that a plane-wave expansion [Eq. (12.29)] is unique and complete. In addition, since $U_n(z)$ are periodic functions, they can be written as

$$U_n(z) = \sum_G U_{nG} \mathrm{e}^{\mathrm{i}Gz}, \tag{12.31}$$

where (with u.c. standing for unit cell)

$$U_{nG} = \int_{\mathrm{u.c.}} \frac{\mathrm{d}z}{a} U_n(z) \mathrm{e}^{-\mathrm{i}Gz}. \tag{12.32}$$

If the $U_n(z)$ are chosen complete and linearly independent, plane waves can be uniquely expanded in terms of them:

$$\mathrm{e}^{\mathrm{i}Gz} = \sum_n \left(U^{-1}\right)_{Gn} U_n(z), \tag{12.33}$$

where

$$\sum_G U_{nG} \left(U^{-1}\right)_{Gm} = \delta_{nm}, \tag{12.34}$$

and

$$\sum_n \left(U^{-1}\right)_{Gn} U_{nG'} = \delta_{GG'}. \tag{12.35}$$

We can now substitute Eq. (12.33) into Eq. (12.29):

$$\begin{aligned}\psi(z) &= \sum_{kG} \widetilde{\psi}_G(k)\mathrm{e}^{\mathrm{i}(k+G)z} = \sum_{kGn} \widetilde{\psi}_G(k)\mathrm{e}^{\mathrm{i}kz} \left(U^{-1}\right)_{Gn} U_n(z) \\ &\equiv \sum_n F_n(z)\, U_n(z),\end{aligned}$$

if

$$F_n(z) = \sum_{kG} \widetilde{\psi}_G(k)\mathrm{e}^{\mathrm{i}kz} \left(U^{-1}\right)_{Gn}. \tag{12.36}$$

Thus, $F_n(z)$ is uniquely determined via Eq. (12.36) and its expansion only consists of plane waves in the FBZ.

12.4.2.2 Uniqueness of $F_n(z)$

We now demonstrate the uniqueness of $F_n(z)$ explicitly. Assume there exists two different envelope-function expansions for the same wave function ψ:

$$\psi = \sum_n F_n^{(1)}(z)\, U_n(z) = \sum_n F_n^{(2)}(z)\, U_n(z).$$

Then

$$\sum_n \left[F_n^{(1)} - F_n^{(2)}\right] U_n = 0. \tag{12.37}$$

We will see below that one cannot yet take $F_n^{(1)}(z) = F_n^{(2)}(z)$. Since we have already established that the expansion of $F_n(z)$ only consists of plane waves in the FBZ,

$$F_n^{(i)}(z) = \sum_k \widetilde{F}_n^{(i)}(k)\mathrm{e}^{\mathrm{i}kz}, \tag{12.38}$$

substituting Eqs. (12.38) and (12.31) into Eq. (12.37):

$$\begin{aligned}&\sum_{kGn} \left[\widetilde{F}_n^{(1)}(k) - \widetilde{F}_n^{(2)}(k)\right] U_{nG}\mathrm{e}^{\mathrm{i}(k+G)z} = 0, \\ \Longrightarrow \quad &\sum_n \left[\widetilde{F}_n^{(1)}(k) - \widetilde{F}_n^{(2)}(k)\right] U_{nG} = 0,\end{aligned}$$

from the linear independence and completeness of all the plane waves. We now $\times \sum_G \left(U^{-1}\right)_{Gm}$ to give

$$\widetilde{F}_m^{(1)}(k) = \widetilde{F}_m^{(2)}(k), \ \forall\, m.$$

Hence, $F_m^{(1)}(z) = F_m^{(2)}(z)$.

Note that, in Eq. (12.37) we did not assume that $F_m^{(1)}(z) = F_m^{(2)}(z)$. The reason is because U_n is only a complete set of linearly-independent *periodic* functions. To wit, Eq. (12.37) becomes

$$\sum_n \left[F_n^{(1)} - F_n^{(2)}\right] U_{nG} \mathrm{e}^{\mathrm{i}Gz} = 0,$$

and $\mathrm{e}^{\mathrm{i}Gz}$ does not form a complete set of linearly-independent functions. If F_n is not restricted to the FBZ,

$$F_n^{(i)}(z) = \sum_{kG} \widetilde{F}_{nG}^{(i)}(k) \mathrm{e}^{\mathrm{i}(k+G)z},$$

and

$$\sum_{nkGG'} \left[\widetilde{F}_{nG}^{(1)}(k) - \widetilde{F}_{nG}^{(2)}(k)\right] U_{nG'} \mathrm{e}^{\mathrm{i}(k+G+G')z} = 0.$$

The coefficients are not necessarily zero because it is an overcomplete expansion. For example, let $G + G' = G''$, then

$$\sum_{nkGG''} \left[\widetilde{F}_{nG}^{(1)}(k) - \widetilde{F}_{nG}^{(2)}(k)\right] U_{nG''-G} \mathrm{e}^{\mathrm{i}(k+G'')z} = 0,$$
$$\Longrightarrow \quad \sum_{nG} \left[\widetilde{F}_{nG}^{(1)}(k) - \widetilde{F}_{nG}^{(2)}(k)\right] U_{nG''-G} = 0.$$

Burt [153] provides the following corollary:

Corollary 12.1. *If*

$$\sum_n \phi_n(z) U_n(z) = 0, \tag{12.39}$$

where $U_n(z)$ is a set of complete linearly independent periodic functions, then

$$\phi_n(z) = 0 \ \forall\, n \tag{12.40}$$

provided *the plane-wave expansion of $\phi_n(z)$ is restricted to the FBZ.*

12.4.2.3 Properties of Periodic Functions $U_n(z)$

It is often useful to impose a further constraint on the $U_n(z)$, that they be orthonormal:

$$\int_{\text{u.c.}} \frac{dz}{a} U_n^*(z) U_m(z) = \delta_{nm}. \tag{12.41}$$

Then, U_{nG} is unitary and, e.g., $\left(U^{-1}\right)_{Gn}$ may be replaced by U_{nG}^*:

$$\begin{aligned}\sum_n \left(U^\dagger\right)_{Gn} U_{nG'} = \sum_n U_{nG}^* U_{nG'} &= \int_{\text{u.c.}} \frac{dz}{a} \int_{\text{u.c.}} \frac{dz'}{a} e^{i(Gz-G'z')} \sum_n U_n^*(z) U_n(z') \\ &= \int_{\text{u.c.}} \frac{dz}{a} e^{i(G-G')z} = \delta_{GG'},\end{aligned}$$

and

$$U^\dagger = U^{-1}.$$

In the penultimate line, we used the completeness result

$$\sum_n U_n^*(z) U_n(z') = \delta(z - z').$$

12.4.3 Envelope-Function Equation

Starting from the Schrödinger equation,

$$-\frac{\hbar^2}{2m_0} \frac{d^2\psi}{dz^2} + V(z)\psi = E\psi, \tag{12.42}$$

the goal is to derive a new equation for the F_n's.

12.4.3.1 Kinetic Energy

Making use of the envelope-function expansion, Eq. (12.28), we note first that the kinetic-energy term becomes

$$-\frac{\hbar^2}{2m_0} \frac{d^2\psi}{dz^2} = -\frac{\hbar^2}{2m_0} \sum_n \left[F_n'' U_n + 2F_n' U_n' + F_n U_n''\right]. \tag{12.43}$$

Since $U_n(z)$ is periodic with period a, so are U_n' and U_n'':

$$U_n'(z+a) = U_n'(z), \; U_n''(z+a) = U_n''(z),$$

and they can, therefore, both be expanded in terms of $U_n(z)$. For example,

$$-\mathrm{i}\hbar\frac{dU_n}{\mathrm{d}z} = \sum_m p_{mn}U_m. \tag{12.44}$$

One can now obtain the coefficients p_{mn} in the same fashion as one would with Fourier coefficients: multiply each side of equation by U_n^*, integrate over z, and use orthonormality. Alternately, one can also state that the following coefficients

$$p_{mn} \equiv \int \frac{\mathrm{d}z}{a}\, U_m^*(z)\left(-\mathrm{i}\hbar\frac{d}{\mathrm{d}z}\right)U_n(z) \tag{12.45}$$

satisfy Eq. (12.44). To wit,

$$\begin{aligned}\sum_m p_{mn}U_m &= \int \frac{\mathrm{d}z'}{a}\sum_m U_m^*(z')\left(-\mathrm{i}\hbar\frac{\mathrm{d}U_n(z')}{\mathrm{d}z'}\right)U_m(z)\\ &= \int \frac{\mathrm{d}z'}{a}\left(-\mathrm{i}\hbar\frac{\mathrm{d}U_n(z')}{\mathrm{d}z'}\right)\sum_m U_m^*(z')U_m(z)\\ &= \int \frac{\mathrm{d}z'}{a}\left(-\mathrm{i}\hbar\frac{\mathrm{d}U_n(z')}{\mathrm{d}z'}\right)\delta(z-z') = -\mathrm{i}\hbar\frac{dU_n}{\mathrm{d}z}.\end{aligned}$$

Similarly, one can define

$$T_{mn} \equiv \int \frac{\mathrm{d}z}{a}\, U_m^*(z)\left(-\frac{\hbar^2}{2m_0}\frac{\mathrm{d}^2}{\mathrm{d}z^2}\right)U_n(z), \tag{12.46}$$

and show that

$$\sum_m T_{mn}U_m = -\frac{\hbar^2}{2m_0}\frac{\mathrm{d}^2U_n}{\mathrm{d}z^2}. \tag{12.47}$$

Putting the two terms together gives the kinetic energy as

$$-\frac{\hbar^2}{2m_0}\frac{\mathrm{d}^2\psi}{\mathrm{d}z^2} = \sum_n\left[-\frac{\hbar^2}{2m_0}\frac{\mathrm{d}^2F_n}{\mathrm{d}z^2} - \frac{\mathrm{i}\hbar}{m_0}\sum_m p_{nm}\frac{\mathrm{d}F_m}{\mathrm{d}z} + \sum_m T_{nm}F_m\right]U_n. \tag{12.48}$$

This is seen to be already in the envelope-function expansion form.

12.4.3.2 Potential Energy

The expression to be rewritten is $V\psi$. Despite its apparent simplicity, there is a problem due to the fact that a Fourier transform of the two functions can lead to terms outside the FBZ.

We start by Fourier transforming $V(z)$ and expanding $\psi(z)$ in the envelope-function expansion; in the latter, we will also replace the envelope functions $F_n(z)$ and the periodic functions $U_n(z)$ by their Fourier transforms. Similar to Eq. (12.30), the Fourier components of the potential are given by

$$\widetilde{V}_G(k) = \int_L \frac{dz}{L} V(z) e^{-i(k+G)z}. \tag{12.49}$$

Thus,

$$\begin{aligned} V(z)\psi(z) &= \sum_{kGn} \widetilde{V}_G(k) e^{i(k+G)z} F_n(z) U_n(z) \\ &= \sum_{kGn} \sum_{k'G'} \widetilde{V}_G(k) e^{i(k+G)z} \widetilde{F}_n(k') e^{ik'z} U_{nG'} e^{iG'z} \\ &= \sum_n \sum_{kk'} \sum_{GG'} \widetilde{V}_G(k) \widetilde{F}_n(k') U_{nG'} e^{i(k+k'+G+G')z}. \end{aligned} \tag{12.50}$$

Let $G \to G - G'$. Then,

$$V(z)\psi(z) = \sum_n \sum_{kk'} \sum_{GG'} \widetilde{V}_{G-G'}(k) \widetilde{F}_n(k') U_{nG'} e^{i(k+k'+G)z}. \tag{12.51}$$

The goal is to express the RHS as an envelope-function expansion, i.e., in an expansion in terms of plane waves in the FBZ only. The sum $k + k'$ can be outside the FBZ; to make this explicit, write

$$k + k' = k_1 + G_1, \tag{12.52}$$

where $k_1 \in$ FBZ. Then, Eq. (12.51) becomes

$$V(z)\psi(z) = \sum_n \sum_{kk'} \sum_{GG'} \widetilde{V}_{G-G'}(k) \widetilde{F}_n(k') U_{nG'} e^{i(k_1+G+G_1)z}. \tag{12.53}$$

But, from Eq. (12.33),

$$\begin{aligned} e^{iGz} &= \sum_n \left(U^{-1}\right)_{Gn} U_n(z) = \sum_n U^*_{nG} U_n(z), \\ \Longrightarrow \quad e^{i(G+G_1)z} &= \sum_m U^*_{m,G+G_1} U_m(z), \end{aligned} \tag{12.54}$$

and Eq. (12.53) becomes

$$V(z)\psi(z) = \sum_m \left[\sum_n \sum_{kk'} \sum_{GG'} U^*_{m,G+G_1} \widetilde{V}_{G-G'}(k) \widetilde{F}_n(k') U_{nG'} e^{ik_1 z} \right] U_m(z). \tag{12.55}$$

This is already in the envelope-function expansion form since the function inside the square brackets has a plane-wave expansion restricted to the FBZ. We, nevertheless, need it in terms of the envelope function F. This requires replacing the Fourier transform of the envelope function by its real space function [inverse of Eq. (12.36)]:

$$\widetilde{F}_n(k') = \int_L \frac{\mathrm{d}z'}{L} F_n(z')\mathrm{e}^{-\mathrm{i}k'z'}, \tag{12.56}$$

giving

$$\begin{aligned} V(z)\psi(z) &= \sum_m \left[\sum_n \sum_k \sum_{GG'} U^*_{m,G+G_1} \widetilde{V}_{G-G'}(k) U_{nG'} \int_L \frac{\mathrm{d}z'}{L} \sum_{k'} F_n(z')\mathrm{e}^{-\mathrm{i}k'z'+\mathrm{i}k_1 z} \right] U_m(z) \\ &= \sum_{n,m} \int_L \frac{\mathrm{d}z'}{L} \sum_{GG'} \sum_{kk'} U^*_{n,G+G_1} \widetilde{V}_{G-G'}(k) U_{mG'} \mathrm{e}^{\mathrm{i}(k_1 z - k'z')} F_m(z') U_n(z) \\ &\equiv \sum_n \left[\sum_m \int \mathrm{d}z' \, V_{nm}(z,z') F_m(z') \right] U_n(z), \end{aligned} \tag{12.57}$$

where

$$V_{nm}(z,z') \equiv \frac{1}{L} \sum_{kk'} \sum_{GG'} U^*_{n,G+G_1} \widetilde{V}_{G-G'}(k) U_{mG'} \mathrm{e}^{\mathrm{i}(k_1 z - k'z')}. \tag{12.58}$$

12.4.3.3 Envelope-Function Equation

Using Eqs. (12.28), (12.48) and (12.57), Eq. (12.42) becomes

$$\sum_n \left[-\frac{\hbar^2}{2m_0} \frac{\mathrm{d}^2 F_n}{\mathrm{d}z^2} - \frac{\mathrm{i}\hbar}{m_0} \sum_m p_{nm} \frac{\mathrm{d}F_m}{\mathrm{d}z} + \sum_m T_{nm} F_m + \sum_m \int \mathrm{d}z' \, V_{nm}(z,z') F_m(z') \right] U_n = E \sum_n F_n(z) \, U_n(z). \tag{12.59}$$

Equating the coefficients of $U_n(z)$ gives the exact envelope-function equation:

$$-\frac{\hbar^2}{2m_0} \frac{\mathrm{d}^2 F_n}{\mathrm{d}z^2} - \frac{\mathrm{i}\hbar}{m_0} \sum_m p_{nm} \frac{\mathrm{d}F_m}{\mathrm{d}z} + \sum_m \int \mathrm{d}z' \, H_{nm}(z,z') F_m(z') = E F_n(z), \tag{12.60}$$

where

$$H_{nm}(z,z') = T_{nm}\delta(z-z') + V_{nm}(z,z'). \tag{12.61}$$

12.4.3.4 Generalization to Three Dimension

It is trivial to generalize the above derivation to three dimension [185]. We will only summarize the main equations.

The envelope-function expansion becomes

$$\psi(\mathbf{R}) = \sum_n F_n(\mathbf{R})\, U_n(\mathbf{R}). \tag{12.62}$$

The envelope-function equation is

$$-\frac{\hbar^2}{2m_0}\nabla^2 F_n(\mathbf{R}) - \frac{i\hbar}{m_0}\sum_m \mathbf{p}_{nm}\cdot\nabla F_m(\mathbf{R}) + \sum_m \int d^3\mathbf{R}'\, H_{nm}(\mathbf{R},\mathbf{R}')F_m(\mathbf{R}) = E F_n(\mathbf{R}), \tag{12.63}$$

where

$$\begin{aligned}
\mathbf{p}_{nm} &= \frac{1}{\Omega}\int d^3\mathbf{R}'\, U_n^* \mathbf{p} U_m,\\
T_{nm} &= \frac{1}{\Omega}\int d^3\mathbf{R}'\, U_n^* T U_m,\\
H_{nm}(\mathbf{R},\mathbf{R}') &= T_{nm}\Delta(\mathbf{R}-\mathbf{R}') + V_{nm}(\mathbf{R},\mathbf{R}')),\\
\Delta(\mathbf{R}-\mathbf{R}') &= \frac{1}{V}\sum_{\mathbf{k}\in\,\mathrm{FBZ}} e^{i\mathbf{k}\cdot(\mathbf{R}-\mathbf{R}')},
\end{aligned} \tag{12.64}$$

and

$$\frac{1}{\Omega}\int d^3\mathbf{R}'\, U_n^* U_m = \delta_{nm}. \tag{12.65}$$

As before, Ω (V) is the unit-cell (crystal) volume.

12.4.3.5 Alternative Derivation

The envelope-function representation can be identified as a unitary transformation. In order to make this connection, we now provide another derivation of the envelope-function equation [185]. We will do so directly in three-dimension.

The starting point is the Schrödinger equation in a plane-wave representation:

$$\frac{\hbar^2}{2m_0}(\mathbf{k}+\mathbf{G})^2\widetilde{\psi}_{\mathbf{G}}(\mathbf{k}) + \sum_{\mathbf{k}',\mathbf{G}'}\langle\mathbf{k}+\mathbf{G}|V|\mathbf{k}'+\mathbf{G}'\rangle\widetilde{\psi}_{\mathbf{G}'}(\mathbf{k}') = E\widetilde{\psi}_{\mathbf{G}}(\mathbf{k}), \tag{12.66}$$

where

$$\widetilde{\psi}_{\mathbf{G}}(\mathbf{k}) = \frac{1}{V}\int d^3\mathbf{R}\,\psi(\mathbf{R})e^{-i(\mathbf{k}+\mathbf{G})\cdot\mathbf{R}}, \tag{12.67}$$

$$\psi(\mathbf{R}) = \sum_{\mathbf{k},\mathbf{G}}\widetilde{\psi}_{\mathbf{G}}(\mathbf{k})e^{i(\mathbf{k}+\mathbf{G})\cdot\mathbf{R}}. \tag{12.68}$$

Now introduce a complete set of periodic functions:

$$U_n(\mathbf{R}) = \sum_{\mathbf{G}} U_{n\mathbf{G}}e^{i\mathbf{G}\cdot\mathbf{R}}, \tag{12.69}$$

giving

$$e^{i\mathbf{G}\cdot\mathbf{R}} = \sum_n (U^{-1})_{\mathbf{G}n} U_n(\mathbf{R}), \tag{12.70}$$

and

$$\sum_n (U^{-1})_{\mathbf{G}n} U_{n\mathbf{G}'} = \delta_{\mathbf{G}\mathbf{G}'}. \tag{12.71}$$

Substituting Eq. (12.70) into Eq. (12.68),

$$\begin{aligned}\psi(\mathbf{R}) &= \sum_{\mathbf{k},\mathbf{G}} \widetilde{\psi}_{\mathbf{G}}(\mathbf{k}) e^{i\mathbf{k}\cdot\mathbf{R}} e^{i\mathbf{G}\cdot\mathbf{R}} = \sum_{\mathbf{k},\mathbf{G}} \widetilde{\psi}_{\mathbf{G}}(\mathbf{k}) e^{i\mathbf{k}\cdot\mathbf{R}} \sum_n (U^{-1})_{\mathbf{G}n} U_n(\mathbf{R}) \\ &\equiv \sum_n F_n(\mathbf{R}) U_n(\mathbf{R}),\end{aligned} \tag{12.72}$$

where

$$F_n(\mathbf{R}) = \sum_{\mathbf{k}} \widetilde{F}_n(\mathbf{k}) e^{i\mathbf{k}\cdot\mathbf{R}}, \tag{12.73}$$

$$\widetilde{F}_n(\mathbf{k}) = \sum_{\mathbf{G}} \widetilde{\psi}_{\mathbf{G}}(\mathbf{k}) (U^{-1})_{\mathbf{G}n} \equiv \frac{1}{V}\int d^3\mathbf{R}\, F_n(\mathbf{R}) e^{-i\mathbf{k}\cdot\mathbf{R}}. \tag{12.74}$$

From the latter equation, one can also write the inverse:

$$\widetilde{\psi}_{\mathbf{G}'}(\mathbf{k}') = \sum_n U_{n\mathbf{G}'} \widetilde{F}_n(\mathbf{k}'). \tag{12.75}$$

Now multiplying Eq. (12.66) by $(U^{-1})_{\mathbf{G}n}$ and summing over $\mathbf{G}$ gives

$$\begin{aligned}&\sum_{\mathbf{G}} (U^{-1})_{\mathbf{G}n} \left[\frac{\hbar^2}{2m_0} (\mathbf{k}+\mathbf{G})^2 \widetilde{\psi}_{\mathbf{G}}(\mathbf{k}) + \sum_{\mathbf{k}',\mathbf{G}'} \langle \mathbf{k}+\mathbf{G}|V|\mathbf{k}'+\mathbf{G}'\rangle \widetilde{\psi}_{\mathbf{G}'}(\mathbf{k}') \right] \\ &= E \sum_{\mathbf{G}} (U^{-1})_{\mathbf{G}n} \widetilde{\psi}_{\mathbf{G}}(\mathbf{k}),\end{aligned}$$

and using Eq. (12.75),

$$\begin{aligned}&\sum_{n'\mathbf{G}} \frac{\hbar^2}{2m_0} (\mathbf{k}+\mathbf{G})^2 (U^{-1})_{\mathbf{G}n} U_{n'\mathbf{G}} \widetilde{F}_{n'}(\mathbf{k}) + \sum_{\substack{\mathbf{k}',\mathbf{G},\mathbf{G}'\\ n'}} \langle \mathbf{k}+\mathbf{G}|V|\mathbf{k}'+\mathbf{G}'\rangle (U^{-1})_{\mathbf{G}n} U_{n'\mathbf{G}'} \widetilde{F}_{n'}(\mathbf{k}') \\ &= E \sum_{\mathbf{G}n'} (U^{-1})_{\mathbf{G}n} U_{n'\mathbf{G}} \widetilde{F}_{n'}(\mathbf{k}).\end{aligned}$$

Then

$$\sum_{n'\mathbf{G}}\left[\frac{\hbar^2}{2m_0}(k^2+G^2+2\mathbf{k}\cdot\mathbf{G})(U^{-1})_{\mathbf{G}n}U_{n'\mathbf{G}}\widetilde{F}_{n'}(\mathbf{k})\right]$$
$$+\sum_{\substack{\mathbf{k}',\mathbf{G},\mathbf{G}'\\ n'}}\langle\mathbf{k}+\mathbf{G}|V|\mathbf{k}'+\mathbf{G}'\rangle(U^{-1})_{\mathbf{G}n}U_{n'\mathbf{G}'}\widetilde{F}_{n'}(\mathbf{k}')=E\sum_{\mathbf{G}n'}(U^{-1})_{\mathbf{G}n}U_{n'\mathbf{G}}\widetilde{F}_{n'}(\mathbf{k}),$$

or

$$\frac{\hbar^2k^2}{2m_0}\widetilde{F}_n(\mathbf{k})+\frac{\hbar^2}{m_0}\sum_{\mathbf{G}n'}\mathbf{k}\cdot\mathbf{G}(U^{-1})_{\mathbf{G}n}U_{n'\mathbf{G}}\widetilde{F}_{n'}(\mathbf{k})$$
$$+\sum_{\mathbf{G}n'}\left[\frac{\hbar^2G^2}{2m_0}(U^{-1})_{\mathbf{G}n}U_{n'\mathbf{G}}\widetilde{F}_{n'}(\mathbf{k})+\sum_{\mathbf{k}',\mathbf{G}'}\langle\mathbf{k}+\mathbf{G}|V|\mathbf{k}'+\mathbf{G}'\rangle(U^{-1})_{\mathbf{G}n}U_{n'\mathbf{G}'}\widetilde{F}_{n'}(\mathbf{k}')\right]$$
$$=E\widetilde{F}_n(\mathbf{k}).$$

Let

$$\hbar\sum_{\mathbf{G}}\mathbf{G}(U^{-1})_{\mathbf{G}n}U_{n'\mathbf{G}}\equiv\mathbf{p}_{nn'},\tag{12.76}$$

$$\sum_{\mathbf{G}}\frac{\hbar^2}{2m_0}G^2(U^{-1})_{\mathbf{G}n}U_{n'\mathbf{G}}\equiv T_{nn'},\tag{12.77}$$

$$\sum_{\mathbf{G}\mathbf{G}'}(U^{-1})_{\mathbf{G}n}\langle\mathbf{k}+\mathbf{G}|V|\mathbf{k}'+\mathbf{G}'\rangle U_{n'\mathbf{G}'}\equiv\widetilde{V}_{nn'}(\mathbf{k},\mathbf{k}'),\tag{12.78}$$

$$T_{nn'}\delta_{\mathbf{k}\mathbf{k}'}+\widetilde{V}_{nn'}(\mathbf{k},\mathbf{k}')\equiv\widetilde{H}_{nn'}(\mathbf{k},\mathbf{k}').\tag{12.79}$$

This gives the envelope function in k space:

$$\frac{\hbar^2k^2}{2m_0}\widetilde{F}_n(\mathbf{k})+\frac{\hbar}{m_0}\sum_{n'}\mathbf{k}\cdot\mathbf{p}_{nn'}\widetilde{F}_{n'}(\mathbf{k})+\sum_{n'\mathbf{k}'}\widetilde{H}_{nn'}(\mathbf{k},\mathbf{k}')\widetilde{F}_{n'}(\mathbf{k}')=E\widetilde{F}_n(\mathbf{k}).\tag{12.80}$$

To relate to the previous derivation, one now transforms to real space by multiplying by $e^{i\mathbf{k}\cdot\mathbf{R}}$ and summing over $\mathbf{k}$. With

$$\sum_{\mathbf{k}}\frac{\hbar^2k^2}{2m_0}\widetilde{F}_n(\mathbf{k})e^{i\mathbf{k}\cdot\mathbf{R}}=-\sum_{\mathbf{k}}\frac{\hbar^2}{2m_0}\nabla^2\widetilde{F}_n(\mathbf{k})e^{i\mathbf{k}\cdot\mathbf{R}}=-\frac{\hbar^2\nabla^2}{2m_0}F_n(\mathbf{R}),$$

$$\frac{\hbar}{m_0}\sum_{n'}\mathbf{p}_{nn'}\cdot\sum_{\mathbf{k}}\mathbf{k}\widetilde{F}_{n'}(\mathbf{k})e^{i\mathbf{k}\cdot\mathbf{R}}=-\frac{i\hbar}{m_0}\sum_{n'}\mathbf{p}_{nn'}\cdot\nabla F_{n'}(\mathbf{R}),$$

$$\sum_{n'\mathbf{k}'\mathbf{k}} \mathrm{e}^{\mathrm{i}\mathbf{k}\cdot\mathbf{R}} \widetilde{H}_{nn'}(\mathbf{k},\mathbf{k}')\widetilde{F}_{n'}(\mathbf{k}') = \frac{1}{V}\int \mathrm{d}^3\mathbf{R}' \sum_{n'\mathbf{k}'\mathbf{k}} \mathrm{e}^{\mathrm{i}\mathbf{k}\cdot\mathbf{R}} \widetilde{H}_{nn'}(\mathbf{k},\mathbf{k}')F_{n'}(\mathbf{R}')\mathrm{e}^{-\mathrm{i}\mathbf{k}'\cdot\mathbf{R}'}$$

$$\equiv \sum_{n'} \int \mathrm{d}^3\mathbf{R}'\, F_{n'}(\mathbf{R}')\frac{1}{V}\sum_{\mathbf{k}'\mathbf{k}} \mathrm{e}^{\mathrm{i}\mathbf{k}\cdot\mathbf{R}} \widetilde{H}_{nn'}(\mathbf{k},\mathbf{k}')\mathrm{e}^{-\mathrm{i}\mathbf{k}'\cdot\mathbf{R}'},$$

using $\nabla = \nabla_{\mathbf{R}}$, Eq. (12.80) becomes

$$-\frac{\hbar^2\nabla^2}{2m_0}F_n(\mathbf{R}) - \frac{\mathrm{i}\hbar}{m_0}\sum_{n'}\mathbf{p}_{nn'}\cdot\nabla F_{n'}(\mathbf{R}) + \sum_{n'}\int \mathrm{d}^3\mathbf{R}'\, H_{nn'}(\mathbf{R},\mathbf{R}')F_{n'}(\mathbf{R}'),$$

which is the same as the exact envelope-function equation, Eq. (12.60). Note, however, that in order to demonstrate complete equivalence of the two approaches, one remains to demonstrate that the newly-defined $\mathbf{p}_{nn'}$, $T_{nn'}$, and $\widetilde{V}_{nn'}(\mathbf{k},\mathbf{k}')$ are identically defined. Consider $\mathbf{p}_{nn'}$:

$$\mathbf{p}_{nn'} = \hbar\sum_{\mathbf{G}}\mathbf{G}(U^{-1})_{\mathbf{G}n}U_{n'\mathbf{G}}.$$

In the previous derivation,

$$\mathbf{p}_{nn'} = \frac{1}{\Omega}\int_{\text{u.c.}} \mathrm{d}^3\mathbf{R}'\, U_n^*\mathbf{p}U_{n'} = \sum_{\mathbf{G}\mathbf{G}'} U_{n\mathbf{G}}^* U_{n'\mathbf{G}'}\frac{1}{\Omega}\int_{\text{u.c.}} \mathrm{d}^3\mathbf{R}'\, \mathrm{e}^{-\mathrm{i}\mathbf{G}\cdot\mathbf{R}}\mathbf{p}\mathrm{e}^{\mathrm{i}\mathbf{G}'\cdot\mathbf{R}}$$

$$= \hbar\sum_{\mathbf{G}\mathbf{G}'}\mathbf{G}'U_{n\mathbf{G}}^*U_{n'\mathbf{G}'}\frac{1}{\Omega}\int_{\text{u.c.}} \mathrm{d}^3\mathbf{R}'\, \mathrm{e}^{\mathrm{i}(\mathbf{G}'-\mathbf{G})\cdot\mathbf{R}} = \hbar\sum_{\mathbf{G}}\mathbf{G}U_{n\mathbf{G}}^*U_{n'\mathbf{G}},$$

and this equals the other definition, Eq. (12.76), if

$$U_{n\mathbf{G}}^* = (U^{-1})_{\mathbf{G}n},$$

i.e., if U is unitary.

12.4.4 Potential-Energy Term

The reason for the nonlocality of the exact envelope-function equation, Eq. (12.60), is because of the potential energy term. The latter indeed holds clues to the difference between the exact envelope-function equation and the conventional one. We, therefore, need to study Eq. (12.58),

$$V_{nm}(z,z') = \frac{1}{L}\sum_{kk'}\sum_{GG'}U_{n,G+G_1}^*\widetilde{V}_{G-G'}(k)U_{mG'}\mathrm{e}^{\mathrm{i}(k_1z-k'z')},$$

where $k+k' = k_1+G_1$, more carefully [153].

12.4.4.1 Separation of $V_{nm}(z, z')$

Remember that k and k' came from the expansion of $V\psi$. We now show that the reason $V_{nm}(z, z')$ is nonlocal is because $k + k'$ can be outside the FBZ. If that were not the case, $G_1 = 0$ and

$$V_{nm}(z, z') = \sum_k \sum_{GG'} U^*_{n,G} \widetilde{V}_{G-G'}(k) U_{mG'} \frac{\mathrm{e}^{\mathrm{i}kz}}{L} \sum_{k'} \mathrm{e}^{\mathrm{i}k'(z-z')}$$
$$\equiv \sum_k \sum_{GG'} U^*_{n,G} \widetilde{V}_{G-G'}(k) U_{mG'} \mathrm{e}^{\mathrm{i}kz} \Delta(z - z'), \tag{12.81}$$

where

$$\Delta(z - z') \equiv \frac{1}{L} \sum_{k' \in \text{ FBZ}} \mathrm{e}^{\mathrm{i}k'(z-z')}. \tag{12.82}$$

Note that the summation is not strictly an orthonormal relation but if we assume Δ to be sufficiently like the delta function,

$$\sum_m \int \mathrm{d}z' \, V_{nm}(z, z') F_m(z') \approx \sum_m V_{nm}(z) F_m(z), \tag{12.83}$$

with

$$V_{nm}(z) = \sum_k \sum_{GG'} U^*_{n,G} \widetilde{V}_{G-G'}(k) U_{mG'} \mathrm{e}^{\mathrm{i}kz}, \tag{12.84}$$

and the term becomes local. This suggest writing

$$V_{nm}(z, z') = V_{nm}(z) \Delta(z - z') + V^{(\mathrm{nl})}_{nm}(z, z'), \tag{12.85}$$

where

$$V^{(\mathrm{nl})}_{nm}(z, z') = \frac{1}{L} \sum_{kk'} \sum_{GG'} \left[U^*_{n,G+G_1} \mathrm{e}^{\mathrm{i}k_1 z} - U^*_{n,G} \mathrm{e}^{\mathrm{i}(k+k')z} \right] \widetilde{V}_{G-G'}(k) U_{mG'} \mathrm{e}^{-\mathrm{i}k'z'}. \tag{12.86}$$

In the above, only sum over k, k' values such that $k + k'$ lies outside the FBZ.

A useful property of $V^{(\mathrm{nl})}_{nm}$ is

$$\int_L \mathrm{d}z' \, V^{(\mathrm{nl})}_{nm}(z, z') = 0. \tag{12.87}$$

One can prove it using Eq. (12.58):

$$\int_L \mathrm{d}z' \, V^{(\mathrm{nl})}_{nm}(z, z') = \frac{1}{L} \int_L \mathrm{d}z' \sum_{kk'} \sum_{GG'} U^*_{n,G+G_1} \widetilde{V}_{G-G'}(k) U_{mG'} \mathrm{e}^{\mathrm{i}(k_1 z - k'z')}$$
$$= \sum_k \sum_{GG'} U^*_{n,G+G_1} \widetilde{V}_{G-G'}(k) U_{mG'} \mathrm{e}^{\mathrm{i}k_1 z}.$$

But $k + k' = k_1 + G$. If $k' = 0$, then $G_1 = 0$, $k_1 = k$, and

$$\int_L \mathrm{d}z'\, V_{nm}^{(\mathrm{nl})}(z, z') = \sum_k \sum_{GG'} U_{n,G}^* \widetilde{V}_{G-G'}(k) U_{mG'} \mathrm{e}^{\mathrm{i}kz} = V_{nm}(z),$$

from Eq. (12.84). Comparing the latter and Eq. (12.85), one gets

$$\int_L \mathrm{d}z'\, V_{nm}^{(\mathrm{nl})}(z, z') = 0.$$

One use of this equation is to show that for slowly-varying envelope functions, the nonlocal part does not contribute.

12.4.4.2 Asymptotic Value of $V_{nm}(z)$

We will now derive the asymptotic value of $V_{nm}(z)$ far from a single heterojunction. We have Eq. (12.84):

$$\begin{aligned} V_{nm}(z) &= \sum_k \sum_{GG'} U_{n,G}^* \widetilde{V}_{G-G'}(k) U_{mG'} \mathrm{e}^{\mathrm{i}kz} \\ &= \sum_{GG'} \sum_k U_{n,G}^* \int \frac{\mathrm{d}z'}{L} V(z') \mathrm{e}^{-\mathrm{i}(k+G-G')z'} U_{mG'} \mathrm{e}^{\mathrm{i}kz} \\ &= \sum_{GG'} U_{n,G}^* U_{mG'} \int \frac{\mathrm{d}z'}{L} V(z') \mathrm{e}^{-\mathrm{i}(G-G')z'} \sum_k \mathrm{e}^{\mathrm{i}k(z-z')} \\ &= \sum_{GG'} U_{n,G}^* U_{mG'} \int \mathrm{d}z'\, V(z') \mathrm{e}^{-\mathrm{i}(G-G')z'} \Delta(z - z'). \end{aligned} \qquad (12.88)$$

Now, assume that, very far away from the interface, the potential is periodic. Consider the potential on the RHS. Write

$$V^{(+)}(z') = \sum_G \widetilde{V}_G^{(+)} \mathrm{e}^{\mathrm{i}Gz'}. \qquad (12.89)$$

For large z, $\Delta(z - z')$ requires z' to be large too. Thus, in the integration, one can replace $V(z')$ by $V^{(+)}(z')$. Going back to Eq. (12.88),

$$\begin{aligned} V_{nm}(z) &\approx \sum_{GG'} U_{n,G}^* U_{mG'} \sum_{G''k} \widetilde{V}_{G''}^{(+)} \int \frac{\mathrm{d}z'}{L} \mathrm{e}^{-\mathrm{i}(k+G-G'-G'')z'} \mathrm{e}^{\mathrm{i}kz} \\ &= \sum_{GG'} U_{n,G}^* U_{mG'} \sum_{G''k} \widetilde{V}_{G''}^{(+)} \delta_{k+G-G'-G'',0} \mathrm{e}^{\mathrm{i}kz}. \end{aligned}$$

Since $k \in$ FBZ, then $k = 0$, $G'' = G - G'$ and

$$V_{nm}(z) \approx \sum_{GG'} U_{n,G}^* U_{mG'} \widetilde{V}_{G-G'}^{(+)} = \sum_{GG'} U_{n,G}^* \widetilde{V}_{G-G'}^{(+)} U_{mG'}$$
$$= \int \frac{dz'}{a} U_n^*(z') V^{(+)}(z') U_m(z'), \tag{12.90}$$

i.e., for large z, $V_{nm}(z)$ tends to a constant $V_{nm}^{(+)}$ which is the matrix element of the local periodic potential.

12.4.4.3 $V_{nm}(z)$ Near an Interface

We now wish to see if $V_{nm}(z)$ near an interface is different from the asymptotic expression. Again, in the previous subsection, we used a periodic form of $V(z)$ far from the interface. Here, we will consider an abrupt interface:

$$V(z) = \Theta^{(+)}(z) V^{(+)}(z) + \Theta^{(-)}(z) V^{(-)}(z), \tag{12.91}$$

where

$$V^{(\pm)}(z) = \sum_G V_G^{(\pm)} e^{iGz},$$

and $\Theta^{(+)}$ is 1 (0) at the RHS (LHS) of the interface; similarly, $\Theta^{(-)}$ is 0 (1) at the RHS (LHS) of the interface. Equation (12.84) for $V_{nm}(z)$ shows that $\widetilde{V}_G(k)$ is required. Fourier transforming the potential:

$$\widetilde{V}_G(k) = \int_{-L/2}^{L/2} \frac{dz}{L} V(z) e^{-i(k+G)z}$$
$$= \sum_{G'} \widetilde{V}_{G'}^{(+)} \int_0^{L/2} \frac{dz}{L} e^{-i(k+G-G')z} + \sum_{G'} \widetilde{V}_{G'}^{(-)} \int_{-L/2}^{0} \frac{dz}{L} e^{-i(k+G-G')z}. \tag{12.92}$$

If $k + G - G' \neq 0$,

$$\int_0^{L/2} \frac{dz}{L} V(z) e^{-i(k+G-G')z} = \frac{1 - e^{-i(k+G-G')L/2}}{i(k+G-G')}.$$

It will be assumed here that the second term in the numerator can be dropped. This is valid if, e.g., $L = 2na$ when $e^{-iGL/2} = e^{-i\frac{2\pi}{a}m2na} = 1$ and $e^{-ikL/2} = e^{-i\frac{2\pi}{L}m\frac{L}{2}} = e^{-i\pi m} = \pm 1$ but the summation over k leads to zero. Hence, for $k \neq 0$,

$$\widetilde{V}_G(k) = \frac{1}{iL} \sum_{G'} \frac{\widetilde{V}_{G'}^{(+)} - \widetilde{V}_{G'}^{(-)}}{k + G - G'},$$

while for $k = 0$,

$$\widetilde{V}_G(0) = \frac{1}{2} \left(\widetilde{V}_G^{(+)} + \widetilde{V}_G^{(-)} \right).$$

Therefore,

$$V_{nm}(z) = \sum_{GG'} U_{n,G}^* \frac{1}{2}\left(\widetilde{V}_{G-G'}^{(+)} + \widetilde{V}_{G-G'}^{(-)}\right) U_{mG'} + \sum_{\substack{GG'\\k\neq 0}} U_{n,G}^* \frac{1}{\mathrm{i}L} \sum_{G''} \frac{\widetilde{V}_{G''}^{(+)} - \widetilde{V}_{G''}^{(-)}}{k+G-G'-G''} U_{mG'}\mathrm{e}^{\mathrm{i}kz}. \tag{12.93}$$

The first term of Eq. (12.93) is simply

$$\frac{1}{2}\left(V_{nm}^{(+)}(z) + V_{nm}^{(-)}(z)\right).$$

For the second term, distinguish $G'' = G - G'$ and $G'' \neq G - G'$. Then

$$V_{nm}(z) = \frac{1}{2}\left[V_{nm}^{(+)}(z) + V_{nm}^{(-)}(z)\right] + \sum_{k\neq 0} \frac{V_{nm}^{(+)} - V_{nm}^{(-)}}{\mathrm{i}kL}\mathrm{e}^{\mathrm{i}kz}$$
$$+ \sum_{\substack{GG'\\k\neq 0}} U_{n,G}^* \left[\sum_{G''\neq G-G'} \frac{\widetilde{V}_{G''}^{(+)} - \widetilde{V}_{G''}^{(-)}}{\mathrm{i}(k+G-G'-G'')L}\right] U_{mG'}\mathrm{e}^{\mathrm{i}kz}. \tag{12.94}$$

One can write the second term as

$$\sum_{k\neq 0} \frac{V_{nm}^{(+)} - V_{nm}^{(-)}}{\mathrm{i}kL}\mathrm{e}^{\mathrm{i}kz} \equiv \sum_{k\neq 0}\sum_{G} \frac{V_{nm}^{(+)} - V_{nm}^{(-)}}{\mathrm{i}(k+G)L}\mathrm{e}^{\mathrm{i}(k+G)z} - \sum_{k\neq 0}\sum_{G\neq 0} \frac{V_{nm}^{(+)} - V_{nm}^{(-)}}{\mathrm{i}(k+G)L}\mathrm{e}^{\mathrm{i}(k+G)z}.$$

Also, given the plane-wave expansion of the step function,

$$\widetilde{\Theta}_G^{(+)}(k) = \int_{-L/2}^{L/2} \frac{\mathrm{d}z}{L}\Theta(z)\mathrm{e}^{-\mathrm{i}(k+G)z} = \int_0^{L/2} \frac{\mathrm{d}z}{L}\mathrm{e}^{-\mathrm{i}(k+G)z} = \frac{\mathrm{e}^{-\mathrm{i}(k+G)L/2} - 1}{-\mathrm{i}(k+G)L},$$
$$\Theta^{(+)}(z) = \sum_{k,G} \widetilde{\Theta}_G^{(+)}(k)\mathrm{e}^{\mathrm{i}(k+G)z},$$

Eq. (12.94) can be rewritten as

$$V_{nm}(z) = \left(V_{nm}^{(+)}\Theta^{(+)}(z) + V_{nm}^{(-)}\Theta^{(-)}(z)\right) - \sum_{k,G\neq 0} \frac{V_{nm}^{(+)} - V_{nm}^{(-)}}{\mathrm{i}(k+G)L}\mathrm{e}^{\mathrm{i}(k+G)z} \tag{12.95}$$
$$+ \sum_{\substack{GG'\\k\neq 0}} U_{n,G}^* \left[\sum_{G''\neq G-G'} \frac{\widetilde{V}_{G''}^{(+)} - \widetilde{V}_{G''}^{(-)}}{\mathrm{i}(k+G-G'-G'')L}\right] U_{mG'}\mathrm{e}^{\mathrm{i}kz}. \tag{12.96}$$

The first term in Eq. (12.96) is the potential term in conventional envelope-function theories. Hence, it is seen that the Burt theory leads to two correction terms to the potential function. A model calculation of the magnitude of such a contribution was given by Burt [153].

12.4.5 Conventional Results

The exact envelope-function equation, Eq. (12.60), will now be made to reproduce well-known theories.

12.4.5.1 Bulk Crystal

In a bulk crystal, the Fourier transform of the potential only has reciprocal lattice components:

$$\widetilde{V}_G(k) = \widetilde{V}_G \delta_{k,0}. \tag{12.97}$$

In Eq. (12.58),

$$V_{nm}(z, z') = \frac{1}{L} \sum_{kk'} \sum_{GG'} U^*_{n,G+G_1} \widetilde{V}_{G-G'}(k) U_{mG'} \mathrm{e}^{\mathrm{i}(k_1 z - k' z')},$$

we then have $k = 0$, which implies $k' = k_1 + G_1$, i.e., $G_1 = 0$. Thus,

$$\begin{aligned} V_{nm}(z, z') &= \frac{1}{L} \sum_{k'} \sum_{GG'} U^*_{n,G} \widetilde{V}_{G-G'} U_{mG'} \mathrm{e}^{\mathrm{i}k'(z-z')} = \sum_{GG'} U^*_{n,G} \widetilde{V}_{G-G'} U_{mG'} \Delta(z - z') \\ &\equiv \Delta(z - z') V^{\text{bulk}}_{nm}. \end{aligned} \tag{12.98}$$

Now Eq. (12.60) becomes

$$-\frac{\hbar^2}{2m_0} \frac{\mathrm{d}^2 F_n}{\mathrm{d}z^2} - \frac{\mathrm{i}\hbar}{m_0} \sum_m p_{nm} \frac{\mathrm{d}F_m}{\mathrm{d}z} + \sum_m H^{\text{bulk}}_{nm} F_m = E F_n, \tag{12.99}$$

where

$$H^{\text{bulk}}_{nm} = T_{nm} + V^{\text{bulk}}_{nm}. \tag{12.100}$$

We note that Eq. (12.99) is exact. Nevertheless, two more steps are needed to recast it into the $k \cdot p$ equation. First, we choose the U_n's to be zone-center solutions. This implies the Hamiltonian is diagonalized in this basis with the diagonal elements being the zone-center energies:

$$H^{\text{bulk}}_{nm} = E_n \delta_{nm}, \tag{12.101}$$

and

$$-\frac{\hbar^2}{2m_0} \frac{\mathrm{d}^2 F_n}{\mathrm{d}z^2} - \frac{\mathrm{i}\hbar}{m_0} \sum_m p_{nm} \frac{\mathrm{d}F_m}{\mathrm{d}z} + E_n F_n = E F_n. \tag{12.102}$$

Now, Eq. (12.102) is a differential equation with constant coefficients. It can be solved by using an exponential form:

$$F_n = A_n e^{ikz}. \tag{12.103}$$

Then, Eq. (12.102) leads to an algebraic equation for the A_n coefficients:

$$\sum_m \left[\left(\frac{\hbar^2 k^2}{2m_0} + E_n - E \right) \delta_{nm} + \frac{\hbar k}{m_0} p_{nm} \right] = 0, \tag{12.104}$$

the well-known $k \cdot p$ equation. Note that, with the choice of Eq. (12.103), the envelope-function expansion, Eq. (12.28) becomes

$$\psi = \sum_n F_n U_n = \sum_n A_n e^{ikz} u_{n0} \equiv \sum_n A_n \chi_{nk}, \tag{12.105}$$

which is the expansion of Luttinger and Kohn [6].

12.4.5.2 Shallow Impurities

We now consider the problem of shallow impurities. The potential energy consists of two terms:

$$V(z) = V^{\text{bulk}}(z) + V^{\text{imp}}(z). \tag{12.106}$$

For a shallow impurity, the definition is that the impurity potential is smooth, i.e., that it only has appreciable Fourier components close to $k = 0$. Thus, the plane-wave expansion of the potential can be written as

$$V_G(k) = \widetilde{V}_G^{\text{bulk}} \delta_{k,0} + \widetilde{V}^{\text{imp}}(k) \delta_{G,0}. \tag{12.107}$$

Substituting into Eq. (12.58),

$$\begin{aligned} V_{nm}(z,z') &= \frac{1}{L} \sum_{kk'} \sum_{GG'} U^*_{n,G+G_1} \widetilde{V}_{G-G'}(k) U_{mG'} e^{i(k_1 z - k' z')} \\ &\approx V_{nm}^{\text{bulk}} \Delta(z-z') + \frac{1}{L} \sum_{kk'} \sum_{GG'} U^*_{n,G+G_1} \widetilde{V}^{\text{imp}}(k) \delta_{G,G'} U_{mG'} e^{i(k_1 z - k' z')} \\ &= V_{nm}^{\text{bulk}} \Delta(z-z') + \frac{1}{L} \sum_{kk'} \sum_{G} U^*_{n,G+G_1} \widetilde{V}^{\text{imp}}(k) U_{mG} e^{i(k_1 z - k' z')} \\ &= \left[V_{nm}^{\text{bulk}} + \delta_{nm} \widetilde{V}^{\text{imp}}(0) \right] \Delta(z-z'), \end{aligned} \tag{12.108}$$

since, for the impurity potential, $k \approx 0$ leads to $k' = k_1 + G_1$ which implies $G_1 = 0$. Hence, the envelope-function equation can be written as (when using a basis of zone-center functions)

$$-\frac{\hbar^2}{2m_0}\frac{\mathrm{d}^2F_n}{\mathrm{d}z^2}-\frac{\mathrm{i}\hbar}{m_0}\sum_m p_{nm}\frac{\mathrm{d}F_m}{\mathrm{d}z}+V^{\mathrm{imp}}(z)F_n=(E-E_n)F_n. \tag{12.109}$$

This is not yet in conventional envelope-function form. To convert into the latter is similar to what is done in conventional theory: it is necessary to remove the interband terms. We will establish this for the one-band case. Assume that V^{imp} is weak so that the bound states are close to the zone-center eigenstates E_n. Here, we choose $n = s$. The envelope-function equation for the latter is

$$-\frac{\hbar^2}{2m_0}\frac{\mathrm{d}^2F_s}{\mathrm{d}z^2}-\frac{\mathrm{i}\hbar}{m_0}\sum_{m\neq s} p_{sm}\frac{\mathrm{d}F_m}{\mathrm{d}z}+V^{\mathrm{imp}}(z)F_s=(E-E_s)F_s. \tag{12.110}$$

The restriction on the summation in the second term is due to the vanishing of p_{ss} in one-dimension. For any other state $m \neq s$, we can write

$$(E-V^{\mathrm{imp}}(z)-E_m)F_m=-\frac{\hbar^2}{2m_0}\frac{\mathrm{d}^2F_m}{\mathrm{d}z^2}-\frac{\mathrm{i}\hbar}{m_0}\sum_{l\neq s} p_{ml}\frac{\mathrm{d}F_l}{\mathrm{d}z}.$$

To eliminate the interband term, we write down an approximate (smooth) solution for the states $m \neq s$ by dropping the second-differential term and approximating $(E-V^{\mathrm{imp}}(z)-E_m)\approx(E_s-E_m)$. Then

$$F_m\approx-\frac{\mathrm{i}\hbar}{m_0}\frac{p_{ms}}{E_s-E_m}\frac{\mathrm{d}F_s}{\mathrm{d}z}.$$

Plugging into Eq. (12.110) gives

$$-\frac{\hbar^2}{2m_0}\frac{\mathrm{d}^2F_s}{\mathrm{d}z^2}-\frac{\mathrm{i}\hbar}{m_0}\sum_{m\neq s} p_{sm}\left[-\frac{\mathrm{i}\hbar}{m_0}\frac{p_{ms}}{E_s-E_m}\frac{\mathrm{d}^2F_s}{\mathrm{d}z^2}\right]+V^{\mathrm{imp}}(z)F_s=(E-E_s)F_s,$$

or

$$-\frac{\hbar^2}{2m_s^*}\frac{\mathrm{d}^2F_s}{\mathrm{d}z^2}+V^{\mathrm{imp}}(z)F_s=(E-E_s)F_s, \tag{12.111}$$

where

$$\frac{1}{m_s^*}=\frac{1}{m_0}+2\left(\frac{1}{m_0}\right)^2\sum_{m\neq s}\frac{|p_{ms}|^2}{E_s-E_m}. \tag{12.112}$$

This is the conventional effective-mass equation.

12.4.5.3 Effective-Mass Equation for Multilayers

We now reproduce the conventional effective-mass equation for a multilayer. A number of approximations are needed. Assume

- the zone-center energies do not change much with respect to position.
- E is close to $E_s(z)$ everywhere. This is similar to the corresponding approximation for the shallow impurity problem.
- F_n is slowly varying.
- the potential change is abrupt (the abrupt-step approximation). Then

$$H_{nm} = T_{nm} + \sum_{GG'} U^*_{nG} \widetilde{V}_{G-G'} U_{mG'}.$$

- same U_n in all layers; i.e., we have only one zone-center function. Then H_{nm} is approximately diagonal throughout the structure and the diagonal elements are the appropriate zone-center energies.

Thus, we have

$$-\frac{\hbar^2}{2m_0}\frac{\mathrm{d}^2 F_n}{\mathrm{d}z^2} - \frac{\mathrm{i}\hbar}{m_0}\sum_m p_{nm}\frac{\mathrm{d}F_m}{\mathrm{d}z} = [E - E_n(z)]F_n. \tag{12.113}$$

The rest of the derivation is similar to the shallow-impurity problem. The equation for the one band is

$$-\frac{\hbar^2}{2m_0}\frac{\mathrm{d}^2 F_s}{\mathrm{d}z^2} - \frac{\mathrm{i}\hbar}{m_0}\sum_{m\neq s} p_{sm}\frac{\mathrm{d}F_m}{\mathrm{d}z} = [E - E_s(z)]F_s,$$

the remote envelopes are approximately

$$F_m \approx -\frac{\mathrm{i}\hbar}{m_0}\frac{p_{ms}}{E_s(z) - E_m(z)}\frac{\mathrm{d}F_s}{\mathrm{d}z},$$

giving

$$-\frac{\hbar^2}{2m_0}\frac{\mathrm{d}^2 F_s}{\mathrm{d}z^2} - \frac{\mathrm{i}\hbar}{m_0}\sum_{m\neq s}\left(-\frac{\mathrm{i}\hbar}{m_0}\right) p_{sm}\frac{\mathrm{d}}{\mathrm{d}z}\left(\frac{p_{ms}}{E_s(z) - E_m(z)}\frac{\mathrm{d}F_s}{\mathrm{d}z}\right) = [E - E_s(z)]F_s,$$

or

$$-\frac{\hbar^2}{2}\frac{\mathrm{d}}{\mathrm{d}z}\left(\frac{1}{m_s(z)}\frac{\mathrm{d}F_s}{\mathrm{d}z}\right) + E_s(z)F_s = EF_s, \tag{12.114}$$

where

$$\frac{1}{m_s(z)} = \frac{1}{m_0} + 2\frac{1}{m_0^2}\sum_{m\neq s}\frac{|p_{ms}|^2}{E_s(z) - E_m(z)}. \tag{12.115}$$

Equation (12.114) has the form of the Ben Daniel-Duke equation. This, therefore, represents a derivation of the latter operator ordering.

One surprising result is the use of Eq. (12.114) suggests the following boundary conditions:

$$F_s \text{ and } \frac{1}{m}\frac{\mathrm{d}F_s}{\mathrm{d}z} \text{ continuous.}$$

However, the exact envelope-function expansion requires

$$F_s \text{ and } \frac{\mathrm{d}F_s}{\mathrm{d}z} \text{ continuous.}$$

The difference is due to the fact that Eq. (12.114) is only approximate.

12.4.5.4 Multiband Model for Multilayers

The final consideration is to derive a multiband $k \cdot p$ theory for multilayers. We can start from the full multilayer equation, Eq. (12.113). Previously, the one-band approximation was made. Here we divide the bands into two subsets à la Löwdin: those close to s, labeled s' (subset S) and those distant labeled r (subset R). Then, rewriting Eq. (12.113), we have for the S states

$$-\frac{\hbar^2}{2m_0}\frac{\mathrm{d}^2 F_s}{\mathrm{d}z^2} - \frac{\mathrm{i}\hbar}{m_0}\sum_{s'} p_{ss'}\frac{\mathrm{d}F_{s'}}{\mathrm{d}z} - \frac{\mathrm{i}\hbar}{m_0}\sum_{r} p_{sr}\frac{\mathrm{d}F_r}{\mathrm{d}z} = [E - E_s(z)]F_s.$$

Eliminating the remote states using

$$F_r \approx -\frac{\mathrm{i}\hbar}{m_0}\sum_{s'}\frac{p_{rs'}}{E - E_r(z)}\frac{\mathrm{d}F_{s'}}{\mathrm{d}z}$$

gives

$$-\frac{\hbar^2}{2m_0}\frac{\mathrm{d}^2 F_s}{\mathrm{d}z^2} - \frac{\mathrm{i}\hbar}{m_0}\sum_{s'} p_{ss'}\frac{\mathrm{d}F_{s'}}{\mathrm{d}z} - \frac{\hbar^2}{m_0^2}\frac{\mathrm{d}}{\mathrm{d}z}\sum_{r}\frac{p_{sr}p_{rs'}}{E - E_r(z)}\frac{\mathrm{d}F_{s'}}{\mathrm{d}z} = [E - E_s(z)]F_s(z). \tag{12.116}$$

12.4.5.5 Multiband $k \cdot p$ Equations for Arbitrary Nanostructures

The exact envelope-function equation, Eq. (12.63), can be simplified to reproduce multiband Hamiltonians for arbitrary nanostructures. These can then be compared to those obtained starting from the bulk Luttinger–Kohn Hamiltonian. Due to the importance of the result, we repeat here the steps in the derivation even though they are similar to the one-dimensional case treated in an earlier section.

The starting point is the approximate, local form of the exact envelope-function equation:

$$-\frac{\hbar^2}{2m_0}\nabla^2 F_n(\mathbf{R}) - \frac{i\hbar}{m_0}\sum_m \mathbf{p}_{nm}\cdot\nabla F_m(\mathbf{R}) + \sum_m H_{nm}(\mathbf{R})F_m(\mathbf{R}) = EF_n(\mathbf{R}). \quad (12.117)$$

In deriving a multiband form of the equations, the conventional technique is to define two sets of states à la Löwdin. The states of immediate interest (so-called dominant states, e.g., the Γ_{15} valence-band complex for zincblende semiconductors), will be denoted by the label $n = s$ (S states) and the remote bands (R states) will be denoted by the label $n = r$. One can now write down separate envelope-function equations, Eq. (12.117), for the two groups:

$$-\frac{\hbar^2}{2m_0}\nabla^2 F_r - \frac{i\hbar}{m_0}\sum_m \mathbf{p}_{rm}\cdot\nabla F_m + \sum_m H_{rm}F_m = EF_r, \quad (12.118)$$

$$-\frac{\hbar^2}{2m_0}\nabla^2 F_s - \frac{i\hbar}{m_0}\sum_m \mathbf{p}_{sm}\cdot\nabla F_m + \sum_m H_{sm}F_m = EF_s. \quad (12.119)$$

Equation (12.118) for the remote states can be simplified by taking into account the smoothness of the envelope function and, thereby, throwing away the curvature term. This leads to

$$-\frac{i\hbar}{m_0}\sum_m \mathbf{p}_{rm}\cdot\nabla F_m + \sum_{r'} H_{rr'}F_{r'} + \sum_{s'} H_{rs'}F_{s'} \approx EF_r,$$

$$F_r \approx (E - H_{rr})^{-1}\sum_{s'}\left[-\frac{i\hbar}{m_0}\mathbf{p}_{rs'}\cdot\nabla F_{s'} + H_{rs'}F_{s'}\right]. \quad (12.120)$$

Equation (12.119) is now written to show the explicit dependence on the R envelope functions:

$$-\frac{\hbar^2}{2m_0}\nabla^2 F_s - \frac{i\hbar}{m_0}\left[\sum_{s'} \mathbf{p}_{ss'}\cdot\nabla F_{s'} + \sum_r \mathbf{p}_{sr}\cdot\nabla F_r\right] + \sum_{s'} H_{ss'}F_{s'} + \sum_r H_{sr}F_r = EF_s,$$

which are then eliminated:

$$\begin{aligned}
&-\frac{\hbar^2}{2m_0}\nabla^2 F_s - \frac{i\hbar}{m_0}\sum_{s'} \mathbf{p}_{ss'}\cdot\nabla F_{s'} + \sum_{s'} H_{ss'}F_{s'} \\
&-\frac{i\hbar}{m_0}\sum_r \mathbf{p}_{sr}\cdot\nabla\left\{(E - H_{rr})^{-1}\sum_{s'}\left[-\frac{i\hbar}{m_0}\mathbf{p}_{rs'}\cdot\nabla F_{s'} + H_{rs'}F_{s'}\right]\right\} \\
&+\sum_r \frac{H_{sr}}{(E - H_{rr})}\sum_{s'}\left[-\frac{i\hbar}{m_0}\mathbf{p}_{rs'}\cdot\nabla F_{s'} + H_{rs'}F_{s'}\right] = EF_s.
\end{aligned}$$

Simplifying the latter gives

$$\begin{aligned}
&-\frac{\hbar^2}{2m_0}\nabla^2 F_s - \frac{\hbar^2}{m_0^2}\sum_{rs'} \mathbf{p}_{sr}\cdot\nabla\frac{1}{(E-H_{rr})}\mathbf{p}_{rs'}\cdot\nabla F_{s'} - \frac{\mathrm{i}\hbar}{m_0}\sum_{s'}\mathbf{p}_{ss'}\cdot\nabla F_{s'}\\
&+\sum_{s'} H_{ss'}F_{s'} + \sum_{rs'}\frac{H_{sr}H_{rs'}}{(E-H_{rr})}F_{s'} - \frac{\mathrm{i}\hbar}{m_0}\sum_{r}\mathbf{p}_{sr}\cdot\nabla\left[(E-H_{rr})^{-1}H_{rs'}F_{s'}\right]\\
&-\frac{\mathrm{i}\hbar}{m_0}\sum_{s'r}(E-H_{rr})^{-1}H_{sr}\mathbf{p}_{rs'}\cdot\nabla F_{s'} = EF_s.
\end{aligned} \tag{12.121}$$

This is the multiband envelope-function equation as given by Burt [185].

12.4.6 Boundary Conditions

We have already derived boundary conditions in the context of the effective-mass equation. Here, we present a more general derivation based upon the local form of the envelope-function equation in three-dimensional, Eq. (12.117). Consider a planar interface with translational symmetry. Let

$$F_n(\mathbf{R}) = \frac{1}{\sqrt{S}}\mathrm{e}^{\mathrm{i}\mathbf{k}_{\|}\cdot\mathbf{r}}F_n(z). \tag{12.122}$$

Then the equation for $F_n(z)$ is

$$\begin{aligned}
&-\frac{\hbar^2}{2m_0}\frac{\mathrm{d}^2F_n}{\mathrm{d}z^2} - \frac{\mathrm{i}\hbar}{m_0}\sum_{n'}p^z_{nn'}\frac{\mathrm{d}F_{n'}}{\mathrm{d}z} + \frac{\hbar^2k_{\|}^2}{2m_0}F_n + \frac{\hbar}{m_0}\sum_{n'}\mathbf{k}_{\|}\cdot\mathbf{p}_{nn'}F_{n'} + \sum_{n'}H_{nn'}F_{n'}\\
&= EF_n.
\end{aligned} \tag{12.123}$$

In general, Eq. (12.117) is only valid far from the interfaces. However, we have already argued that the interfacial terms are small. If they are dropped, Eq. (12.117) is (approximately) valid everywhere and can be integrated across the interface.

First, we know that F_n is continuous, otherwise F_n'' blows up. Integrating across an interface (say at $z = 0$),

$$\begin{aligned}
&-\frac{\hbar^2}{2m_0}\int_{-\varepsilon}^{\varepsilon}\mathrm{d}z\,\frac{\mathrm{d}^2F_n}{\mathrm{d}z^2} - \frac{\mathrm{i}\hbar}{m_0}\sum_{n'}p^z_{nn'}\int_{-\varepsilon}^{\varepsilon}\mathrm{d}z\,\frac{\mathrm{d}F_{n'}}{\mathrm{d}z} + \frac{\hbar^2k_{\|}^2}{2m_0}\int_{-\varepsilon}^{\varepsilon}\mathrm{d}z\,F_n\\
&+\frac{\hbar}{m_0}\sum_{n'}\mathbf{k}_{\|}\cdot\mathbf{p}_{nn'}\int_{-\varepsilon}^{\varepsilon}\mathrm{d}z\,F_{n'} + \sum_{n'}H_{nn'}\int_{-\varepsilon}^{\varepsilon}\mathrm{d}z\,F_{n'} = E\int_{-\varepsilon}^{\varepsilon}\mathrm{d}z\,F_n.
\end{aligned}$$

Note that $\mathbf{p}_{nn'}$ and $H_{nn'}$ are continuous. Then

$$-\frac{\hbar^2}{2m_0}\left[\frac{\mathrm{d}F_n}{\mathrm{d}z}\right]_{-\varepsilon}^{\varepsilon} - \frac{\mathrm{i}\hbar}{m_0}\sum_{n'}p^z_{nn'}\left[F_{n'}\right]_{-\varepsilon}^{\varepsilon} = 0,$$

or

$$-i\hbar\frac{\mathrm{d}F_n}{\mathrm{d}z} + 2\sum_{n'} \mathbf{p}^z_{nn'} F_{n'} \tag{12.124}$$

is continuous; i.e., F'_n is. So far, we have made the abrupt-step approximation and used a local $H_{nn'}$. If one further makes the effective-mass approximation, then one requires $\frac{1}{m}\frac{\mathrm{d}F_n}{\mathrm{d}z}$ continuous.

12.4.7 Burt–Foreman Hamiltonian

Foreman [34], in 1993, was the first to derive an explicit form of the valence-band Hamiltonian on the basis of Burt's theory. In 1997, he extended it to an eight-band model [228].

In his 1992 review article, Burt had already indicated how to simplify Eq. (12.121). All the terms that are only significant near an interface will be dropped. Considering the terms on the LHS of Eq. (12.121), it turns out such terms are the last two of the seven. This is obvious for the sixth term since it involves the derivative of H_{rs} and the latter is a constant in a bulk layer. The last term would correspond, in the bulk, to a term linear-in-k; such terms are absent in the bulk valence-band Hamiltonian. Thus, they can only arise from the interfaces and are also dropped. Finally, the third term involves momentum matrix elements within the subspace; they are known to be zero from symmetry arguments. An additional simplification is to recognize the two terms on the second line to correspond to the energy correct to second order in perturbation theory; we can denote the latter as $H^{(2)}_{ss'}$.

What remains looks very much like the second-order $k \cdot p$ term since the differential operator transforms just like the wave vector:

$$-\frac{\hbar^2}{2m_0}\nabla^2 F_s - \frac{\hbar^2}{m_0^2}\sum_{rs'} \mathbf{p}_{sr} \cdot \nabla \frac{1}{(E - H_{rr})} \mathbf{p}_{rs'} \cdot \nabla F_{s'} + \sum_{s'} H^{(2)}_{ss'} F_{s'} = E F_s. \tag{12.125}$$

The main difference lies in the noncommutativity of the differential operator and the band parameters since the latter are position dependent in a heterostructure. Thus, one can expect additional terms in the BF Hamiltonian. We now obtain the latter explicitly. We will first do so for the zincblende structure without spin-orbit coupling. We will refer extensively to the derivations of the DKK and Kane parameters in earlier chapters.

12.4.7.1 H_{SS}

The electron mass arises from the interaction of the conduction S state with remote bands. We treated this both within a one-band and the Kane models. The largest contribution stems from the P interaction with the top of the valence band Γ_{15v} states. In the bulk, the latter was written as

$$A'k^2.$$

Now we have

$$H_{SS} = \frac{\hbar^2}{m_0^2}\sum_{l\in\Gamma_{15}}\sum_{i=x,y,z}\widehat{k}_i\frac{\langle S|p_\alpha|l\alpha\rangle\langle l\alpha|p_\alpha|S\rangle}{E-H_l(\mathbf{r})}\widehat{k}_i$$
$$= \widehat{k}_x A'\widehat{k}_x + \widehat{k}_y A'\widehat{k}_y + \widehat{k}_z A'\widehat{k}_z, \qquad (12.126)$$

where $\widehat{k}_i = -\mathrm{i}\nabla_i$.

12.4.7.2 H_{XX}

Again, we give the bulk result as a reminder:

$$L'k_x^2 + M(k_y^2 + k_z^2).$$

Now,

$$H_{XX}^{\Gamma_1} = \widehat{k}_x\frac{\hbar^2}{m_0^2}\sum_{l\in\Gamma_1}\frac{\langle X|p_x|lS\rangle\langle lS|p_x|X\rangle}{E-H_l(\mathbf{r})}\widehat{k}_x = \widehat{k}_x F'\widehat{k}_x,$$

$$H_{XX}^{\Gamma_{12}} = \widehat{k}_x\frac{\hbar^2}{m_0^2}\sum_{l\in\Gamma_{12}}\frac{\langle X|p_x|l\rangle\langle l|p_x|X\rangle}{E-H_l(\mathbf{r})}\widehat{k}_x = \widehat{k}_x 2G\widehat{k}_x,$$

$$H_{XX}^{\Gamma_{15}} = \widehat{k}_y\frac{\hbar^2}{m_0^2}\sum_{l\in\Gamma_{15}}\frac{\langle X|p_y|lZ\rangle\langle lZ|p_y|X\rangle}{E-H_l(\mathbf{r})}\widehat{k}_y + \widehat{k}_z\frac{\hbar^2}{m_0^2}\sum_{l\in\Gamma_{15}}\frac{\langle X|p_z|lY\rangle\langle lY|p_z|X\rangle}{E-H_l(\mathbf{r})}\widehat{k}_z$$
$$= \widehat{k}_y H_1\widehat{k}_y + \widehat{k}_z H_1\widehat{k}_z,$$

$$H_{XX}^{\Gamma_{25}} = \widehat{k}_y\frac{\hbar^2}{m_0^2}\sum_{l\in\Gamma_{25}}\frac{\langle X|p_y|lxy\rangle\langle lxy|p_y|X\rangle}{E-H_l(\mathbf{r})}\widehat{k}_y + \widehat{k}_z\frac{\hbar^2}{m_0^2}\sum_{l\in\Gamma_{25}}\frac{\langle X|p_z|lzx\rangle\langle lzx|p_z|X\rangle}{E-H_l(\mathbf{r})}\widehat{k}_z$$
$$= \widehat{k}_y H_2\widehat{k}_y + \widehat{k}_z H_2\widehat{k}_z,$$

giving

$$H_{XX} = \widehat{k}_x(F' + 2G)\widehat{k}_x + \widehat{k}_y(H_1 + H_2)\widehat{k}_y + \widehat{k}_z(H_1 + H_2)\widehat{k}_z. \qquad (12.127)$$

12.4.7.3 H_{XY}

The bulk result is:

$$Nk_xk_y,$$

where

$$N = F - G + H_1 - H_2.$$

Now,

$$H_{XY}^{\Gamma_1} = \widehat{k}_x \frac{\hbar^2}{m_0^2} \sum_{l \in \Gamma_1} \frac{\langle X|p_x|lS\rangle\langle lS|p_y|Y\rangle}{E - H_l(\mathbf{r})} \widehat{k}_y = \widehat{k}_x F' \widehat{k}_y,$$

$$H_{XY}^{\Gamma_{12}} = \widehat{k}_x \frac{\hbar^2}{m_0^2} \sum_{l\gamma \in \Gamma_{12}} \frac{\langle X|p_x|l\gamma\rangle\langle l\gamma|p_y|Y\rangle}{E - H_l(\mathbf{r})} \widehat{k}_y = -\widehat{k}_x G \widehat{k}_y,$$

$$H_{XY}^{\Gamma_{15}} = \widehat{k}_y \frac{\hbar^2}{m_0^2} \sum_{l \in \Gamma_{15}} \frac{\langle X|p_y|l\delta_1\rangle\langle l\delta_1|p_x|Y\rangle}{E - H_l(\mathbf{r})} \widehat{k}_x = \widehat{k}_y H_1 \widehat{k}_x,$$

$$H_{XY}^{\Gamma_{25}} = \widehat{k}_y \frac{\hbar^2}{m_0^2} \sum_{l \in \Gamma_{25}} \frac{\langle X|p_y|l\epsilon_3\rangle\langle l\epsilon_3|p_x|Y\rangle}{E - H_l(\mathbf{r})} \widehat{k}_x = -\widehat{k}_y H_2 \widehat{k}_x,$$

giving

$$H_{XY} = \widehat{k}_x (F' - G) \widehat{k}_y + \widehat{k}_y (H_1 - H_2) \widehat{k}_x. \tag{12.128}$$

12.4.7.4 H_{SX}

The linear-in-k term is unchanged. For the quadratic term, in the bulk result,

$$B k_y k_z,$$

where

$$B = \frac{2\hbar^2}{m_0^2} \sum_{l \in \Gamma_{15}}{}' \frac{\langle s|p_x|u_l\rangle\langle u_l|p_y|z\rangle}{\frac{E_0}{2} - E_l}.$$

Now,

$$\begin{aligned} H_{SX}^{\Gamma_{15}} &= \widehat{k}_y \frac{\hbar^2}{m_0^2} \sum_{l \in \Gamma_{15}} \frac{\langle S|p_y|lY\rangle\langle lY|p_z|X\rangle}{E - H_l(\mathbf{r})} \widehat{k}_z + \widehat{k}_z \frac{\hbar^2}{m_0^2} \sum_{l \in \Gamma_{15}} \frac{\langle S|p_z|lZ\rangle\langle lZ|p_y|X\rangle}{E - H_l(\mathbf{r})} \widehat{k}_y \\ &= \widehat{k}_z \frac{B}{2} \widehat{k}_y + \widehat{k}_y \frac{B}{2} \widehat{k}_z. \end{aligned} \tag{12.129}$$

The resulting Hamiltonian is given in Table 12.2. The three-band version (moving the S state into the remote class) is easily obtained by removing the S state and changing the primed DKK parameters to unprimed. This Hamiltonian was not explicitly written down by Foreman; an explicit form was first given by Stravinou and van Dalen [232] and we give their form of it in Table 12.3.

Table 12.2 Four-band BF Hamiltonian in LS basis for zincblende

$$
H_{\mathrm{BF}}(\mathbf{k}) = \begin{pmatrix}
|S\rangle & |X\rangle & |Y\rangle & |Z\rangle \\
\widehat{k}_x A' \widehat{k}_x + \widehat{k}_y A' \widehat{k}_y + \widehat{k}_z A' \widehat{k}_z & \begin{array}{c} \mathrm{i}\frac{\hbar}{m_0}\widehat{k}_x P \\ \widehat{k}_y \frac{B}{2}\widehat{k}_z + \widehat{k}_z \frac{B}{2}\widehat{k}_y \end{array} & \begin{array}{c} \mathrm{i}\frac{\hbar}{m_0}\widehat{k}_y P \\ \widehat{k}_z \frac{B}{2}\widehat{k}_x + \widehat{k}_x \frac{B}{2}\widehat{k}_z \end{array} & \begin{array}{c} \mathrm{i}\frac{\hbar}{m_0}\widehat{k}_z P \\ \widehat{k}_x \frac{B}{2}\widehat{k}_y + \widehat{k}_x \frac{B}{2}\widehat{k}_y \end{array} \\
 & \begin{array}{c} \widehat{k}_x (F' + 2G)\widehat{k}_x \\ +\widehat{k}_y (H_1 + H_2)\widehat{k}_y + \widehat{k}_z (H_1 + H_2)\widehat{k}_z \end{array} & \widehat{k}_x (F' - G)\widehat{k}_y + \widehat{k}_y (H_1 - H_2)\widehat{k}_x & \widehat{k}_x (F' - G)\widehat{k}_z + \widehat{k}_z (H_1 - H_2)\widehat{k}_x \\
 & & \begin{array}{c} \widehat{k}_y (F' + 2G)\widehat{k}_y \\ +\widehat{k}_x (H_1 + H_2)\widehat{k}_x + \widehat{k}_z (H_1 + H_2)\widehat{k}_z \end{array} & \widehat{k}_y (F' - G)\widehat{k}_z + \widehat{k}_z (H_1 - H_2)\widehat{k}_y \\
 & \dagger & & \begin{array}{c} \widehat{k}_z (F' + 2G)\widehat{k}_z \\ +\widehat{k}_x (H_1 + H_2)\widehat{k}_x + \widehat{k}_y (H_1 + H_2)\widehat{k}_y \end{array}
\end{pmatrix}
$$

Table 12.3 Three-band Stravinou-van Dalen Hamiltonian for zincblende. This includes the free-electron term

$$H_{SV}(\mathbf{k}) = \frac{\hbar^2}{2m_0}\begin{bmatrix} |X\rangle & |Y\rangle & |Z\rangle \\ \widehat{k}_x A\widehat{k}_x + \widehat{k}_y B\widehat{k}_y + \widehat{k}_z B\widehat{k}_z & \widehat{k}_x C_1\widehat{k}_y - \widehat{k}_y C_2\widehat{k}_x & \widehat{k}_x C_1\widehat{k}_z - \widehat{k}_z C_2\widehat{k}_x \\ \widehat{k}_y C_1\widehat{k}_x - \widehat{k}_x C_2\widehat{k}_y & \widehat{k}_y A\widehat{k}_y + \widehat{k}_x B\widehat{k}_x + \widehat{k}_z B\widehat{k}_z & \widehat{k}_y C_1\widehat{k}_z - \widehat{k}_z C_2\widehat{k}_y \\ \widehat{k}_z C_1\widehat{k}_x - \widehat{k}_x C_2\widehat{k}_z & \widehat{k}_z C_1\widehat{k}_y - \widehat{k}_y C_2\widehat{k}_z & \widehat{k}_z A\widehat{k}_z + \widehat{k}_x B\widehat{k}_x + \widehat{k}_y B\widehat{k}_y \end{bmatrix},$$

with

$$A = \frac{2m_0}{\hbar^2}(F + 2G) + 1,$$

$$B = \frac{2m_0}{\hbar^2}(H_1 + H_2) + 1,$$

$$C_1 = \frac{2m_0}{\hbar^2}(F - G),$$

$$C_2 = -\frac{2m_0}{\hbar^2}(H_1 - H_2).$$

12.4.7.5 Foreman Parameters

Foreman introduced a new notation for the symmetry expansion of the Luttinger parameters. He introduced dimensionless

$$\begin{aligned} \sigma &= -\frac{1}{3m_0}\sum_{l\in\Gamma_1}\frac{|\langle X|p_x|lS\rangle|^2}{E - E_l} = -\frac{1}{3}\frac{m_0}{\hbar^2}F, \\ \pi &= -\frac{1}{3m_0}\sum_{l\in\Gamma_{15}}\frac{|\langle X|p_y|l\delta_1\rangle|^2}{E - E_l} = -\frac{1}{3}\frac{m_0}{\hbar^2}H_1, \\ \delta &= -\frac{1}{3m_0}\sum_{l\in\Gamma_{12}}\frac{|\langle X|p_x|l\gamma_1\rangle|^2}{E - E_l} = -\frac{1}{3}\frac{m_0}{\hbar^2}G. \end{aligned} \tag{12.130}$$

In terms of the latter, the Luttinger parameters become [making use of Eqs. (5.77) and (5.78) and neglecting H_2 since the Γ_{25} states are distant],

$$\begin{aligned} \gamma_1 &= -\frac{2m_0}{3\hbar^2}(L + 2M) - 1 \approx -1 - \frac{2m_0}{3\hbar^2}(F + 2G + 2H_1) = -1 + 2\sigma + 4\pi + 4\delta, \\ \gamma_2 &= -\frac{2m_0}{6\hbar^2}(L - M) \approx -\frac{2m_0}{6\hbar^2}(F + 2G - H_1) = \sigma - \pi + 2\delta, \\ \gamma_3 &= -\frac{2m_0}{6\hbar^2}N \approx -\frac{2m_0}{6\hbar^2}(F - G + H_1) = \sigma + \pi - \delta. \end{aligned}$$

Writing the latter as a matrix equation easily leads to the inverted equations:

$$\begin{pmatrix} \sigma \\ \pi \\ \delta \end{pmatrix} = \frac{1}{9} \begin{pmatrix} -1 & 8 & 12 \\ 3 & -6 & 0 \\ 2 & 2 & -6 \end{pmatrix} \begin{pmatrix} \gamma_1 + 1 \\ \gamma_2 \\ \gamma_3 \end{pmatrix}.$$

Writing

$$\overline{\gamma} = \frac{1}{2}(\gamma_3 + \gamma_2), \mu = \frac{1}{2}(\gamma_3 - \gamma_2), \tag{12.131}$$

we have

$$\begin{aligned} \delta &= \frac{1}{9}(1 + \gamma_1 + \gamma_2 - 3\gamma_3), \\ \pi &= \frac{3}{2}\delta + \mu, \\ \sigma &= -\frac{1}{2}\delta + \overline{\gamma}. \end{aligned} \tag{12.132}$$

For completeness, we list the connection between Foreman's parameters and those of Stravinou and van Dalen, and the Luttinger parameters:

$$\begin{aligned} A &\equiv \frac{2m_0}{\hbar^2}(F + 2G) + 1 = 1 - 6\sigma - 12\delta = -(\gamma_1 + 4\gamma_2), \\ B &\equiv \frac{2m_0}{\hbar^2}(H_1 + H_2) + 1 = 1 - 6\pi = -(\gamma_1 - 2\gamma_2), \\ C_1 &\equiv \frac{2m_0}{\hbar^2}(F - G) - 6\delta - 6\sigma = 1 + \gamma_1 - 2\gamma_2 - 6\gamma_3, \\ C_2 &\equiv -\frac{2m_0}{\hbar^2}(H_1 - H_2) = 6\pi = 1 + \gamma_1 - 2\gamma_2. \end{aligned} \tag{12.133}$$

12.4.7.6 Spin

We now consider what happens when spin is included. We first derive the six-band Hamiltonian corresponding to the bulk LK one. The procedure is similar to adding spin to the DKK matrix. However, we have to repeat the derivation since there are terms that might not commute nor reduce to zero. At this point, we will leave the Hamiltonian as general as possible so that it is applicable to arbitrary nanostructures; we will afterwards consider the simplications resulting for a QW. To simplify the notation, we will not write out explicitly the free-electron term $\hbar^2 k^2/(2m_0)$. Thus, it is implied to be added to the Hamiltonian when the latter is given in terms of the DKK parameters. However, when the latter are converted to the Luttinger ones, the presence of the free-electron term is explicit. We also note that the overall structure of the Hamiltonian matrix is actually determined by the Clebsch-Gordan coefficients. Hence, it is possible to write down a general form of the matrix that is the same independent of the underlying $k \cdot p$ model.

Refering to Sect. 3.3, we have starting with the diagonal elements

$$
\begin{aligned}
P' \equiv \left\langle \frac{3}{2}\ \frac{3}{2} \right| H_{k\cdot p} \left| \frac{3}{2}\ \frac{3}{2} \right\rangle &= \frac{1}{2}\langle (x+\mathrm{i}y)\uparrow |H_{k\cdot p}| (x+\mathrm{i}y)\uparrow \rangle \\
&= \frac{1}{2}(H_{11}+H_{22}) + \frac{i}{2}(H_{12}-H_{21}) \\
&= \frac{1}{2}\left\{\widehat{k}_x(L+M)\widehat{k}_x + \widehat{k}_y(L+M)\widehat{k}_y + \widehat{k}_z 2M\widehat{k}_z\right\} \\
&\quad + \frac{\mathrm{i}}{2}\left\{\widehat{k}_x(F-G-H_1+H_2)\widehat{k}_y - \widehat{k}_y(F-G-H_1+H_2)\widehat{k}_x\right\}, \quad (12.134)
\end{aligned}
$$

$$
\begin{aligned}
P'' \equiv \left\langle \frac{3}{2}\ \frac{1}{2} \right| H_{k\cdot p} \left| \frac{3}{2}\ \frac{1}{2} \right\rangle &= \frac{1}{6}(H_{11}+H_{22}) + \frac{2}{3}H_{33} + \frac{\mathrm{i}}{6}(H_{12}-H_{21}) \\
&= \frac{1}{6}\left\{\widehat{k}_x(L+5M)\widehat{k}_x + \widehat{k}_y(L+5M)\widehat{k}_y + 2\widehat{k}_z(2L+M)\widehat{k}_z\right\} \\
&\quad + \frac{\mathrm{i}}{6}\left\{\widehat{k}_x(F-G-H_1+H_2)\widehat{k}_y - \widehat{k}_y(F-G-H_1+H_2)\widehat{k}_x\right\}, \quad (12.135)
\end{aligned}
$$

$$
\begin{aligned}
P''' \equiv \left\langle \frac{1}{2}\ \frac{1}{2} \right| H_{k\cdot p} \left| \frac{1}{2}\ \frac{1}{2} \right\rangle &= \frac{1}{3}(H_{11}+H_{22}+H_{33}) + \frac{\mathrm{i}}{3}(H_{12}-H_{21}) \\
&= \frac{1}{3}\left\{\widehat{k}_x(L+2M)\widehat{k}_x + \widehat{k}_y(L+2M)\widehat{k}_y + \widehat{k}_z(L+2M)\widehat{k}_z\right\} \\
&\quad + \frac{\mathrm{i}}{3}\left\{\widehat{k}_x(F-G-H_1+H_2)\widehat{k}_y - \widehat{k}_y(F-G-H_1+H_2)\widehat{k}_x\right\}. \quad (12.136)
\end{aligned}
$$

Now for the distinct off-diagonal matrix elements (writing $\widehat{k}_- = \widehat{k}_x - \mathrm{i}\widehat{k}_y$),

$$
S_- \equiv \left\langle \frac{3}{2}\ \frac{3}{2} \right| H_{k\cdot p} \left| \frac{3}{2}\ \frac{1}{2} \right\rangle = -\frac{1}{\sqrt{3}}(H_{13} - \mathrm{i}H_{23}) \quad (12.137)
$$

$$
\begin{aligned}
&= -\frac{1}{\sqrt{3}}\left\{\widehat{k}_x(F-G)\widehat{k}_z + \widehat{k}_z(H_1-H_2)\widehat{k}_x - \mathrm{i}[\widehat{k}_y(F-G)\widehat{k}_z + \widehat{k}_z(H_1-H_2)\widehat{k}_y]\right\} \\
&= -\frac{1}{\sqrt{3}}\left\{\widehat{k}_-(F-G)\widehat{k}_z + \widehat{k}_z(H_1-H_2)\widehat{k}_-\right\}, \quad (12.138)
\end{aligned}
$$

$$
\begin{aligned}
-R \equiv \left\langle \frac{3}{2}\ \frac{3}{2} \right| H_{k\cdot p} \left| \frac{3}{2}\ -\frac{1}{2} \right\rangle &= -\frac{1}{2\sqrt{3}}(H_{11} - H_{22} - \mathrm{i}H_{12} - \mathrm{i}H_{21}) \\
&= -\frac{1}{2\sqrt{3}}\left\{\widehat{k}_x(L-M)\widehat{k}_x - \widehat{k}_y(L-M)\widehat{k}_y - \mathrm{i}[\widehat{k}_x N\widehat{k}_y + \widehat{k}_y N\widehat{k}_x]\right\}, \quad (12.139)
\end{aligned}
$$

$$
\begin{aligned}
-C \equiv \left\langle \frac{3}{2}\ \frac{1}{2} \right| H_{k\cdot p} \left| \frac{3}{2}\ -\frac{1}{2} \right\rangle &= \frac{1}{3}(-H_{13} + H_{31} + \mathrm{i}H_{23} - \mathrm{i}H_{32}) \\
&= \frac{1}{3}\left\{\widehat{k}_z(F-G-H_1+H_2)\widehat{k}_- - \widehat{k}_-(F-G-H_1+H_2)\widehat{k}_z\right\}, \quad (12.140)
\end{aligned}
$$

$$-\sqrt{2}Q \equiv \left\langle \frac{3}{2}\ \frac{1}{2}\right| H_{k\cdot p} \left|\frac{1}{2}\ \frac{1}{2}\right\rangle = \frac{1}{3\sqrt{2}}(H_{11}+H_{22}+\mathrm{i}H_{12}-\mathrm{i}H_{21}) - \frac{\sqrt{2}}{3}H_{33}$$
$$= \frac{1}{3\sqrt{2}}\left\{\widehat{k}_x(L-M)\widehat{k}_x + \widehat{k}_y(L-M)\widehat{k}_y - 2\widehat{k}_z(L-M)\widehat{k}_z\right.$$
$$\left. + \mathrm{i}[\widehat{k}_x(F-G-H_1+H_2)\widehat{k}_y - \widehat{k}_y(F-G-H_1+H_2)\widehat{k}_x]\right\}, \quad (12.141)$$

$$-\sqrt{\frac{3}{2}}\Sigma_- \equiv \left\langle \frac{3}{2}\ \frac{1}{2}\right| H_{k\cdot p} \left|\frac{1}{2}\ -\frac{1}{2}\right\rangle = \frac{1}{3\sqrt{2}}(H_{13}+2H_{31}-\mathrm{i}H_{23}-2\mathrm{i}H_{32})$$
$$= \frac{1}{3\sqrt{2}}\left\{\widehat{k}_-(F-G+2H_1-2H_2)\widehat{k}_z + \widehat{k}_z(2F-2G+H_1-H_2)\widehat{k}_-\right\}. \quad (12.142)$$

Further,

$$\left\langle \frac{3}{2}\ -\frac{1}{2}\right| H_{k\cdot p} \left|\frac{3}{2}\ -\frac{3}{2}\right\rangle = -\frac{1}{\sqrt{3}}(H_{31}-\mathrm{i}H_{32})$$
$$= -\frac{1}{\sqrt{3}}\left\{\widehat{k}_z(F-G)\widehat{k}_x + \widehat{k}_x(H_1-H_2)\widehat{k}_z - \mathrm{i}[\widehat{k}_z(F-G)\widehat{k}_y + \widehat{k}_y(H_1-H_2)\widehat{k}_z]\right\}$$
$$= -\frac{1}{\sqrt{3}}\left\{\widehat{k}_z(F-G)\widehat{k}_- + \widehat{k}_-(H_1-H_2)\widehat{k}_z\right\}.$$

The latter appears to involve the hermitian of S. To verify, we compute instead the transpose matrix element:

$$\left\langle \frac{3}{2}\ -\frac{3}{2}\right| H_{k\cdot p} \left|\frac{3}{2}\ -\frac{1}{2}\right\rangle = -\frac{1}{\sqrt{3}}(H_{13}+\mathrm{i}H_{23}) \equiv S_+. \quad (12.143)$$

Thus

$$\left\langle \frac{3}{2}\ -\frac{1}{2}\right| H_{k\cdot p} \left|\frac{3}{2}\ -\frac{3}{2}\right\rangle = \left\langle \frac{3}{2}\ -\frac{3}{2}\right| H_{k\cdot p} \left|\frac{3}{2}\ -\frac{1}{2}\right\rangle^{\dagger} = S_+^{\dagger}. \quad (12.144)$$

The resulting Hamiltonian is given in Table 12.4.

12.4.7.7 [001] QW

Having obtained the most general Hamiltonian, we now look at the specific case of a QW grown along the [001] direction; this is the case considered by Foreman [34].

In this case, $\widehat{k}_x \to k_x$ and $\widehat{k}_y \to k_y$ and they, therefore, commute. We obtain the following simplifications:

$$\left\langle \frac{3}{2}\ \frac{3}{2}\right| H_{k\cdot p} \left|\frac{3}{2}\ \frac{3}{2}\right\rangle = \frac{1}{2}(L+M)(k_x^2+k_y^2) + \widehat{k}_z M\widehat{k}_z \equiv -(P+Q), \quad (12.145)$$

Table 12.4 Six-band BF Hamiltonian in JM_J basis for arbitrary quantization

$$H_{\mathrm{BF1}}(\mathbf{k}) = \begin{pmatrix} \left|\frac{3}{2}\,\frac{3}{2}\right\rangle & \left|\frac{3}{2}\,\frac{1}{2}\right\rangle & \left|\frac{3}{2}\,-\frac{1}{2}\right\rangle & \left|\frac{3}{2}\,-\frac{3}{2}\right\rangle & \left|\frac{1}{2}\,\frac{1}{2}\right\rangle & \left|\frac{1}{2}\,-\frac{1}{2}\right\rangle \\ P' & S_- & -R & 0 & -\frac{1}{\sqrt{2}}S_- & -\sqrt{2}R \\ S_-^\dagger & P'' & -C & R & -\sqrt{2}Q & -\sqrt{\frac{3}{2}}\Sigma_- \\ -R^\dagger & -C^\dagger & P''^* & S_+^\dagger & \sqrt{\frac{3}{2}}\Sigma_+ & -\sqrt{2}Q^* \\ 0 & R^\dagger & S_+ & P'^* & \sqrt{2}R^\dagger & -\frac{1}{\sqrt{2}}S_+ \\ -\frac{1}{\sqrt{2}}S_-^\dagger & -\sqrt{2}Q^\dagger & \sqrt{\frac{3}{2}}\Sigma_+^\dagger & -\sqrt{2}R & P''' - \Delta_0 & C \\ -\sqrt{2}R^\dagger & -\sqrt{\frac{3}{2}}\Sigma_-^\dagger & -\sqrt{2}Q & -\frac{1}{\sqrt{2}}S_+^\dagger & C^\dagger & P'''^* - \Delta_0 \end{pmatrix}$$

with

$$P' = \frac{1}{2}\left\{\widehat{k}_x(L+M)\widehat{k}_x + \widehat{k}_y(L+M)\widehat{k}_y + \widehat{k}_z 2M\widehat{k}_z\right\}$$
$$+\frac{\mathrm{i}}{2}\left\{\widehat{k}_x(F-G-H_1+H_2)\widehat{k}_y - \widehat{k}_y(F-G-H_1+H_2)\widehat{k}_x\right\},$$

$$P'' = \frac{1}{6}\left\{\widehat{k}_x(L+5M)\widehat{k}_x + \widehat{k}_y(L+5M)\widehat{k}_y + 2\widehat{k}_z(2L+M)\widehat{k}_z\right\}$$
$$+\frac{\mathrm{i}}{6}\left\{\widehat{k}_x(F-G-H_1+H_2)\widehat{k}_y - \widehat{k}_y(F-G-H_1+H_2)\widehat{k}_x\right\},$$

$$P''' = \frac{1}{3}\left\{\widehat{k}_x(L+2M)\widehat{k}_x + \widehat{k}_y(L+2M)\widehat{k}_y + \widehat{k}_z(L+2M)\widehat{k}_z\right\}$$
$$+\frac{\mathrm{i}}{3}\left\{\widehat{k}_x(F-G-H_1+H_2)\widehat{k}_y - \widehat{k}_y(F-G-H_1+H_2)\widehat{k}_x\right\},$$

$$Q = -\frac{1}{6}\left\{\widehat{k}_x(L-M)\widehat{k}_x + \widehat{k}_y(L-M)\widehat{k}_y + 2\widehat{k}_z(L-M)\widehat{k}_z\right.$$
$$\left.+\,\mathrm{i}[\widehat{k}_x(F-G-H_1+H_2)\widehat{k}_y - \widehat{k}_y(F-G-H_1+H_2)\widehat{k}_x]\right\},$$

$$R = \frac{1}{2\sqrt{3}}\left\{\widehat{k}_x(L-M)\widehat{k}_x - \widehat{k}_y(L-M)\widehat{k}_y - \mathrm{i}[\widehat{k}_x N\widehat{k}_y + \widehat{k}_y N\widehat{k}_x]\right\},$$

$$S_\pm = -\frac{1}{\sqrt{3}}\left\{\widehat{k}_\pm(F-G)\widehat{k}_z + \widehat{k}_z(H_1-H_2)\widehat{k}_\pm\right\},$$

$$C = -\frac{1}{3}\left\{\widehat{k}_z(F-G-H_1+H_2)\widehat{k}_- - \widehat{k}_-(F-G-H_1+H_2)\widehat{k}_z\right\},$$

$$\Sigma_\pm = -\frac{1}{3\sqrt{3}}\left\{\widehat{k}_\pm(F-G+2H_1-2H_2)\widehat{k}_z + \widehat{k}_z(2F-2G+H_1-H_2)\widehat{k}_\pm\right\}.$$

where

$$P = \frac{\hbar^2}{2m_0}(\gamma_1 k_{||}^2 + \widehat{k}_z \gamma_1 \widehat{k}_z),$$
$$Q = \frac{\hbar^2}{2m_0}(\gamma_2 k_{||}^2 - 2\widehat{k}_z \gamma_2 \widehat{k}_z), \tag{12.146}$$
$$k_{||} = k_x^2 + k_y^2.$$

Similarly,

$$\left\langle \frac{3}{2}\ \frac{1}{2} \middle| H_{k\cdot p} \middle| \frac{3}{2}\ \frac{1}{2} \right\rangle = \frac{1}{6}\left\{(L+5M)(k_x^2+k_y^2) + 2\widehat{k}_z(2L+M)\widehat{k}_z\right\}$$
$$= -\frac{\hbar^2}{2m_0}\left[(\gamma_1-\gamma_2)k_{||}^2 + \widehat{k}_z(\gamma_1+2\gamma_2)\widehat{k}_z)\right]$$
$$= -(P-Q), \tag{12.147}$$
$$\left\langle \frac{1}{2}\ \frac{1}{2} \middle| H_{k\cdot p} \middle| \frac{1}{2}\ \frac{1}{2} \right\rangle = \frac{1}{3}\left[(L+2M)(k_x^2+k_y^2) + \widehat{k}_z(L+2M)\widehat{k}_z\right] - \Delta_0$$
$$= -P - \Delta_0. \tag{12.148}$$

For the off-diagonal elements,

$$\left\langle \frac{3}{2}\ \frac{3}{2} \middle| H_{k\cdot p} \middle| \frac{3}{2}\ \frac{1}{2} \right\rangle = -\frac{1}{\sqrt{3}}\left[k_-(F-G)\widehat{k}_z + \widehat{k}_z(H_1-H_2)k_-\right]$$
$$= \sqrt{3}k_-\left[(\sigma-\delta)\widehat{k}_z + \widehat{k}_z\pi\right] = S_-, \tag{12.149}$$
$$\left\langle \frac{3}{2}\ \frac{3}{2} \middle| H_{k\cdot p} \middle| \frac{3}{2}\ -\frac{1}{2} \right\rangle = -\frac{1}{2\sqrt{3}}\left\{k_x(L-M)k_x - k_y(L-M)k_y - 2\mathrm{i}Nk_xk_y\right\}$$
$$= \frac{\hbar^2}{2m_0}\sqrt{3}\left[\gamma_2(k_x^2-k_y^2) - 2\mathrm{i}\gamma_3 k_x k_y\right].$$

Now

$$k_-^2 = (k_x - \mathrm{i}k_y)^2 = k_x^2 - k_y^2 - 2\mathrm{i}k_xk_y,$$
$$k_+^2 = (k_x + \mathrm{i}k_y)^2 = k_x^2 - k_y^2 + 2\mathrm{i}k_xk_y.$$

Therefore

$$\left\langle \frac{3}{2}\ \frac{3}{2} \middle| H_{k\cdot p} \middle| \frac{3}{2}\ -\frac{1}{2} \right\rangle = \frac{\hbar^2}{2m_0}\sqrt{3}(\overline{\gamma}k_-^2 - \mu k_+^2) \equiv -R. \tag{12.150}$$

Continuing

$$\left\langle \frac{3}{2}\ \frac{1}{2} \right| H_{k\cdot p} \left| \frac{3}{2}\ -\frac{1}{2} \right\rangle = \frac{1}{3}\left\{ \widehat{k}_z(F-G-H_1+H_2)k_- - k_-(F-G-H_1+H_2)\widehat{k}_z \right\}$$
$$= -\left[\widehat{k}_z(\sigma-\delta-\pi)k_- - k_-(\sigma-\delta-\pi)\widehat{k}_z \right]$$
$$= -k_-\left[\widehat{k}_z(\sigma-\delta-\pi) - (\sigma-\delta-\pi)\widehat{k}_z \right] \equiv -C, \qquad (12.151)$$

$$\left\langle \frac{3}{2}\ \frac{1}{2} \right| H_{k\cdot p} \left| \frac{1}{2}\ \frac{1}{2} \right\rangle = \frac{1}{3\sqrt{2}}\left[(L-M)(k_x^2+k_y^2) - 2\widehat{k}_z(L-M)\widehat{k}_z \right]$$
$$= -\frac{\hbar^2}{m_0}\frac{1}{\sqrt{2}}\left[\gamma_2 k_{\|}^2 - 2\widehat{k}_z\gamma_2\widehat{k}_z \right] \equiv -\sqrt{2}Q, \qquad (12.152)$$

$$\left\langle \frac{3}{2}\ \frac{1}{2} \right| H_{k\cdot p} \left| \frac{1}{2}\ -\frac{1}{2} \right\rangle = \frac{1}{3\sqrt{2}}k_-\left\{ (F-G+2H_1-2H_2)\widehat{k}_z + \widehat{k}_z(2F-2G+H_1-H_2) \right\}$$
$$= -\frac{1}{\sqrt{2}}k_-\left\{ (\sigma-\delta+2\pi)\widehat{k}_z + \widehat{k}_z(2\sigma-2\delta+\pi) \right\}$$
$$\equiv -\sqrt{\frac{3}{2}}\Sigma_-. \qquad (12.153)$$

The resulting Hamiltonian is given in Table 12.5. Note that we have left out the band-edge energies for simplicity. Using this Hamiltonian, Foreman showed that the heavy-hole mass for an InGaAs QW can be unphysical if the conventional symmetrized Hamiltonian is used, but is well-behaved using his Hamiltonian (Fig. 12.3).

12.4.8 Beyond Burt–Foreman Theory?

Meney et al. [233] compared the BF Hamiltonian to the LK one and made a number of interesting observations. For the eight-band models, they are very similar. For the six-band models, there is substantial difference (up to 20 meV) between the two at intermediate wave numbers (around 0.1 Å^{-1}) but also they both differ from the eight-band models, with the LK one worst. For the four-band models, there was only a small difference between the two, again at intermediate wave numbers. Similar studies have recently been reported [234]. Other studies of the BF model since were for QW's with arbitrary growth direction [232, 235], QW's with wurtzite constituents [55, 56], and for a spherical quantum dot within the spherical approximation [236, 237]. A study of the optical properties of QW's [238] and of nanowires of arbitrary orientation can be found in [239, 240]. It is, therefore, clear that, by now, the eight-band BF model has been sufficiently studied for a variety of nanostructures.

Table 12.5 Six-band BF Hamiltonian in JM_J basis for [001] quantization. Note the overall negative sign. Foreman's Hamiltonian is the negative of ours because he gives the hole energies as positive; also, he has a different ordering of the basis states

$$
H_{\mathrm{BF2}}(\mathbf{k}) = -\begin{pmatrix}
\left|\frac{3}{2}\,\frac{3}{2}\right\rangle & \left|\frac{3}{2}\,\frac{1}{2}\right\rangle & \left|\frac{3}{2}\,-\frac{1}{2}\right\rangle & \left|\frac{3}{2}\,-\frac{3}{2}\right\rangle & \left|\frac{1}{2}\,\frac{1}{2}\right\rangle & \left|\frac{1}{2}\,-\frac{1}{2}\right\rangle \\
P+Q & -S_- & R & 0 & \frac{1}{\sqrt{2}}S_- & \sqrt{2}R \\
-S_-^\dagger & P-Q & C & -R & \sqrt{2}Q & \sqrt{\frac{3}{2}}\Sigma_- \\
R^\dagger & C^\dagger & P-Q & -S_+^\dagger & -\sqrt{\frac{3}{2}}\Sigma_+ & \sqrt{2}Q^* \\
0 & -R^\dagger & -S_+ & P+Q & -\sqrt{2}R^\dagger & \frac{1}{\sqrt{2}}S_+ \\
\frac{1}{\sqrt{2}}S_-^\dagger & \sqrt{2}Q & -\sqrt{\frac{3}{2}}\Sigma_+^\dagger & -\sqrt{2}R & P+\Delta_0 & -C \\
\sqrt{2}R^\dagger & \sqrt{\frac{3}{2}}\Sigma_-^\dagger & \sqrt{2}Q & \frac{1}{\sqrt{2}}S_+^\dagger & -C^\dagger & P+\Delta_0
\end{pmatrix}
$$

with

$$P = \frac{\hbar^2}{2m_0}\left(\gamma_1 k_{\parallel}^2 + \widehat{k}_z \gamma_1 \widehat{k}_z\right),$$

$$Q = \frac{\hbar^2}{2m_0}\left(\gamma_2 k_{\parallel}^2 - 2\widehat{k}_z \gamma_2 \widehat{k}_z\right),$$

$$R = -\frac{\hbar^2}{2m_0}\sqrt{3}\left(\overline{\gamma} k_-^2 - \mu k_+^2\right),$$

$$S_\pm = \frac{\hbar^2}{m_0}\sqrt{3}k_\pm\left[(\sigma-\delta)\widehat{k}_z + \widehat{k}_z \pi\right],$$

$$C = \frac{\hbar^2}{m_0}k_-\left[\widehat{k}_z(\sigma-\delta-\pi) - (\sigma-\delta-\pi)\widehat{k}_z\right],$$

$$\Sigma_\pm = \frac{\hbar^2}{m_0}\sqrt{3}k_\pm\left\{\left[\frac{1}{3}(\sigma-\delta)+\frac{2}{3}\pi\right]\widehat{k}_z + \widehat{k}_z\left[\left(\frac{2}{3}(\sigma-\delta)+\frac{1}{3}\pi\right]\right\}.$$

Nevertheless, there are difficulties with the BF theory. Takhtamirov and Volkov [241–244] have demonstrated that most heterostructure effective-mass equations do not consistently include all perturbative corrections to the same order. In particular, due to the presence of material interfaces in heterostructures, important contributions arise from couplings between near band-gap states and remote states that must be accounted for (this fact is also pointed out in [20]). The fact that there must be some so-called interface Hamiltonian has been well documented [245–248]. In [249, 250], based on first-principles envelope-function theory, an effective one-electron Hamiltonian is derived including perturbative effects of material inhomogeneity in agreement with the results of Takhtamirov and Volkov. An important driving mechanism behind formulating effective-mass one-electron Hamiltonian models as approximations to the real semiconductor many-particle problem is the

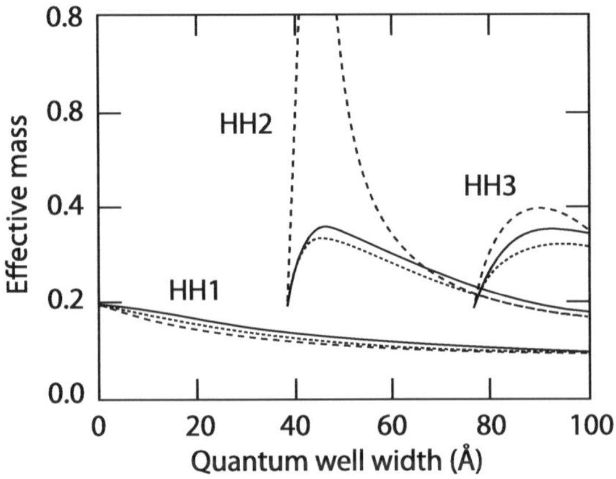

Fig. 12.3 Heavy-hole mass of subband of InGaAs quantum well – comparison of different models. Solid line is BF Hamiltonian, dashed is symmetrized, and dotted is uncoupled. Reprinted with permission from [34]. ©1993 by the American Physical Society

possibility to obtain accuracy while maintaining fast computational speed in the determination of electronic states and energies. This is essential for solving complicated structures such as quantum wires and dots where ab initio atomic-based methods are still too slow.

12.5 Sercel–Vahala Theory

In 1990–1991, Sercel and Vahala wrote a series of papers expounding a new formulation of the multiband envelope-function theory and applied it to spherical quantum dots and cylindrical quantum wires [36, 251–253]. The method is based upon a similar treatment of the acceptor problem by Altarelli, Baldereschi, and Lipari in the 1970s [30, 92]. The theory (within the LK framework) has subsequently been applied to model quantum rods [254], quantum rings [255–257], and quantum dots [258]. An extension to the Burt–Foreman theory and application to wurtzite nanostructures [240, 259] and nanowire superlattices have been carried out [260].

12.5.1 Overview

The basic idea of the Sercel–Vahala (SV) theory is to re-express the zincblende $k \cdot p$ Hamiltonian in a basis of eigenstates of total angular momentum $\mathbf{F}$ or of its z-projection F_z. This results from the vector addition of the envelope angular momentum $\mathbf{L}$ and the zone-center Bloch angular momentum $\mathbf{J}$ (in direct analogy to the coupling of atomic orbitals and electron spin in atomic physics). The envelope angular momentum, being a spatial degree of freedom, is integral. The Bloch angular momentum corresponds to a spin-$\frac{3}{2}$ particle. A formal feature of the theory is the treatment of the envelope function and of the underlying Bloch functions as functions of spatial coordinates corresponding to different state spaces. What makes the theory really useful is the simplifying assumption that the total angular momentum (for the spherical problem) or of its z-projection (for the cylindrical problem) is a constant of motion. This is certainly true when anisotropic coupling to remote bands are neglected resulting in either the spherical or axial approximation. Furthermore, Kramers degeneracy leads to degeneracy of the $\pm F_z$ solutions (this can be most directly proven by observing that the $-F_z$ states can be obtained from the $+F_z$ ones by applying the Kramers operator). The SV representation also reduces the three-dimensional problem to a two-dimensional one when cylindrical polar coordinates are used.

We now detail the theory for each of the two symmetries considered by Sercel and Vahala, followed by application to wurtzite structures. For the latter case, the reduction is *exact*, i.e., no axial approximation is needed. That this should be possible is already clear from work on the bulk band structure since the latter was found to be azimuthal-angle independent. The theory has also been applied to the linear strain Hamiltonian.

12.5.2 Spherical Representation

A complete set of commuting observables is H, $\mathbf{F}^2$, F_z, where

$$\mathbf{F} = \mathbf{L} + \mathbf{J}. \tag{12.154}$$

Hence, the Hamiltonian is block diagonal in a basis of eigenstates of $\mathbf{F}^2$ and F_z. This basis can be obtained by coupling those of the envelope and Bloch functions:

$$|\mathbf{k}; F; F_z; J, L\rangle = \sum_{J_z=-J}^{J} \sum_{L_z=-L}^{L} \langle J, J_z; L, L_z|F, F_z\rangle \, |J, J_z\rangle|k; L, , L_z\rangle. \tag{12.155}$$

For the Bloch states $|J, J_z\rangle$, Sercel and Vahala use a slightly different set of functions from what is conventional (see Table C.2). Here, the envelope functions have also been expanded in terms of a basis set. For a spherical problem within the flat-band approximation (i.e., with a piecewise-constant potential profile), it is advantageous to use free spherical waves. In the coordinate representation,

$$\langle r, \theta, \phi|k; L, L_z\rangle = \sqrt{\frac{2}{\pi}} \mathrm{i}^L h_L(kr) Y_L^{L_z}(\theta, \phi), \tag{12.156}$$

where h_L is a spherical Hankel function, $Y_L^{L_z}$ is a spherical harmonic, and k is the radial wave number. The envelope functions form a complete set in envelope state space, with orthonormality

$$\langle k; L, L_z | k'; L', L'_z \rangle = \frac{\delta(k-k')}{k^2} \delta_{L,L'} \delta_{L_z,L'_z}, \tag{12.157}$$

and closure

$$\mathcal{I}_E = \sum_{L=0}^{\infty} \sum_{L_z=-L}^{L} \int_0^{\infty} \mathrm{d}k\, k^2 |k; L, L_z\rangle\langle k; L, L_z|. \tag{12.158}$$

The conventional Bloch plane-wave basis can be written as

$$|\mathbf{k}; J, J_z\rangle = |\mathbf{k}\rangle |J, J_z\rangle. \tag{12.159}$$

The Hamiltonian in this basis is related to the new one in the total-angular-momentum basis via a unitary transformation. In order to establish the transformation, we need to express the plane-wave envelope states in terms of the spherical-wave ones:

$$|\mathbf{k}\rangle = \sum_{L,L_z} \left[Y_L^{L_z}(\Omega_{\mathbf{k}}) \right]^* |k; L, L_z\rangle, \tag{12.160}$$

where $\Omega_{\mathbf{k}} = (\theta_{\mathbf{k}}, \phi_{\mathbf{k}})$ is the orientation of the wave vector with respect to the axes. Then, the unitary transformation between the two envelope functions is

$$\langle k'; L', L'_z | \mathbf{k}\rangle = \sum_{L,L_z} \left[Y_L^{L_z}(\Omega_{\mathbf{k}}) \right]^* \langle k'; L', L'_z | k; L, L_z\rangle = \left[Y_L^{L_z}(\Omega_{\mathbf{k}}) \right]^* \frac{\delta(k-k')}{k^2}. \tag{12.161}$$

Using the closure relation for the old product Bloch plane-wave basis,

$$\mathcal{I} = \mathcal{I}_B \otimes \mathcal{I}_E = \left(\sum_{J,J_z} |J, J_z\rangle\langle J, J_z| \right) \otimes \left(\int_0^{\infty} \mathrm{d}k\, k^2 \int \mathrm{d}\Omega_{\mathbf{k}} |\mathbf{k}\rangle\langle\mathbf{k}| \right), \tag{12.162}$$

a wave function ψ in the new total-angular-momentum basis can be written as

$$\begin{aligned} &\langle \mathbf{k}; F; F_z; J, L | \psi\rangle \\ &= \langle \mathbf{k}; F; F_z; J, L | \left(\sum_{J',J'_z} |J', J'_z\rangle\langle J', J'_z| \right) \otimes \left(\int_0^{\infty} \mathrm{d}k'\, (k')^2 \int \mathrm{d}\Omega_{\mathbf{k}'} |\mathbf{k}'\rangle\langle\mathbf{k}'| \right) |\psi\rangle. \end{aligned} \tag{12.163}$$

This can be simplified using Eqs. (12.155) and (12.161):

$$\begin{aligned} \langle \mathbf{k}; F; F_z; J, L | \psi\rangle = &\sum_{J',J'_z} \sum_{J_z=-J}^{J} \sum_{L_z=-L}^{L} \langle J, J_z; L, L_z | F, F_z\rangle^* \langle J, J_z | J', J'_z\rangle \langle k; L,, L_z | \mathbf{k}'\rangle \\ &\times \int_0^{\infty} \mathrm{d}k'\, (k')^2 \int \mathrm{d}\Omega_{\mathbf{k}'} \langle J', J'_z | \langle \mathbf{k}' | \psi\rangle \end{aligned}$$

$$= \sum_{J_z=-J}^{J} \sum_{L_z=-L}^{L} \langle J, J_z; L, L_z | F, F_z \rangle^*$$
$$\times \int_0^\infty \mathrm{d}k' \, (k')^2 \int \mathrm{d}\Omega_{\mathbf{k}'} \left[Y_L^{L_z}(\Omega_{\mathbf{k}'}) \right]^* \frac{\delta(k-k')}{k^2} \langle \mathbf{k}'; J, J_z | \psi \rangle$$
$$= \sum_{J_z=-J}^{J} \sum_{L_z=-L}^{L} \langle J, J_z; L, L_z | F, F_z \rangle^* \int \mathrm{d}\Omega_{\mathbf{k}} \left[Y_L^{L_z}(\Omega_{\mathbf{k}}) \right]^* \langle \mathbf{k}; J, J_z | \psi \rangle. \tag{12.164}$$

This is, of course, the unitary transformation connecting the wave function in the two representations. The Hamiltonian in the new total-angular-momentum basis is, therefore, given by

$$\langle \mathbf{k}; F; F_z; J, L | H | \mathbf{k}'; F'; F_z'; J', L' \rangle$$
$$= \sum_{J_z=-J}^{J} \sum_{L_z=-L}^{L} \sum_{J_z'=-J'}^{J'} \sum_{L_z'=-L'}^{L'} \langle J, J_z; L, L_z | F, F_z \rangle^* \langle J', J_z'; L', L_z' | F', F_z' \rangle$$
$$\times \int \mathrm{d}\Omega_{\mathbf{k}} \int \mathrm{d}\Omega_{\mathbf{k}'} \left[Y_L^{L_z}(\Omega_{\mathbf{k}}) \right]^* \left[Y_{L'}^{L_z'}(\Omega_{\mathbf{k}'}) \right] \langle \mathbf{k}; J, J_z | H | \mathbf{k}'; J', J_z' \rangle.$$

The old Hamiltonian is diagonal in **k**:

$$\langle \mathbf{k}; F; F_z; J, L | H | \mathbf{k}'; F'; F_z'; J', L' \rangle$$
$$= \sum_{J_z=-J}^{J} \sum_{L_z=-L}^{L} \sum_{J_z'=-J'}^{J'} \sum_{L_z'=-L'}^{L'} \langle J, J_z; L, L_z | F, F_z \rangle^* \langle J', J_z'; L', L_z' | F', F_z' \rangle$$
$$\times \int \mathrm{d}\Omega_{\mathbf{k}} \int \mathrm{d}\Omega_{\mathbf{k}'} \left[Y_L^{L_z}(\Omega_{\mathbf{k}}) \right]^* \left[Y_{L'}^{L_z'}(\Omega_{\mathbf{k}'}) \right] \langle \mathbf{k}; J, J_z | H | \mathbf{k}'; J', J_z' \rangle \delta(k-k') \delta(\Omega_{\mathbf{k}} - \Omega_{\mathbf{k}'})$$
$$= \sum_{J_z=-J}^{J} \sum_{L_z=-L}^{L} \sum_{J_z'=-J'}^{J'} \sum_{L_z'=-L'}^{L'} \langle J, J_z; L, L_z | F, F_z \rangle^* \langle J', J_z'; L', L_z' | F', F_z' \rangle$$
$$\times \int \mathrm{d}\Omega_{\mathbf{k}} \left[Y_L^{L_z}(\Omega_{\mathbf{k}}) \right]^* \left[Y_{L'}^{L_z'}(\Omega_{\mathbf{k}}) \right] \langle \mathbf{k}; J, J_z | H | \mathbf{k}; J', J_z' \rangle \delta(k-k').$$

Hence,

$$\langle \mathbf{k}; F; F_z; J, L | H | \mathbf{k}; F'; F_z'; J', L' \rangle$$
$$= \sum_{J_z=-J}^{J} \sum_{L_z=-L}^{L} \sum_{J_z'=-J'}^{J'} \sum_{L_z'=-L'}^{L'} \langle J, J_z; L, L_z | F, F_z \rangle^* \langle J', J_z'; L', L_z' | F', F_z' \rangle$$
$$\times \int \mathrm{d}\Omega_{\mathbf{k}} \left[Y_L^{L_z}(\Omega_{\mathbf{k}}) \right]^* \left[Y_{L'}^{L_z'}(\Omega_{\mathbf{k}}) \right] \langle \mathbf{k}; J, J_z | H | \mathbf{k}; J', J_z' \rangle. \tag{12.165}$$

We know from angular-momentum theory that $L_z = F_z - J_z$ and $L'_z = F'_z - J_z$. The key result here is that the old Hamiltonian is not diagonal in J. This implies the Sercel–Vahala model goes beyond the $J = \frac{3}{2}$ subspace and includes J-subspace coupling (e.g., as in Kane's eight-band model). Note also, though, that Eq. (12.165) is really diagonal in F and F_z. Finally, Sercel and Vahala used real Clebsch-Gordan coefficients; we have left our expressions in terms of more general complex Clebsch-Gordan coefficients.

12.5.2.1 $F = \frac{1}{2}$, $F_z = \pm\frac{1}{2}$

We can now start looking at individual blocks of the Hamiltonian. We start with $F = \frac{1}{2}$, $F_z = \pm\frac{1}{2}$. The two $F_z = \pm\frac{1}{2}$ subspaces are degenerate due to Kramers degeneracy. Taking the old Hamiltonian to be an eight-band one, we have, using

$$\mathbf{F} = \mathbf{L} + \mathbf{J}, \mathbf{F}_z = \mathbf{L}_z + \mathbf{J}_z,$$

that the only allowed coupling of J and L are

$$(J, L) = \left(\frac{1}{2}, 0\right), \left(\frac{1}{2}, 1\right), \left(\frac{3}{2}, 1\right), \left(\frac{3}{2}, 2\right);$$

i.e., there are six ways to get the required ($\mathbf{F}^2$, F_z). These combinations, together with the Clebsch-Gordan coefficients for the uncoupled states, are given in Table 12.6. The absence of hh states will be explained later. We now show how the matrix elements $\langle J, L|H|J', L'\rangle$ are obtained. For example,

$$\begin{aligned}
\left\langle \frac{1}{2}, 1 \middle| H \middle| \frac{1}{2}, 1 \right\rangle &= \left\langle \frac{1}{2}, \frac{1}{2}; 1, 0 \middle| \frac{1}{2}, 1 \right\rangle \left\langle \frac{1}{2}, \frac{1}{2}; 1, 0 \middle| \frac{1}{2}, 1 \right\rangle \int \mathrm{d}\Omega \, (Y_1^0)^* (Y_1^0) H_{cc} \\
&+ \left\langle \frac{1}{2}, -\frac{1}{2}; 1, 1 \middle| \frac{1}{2}, 1 \right\rangle \left\langle \frac{1}{2}, -\frac{1}{2}; 1, 1 \middle| \frac{1}{2}, 1 \right\rangle \int \mathrm{d}\Omega \, (Y_1^1)^* (Y_1^1) H_{cc} \\
&+ \left\langle \frac{1}{2}, \frac{1}{2}; 1, 0 \middle| \frac{1}{2}, 1 \right\rangle \left\langle \frac{1}{2}, -\frac{1}{2}; 1, 1 \middle| \frac{1}{2}, 1 \right\rangle \int \mathrm{d}\Omega \, (Y_1^0)^* (Y_1^1) H_{cc} \\
&+ \left\langle \frac{1}{2}, -\frac{1}{2}; 1, 1 \middle| \frac{1}{2}, 1 \right\rangle \left\langle \frac{1}{2}, \frac{1}{2}; 1, 0 \middle| \frac{1}{2}, 1 \right\rangle \int \mathrm{d}\Omega \, (Y_1^1)^* (Y_1^0) H_{cc} \\
&= \left(\frac{1}{3} + \frac{2}{3}\right) H_{cc} = H_{cc} = E_c + \frac{\hbar^2 k^2}{2m_0}.
\end{aligned}$$

The 6×6 block diagonalizes into two identical subblocks. Sercel and Vahala obtained, e.g.,

Table 12.6 Total-angular-momentum states for $F = \frac{1}{2}$, $F_z = \frac{1}{2}$ [36]

Band	$\lvert J, L\rangle$	$\lvert J, J_z\rangle \otimes \lvert k; L, L_z\rangle$
e	$\left\lvert \frac{1}{2}, 0\right\rangle$	$\left\lvert \frac{1}{2}, \frac{1}{2}\right\rangle \otimes \lvert k; 0, 0\rangle$
	$\left\lvert \frac{1}{2}, 1\right\rangle$	$\sqrt{\frac{1}{3}} \left\lvert \frac{1}{2}, \frac{1}{2}\right\rangle \otimes \lvert k; 1, 0\rangle - \sqrt{\frac{2}{3}} \left\lvert \frac{1}{2}, -\frac{1}{2}\right\rangle \otimes \lvert k; 1, 1\rangle$
lh	$\left\lvert \frac{3}{2}, 1\right\rangle$	$\sqrt{\frac{1}{2}} \left\lvert \frac{3}{2}, \frac{3}{2}\right\rangle \otimes \lvert k; 1, -1\rangle - \sqrt{\frac{1}{3}} \left\lvert \frac{3}{2}, \frac{1}{2}\right\rangle \otimes \lvert k; 1, 0\rangle + \sqrt{\frac{1}{6}} \left\lvert \frac{3}{2}, -\frac{1}{2}\right\rangle \otimes \lvert k; 1, 1\rangle$
	$\left\lvert \frac{3}{2}, 2\right\rangle$	$\sqrt{\frac{1}{10}} \left\lvert \frac{3}{2}, \frac{3}{2}\right\rangle \otimes \lvert k; 2, -1\rangle - \sqrt{\frac{1}{5}} \left\lvert \frac{3}{2}, \frac{1}{2}\right\rangle \otimes \lvert k; 2, 0\rangle + \sqrt{\frac{3}{10}} \left\lvert \frac{3}{2}, -\frac{1}{2}\right\rangle \otimes \lvert k; 2, 1\rangle - \sqrt{\frac{2}{5}} \left\lvert \frac{3}{2}, -\frac{3}{2}\right\rangle \otimes \lvert k; 2, 2\rangle$
sh	$\left\lvert \frac{1}{2}, 0\right\rangle$	$\left\lvert \frac{1}{2}, \frac{1}{2}\right\rangle \otimes \lvert k; 0, 0\rangle$
	$\left\lvert \frac{1}{2}, 1\right\rangle$	$\sqrt{\frac{1}{3}} \left\lvert \frac{1}{2}, \frac{1}{2}\right\rangle \otimes \lvert k; 1, 0\rangle - \sqrt{\frac{2}{3}} \left\lvert \frac{1}{2}, -\frac{1}{2}\right\rangle \otimes \lvert k; 1, 1\rangle$

$$H^{F=\frac{1}{2}} = \begin{pmatrix} |\frac{1}{2},0\rangle & |\frac{3}{2},1\rangle & |\frac{1}{2},1\rangle \\ E_c + \frac{\hbar^2 k^2}{2m_0} & -\mathrm{i}\sqrt{\frac{2}{3}}Pk & -\mathrm{i}\sqrt{\frac{1}{3}}Pk \\ \mathrm{i}\sqrt{\frac{2}{3}}Pk & E_v - \frac{\hbar^2}{2m_0}(\gamma_1 + 2\gamma_2)k^2 & -2\sqrt{2}\frac{\hbar^2}{2m_0}\gamma_2 k^2 \\ \mathrm{i}\sqrt{\frac{1}{3}}Pk & -2\sqrt{2}\frac{\hbar^2}{2m_0}\gamma_2 k^2 & E_v - \Delta_0 - \frac{\hbar^2}{2m_0}\gamma_1 k^2 \end{pmatrix}, \tag{12.166}$$

where the three states are, respectively, the conduction, light-hole (lh) and split-hole states. Note that, in general, the lh contains a mixture of J_z states (Table 12.6). Heavy-hole states do not appear until $F = \frac{3}{2}$ [36]. Sercel and Vahala now solved the spherical-dot problem. We will instead focus on the solutions for the cylindrical problem.

12.5.3 Cylindrical Representation

A complete set of commuting observables is H, F_z, P_z, where P_z is the component of the envelope linear momentum along the z-axis. F_z (but not F) is a good quantum number.

12.5.3.1 Hamiltonian

The Hamiltonian is block diagonal in this basis:

$$H = \sum_{\oplus F_z} H_{F_z}. \tag{12.167}$$

Because F is not a good quantum number, it is now better to work in an uncoupled angular momentum representation:

$$|k_z, k_{||}, F_z, J, J_z\rangle \equiv |J, J_z\rangle |k_z, k_{||}, L_z = F_z - J_z\rangle. \tag{12.168}$$

Hence, Clebsch-Gordan coefficients are not required for this step. The envelope function $|k_z, k_{||}, L_z\rangle$ is an eigenstate of L_z and P_z. The cylindrical coordinate representation is

$$\langle \rho, \phi, z | k_z, k_{||}, L_z\rangle = \frac{\mathrm{i}^{L_z}}{2\pi} H_{L_z}(k_{||}\rho) \mathrm{e}^{\mathrm{i}L_z\phi} \mathrm{e}^{\mathrm{i}k_z z}. \tag{12.169}$$

The basis has cylindrical symmetry and is orthonormal:

$$\langle k_z', k_{||}', F_z', J', J_z' | k_z, k_{||}, F_z, J, J_z\rangle = \delta_{F_z,F_z'} \delta_{J,J'} \delta_{J_z,J_z'} \frac{\delta(k_{||} - k_{||}')}{k_{||}} \delta(k_z - k_z'). \tag{12.170}$$

We still have the conventional Bloch plane-wave basis written as

$$|\mathbf{k}; J, J_z\rangle = |\mathbf{k}\rangle |J, J_z\rangle. \tag{12.171}$$

Here, however,

$$|\mathbf{k}\rangle = |k_z\rangle |\mathbf{k}_{||}\rangle = \mathrm{e}^{\mathrm{i}k_z z} \sum_{L_z} \mathrm{e}^{-\mathrm{i}L_z \phi_{\mathbf{k}_{||}}} |k_{||}, L_z\rangle, \tag{12.172}$$

giving

$$\begin{aligned}\langle k'_{||}, L'_z|\mathbf{k}\rangle &= \mathrm{e}^{\mathrm{i}k_z z} \sum_{L_z} \mathrm{e}^{-\mathrm{i}L_z \phi_{\mathbf{k}_{||}}} \langle k'_{||}, L'_z|k_{||}, L_z\rangle = \mathrm{e}^{\mathrm{i}k_z z} \mathrm{e}^{-\mathrm{i}L'_z \phi_{\mathbf{k}_{||}}} \frac{\delta(k_{||} - k'_{||})}{k_{||}} \\ &\equiv |k_z\rangle \, \mathrm{e}^{-\mathrm{i}L'_z \phi_{\mathbf{k}_{||}}} \frac{\delta(k_{||} - k'_{||})}{k_{||}}. \end{aligned} \tag{12.173}$$

The closure relation for the old product Bloch plane-wave basis is

$$\begin{aligned}\mathcal{I} &= \mathcal{I}_B \otimes \mathcal{I}_E \\ &= \left(\sum_{J, J_z} |J, J_z\rangle \langle J, J_z| \right) \otimes \left(\int_0^\infty \mathrm{d}k_{||}\, k_{||} \int \mathrm{d}\phi_{\mathbf{k}_{||}} |\mathbf{k}_{||}\rangle \langle \mathbf{k}_{||}| \right) \otimes \left(\int_0^\infty \mathrm{d}k_z\, |k_z\rangle \langle k_z| \right), \end{aligned} \tag{12.174}$$

and a wave function ψ in the product basis can be written as

$$\begin{aligned}\langle k_z, k_{||}, F_z, J, J_z|\psi\rangle &= \langle k_z, k_{||}, F_z, J, J_z| \\ &\times \left(\sum_{J', J'_z} |J', J'_z\rangle \langle J', J'_z| \right) \otimes \left(\int_0^\infty \mathrm{d}k_{||}\, k_{||} \int \mathrm{d}\phi_{\mathbf{k}_{||}} |\mathbf{k}_{||}\rangle \langle \mathbf{k}_{||}| \right) \otimes \left(\int_0^\infty \mathrm{d}k_z\, |k_z\rangle \langle k_z| \right) |\psi\rangle. \end{aligned} \tag{12.175}$$

This can be simplified using Eqs. (12.168) and (12.173):

$$\begin{aligned}&\langle k_z, k_{||}, F_z, J, J_z|\psi\rangle \\ &= \Big(\sum_{J', J'_z} \langle J, J_z|J', J'_z\rangle \langle J', J'_z| \Big) \Big(\int_0^\infty \mathrm{d}k'_{||}\, k'_{||} \int \mathrm{d}\phi_{\mathbf{k}'_{||}} \int_0^\infty \mathrm{d}k'_z\, \langle k_z| \langle k_{||}, L_z|\mathbf{k}'_{||}\rangle |k'_z\rangle \Big) \\ &\quad \times \Big(\langle \mathbf{k}'_{||}| \otimes \langle k'_z| \Big) |\psi\rangle \\ &= \int_0^\infty \mathrm{d}k'_{||}\, k'_{||} \int \mathrm{d}\phi_{\mathbf{k}'_{||}} \int_0^\infty \mathrm{d}k'_z\, \langle k_z| \langle k_{||}, L_z|\mathbf{k}'\rangle \left(\langle J, J_z| \otimes \langle \mathbf{k}'_{||}| \otimes \langle k'_z| \right) |\psi\rangle \\ &= \int_0^\infty \mathrm{d}k'_{||}\, k'_{||} \frac{\delta(k_{||} - k'_{||})}{k_{||}} \int \mathrm{d}\phi_{\mathbf{k}'_{||}} \mathrm{e}^{-\mathrm{i}L_z \phi_{\mathbf{k}_{||}}} \int_0^\infty \mathrm{d}k'_z\, \langle k_z|k'_z\rangle \langle \mathbf{k}, J, J_z|\psi\rangle \\ &= \int \mathrm{d}\phi_{\mathbf{k}_{||}} \mathrm{e}^{-\mathrm{i}(F_z - J_z)\phi_{\mathbf{k}_{||}}} \langle \mathbf{k}, J, J_z|\psi\rangle. \end{aligned} \tag{12.176}$$

Finally, the Hamiltonian is given by

$$\begin{aligned}&\langle k_z', k_{||}', F_z', J', J_z'|H|k_z, k_{||}, F_z, J, J_z\rangle\\&= \int \mathrm{d}\phi_{\mathbf{k}_{||}'} \int \mathrm{d}\phi_{\mathbf{k}_{||}} \mathrm{e}^{\mathrm{i}(F_z - J_z)\phi_{\mathbf{k}_{||}} - \mathrm{i}(F_z' - J_z')\phi_{\mathbf{k}_{||}'}} \langle \mathbf{k}', J', J_z'|H|\mathbf{k}, J, J_z\rangle.\end{aligned}$$

Again, the old Hamiltonian is diagonal in **k** and one angular integral becomes unity (note that the measure of 2π was not explicit written down):

$$\begin{aligned}&\langle k_z', k_{||}', F_z', J', J_z'|H|k_z, k_{||}, F_z, J, J_z\rangle\\&= \int \mathrm{d}\phi_{\mathbf{k}_{||}'} \int \mathrm{d}\phi_{\mathbf{k}_{||}} \mathrm{e}^{\mathrm{i}(F_z - J_z)\phi_{\mathbf{k}_{||}} - \mathrm{i}(F_z' - J_z')\phi_{\mathbf{k}_{||}'}} \langle \mathbf{k}, J', J_z'|H|\mathbf{k}, J, J_z\rangle\\&\quad\times \delta(k_z - k_z')\delta(\mathbf{k}_{||} - \mathbf{k}_{|}|')\delta(\phi_{\mathbf{k}_{||}} - \phi_{\mathbf{k}_{||}'})\\&= \int \mathrm{d}\phi_{\mathbf{k}_{||}} \mathrm{e}^{\mathrm{i}[(F_z - J_z) - (F_z' - J_z')]\phi_{\mathbf{k}_{||}}} \langle \mathbf{k}, J', J_z'|H|\mathbf{k}, J, J_z\rangle \delta(k_z - k_z')\delta(\mathbf{k}_{||} - \mathbf{k}_{||}').\end{aligned}$$

Therefore,

$$\begin{aligned}&\langle k_z, k_{||}, F_z', J', J_z'|H|k_z, k_{||}, F_z, J, J_z\rangle\\&= \frac{1}{2\pi} \int_0^{2\pi} \mathrm{d}\phi_{\mathbf{k}_{||}} \mathrm{e}^{\mathrm{i}[(F_z - F_z') - (J_z - J_z')]\phi_{\mathbf{k}_{||}}} \langle \mathbf{k}, J', J_z'|H|\mathbf{k}, J, J_z\rangle. \qquad (12.177)\end{aligned}$$

In the old basis,

$$\langle \mathbf{k}, J', J_z'|H|\mathbf{k}, J, J_z\rangle \sim \mathrm{e}^{\mathrm{i}(J_z - J_z')\phi_{\mathbf{k}_{||}}},$$

and the angular dependence in J_z goes away; for the integral over ϕ not to be zero we have $F_z = F_z'$.

12.5.3.2 Four-Band Model

Within the axial approximation, the four-band model is exactly solvable for a cylindrical quantum wire [252]. The method of Sercel and Vahala is to expand the solutions inside and outside the quantum wire in terms of the cylindrical functions and to match boundary conditions at the interface.

12.5.3.3 Bulk Solutions

In the product space of well-defined F_z, the four-band Hamiltonian looks exactly the same as the one in the conventional basis except for the replacement $k_x = k_y = k_{||}$. This can be shown by using Eq. (12.177) and the conventional Hamiltonian (e.g., Table 5.3). The new Sercel–Vahala Hamiltonian is given in Table 12.7. It is now

Table 12.7 Four-band Sercel–Vahala Hamiltonian. Also note the conventional negative sign outside the matrix

$$H_{SV} = -\begin{pmatrix} \left|\frac{3}{2}\,\frac{3}{2}\right\rangle\left|F_z-\frac{3}{2}\right\rangle & \left|\frac{3}{2}\,\frac{1}{2}\right\rangle\left|F_z-\frac{1}{2}\right\rangle & \left|\frac{3}{2}\,-\frac{1}{2}\right\rangle\left|F_z+\frac{1}{2}\right\rangle & \left|\frac{3}{2}\,-\frac{3}{2}\right\rangle\left|F_z+\frac{3}{2}\right\rangle \\ P' & -S & R & 0 \\ -S^* & P'' & 0 & -R \\ R^* & 0 & P'' & -S \\ 0 & -R^* & -S^* & P' \end{pmatrix}$$

with

$$P' = \frac{\hbar^2}{2m_0}\left\{(\gamma_1+\gamma_2)k_{||}^2 + (\gamma_1-2\gamma_2)k_z^2\right\},$$

$$P'' = \frac{\hbar^2}{2m_0}\left\{(\gamma_1-\gamma_2)k_{||}^2 + (\gamma_1+2\gamma_2)k_z^2\right\},$$

$$S = \frac{\hbar^2}{2m_0}2\sqrt{3}\tilde{\gamma}k_{||}k_z,$$

$$R = -\frac{\hbar^2}{2m_0}\sqrt{3}\tilde{\gamma}k_{||}^2.$$

necessary to obtain the bulk solutions. It is clear there are analytic solutions since the four-band is exactly solvable even without making the axial approximation. By direct substitution, one can show that the four eigenvectors are

$$|hh_1\rangle = \begin{pmatrix} \frac{k_{||}^2+4k_z^2}{\sqrt{3}k_{||}^2} \\ \frac{2k_z}{k_{||}} \\ 1 \\ 0 \end{pmatrix}, |hh_2\rangle = \begin{pmatrix} \frac{2k_z}{k_{||}} \\ \sqrt{3} \\ 0 \\ -1 \end{pmatrix}, |lh_1\rangle = \begin{pmatrix} -\sqrt{3} \\ \frac{2k_z}{k_{||}} \\ 1 \\ 0 \end{pmatrix}, |lh_2\rangle = \begin{pmatrix} \frac{2k_z}{k_{||}} \\ -\frac{k_{||}^2+4k_z^2}{\sqrt{3}k_{||}^2} \\ 0 \\ -1 \end{pmatrix}, \tag{12.178}$$

with eigenvalues $-\hbar^2(\gamma_1-2\gamma)k^2/(2m_0)$ (twice) and $-\hbar^2(\gamma_1+2\gamma)k^2/(2m_0)$ (twice).

If $S = 0$ in Table 12.7 (occurs when $k_z = 0$), the matrix decouples into two 2×2 subblocks and the coupled states are $\left|\frac{3}{2}\frac{3}{2}\right\rangle$, $\left|\frac{3}{2}-\frac{1}{2}\right\rangle$ and $\left|\frac{3}{2}\frac{1}{2}\right\rangle$, $\left|\frac{3}{2}-\frac{3}{2}\right\rangle$:

$$|hh_1\rangle = \begin{pmatrix} \frac{1}{\sqrt{3}} \\ 0 \\ 1 \\ 0 \end{pmatrix}, |hh_2\rangle = \begin{pmatrix} 0 \\ \sqrt{3} \\ 0 \\ -1 \end{pmatrix}, |lh_1\rangle = \begin{pmatrix} -\sqrt{3} \\ 0 \\ 1 \\ 0 \end{pmatrix}, |lh_2\rangle = \begin{pmatrix} 0 \\ -\frac{1}{\sqrt{3}} \\ 0 \\ -1 \end{pmatrix}. \tag{12.179}$$

The corresponding values of L_z for the above four states are, respectively, $L_z = F_z-\frac{3}{2}, F_z-\frac{1}{2}, F_z+\frac{1}{2}, F_z+\frac{3}{2}$ and, depending upon the actual F_z value, they have opposite parity.

12.5.3.4 Envelope Functions

The quantum-wire envelope functions can, therefore, be written as

$$\phi_{F_z}^{hh1} = \begin{pmatrix} \frac{k_{\|}^2+4k_z^2}{\sqrt{3}k_{\|}^2} J_{F_z-\frac{3}{2}}(k_{hh}\rho) \\ \frac{2k_z}{k_{\|}} J_{F_z-\frac{1}{2}}(k_{hh}\rho) \\ J_{F_z+\frac{1}{2}}(k_{hh}\rho) \\ 0 \end{pmatrix} \mathrm{e}^{\mathrm{i}k_z z},\ \phi_{F_z}^{hh2} = \begin{pmatrix} \frac{2k_z}{k_{\|}} J_{F_z-\frac{3}{2}}(k_{hh}\rho) \\ \sqrt{3} J_{F_z-\frac{1}{2}}(k_{hh}\rho) \\ 0 \\ J_{F_z+\frac{1}{2}}(k_{hh}\rho) \end{pmatrix} \mathrm{e}^{\mathrm{i}k_z z},$$

$$\phi_{F_z}^{lh1} = \begin{pmatrix} -\sqrt{3} J_{F_z-\frac{3}{2}}(k_{lh}\rho) \\ \frac{2k_z}{k_{\|}} J_{F_z-\frac{1}{2}}(k_{lh}\rho) \\ J_{F_z+\frac{1}{2}}(k_{lh}\rho) \\ 0 \end{pmatrix} \mathrm{e}^{\mathrm{i}k_z z},\ \phi_{F_z}^{lh2} = \begin{pmatrix} \frac{2k_z}{k_{\|}} J_{F_z-\frac{3}{2}}(k_{lh}\rho) \\ -\frac{k_{\|}^2+4k_z^2}{\sqrt{3}k_{\|}^2} J_{F_z-\frac{1}{2}}(k_{lh}\rho) \\ J_{F_z+\frac{3}{2}}(k_{lh}\rho) \\ 0 \end{pmatrix} \mathrm{e}^{\mathrm{i}k_z z}, \tag{12.180}$$

where the angular dependence is left out. Note that not all the Bessel functions will go to zero at the origin. A quantum-wire eigenfunction is then

$$\psi_{F_z} = A|hh_1\rangle + B|hh_2\rangle + C|lh_1\rangle + D|lh_2\rangle. \tag{12.181}$$

12.5.3.5 Infinite-Barrier Solutions

For a quantum wire of radius R with an infinite barrier, the energies are given by the solutions to the following determinantal equation [36]:

$$\begin{vmatrix} \frac{k_{hh}^2+4k_z^2}{\sqrt{3}k_{hh}^2} J_{F_z-\frac{1}{2}}(k_{hh}R) & \frac{2k_z}{k_{hh}} J_{F_z-\frac{3}{2}}(k_{hh}R) & -\sqrt{3} J_{F_z-\frac{3}{2}}(k_{lh}R) & \frac{2k_z}{k_{lh}} J_{F_z-\frac{3}{2}}(k_{lh}R) \\ \frac{2k_z}{k_{hh}} J_{F_z-\frac{1}{2}}(k_{hh}R) & \sqrt{3} J_{F_z-\frac{1}{2}}(k_{hh}R) & \frac{2k_z}{k_{lh}} J_{F_z-\frac{1}{2}}(k_{lh}R) & \\ & -\frac{k_{lh}^2+4k_z^2}{\sqrt{3}k_{lh}^2} J_{F_z-\frac{1}{2}}(k_{lh}R) & & \\ J_{F_z+\frac{1}{2}}(k_{hh}R) & 0 & J_{F_z+\frac{1}{2}}(k_{lh}R) & 0 \\ 0 & J_{F_z+\frac{3}{2}}(k_{hh}R) & 0 & J_{F_z+\frac{3}{2}}(k_{lh}R) \end{vmatrix} = 0. \tag{12.182}$$

Note that the determinant is not symmetrical.

For $k_z = 0$, only hhi and lhi are coupled:

$$F_z = +\frac{1}{2}:$$

$$|hh_1\rangle = \begin{pmatrix} \frac{1}{\sqrt{3}} J_{-1} \\ 0 \\ J_{+1} \\ 0 \end{pmatrix}, |hh_2\rangle = \begin{pmatrix} 0 \\ \sqrt{3} J_0 \\ 0 \\ -J_2 \end{pmatrix}, |lh_1\rangle = \begin{pmatrix} -\sqrt{3} J_{-1} \\ 0 \\ J_{+1} \\ 0 \end{pmatrix}, |lh_2\rangle = \begin{pmatrix} 0 \\ -\frac{1}{\sqrt{3}} J_0 \\ 0 \\ -J_2 \end{pmatrix},$$

$$F_z = -\frac{1}{2}: \tag{12.183}$$

$$|hh_1\rangle = \begin{pmatrix} \frac{1}{\sqrt{3}} J_{-2} \\ 0 \\ J_0 \\ 0 \end{pmatrix}, |hh_2\rangle = \begin{pmatrix} 0 \\ \sqrt{3} J_{-1} \\ 0 \\ -J_{+1} \end{pmatrix}, |lh_1\rangle = \begin{pmatrix} -\sqrt{3} J_{-2} \\ 0 \\ J_0 \\ 0 \end{pmatrix}, |lh_2\rangle = \begin{pmatrix} 0 \\ -\frac{1}{\sqrt{3}} J_{-1} \\ 0 \\ -J_{+1} \end{pmatrix},$$

$$F_z = +\frac{3}{2}:$$

$$|hh_1\rangle = \begin{pmatrix} \frac{1}{\sqrt{3}} J_0 \\ 0 \\ J_2 \\ 0 \end{pmatrix}, |hh_2\rangle = \begin{pmatrix} 0 \\ \sqrt{3} J_1 \\ 0 \\ -J_3 \end{pmatrix}, |lh_1\rangle = \begin{pmatrix} -\sqrt{3} J_0 \\ 0 \\ J_2 \\ 0 \end{pmatrix}, |lh_2\rangle = \begin{pmatrix} 0 \\ -\frac{1}{\sqrt{3}} J_1 \\ 0 \\ -J_3 \end{pmatrix},$$

$$F_z = -\frac{3}{2}:$$

$$|hh_1\rangle = \begin{pmatrix} \frac{1}{\sqrt{3}} J_{-3} \\ 0 \\ J_{-1} \\ 0 \end{pmatrix}, |hh_2\rangle = \begin{pmatrix} 0 \\ \sqrt{3} J_{-2} \\ 0 \\ -J_0 \end{pmatrix}, |lh_1\rangle = \begin{pmatrix} -\sqrt{3} J_{-3} \\ 0 \\ J_{-1} \\ 0 \end{pmatrix}, |lh_2\rangle = \begin{pmatrix} 0 \\ -\frac{1}{\sqrt{3}} J_{-2} \\ 0 \\ -J_0 \end{pmatrix}.$$

Next, we note that

$$J_{-n}(k\rho) = (-1)^n J_n(k\rho).$$

Thus, we can label the states as being either even or odd depending upon the parity of the Bessel function. For $F_z = \pm\frac{1}{2}$, we see from Eq. (12.183) that the odd states can satisfy the boundary condition independently for the hh and lh components, i.e., hh and lh are uncoupled. For all other states, the Bessel functions cannot be simultaneously zero; hence, a linear combination is required in order to satisfy the boundary condition, leading the hh–lh mixing.

12.5.4 Four-Band Hamiltonian in Cylindrical Polar Coordinates

We now derive the four-band BF Hamiltonian in cylindrical polar coordinates for zincblende. This has application to problems with axial symmetry, e.g., cylindrical quantum wires and circular quantum rings. Both problems have been studied within the LK model [36, 252, 253, 255–257].

We first give the Hamiltonian (Table 12.8) before the derivation. The Hamiltonian in Table 12.8 is similar to the one given by, e.g., Planelles et al. [255] in the limit of LK, constant Luttinger parameters, and zero magnetic field (except for differences that can be accounted for by interchanging the sign of the second row and column). Note that they set $\hbar = m_0 = 1$ and their factor of $1/2$ is inside their matrix elements.

Table 12.8 Four-band BF Hamiltonian in cylindrical polar coordinates for a zincblende quantum wire. $E_v(z)$ is the valence-band offset

$$H_{\rm BF}(F_z) = \frac{\hbar^2}{2m_0}\begin{pmatrix} |\frac{3}{2}\,\frac{3}{2}\rangle & |\frac{3}{2}\,\frac{1}{2}\rangle & |\frac{3}{2}\,-\frac{1}{2}\rangle & |\frac{3}{2}\,-\frac{3}{2}\rangle \\ P_1 & S_1 & -R_1 & 0 \\ S_1^\dagger & P_2 & -C & R_2 \\ -R_1^\dagger & -C^\dagger & P_3 & S_2 \\ 0 & R_2^\dagger & S_2^\dagger & P_4 \end{pmatrix} + E_v(z)\mathcal{I},$$

where

$$P_1 = \frac{\partial}{\partial\rho}(\gamma_1+\gamma_2)\frac{\partial}{\partial\rho} + \frac{(\gamma_1+\gamma_2)}{\rho}\frac{\partial}{\partial\rho} + \frac{\partial}{\partial z}(\gamma_1-2\gamma_2)\frac{\partial}{\partial z} - \frac{(\gamma_1+\gamma_2)}{\rho^2}\left(F_z-\frac{3}{2}\right)^2 + \frac{(F_z-\frac{3}{2})}{2\rho}\left[\frac{\partial}{\partial\rho}(C_1+C_2)-(C_1+C_2)\frac{\partial}{\partial\rho}\right],$$

$$P_2 = \frac{\partial}{\partial\rho}(\gamma_1-\gamma_2)\frac{\partial}{\partial\rho} + \frac{(\gamma_1-\gamma_2)}{\rho}\frac{\partial}{\partial\rho} + \frac{\partial}{\partial z}(\gamma_1+2\gamma_2)\frac{\partial}{\partial z} - \frac{(\gamma_1-\gamma_2)}{\rho^2}\left(F_z-\frac{1}{2}\right)^2 + \frac{(F_z-\frac{1}{2})}{6\rho}\left[\frac{\partial}{\partial\rho}(C_1+C_2)-(C_1+C_2)\frac{\partial}{\partial\rho}\right],$$

$$P_3 = \frac{\partial}{\partial\rho}(\gamma_1-\gamma_2)\frac{\partial}{\partial\rho} + \frac{(\gamma_1-\gamma_2)}{\rho}\frac{\partial}{\partial\rho} + \frac{\partial}{\partial z}(\gamma_1+2\gamma_2)\frac{\partial}{\partial z} - \frac{(\gamma_1-\gamma_2)}{\rho^2}\left(F_z+\frac{1}{2}\right)^2 - \frac{(F_z+\frac{1}{2})}{6\rho}\left[\frac{\partial}{\partial\rho}(C_1+C_2)-(C_1+C_2)\frac{\partial}{\partial\rho}\right],$$

$$P_4 = \frac{\partial}{\partial\rho}(\gamma_1+\gamma_2)\frac{\partial}{\partial\rho} + \frac{(\gamma_1+\gamma_2)}{\rho}\frac{\partial}{\partial\rho} + \frac{\partial}{\partial z}(\gamma_1-2\gamma_2)\frac{\partial}{\partial z} - \frac{(\gamma_1+\gamma_2)}{\rho^2}\left(F_z+\frac{3}{2}\right)^2 - \frac{(F_z+\frac{3}{2})}{2\rho}\left[\frac{\partial}{\partial\rho}(C_1+C_2)-(C_1+C_2)\frac{\partial}{\partial\rho}\right],$$

$$S_1 = \frac{1}{\sqrt{3}}\left\{\frac{\partial}{\partial\rho}C_1\frac{\partial}{\partial z} - \frac{\partial}{\partial z}C_2\frac{\partial}{\partial\rho} + \frac{(F_z-\frac{1}{2})}{\rho}\left[C_1\frac{\partial}{\partial z}-\frac{\partial}{\partial z}C_2\right]\right\},$$

$$S_1^\dagger = \frac{1}{\sqrt{3}}\left\{\frac{\partial}{\partial z}C_1\frac{\partial}{\partial\rho} - \frac{\partial}{\partial\rho}C_2\frac{\partial}{\partial z} - \frac{(F_z-\frac{3}{2})}{\rho}\left[\frac{\partial}{\partial z}C_1 - C_2\frac{\partial}{\partial z}\right]\right\},$$

$$S_2 = \frac{1}{\sqrt{3}}\left\{\frac{\partial}{\partial z}C_1\frac{\partial}{\partial\rho} - \frac{\partial}{\partial\rho}C_2\frac{\partial}{\partial z} + \frac{(F_z+\frac{3}{2})}{\rho}\left[\frac{\partial}{\partial z}C_1 - C_2\frac{\partial}{\partial z}\right]\right\},$$

$$S_2^\dagger = \frac{1}{\sqrt{3}}\left\{\frac{\partial}{\partial\rho}C_1\frac{\partial}{\partial z} - \frac{\partial}{\partial z}C_2\frac{\partial}{\partial\rho} - \frac{(F_z+\frac{1}{2})}{\rho}\left[C_1\frac{\partial}{\partial z}-\frac{\partial}{\partial z}C_2\right]\right\},$$

$$R_1 = \sqrt{3}\left\{\frac{\partial}{\partial\rho}\widetilde{\gamma}\frac{\partial}{\partial\rho} + \frac{(F_z+\frac{1}{2})}{\rho}\frac{\partial}{\partial\rho}\widetilde{\gamma} + \frac{(F_z-\frac{1}{2})}{\rho}\widetilde{\gamma}\frac{\partial}{\partial\rho} + \frac{F_z(F_z-1)-\frac{3}{4}}{\rho^2}\widetilde{\gamma}\right\},$$

Table 12.8 (continued)

$$R_1^\dagger = \sqrt{3}\left\{\frac{\partial}{\partial\rho}\tilde{\gamma}\frac{\partial}{\partial\rho} - \frac{\left(F_z-\frac{3}{2}\right)}{\rho}\frac{\partial}{\partial\rho}\tilde{\gamma} - \frac{\left(F_z-\frac{1}{2}\right)}{\rho}\tilde{\gamma}\frac{\partial}{\partial\rho} + \frac{F_z(F_z-1)-\frac{3}{4}}{\rho^2}\tilde{\gamma}\right\},$$

$$R_2 = \sqrt{3}\left\{\frac{\partial}{\partial\rho}\tilde{\gamma}\frac{\partial}{\partial\rho} + \frac{\left(F_z+\frac{3}{2}\right)}{\rho}\frac{\partial}{\partial\rho}\tilde{\gamma} + \frac{\left(F_z+\frac{1}{2}\right)}{\rho}\tilde{\gamma}\frac{\partial}{\partial\rho} + \frac{F_z(F_z+1)-\frac{3}{4}}{\rho^2}\tilde{\gamma}\right\},$$

$$R_2^\dagger = \sqrt{3}\left\{\frac{\partial}{\partial\rho}\tilde{\gamma}\frac{\partial}{\partial\rho} - \frac{\left(F_z-\frac{1}{2}\right)}{\rho}\frac{\partial}{\partial\rho}\tilde{\gamma} - \frac{\left(F_z+\frac{1}{2}\right)}{\rho}\tilde{\gamma}\frac{\partial}{\partial\rho} + \frac{F_z(F_z+1)-\frac{3}{4}}{\rho^2}\tilde{\gamma}\right\},$$

$$C = \frac{1}{\sqrt{3}}\left\{\frac{\partial}{\partial z}(C_1+C_2)\frac{\partial}{\partial\rho} - \frac{\partial}{\partial\rho}(C_1+C_2)\frac{\partial}{\partial z} + \frac{\left(F_z+\frac{1}{2}\right)}{\rho}\left[\frac{\partial}{\partial z}(C_1+C_2) - (C_1+C_2)\frac{\partial}{\partial z}\right]\right\},$$

$$C^\dagger = \frac{1}{\sqrt{3}}\left\{\frac{\partial}{\partial\rho}(C_1+C_2)\frac{\partial}{\partial z} - \frac{\partial}{\partial z}(C_1+C_2)\frac{\partial}{\partial\rho} + \frac{\left(F_z-\frac{1}{2}\right)}{\rho}\left[\frac{\partial}{\partial z}(C_1+C_2) - (C_1+C_2)\frac{\partial}{\partial z}\right]\right\}.$$

12.5.4.1 Conventional Basis

The procedure involves first taking the BF Hamiltonian in the conventional basis and replacing all the cartesian differential operators by their cylindrical counterparts. Here, we will do so explicitly for a quantum wire with [001] axis. We will need the following results:

$$\begin{aligned} x &= \rho\cos\phi,\ y = \rho\sin\phi,\ z = z, \\ \frac{\partial}{\partial x} &= \cos\phi\frac{\partial}{\partial\rho} - \frac{\sin\phi}{\rho}\frac{\partial}{\partial\phi}, \\ \frac{\partial}{\partial y} &= \sin\phi\frac{\partial}{\partial\rho} + \frac{\cos\phi}{\rho}\frac{\partial}{\partial\phi}, \end{aligned} \tag{12.184}$$

and, therefore,

$$\begin{aligned} \frac{\partial}{\partial x}\left[A(\mathbf{r})\frac{\partial}{\partial x}\right] &= \cos^2\phi\frac{\partial}{\partial\rho}\left[A(\mathbf{r})\frac{\partial}{\partial\rho}\right] + \frac{\sin\phi}{\rho^2}\frac{\partial}{\partial\phi}\left[A(\mathbf{r})\sin\phi\frac{\partial}{\partial\phi}\right] \\ &\quad - \sin\phi\cos\phi\frac{\partial}{\partial\rho}\left[\frac{A(\mathbf{r})}{\rho}\frac{\partial}{\partial\phi}\right] - \frac{\sin\phi}{\rho}\frac{\partial}{\partial\phi}\left[A(\mathbf{r})\cos\phi\frac{\partial}{\partial\rho}\right], \\ \frac{\partial}{\partial y}\left[A(\mathbf{r})\frac{\partial}{\partial y}\right] &= \sin^2\phi\frac{\partial}{\partial\rho}\left[A(\mathbf{r})\frac{\partial}{\partial\rho}\right] + \frac{\cos\phi}{\rho^2}\frac{\partial}{\partial\phi}\left[A(\mathbf{r})\cos\phi\frac{\partial}{\partial\phi}\right] \\ &\quad + \sin\phi\cos\phi\frac{\partial}{\partial\rho}\left[\frac{A(\mathbf{r})}{\rho}\frac{\partial}{\partial\phi}\right] + \frac{\cos\phi}{\rho}\frac{\partial}{\partial\phi}\left[A(\mathbf{r})\sin\phi\frac{\partial}{\partial\rho}\right], \end{aligned}$$

$$
\begin{aligned}
\frac{\partial}{\partial x}\left[A(\mathbf{r})\frac{\partial}{\partial y}\right] &= \sin\phi\cos\phi\frac{\partial}{\partial\rho}\left[A(\mathbf{r})\frac{\partial}{\partial\rho}\right] - \frac{\sin\phi}{\rho^2}\frac{\partial}{\partial\phi}\left[A(\mathbf{r})\cos\phi\frac{\partial}{\partial\phi}\right] \\
&\quad + \cos^2\phi\frac{\partial}{\partial\rho}\left[\frac{A(\mathbf{r})}{\rho}\frac{\partial}{\partial\phi}\right] - \frac{\sin\phi}{\rho}\frac{\partial}{\partial\phi}\left[A(\mathbf{r})\sin\phi\frac{\partial}{\partial\rho}\right], \\
\frac{\partial}{\partial y}\left[A(\mathbf{r})\frac{\partial}{\partial x}\right] &= \sin\phi\cos\phi\frac{\partial}{\partial\rho}\left[A(\mathbf{r})\frac{\partial}{\partial\rho}\right] - \frac{\cos\phi}{\rho^2}\frac{\partial}{\partial\phi}\left[A(\mathbf{r})\sin\phi\frac{\partial}{\partial\phi}\right] \\
&\quad - \sin^2\phi\frac{\partial}{\partial\rho}\left[\frac{A(\mathbf{r})}{\rho}\frac{\partial}{\partial\phi}\right] + \frac{\cos\phi}{\rho}\frac{\partial}{\partial\phi}\left[A(\mathbf{r})\cos\phi\frac{\partial}{\partial\rho}\right], \\
\frac{\partial}{\partial x}\left[A(\mathbf{r})\frac{\partial}{\partial z}\right] &= \cos\phi\frac{\partial}{\partial\rho}\left[A(\mathbf{r})\frac{\partial}{\partial z}\right] - \frac{\sin\phi}{\rho}\frac{\partial}{\partial\phi}\left[A(\mathbf{r})\frac{\partial}{\partial z}\right], \qquad (12.185) \\
\frac{\partial}{\partial z}\left[A(\mathbf{r})\frac{\partial}{\partial x}\right] &= \cos\phi\frac{\partial}{\partial z}\left[A(\mathbf{r})\frac{\partial}{\partial\rho}\right] - \frac{\sin\phi}{\rho}\frac{\partial}{\partial z}\left[A(\mathbf{r})\frac{\partial}{\partial\phi}\right], \\
\frac{\partial}{\partial y}\left[A(\mathbf{r})\frac{\partial}{\partial z}\right] &= \sin\phi\frac{\partial}{\partial\rho}\left[A(\mathbf{r})\frac{\partial}{\partial z}\right] + \frac{\cos\phi}{\rho}\frac{\partial}{\partial\phi}\left[A(\mathbf{r})\frac{\partial}{\partial z}\right], \\
\frac{\partial}{\partial z}\left[A(\mathbf{r})\frac{\partial}{\partial y}\right] &= \sin\phi\frac{\partial}{\partial z}\left[A(\mathbf{r})\frac{\partial}{\partial\rho}\right] + \frac{\cos\phi}{\rho}\frac{\partial}{\partial z}\left[A(\mathbf{r})\frac{\partial}{\partial\phi}\right].
\end{aligned}
$$

We now apply the above transformation to the Hamiltonian in Table 12.4. Note that

$$
\widehat{k}_i\left[A(\mathbf{r})\widehat{k}_j\right] = -\frac{\partial}{\partial x_i}\left[A(\mathbf{r})\frac{\partial}{\partial x_j}\right],
$$

and we will convert from the DKK parameters to the Stravinou-van Dalen ones, which brings out a factor of $\hbar^2/(2m_0)$ but we will leave it out of all the matrix elements, except for the final Hamiltonian of course. We have

$$
\begin{aligned}
P' &= \frac{1}{2}\left\{\widehat{k}_x(A+B)\widehat{k}_x + \widehat{k}_y(A+B)\widehat{k}_y + \widehat{k}_z 2B\widehat{k}_z\right\} + \frac{\mathrm{i}}{2}\left\{\widehat{k}_x(C_1+C_2)\widehat{k}_y - \widehat{k}_y(C_1+C_2)\widehat{k}_x\right\} \\
&= -\frac{1}{2}\left\{\frac{\partial}{\partial x}(A+B)\frac{\partial}{\partial x} + \frac{\partial}{\partial y}(A+B)\frac{\partial}{\partial y} + \frac{\partial}{\partial z}2B\frac{\partial}{\partial z}\right\} \\
&\quad -\frac{\mathrm{i}}{2}\left\{\frac{\partial}{\partial x}(C_1+C_2)\frac{\partial}{\partial y} - \frac{\partial}{\partial y}(C_1+C_2)\frac{\partial}{\partial x}\right\} \\
&= -\frac{1}{2}\left\{\frac{\partial}{\partial\rho}\left[(A+B)\frac{\partial}{\partial\rho}\right] + \frac{\sin\phi}{\rho^2}\frac{\partial}{\partial\phi}\left[(A+B)\sin\phi\frac{\partial}{\partial\phi}\right]\right. \\
&\quad + \frac{\cos\phi}{\rho^2}\frac{\partial}{\partial\phi}\left[(A+B)\cos\phi\frac{\partial}{\partial\phi}\right] - \frac{\sin\phi}{\rho}\frac{\partial}{\partial\phi}\left[(A+B)\cos\phi\frac{\partial}{\partial\rho}\right] \\
&\quad \left. + \frac{\cos\phi}{\rho}\frac{\partial}{\partial\phi}\left[(A+B)\sin\phi\frac{\partial}{\partial\rho}\right] + 2\frac{\partial}{\partial z}B\frac{\partial}{\partial z}\right\} - \frac{\mathrm{i}}{2}\left\{\frac{\partial}{\partial\rho}\left[\frac{(C_1+C_2)}{\rho}\frac{\partial}{\partial\phi}\right]\right.
\end{aligned}
$$

$$-\frac{\sin\phi}{\rho}\frac{\partial}{\partial\phi}\left[(C_1+C_2)\sin\phi\frac{\partial}{\partial\rho}\right]-\frac{\cos\phi}{\rho}\frac{\partial}{\partial\phi}\left[(C_1+C_2)\cos\phi\frac{\partial}{\partial\rho}\right]$$
$$-\frac{\sin\phi}{\rho^2}\frac{\partial}{\partial\phi}\left[(C_1+C_2)\cos\phi\frac{\partial}{\partial\phi}\right]+\frac{\cos\phi}{\rho^2}\frac{\partial}{\partial\phi}\left[(C_1+C_2)\sin\phi\frac{\partial}{\partial\phi}\right]\Bigg\}$$
$$=-\frac{1}{2}\left\{\frac{\partial}{\partial\rho}\left[(A+B)\frac{\partial}{\partial\rho}\right]+\frac{1}{\rho^2}\frac{\partial}{\partial\phi}\left[(A+B)\frac{\partial}{\partial\phi}\right]+\frac{(A+B)}{\rho}\frac{\partial}{\partial\rho}+2\frac{\partial}{\partial z}B\frac{\partial}{\partial z}\right\}$$
$$-\frac{\mathrm{i}}{2}\left\{\frac{\partial}{\partial\rho}\left[\frac{(C_1+C_2)}{\rho}\frac{\partial}{\partial\phi}\right]-\frac{1}{\rho}\frac{\partial}{\partial\phi}\left[(C_1+C_2)\frac{\partial}{\partial\rho}\right]+\frac{(C_1+C_2)}{\rho^2}\frac{\partial}{\partial\phi}\right\}. \tag{12.186}$$

The A, B, C_1, C_2 are the parameters of Stravinou and van Dalen; we will re-express them in terms of the Luttinger parameters at the end. Equation (12.186) is at this point quite general in the sense that the material parameters have arbitrary spatial dependence. We now restrict to the case of a cylindrical quantum wire with possible modulation in the z-direction (e.g., a modulated nanowire [261]), and we will, therefore, also give the results for when the parameters only have radial and z dependence. This gives

$$P'=-\frac{1}{2}\left\{\frac{\partial}{\partial\rho}\left[(A+B)\frac{\partial}{\partial\rho}\right]+\frac{(A+B)}{\rho^2}\frac{\partial^2}{\partial\phi^2}+\frac{(A+B)}{\rho}\frac{\partial}{\partial\rho}+2\frac{\partial}{\partial z}B\frac{\partial}{\partial z}\right\}$$
$$-\frac{\mathrm{i}}{2}\left\{\frac{\partial}{\partial\rho}\left[\frac{(C_1+C_2)}{\rho}\frac{\partial}{\partial\phi}\right]-\frac{\partial}{\partial\phi}\left[\frac{(C_1+C_2)}{\rho}\frac{\partial}{\partial\rho}\right]+\frac{(C_1+C_2)}{\rho^2}\frac{\partial}{\partial\phi}\right\}$$
$$=\frac{\partial}{\partial\rho}(\gamma_1+\gamma_2)\frac{\partial}{\partial\rho}+\frac{(\gamma_1+\gamma_2)}{\rho}\frac{\partial}{\partial\rho}+\frac{\partial}{\partial z}(\gamma_1-2\gamma_2)\frac{\partial}{\partial z}+\frac{(\gamma_1+\gamma_2)}{\rho^2}\frac{\partial^2}{\partial\phi^2}$$
$$-\frac{\mathrm{i}}{2\rho}\left\{\frac{\partial}{\partial\rho}\left[(C_1+C_2)\frac{\partial}{\partial\phi}\right]-\frac{\partial}{\partial\phi}\left[(C_1+C_2)\frac{\partial}{\partial\rho}\right]\right\}.$$

Continuing,

$$P''=\frac{1}{6}\left\{\widehat{k}_x(A+5B)\widehat{k}_x+\widehat{k}_y(A+5B)\widehat{k}_y+2\widehat{k}_z(2A+B)\widehat{k}_z\right\}$$
$$+\frac{\mathrm{i}}{6}\left\{\widehat{k}_x(C_1+C_2)\widehat{k}_y-\widehat{k}_y(C_1+C_2)\widehat{k}_x\right\}$$
$$=-\frac{1}{6}\left\{\frac{\partial}{\partial\rho}\left[(A+5B)\frac{\partial}{\partial\rho}\right]+\frac{1}{\rho^2}\frac{\partial}{\partial\phi}\left[(A+5B)\frac{\partial}{\partial\phi}\right]+\frac{(A+5B)}{\rho}\frac{\partial}{\partial\rho}\right.$$
$$\left.+2\frac{\partial}{\partial z}(2A+B)\frac{\partial}{\partial z}\right\}-\frac{\mathrm{i}}{6}\left\{\frac{\partial}{\partial\rho}\left[\frac{(C_1+C_2)}{\rho}\frac{\partial}{\partial\phi}\right]-\frac{1}{\rho}\frac{\partial}{\partial\phi}\left[(C_1+C_2)\frac{\partial}{\partial\rho}\right]\right.$$
$$\left.+\frac{(C_1+C_2)}{\rho^2}\frac{\partial}{\partial\phi}\right\} \tag{12.187}$$

$$= \frac{\partial}{\partial \rho}(\gamma_1 - \gamma_2)\frac{\partial}{\partial \rho} + \frac{(\gamma_1 - \gamma_2)}{\rho}\frac{\partial}{\partial \rho} + \frac{\partial}{\partial z}(\gamma_1 + 2\gamma_2)\frac{\partial}{\partial z} + \frac{(\gamma_1 - \gamma_2)}{\rho^2}\frac{\partial^2}{\partial \phi^2}$$
$$-\frac{\mathrm{i}}{6\rho}\left\{\frac{\partial}{\partial \rho}\left[(C_1 + C_2)\frac{\partial}{\partial \phi}\right] - \frac{\partial}{\partial \phi}\left[(C_1 + C_2)\frac{\partial}{\partial \rho}\right]\right\}, \quad (12.188)$$

$$S_\pm = -\frac{1}{\sqrt{3}}\left\{\widehat{k}_\pm C_1 \widehat{k}_z - \widehat{k}_z C_2 \widehat{k}_\pm\right\}$$
$$= \frac{1}{\sqrt{3}}\left\{\cos\phi\left[\frac{\partial}{\partial \rho}C_1\frac{\partial}{\partial z} - \frac{\partial}{\partial z}C_2\frac{\partial}{\partial \rho}\right] - \frac{\sin\phi}{\rho}\left[\frac{\partial}{\partial \phi}C_1\frac{\partial}{\partial z} - \frac{\partial}{\partial z}C_2\frac{\partial}{\partial \phi}\right]\right\}$$
$$\pm\frac{\mathrm{i}}{\sqrt{3}}\left\{\sin\phi\left[\frac{\partial}{\partial \rho}C_1\frac{\partial}{\partial z} - \frac{\partial}{\partial z}C_2\frac{\partial}{\partial \rho}\right] + \frac{\cos\phi}{\rho}\left[\frac{\partial}{\partial \phi}C_1\frac{\partial}{\partial z} - \frac{\partial}{\partial z}C_2\frac{\partial}{\partial \phi}\right]\right\}$$
$$= \frac{\mathrm{e}^{\pm \mathrm{i}\phi}}{\sqrt{3}}\left\{\left[\frac{\partial}{\partial \rho}C_1\frac{\partial}{\partial z} - \frac{\partial}{\partial z}C_2\frac{\partial}{\partial \rho}\right] \pm \frac{\mathrm{i}}{\rho}\left[\frac{\partial}{\partial \phi}C_1\frac{\partial}{\partial z} - \frac{\partial}{\partial z}C_2\frac{\partial}{\partial \phi}\right]\right\}, \quad (12.189)$$

$$S_1 \equiv S_-,$$

$$S_1^\dagger = \frac{\mathrm{e}^{\mathrm{i}\phi}}{\sqrt{3}}\left\{\left[\frac{\partial}{\partial z}C_1\frac{\partial}{\partial \rho} - \frac{\partial}{\partial \rho}C_2\frac{\partial}{\partial z}\right] + \frac{\mathrm{i}}{\rho}\left[\frac{\partial}{\partial z}C_1\frac{\partial}{\partial \phi} - \frac{\partial}{\partial \phi}C_2\frac{\partial}{\partial z}\right]\right\},$$

$$S_2 = -\frac{1}{\sqrt{3}}\left\{\widehat{k}_z C_1 \widehat{k}_- - \widehat{k}_- C_2 \widehat{k}_z\right\}$$
$$= \frac{\mathrm{e}^{-\mathrm{i}\phi}}{\sqrt{3}}\left\{\left[\frac{\partial}{\partial z}C_1\frac{\partial}{\partial \rho} - \frac{\partial}{\partial \rho}C_2\frac{\partial}{\partial z}\right] - \frac{\mathrm{i}}{\rho}\left[\frac{\partial}{\partial z}C_1\frac{\partial}{\partial \phi} - \frac{\partial}{\partial \phi}C_2\frac{\partial}{\partial z}\right]\right\}, \quad (12.190)$$

$$S_2^\dagger = -\frac{1}{\sqrt{3}}\left\{\widehat{k}_+ C_1 \widehat{k}_z - \widehat{k}_z C_2 \widehat{k}_+\right\}$$
$$= \frac{\mathrm{e}^{\mathrm{i}\phi}}{\sqrt{3}}\left\{\left[\frac{\partial}{\partial \rho}C_1\frac{\partial}{\partial z} - \frac{\partial}{\partial z}C_2\frac{\partial}{\partial \rho}\right] + \frac{\mathrm{i}}{\rho}\left[\frac{\partial}{\partial \phi}C_1\frac{\partial}{\partial z} - \frac{\partial}{\partial z}C_2\frac{\partial}{\partial \phi}\right]\right\}, \quad (12.191)$$

$$R_1 = \frac{1}{2\sqrt{3}}\left\{\widehat{k}_x(A-B)\widehat{k}_x - \widehat{k}_y(A-B)\widehat{k}_y - \mathrm{i}\left[\widehat{k}_x(C_1 - C_2)\widehat{k}_y + \widehat{k}_y(C_1 - C_2)\widehat{k}_x\right]\right\}$$
$$= -\frac{1}{2\sqrt{3}}\left\{\cos 2\phi\frac{\partial}{\partial \rho}(A-B)\frac{\partial}{\partial \rho} - \frac{\cos 2\phi}{\rho^2}\frac{\partial}{\partial \phi}(A-B)\frac{\partial}{\partial \phi} + \frac{\sin 2\phi}{\rho^2}(A-B)\frac{\partial}{\partial \phi}\right.$$
$$\left. - \sin 2\phi\frac{\partial}{\partial \rho}\frac{(A-B)}{\rho}\frac{\partial}{\partial \phi} - \frac{\sin 2\phi}{\rho}\frac{\partial}{\partial \phi}(A-B)\frac{\partial}{\partial \rho} - \frac{\cos 2\phi}{\rho}(A-B)\frac{\partial}{\partial \rho}\right\}$$
$$+\frac{\mathrm{i}}{2\sqrt{3}}\left\{\sin 2\phi\frac{\partial}{\partial \rho}(C_1 - C_2)\frac{\partial}{\partial \rho} - \frac{\sin 2\phi}{\rho^2}\frac{\partial}{\partial \phi}(C_1 - C_2)\frac{\partial}{\partial \phi}\right.$$
$$-\frac{\cos 2\phi}{\rho^2}(C_1 - C_2)\frac{\partial}{\partial \phi} + \cos 2\phi\frac{\partial}{\partial \rho}\frac{(C_1 - C_2)}{\rho}\frac{\partial}{\partial \phi} + \frac{\cos 2\phi}{\rho}\frac{\partial}{\partial \phi}(C_1 - C_2)\frac{\partial}{\partial \rho}$$

$$
\begin{aligned}
&-\frac{\sin 2\phi}{\rho}(C_1 - C_2)\frac{\partial}{\partial\rho}\Big\} \\
&= \sqrt{3}\Big\{\cos 2\phi\frac{\partial}{\partial\rho}\gamma_2\frac{\partial}{\partial\rho} - \frac{\cos 2\phi}{\rho^2}\frac{\partial}{\partial\phi}\gamma_2\frac{\partial}{\partial\phi} + \frac{\sin 2\phi}{\rho^2}\gamma_2\frac{\partial}{\partial\phi} \\
&\quad - \sin 2\phi\frac{\partial}{\partial\rho}\frac{\gamma_2}{\rho}\frac{\partial}{\partial\phi} - \frac{\sin 2\phi}{\rho}\frac{\partial}{\partial\phi}\gamma_2\frac{\partial}{\partial\rho} - \frac{\cos 2\phi}{\rho}\gamma_2\frac{\partial}{\partial\rho}\Big\} \\
&\quad -\mathrm{i}\sqrt{3}\Big\{\sin 2\phi\frac{\partial}{\partial\rho}\gamma_3\frac{\partial}{\partial\rho} - \frac{\sin 2\phi}{\rho^2}\frac{\partial}{\partial\phi}\gamma_3\frac{\partial}{\partial\phi} - \frac{\cos 2\phi}{\rho^2}\gamma_3\frac{\partial}{\partial\phi} \\
&\quad + \cos 2\phi\frac{\partial}{\partial\rho}\frac{\gamma_3}{\rho}\frac{\partial}{\partial\phi} + \frac{\cos 2\phi}{\rho}\frac{\partial}{\partial\phi}\gamma_3\frac{\partial}{\partial\rho} - \frac{\sin 2\phi}{\rho}\gamma_3\frac{\partial}{\partial\rho}\Big\} \\
&\overset{\gamma_2=\gamma_3}{\approx} \sqrt{3}\mathrm{e}^{-\mathrm{i}2\phi}\Big\{\frac{\partial}{\partial\rho}\widetilde{\gamma}\frac{\partial}{\partial\rho} - \frac{1}{\rho^2}\widetilde{\gamma}\frac{\partial^2}{\partial\phi^2} + \frac{\mathrm{i}}{\rho^2}\widetilde{\gamma}\frac{\partial}{\partial\phi} - \mathrm{i}\frac{\partial}{\partial\rho}\frac{\widetilde{\gamma}}{\rho}\frac{\partial}{\partial\phi} - \frac{\mathrm{i}}{\rho}\frac{\partial}{\partial\phi}\widetilde{\gamma}\frac{\partial}{\partial\rho} \\
&\quad -\frac{1}{\rho}\widetilde{\gamma}\frac{\partial}{\partial\rho}\Big\},
\end{aligned}
\tag{12.192}
$$

$$
R_2 = R_1, \tag{12.193}
$$

$$
\begin{aligned}
R_2^{\dagger} &\overset{\gamma_2=\gamma_3}{\approx} \sqrt{3}\mathrm{e}^{\mathrm{i}2\phi}\Big\{\frac{\partial}{\partial\rho}\widetilde{\gamma}\frac{\partial}{\partial\rho} - \frac{1}{\rho^2}\widetilde{\gamma}\frac{\partial^2}{\partial\phi^2} - \frac{\mathrm{i}}{\rho^2}\widetilde{\gamma}\frac{\partial}{\partial\phi} + \mathrm{i}\frac{\partial}{\partial\rho}\frac{\widetilde{\gamma}}{\rho}\frac{\partial}{\partial\phi} + \frac{\mathrm{i}}{\rho}\frac{\partial}{\partial\phi}\widetilde{\gamma}\frac{\partial}{\partial\rho} \\
&\quad -\frac{1}{\rho}\widetilde{\gamma}\frac{\partial}{\partial\rho}\Big\},
\end{aligned}
\tag{12.194}
$$

$$
\begin{aligned}
C &= -\frac{1}{3}\left\{\widehat{k}_z(C_1 + C_2)\widehat{k}_- - \widehat{k}_-(C_1 + C_2)\widehat{k}_z\right\} \\
&= \frac{1}{3}\Big\{\cos\phi\left[\frac{\partial}{\partial z}(C_1 + C_2)\frac{\partial}{\partial\rho} - \frac{\partial}{\partial\rho}(C_1 + C_2)\frac{\partial}{\partial z}\right] - \frac{\sin\phi}{\rho}\Big[\frac{\partial}{\partial z}(C_1 + C_2)\frac{\partial}{\partial\phi} \\
&\quad -\frac{\partial}{\partial\phi}(C_1 + C_2)\frac{\partial}{\partial z}\Big]\Big\} - \frac{\mathrm{i}}{3}\Big\{\sin\phi\left[\frac{\partial}{\partial z}(C_1 + C_2)\frac{\partial}{\partial\rho} - \frac{\partial}{\partial\rho}(C_1 + C_2)\frac{\partial}{\partial z}\right] \\
&\quad + \frac{\cos\phi}{\rho}\left[\frac{\partial}{\partial z}(C_1 + C_2)\frac{\partial}{\partial\phi} - \frac{\partial}{\partial\phi}(C_1 + C_2)\frac{\partial}{\partial z}\right]\Big\} \\
&= \frac{\mathrm{e}^{-\mathrm{i}\phi}}{3}\Big\{\left[\frac{\partial}{\partial z}(C_1 + C_2)\frac{\partial}{\partial\rho} - \frac{\partial}{\partial\rho}(C_1 + C_2)\frac{\partial}{\partial z}\right] - \frac{\mathrm{i}}{\rho}\Big[\frac{\partial}{\partial z}(C_1 + C_2)\frac{\partial}{\partial\phi} \\
&\quad -\frac{\partial}{\partial\phi}(C_1 + C_2)\frac{\partial}{\partial z}\Big]\Big\},
\end{aligned}
\tag{12.195}
$$

$$
\begin{aligned}
C^{\dagger} &= \frac{\mathrm{e}^{\mathrm{i}\phi}}{3}\Big\{\left[\frac{\partial}{\partial\rho}(C_1 + C_2)\frac{\partial}{\partial z} - \frac{\partial}{\partial z}(C_1 + C_2)\frac{\partial}{\partial\rho}\right] + \frac{\mathrm{i}}{\rho}\Big[\frac{\partial}{\partial\phi}(C_1 + C_2)\frac{\partial}{\partial z} \\
&\quad -\frac{\partial}{\partial z}(C_1 + C_2)\frac{\partial}{\partial\phi}\Big]\Big\}.
\end{aligned}
\tag{12.196}
$$

12.5.4.2 F_z-Basis

For problems with axial symmetry, one can use the following state (i.e., in the F_z basis):

$$\psi(\mathbf{r}) = \sum_{J_z} f_{J_z}(\rho, z) \mathrm{e}^{\mathrm{i}(F_z - J_z)\phi} \left| \frac{3}{2} \, J_z \right\rangle. \tag{12.197}$$

Then, the matrix elements given earlier are obtained.

12.5.5 Wurtzite Structure

For problems with cylindrical symmetry, it is still useful to derive the equivalent Sercel–Vahala theory; this was first done by Lew Yan Voon et al. [58]. Using expressions from above, we have

$$\widehat{k}_x \pm \mathrm{i}\widehat{k}_y \equiv \widehat{k}_\pm = \mathrm{e}^{\pm \mathrm{i}\phi} \left(\frac{\partial}{\partial \rho} \pm \frac{\mathrm{i}}{\rho} \frac{\partial}{\partial \phi} \right),$$

and

$$\begin{aligned}
-\widehat{k}_z A(\mathbf{r}) \widehat{k}_z &= \frac{\partial}{\partial z} A(\mathbf{r}) \frac{\partial}{\partial z}, \\
-\left(\widehat{k}_\pm A(\mathbf{r}) \widehat{k}_\pm \right) &= \mathrm{e}^{\pm \mathrm{i}2\phi} \left\{ -\frac{A(\mathbf{r})}{\rho} \left(\frac{\partial}{\partial \rho} \pm \frac{1}{\rho} \frac{\partial}{\partial \phi} \right) + \frac{\partial}{\partial \rho} \left(A(\mathbf{r}) \frac{\partial}{\partial \rho} \right) \right. \\
&\quad \left. - \frac{1}{\rho^2} \frac{\partial}{\partial \phi} \left(A(\mathbf{r}) \frac{\partial}{\partial \phi} \right) \pm \mathrm{i} \frac{\partial}{\partial \rho} \left(\frac{A(\mathbf{r})}{\rho} \frac{\partial}{\partial \phi} \right) \pm \frac{\mathrm{i}}{\rho} \frac{\partial}{\partial \phi} \left(A(\mathbf{r}) \frac{\partial}{\partial \rho} \right) \right\}, \\
-\widehat{k}_\pm A(\mathbf{r}) \widehat{k}_z &= \mathrm{e}^{\pm \mathrm{i}\phi} \left(\frac{\partial}{\partial \rho} \pm \frac{\mathrm{i}}{\rho} \frac{\partial}{\partial \phi} \right) \left(A(\mathbf{r}) \frac{\partial}{\partial z} \right), \\
-\widehat{k}_z A(\mathbf{r}) \widehat{k}_\pm &= \mathrm{e}^{\pm \mathrm{i}\phi} \frac{\partial}{\partial z} \left[A(\mathbf{r}) \left(\frac{\partial}{\partial \rho} \pm \frac{\mathrm{i}}{\rho} \frac{\partial}{\partial \phi} \right) \right],
\end{aligned}$$

$$\begin{aligned}
-\left(\widehat{k}_x A(\mathbf{r}) \widehat{k}_x + \widehat{k}_y A(\mathbf{r}) \widehat{k}_y \right) &= \frac{\partial}{\partial \rho} \left(A(\mathbf{r}) \frac{\partial}{\partial \rho} \right) + \frac{1}{\rho^2} \left(\frac{\partial A(\mathbf{r})}{\partial \phi} \right) \frac{\partial}{\partial \phi} + \frac{A(\mathbf{r})}{\rho^2} \frac{\partial^2}{\partial \phi^2} \\
&\quad + \frac{A(\mathbf{r})}{\rho} \frac{\partial}{\partial \rho}.
\end{aligned}$$

If $A(\mathbf{r}) = A(\rho, z)$, then

$$-\widehat{k}_z A \widehat{k}_z = \frac{\partial}{\partial z} A \frac{\partial}{\partial z},$$

$$-\left(\widehat{k}_x A\widehat{k}_x + \widehat{k}_y A\widehat{k}_y\right) = \frac{\partial}{\partial \rho}\left(A\frac{\partial}{\partial \rho}\right) + \frac{A}{\rho}\frac{\partial}{\partial \rho} + \frac{A}{\rho^2}\frac{\partial^2}{\partial \phi^2},$$

$$-\widehat{k}_\pm A\widehat{k}_\pm = \mathrm{e}^{\pm \mathrm{i}2\phi}\left\{-\frac{A(\mathbf{r})}{\rho}\left(\frac{\partial}{\partial \rho} \pm \frac{\mathrm{i}}{\rho}\frac{\partial}{\partial \phi}\right) + \frac{\partial}{\partial \rho}\left(A\frac{\partial}{\partial \rho}\right) - \frac{A}{\rho^2}\frac{\partial^2}{\partial \phi^2}\right.$$
$$\left. \pm \mathrm{i}\frac{\partial}{\partial \rho}\left(\frac{A}{\rho}\frac{\partial}{\partial \phi}\right) \pm \mathrm{i}\frac{A}{\rho}\frac{\partial^2}{\partial \phi \partial \rho}\right\},$$

$$-\left(\widehat{k}_\pm A\widehat{k}_z + \widehat{k}_z A'\widehat{k}_\pm\right) = \mathrm{e}^{\pm \mathrm{i}\phi}\left\{\frac{\partial}{\partial \rho}\left(A\frac{\partial}{\partial z}\right) + \frac{\partial}{\partial z}\left(A'\frac{\partial}{\partial \rho}\right) \pm \mathrm{i}\frac{A}{\rho}\frac{\partial}{\partial \phi}\left(\frac{\partial}{\partial z}\right)\right.$$
$$\left. \pm \mathrm{i}\frac{\partial}{\partial z}\left(\frac{A'}{\rho}\frac{\partial}{\partial \phi}\right)\right\}.$$

We see, from Table C.9, that J_z is well-defined for the product states. Similar to the ZB case, the Sercel–Vahala wave function can be written as

$$\psi(\mathbf{r}) = \sum_i f(\mathbf{r})|u_i(J_z)\rangle = \sum_i f_i(\rho, z)\mathrm{e}^{\mathrm{i}(F_z - J_z[u_i])\phi}|u_i(J_z)\rangle \equiv \sum_i f_i(\rho, z)|u_i'(J_z)\rangle. \tag{12.198}$$

Then, the new matrix elements are given by

$$\langle u_i'(J_z)|\widehat{H}|u_j'(J_z)\rangle = \langle u_i(J_z)|\mathrm{e}^{-\mathrm{i}(F_z - J_z)\phi}\widehat{H}\mathrm{e}^{\mathrm{i}(F_z' - J_z')\phi}|u_j(J_z)\rangle. \tag{12.199}$$

The SV transformation can now be applied to different Hamiltonians.

12.5.5.1 Symmetrized RSP Hamiltonian

We start by transforming the symmetrized RSP Hamiltonian. In Table 3.8, the symmetrized Hamiltonian has

$$\begin{aligned}
\lambda &= \frac{\hbar^2}{2m_0}\left[\widehat{k}_z A_1\widehat{k}_z + \widehat{k}_x A_2\widehat{k}_x + \widehat{k}_y A_2\widehat{k}_y\right],\\
\theta &= \frac{\hbar^2}{2m_0}\left[\widehat{k}_z A_3\widehat{k}_z + \widehat{k}_x A_4\widehat{k}_x + \widehat{k}_y A_4\widehat{k}_y\right],\\
K &= \frac{\hbar^2}{2m_0}\left(\widehat{k}_x + \mathrm{i}\widehat{k}_y\right) A_5 \left(\widehat{k}_x + \mathrm{i}\widehat{k}_y\right),\\
H &= \frac{\hbar^2}{2m_0}\frac{1}{2}\left[\left(\widehat{k}_x + \mathrm{i}\widehat{k}_y\right) A_6\widehat{k}_z + \widehat{k}_z A_6 \left(\widehat{k}_x + \mathrm{i}\widehat{k}_y\right)\right].
\end{aligned} \tag{12.200}$$

In the SV representation (and leaving out factors of $\hbar^2/(2m_0)$ for now), we have

$$\lambda = -e^{-i(F_z-J_1)\phi}\left\{\frac{\partial}{\partial z}A_1\frac{\partial}{\partial z}+\frac{\partial}{\partial \rho}\left(A_2\frac{\partial}{\partial \rho}\right)+\frac{A_2}{\rho}\frac{\partial}{\partial \rho}+\frac{A_2}{\rho^2}\frac{\partial^2}{\partial \phi^2}\right\}e^{i(F_z-J_1)\phi}$$

$$= -\left\{\frac{\partial}{\partial z}A_1\frac{\partial}{\partial z}+\frac{\partial}{\partial \rho}\left(A_2\frac{\partial}{\partial \rho}\right)+\frac{A_2}{\rho}\frac{\partial}{\partial \rho}-\frac{(F_z-J_z)^2}{\rho^2}A_2\right\}, \quad (12.201)$$

$$\theta = -\left\{\frac{\partial}{\partial z}A_3\frac{\partial}{\partial z}+\frac{\partial}{\partial \rho}\left(A_4\frac{\partial}{\partial \rho}\right)+\frac{A_4}{\rho}\frac{\partial}{\partial \rho}-\frac{(F_z-J_z)^2}{\rho^2}A_4\right\}, \quad (12.202)$$

$$K = -e^{-i(F_z-J_2)\phi}e^{i2\phi}\left\{-\frac{A_5}{\rho}\left(\frac{\partial}{\partial \rho}+\frac{i}{\rho}\frac{\partial}{\partial \phi}\right)+\frac{\partial}{\partial \rho}\left(A_5\frac{\partial}{\partial \rho}\right)-\frac{A_5}{\rho^2}\frac{\partial^2}{\partial \phi^2}\right.$$

$$\left.+\,i\frac{\partial}{\partial \rho}\left(\frac{A_5}{\rho}\frac{\partial}{\partial \phi}\right)+i\frac{A_5}{\rho}\frac{\partial^2}{\partial \phi \partial \rho}\right\}e^{i(F_z-J_1)\phi}$$

$$= -\left\{\frac{\partial}{\partial \rho}\left(A_5\frac{\partial}{\partial \rho}\right)+\frac{(F_z-J_z)(F_z-J_z-1)}{\rho^2}A_5-(F_z-J_z)\frac{\partial}{\partial \rho}\left(\frac{A_5}{\rho}\cdot\right)\right.$$

$$\left.-(F_z-J_z+1)\frac{A_5}{\rho}\frac{\partial}{\partial \rho}\right\}, \quad (12.203)$$

$$H = -\frac{1}{2}e^{-i(F_z-J_2)\phi}e^{i\phi}\left\{\frac{\partial}{\partial \rho}\left(A_6\frac{\partial}{\partial z}\right)+\frac{\partial}{\partial z}\left(A_6\frac{\partial}{\partial \rho}\right)+i\frac{A_6}{\rho}\frac{\partial}{\partial \phi}\left(\frac{\partial}{\partial z}\right)\right.$$

$$\left.+i\frac{\partial}{\partial z}\left(\frac{A_6}{\rho}\frac{\partial}{\partial \phi}\right)\right\}e^{i(F_z-J_3)\phi} \quad (12.204)$$

$$= -\frac{1}{2}\left\{\frac{\partial}{\partial \rho}\left(A_6\frac{\partial}{\partial z}\right)+\frac{\partial}{\partial z}\left(A_6\frac{\partial}{\partial \rho}\right)-(F_z-J_z)\frac{A_6}{\rho}\frac{\partial}{\partial z}-(F_z-J_z)\frac{\partial}{\partial z}\left(\frac{A_6}{\rho}\cdot\right)\right\}.$$

In the above, we have used $J_i \equiv J_z[u_i]$ and explicit values in the top 3×3 block of the Hamiltonian. The appropriate J_z values should be used in the transformed matrix (i.e., the value J_z need not all be the same) but what is common is that all ϕ dependence goes away.

Indeed, it is found that all ϕ dependence drops out following the SV transformation, without making an axial approximation as for ZB. The validity of this result is due to the axial symmetry already found to be true for the bulk dispersion relation [46]. One difference compared to the bulk case, though, is that the lateral quantization now prevents setting up the unitary transformation that block diagonalized the bulk Hamiltonian. Hence, we here still have to solve a 6×6 matrix. The full matrix is summarized in Table 12.9.

12.5.5.2 MU Hamiltonian

The Mireles–Ulloa (MU) Hamiltonian [55, 56] results from the application of Burt's envelope-function theory to obtain the equivalent of the $D_{3\times3}$ matrix, Eq. (3.71), for WZ heterostructures. Just as for the ZB case, the result is the presence of nonsymmetric terms in the Hamiltonian matrix elements. Thus, the heterostructure $D_{3\times3}$ matrix is given by

Table 12.9 Six-band RSP Hamiltonian in SV representation

$$H = \begin{pmatrix} |u_1'\rangle & |u_2'\rangle & |u_3'\rangle & |u_4'\rangle & |u_5'\rangle & |u_6'\rangle \\ F & -K^\dagger & -H^\dagger & 0 & 0 & 0 \\ -K & G & H & 0 & 0 & \Delta \\ -H & H^\dagger & \lambda & 0 & \Delta & 0 \\ 0 & 0 & 0 & F & -K & H \\ 0 & 0 & \Delta & -K^\dagger & G & -H^\dagger \\ 0 & \Delta & 0 & H^\dagger & -H & \lambda \end{pmatrix},$$

where

$$F = \Delta_1 + \Delta_2 + \lambda + \theta,$$

$$G = \Delta_1 - \Delta_2 + \lambda + \theta,$$

$$\Delta = \sqrt{2}\Delta_3,$$

$$\lambda = -\frac{\hbar^2}{2m_0}\left\{\frac{\partial}{\partial z}A_1\frac{\partial}{\partial z} + \frac{\partial}{\partial \rho}\left(A_2\frac{\partial}{\partial \rho}\right) + \frac{A_2}{\rho}\frac{\partial}{\partial \rho} - \frac{(F_z - J_i)^2}{\rho^2}A_2\right\},$$

$$\theta = -\frac{\hbar^2}{2m_0}\left\{\frac{\partial}{\partial z}A_3\frac{\partial}{\partial z} + \frac{\partial}{\partial \rho}\left(A_4\frac{\partial}{\partial \rho}\right) + \frac{A_4}{\rho}\frac{\partial}{\partial \rho} - \frac{(F_z - J_i)^2}{\rho^2}A_4\right\},$$

$$K = -\frac{\hbar^2}{2m_0}\left\{\frac{\partial}{\partial \rho}\left(A_5\frac{\partial}{\partial \rho}\right) + \frac{(F_z - J_z)(F_z - J_z - 1)}{\rho^2}A_5 - (F_z - J_z)\frac{\partial}{\partial \rho}\left(\frac{A_5}{\rho}\cdot\right)\right.$$
$$\left. - (F_z - J_z + 1)\frac{A_5}{\rho}\frac{\partial}{\partial \rho}\right\},$$

$$H = -\frac{1}{2}\frac{\hbar^2}{2m_0}\left\{\frac{\partial}{\partial \rho}\left(A_6\frac{\partial}{\partial z}\right) + \frac{\partial}{\partial z}\left(A_6\frac{\partial}{\partial \rho}\right) - (F_z - J_3)\frac{A_6}{\rho}\frac{\partial}{\partial z} - (F_z - J_3)\frac{\partial}{\partial z}\left(\frac{A_6}{\rho}\cdot\right)\right\}.$$

$$D = \tag{12.205}$$
$$\begin{pmatrix} |X\rangle & |Y\rangle & |Z\rangle \\ \widehat{k}_x L_1 \widehat{k}_x + \widehat{k}_y M_1 \widehat{k}_y + \widehat{k}_z M_2 \widehat{k}_z & \widehat{k}_x N_1 \widehat{k}_y + \widehat{k}_y N_1' \widehat{k}_x & \widehat{k}_x N_2 \widehat{k}_z + \widehat{k}_z N_2' \widehat{k}_x \\ \widehat{k}_y N_1 \widehat{k}_x + \widehat{k}_x N_1' \widehat{k}_y & \widehat{k}_x M_1 \widehat{k}_x + \widehat{k}_y L_1 \widehat{k}_y + \widehat{k}_z M_2 \widehat{k}_z & \widehat{k}_y N_2 \widehat{k}_z + \widehat{k}_z N_2' \widehat{k}_y \\ \widehat{k}_z N_2 \widehat{k}_x + \widehat{k}_x N_2' \widehat{k}_z & \widehat{k}_z N_2 \widehat{k}_y + \widehat{k}_y N_2' \widehat{k}_z & \widehat{k}_x M_3 \widehat{k}_x + \widehat{k}_y M_3 \widehat{k}_y + \widehat{k}_z L_2 \widehat{k}_z \end{pmatrix}.$$

The new parameters can be related to the RSP ones (Table 12.12). First, with spin and in terms of the $|u_i\rangle$ basis, one gets the 6×6 matrix in Table 12.10. This was obtained as follows:

$$\widetilde{D}_{11} = \langle u_1 | D_{6\times6} | u_1 \rangle = \frac{1}{2}\langle (X + \mathrm{i}Y) \uparrow | D_{6\times6} | (X + \mathrm{i}Y) \uparrow \rangle$$
$$= \frac{1}{2}[D_{11} + D_{22} + \mathrm{i}(D_{12} - D_{21})]$$

Table 12.10 MU Hamiltonian in LS basis for wurtzite heterostructures [55]

$$H = \begin{pmatrix} |u_1\rangle & |u_2\rangle & |u_3\rangle & |u_4\rangle & |u_5\rangle & |u_6\rangle \\ \widetilde{D}_{11}+\Delta_1+\Delta_2 & \widetilde{D}_{12} & \widetilde{D}_{13} & 0 & 0 & 0 \\ \widetilde{D}_{12}^{\dagger} & \widetilde{D}_{22}+\Delta_1-\Delta_2 & \widetilde{D}_{23} & 0 & 0 & \Delta \\ \widetilde{D}_{13}^{\dagger} & \widetilde{D}_{23}^{\dagger} & \widetilde{D}_{33} & 0 & \Delta & 0 \\ 0 & 0 & 0 & \widetilde{D}_{22}+\Delta_1+\Delta_2 & \widetilde{D}_{12}^{\dagger} & \widetilde{D}_{23} \\ 0 & 0 & \Delta & \widetilde{D}_{12} & \widetilde{D}_{11}+\Delta_1-\Delta_2 & \widetilde{D}_{13} \\ 0 & \Delta & 0 & \widetilde{D}_{23}^{\dagger} & \widetilde{D}_{13}^{\dagger} & \widetilde{D}_{33} \end{pmatrix},$$

where

$$\widetilde{D}_{11} = \frac{1}{2}\Big\{\widehat{k}_x(L_1+M_1)\widehat{k}_x + \widehat{k}_y(L_1+M_1)\widehat{k}_y + 2\widehat{k}_z M_2 \widehat{k}_z + \mathrm{i}[\widehat{k}_x(N_1-N_1')\widehat{k}_y - \widehat{k}_y(N_1-N_1')\widehat{k}_x]\Big\},$$

$$\widetilde{D}_{22} = \frac{1}{2}\Big\{\widehat{k}_x(L_1+M_1)\widehat{k}_x + \widehat{k}_y(L_1+M_1)\widehat{k}_y + 2\widehat{k}_z M_2 \widehat{k}_z - \mathrm{i}[\widehat{k}_x(N_1-N_1')\widehat{k}_y - \widehat{k}_y(N_1-N_1')\widehat{k}_x]\Big\},$$

$$\widetilde{D}_{33} = D_{33} = \widehat{k}_x M_3 \widehat{k}_x + \widehat{k}_y M_3 \widehat{k}_y + \widehat{k}_z L_2 \widehat{k}_z,$$

$$\widetilde{D}_{12} = -\frac{1}{2}\Big\{\widehat{k}_x(L_1-M_1)\widehat{k}_x - \widehat{k}_y(L_1-M_1)\widehat{k}_y - \mathrm{i}[\widehat{k}_x(N_1+N_1')\widehat{k}_y + \widehat{k}_y(N_1+N_1')\widehat{k}_x]\Big\},$$

$$\widetilde{D}_{13} = -\frac{1}{\sqrt{2}}\Big\{\widehat{k}_- N_2 \widehat{k}_z + \widehat{k}_z N_2' \widehat{k}_-\Big\},$$

$$\widetilde{D}_{23} = \frac{1}{\sqrt{2}}\Big\{\widehat{k}_+ N_2 \widehat{k}_z + \widehat{k}_z N_2' \widehat{k}_+\Big\},$$

$$\widetilde{D}_{21} = -\frac{1}{2}\widehat{k}_+(L_1-M_1)\widehat{k}_+,$$

$$\widetilde{D}_{31} = -\frac{1}{\sqrt{2}}\Big\{\widehat{k}_z N_2 \widehat{k}_+ + \widehat{k}_+ N_2' \widehat{k}_z\Big\},$$

$$\widetilde{D}_{32} = \frac{1}{\sqrt{2}}\Big\{\widehat{k}_z N_2 \widehat{k}_- + \widehat{k}_- N_2' \widehat{k}_z\Big\}.$$

$$= \frac{1}{2}\Big\{\widehat{k}_x(L_1+M_1)\widehat{k}_x + \widehat{k}_y(L_1+M_1)\widehat{k}_y + 2\widehat{k}_z M_2 \widehat{k}_z$$
$$+\mathrm{i}\left[\widehat{k}_x(N_1-N_1')\widehat{k}_y - \widehat{k}_y(N_1-N_1')\widehat{k}_x\right]\Big\},$$

$$\widetilde{D}_{22} = \langle u_2|D_{6\times6}|u_2\rangle = \frac{1}{2}\langle(X-\mathrm{i}Y)\uparrow|D_{6\times6}|(X-\mathrm{i}Y)\uparrow\rangle$$
$$= \frac{1}{2}[D_{11}+D_{22}-\mathrm{i}(D_{12}-D_{21})] = \widetilde{D}_{11}^{*}$$
$$= \frac{1}{2}\Big\{\widehat{k}_x(L_1+M_1)\widehat{k}_x + \widehat{k}_y(L_1+M_1)\widehat{k}_y + 2\widehat{k}_z M_2 \widehat{k}_z$$

$$-\mathrm{i}\left[\widehat{k}_x(N_1 - N_1')\widehat{k}_y - \widehat{k}_y(N_1 - N_1')\widehat{k}_x\right]\Big\},$$

$$\widetilde{D}_{33} = \langle u_3|D_{6\times 6}|u_3\rangle = \langle Z\uparrow|D_{6\times 6}|Z\uparrow\rangle = D_{33} = \widehat{k}_x M_3 \widehat{k}_x + \widehat{k}_y M_3 \widehat{k}_y + \widehat{k}_z L_2 \widehat{k}_z,$$

$$\widetilde{D}_{12} = \langle u_1|D_{6\times 6}|u_2\rangle = -\frac{1}{2}\langle (X+\mathrm{i}Y)\uparrow|D_{6\times 6}|(X-\mathrm{i}Y)\uparrow\rangle$$

$$= -\frac{1}{2}[D_{11} - D_{22} - \mathrm{i}(D_{12} + D_{21})]$$

$$= -\frac{1}{2}\Big\{\widehat{k}_x(L_1 - M_1)\widehat{k}_x - \widehat{k}_y(L_1 - M_1)\widehat{k}_y - \mathrm{i}\left[\widehat{k}_x(N_1 + N_1')\widehat{k}_y + \widehat{k}_y(N_1 + N_1')\widehat{k}_x\right]\Big\}$$

$$= -\frac{1}{2}(\widehat{k}_x - \mathrm{i}\widehat{k}_y)(L_1 - M_1)\left(\widehat{k}_x - \mathrm{i}\widehat{k}_y\right) \text{ if } L_1 - M_1 = N_1 + N_1',$$

$$\widetilde{D}_{21} = \langle u_2|D_{6\times 6}|u_1\rangle = -\frac{1}{2}\langle (X-\mathrm{i}Y)\uparrow|D_{6\times 6}|(X+\mathrm{i}Y)\uparrow\rangle$$

$$= -\frac{1}{2}[D_{11} - D_{22} + \mathrm{i}(D_{12} + D_{21})] = \widetilde{D}_{12}^{\dagger} = -\frac{1}{2}\widehat{k}_+(L_1 - M_1)\widehat{k}_+,$$

$$\widetilde{D}_{13} = \langle u_1|D_{6\times 6}|u_3\rangle = -\frac{1}{\sqrt{2}}\langle (X+\mathrm{i}Y)\uparrow|D_{6\times 6}|Z\uparrow\rangle = -\frac{1}{\sqrt{2}}(D_{13} - \mathrm{i}D_{23})$$

$$= -\frac{1}{\sqrt{2}}\Big\{\widehat{k}_- N_2 \widehat{k}_z + \widehat{k}_z N_2' \widehat{k}_-\Big\},$$

$$\widetilde{D}_{31} = \langle u_3|D_{6\times 6}|u_1\rangle = -\frac{1}{\sqrt{2}}\langle Z\uparrow|D_{6\times 6}|(X+\mathrm{i}Y)\uparrow\rangle = -\frac{1}{\sqrt{2}}(D_{31} + \mathrm{i}D_{32}) = \widetilde{D}_{13}^{\dagger}$$

$$= -\frac{1}{\sqrt{2}}\Big\{\widehat{k}_z N_2 \widehat{k}_+ + \widehat{k}_+ N_2' \widehat{k}_z\Big\},$$

$$\widetilde{D}_{23} = \langle u_2|D_{6\times 6}|u_3\rangle = \frac{1}{\sqrt{2}}\langle (X-\mathrm{i}Y)\uparrow|D_{6\times 6}|Z\uparrow\rangle = \frac{1}{\sqrt{2}}(D_{13} + \mathrm{i}D_{23}) = -\widetilde{D}_{13}^{\dagger}$$

$$= \frac{1}{\sqrt{2}}\Big\{\widehat{k}_+ N_2 \widehat{k}_z + \widehat{k}_z N_2' \widehat{k}_+\Big\},$$

$$\widetilde{D}_{32} = \langle u_3|D_{6\times 6}|u_2\rangle = \frac{1}{\sqrt{2}}\langle Z\uparrow|D_{6\times 6}|(X-\mathrm{i}Y)\uparrow\rangle = \frac{1}{\sqrt{2}}(D_{31} - \mathrm{i}D_{32}) = \widetilde{D}_{31}^{*} = \widetilde{D}_{23}^{\dagger}$$

$$= \frac{1}{\sqrt{2}}\Big\{\widehat{k}_z N_2 \widehat{k}_- + \widehat{k}_- N_2' \widehat{k}_z\Big\}.$$

We are now ready for the SV transformation. Compared to the LK theory already presented, here we will also need:

$$-\widehat{k}_x A \widehat{k}_y = \left(\cos\phi\frac{\partial}{\partial\rho} - \frac{\sin\phi}{\rho}\frac{\partial}{\partial\phi}\right) A \left(\sin\phi\frac{\partial}{\partial\rho} + \frac{\cos\phi}{\rho}\frac{\partial}{\partial\phi}\right)$$

$$= \sin\phi\cos\phi\frac{\partial}{\partial\rho}\left(A\frac{\partial}{\partial\rho}\right) - \frac{\sin\phi}{\rho^2}\frac{\partial}{\partial\phi}\left(A\cos\phi\frac{\partial}{\partial\phi}\right)$$

$$+\cos^2\phi\frac{\partial}{\partial\rho}\left(\frac{A}{\rho}\frac{\partial}{\partial\phi}\right)-\frac{\sin\phi}{\rho}\frac{\partial}{\partial\phi}\left(A\sin\phi\frac{\partial}{\partial\rho}\right),$$

$$-\widehat{k}_y A\widehat{k}_x = \sin\phi\cos\phi\frac{\partial}{\partial\rho}\left(A\frac{\partial}{\partial\rho}\right)-\frac{\cos\phi}{\rho^2}\frac{\partial}{\partial\phi}\left(A\sin\phi\frac{\partial}{\partial\phi}\right)$$

$$-\sin^2\phi\frac{\partial}{\partial\rho}\left(\frac{A}{\rho}\frac{\partial}{\partial\phi}\right)+\frac{\cos\phi}{\rho}\frac{\partial}{\partial\phi}\left(A\cos\phi\frac{\partial}{\partial\rho}\right),$$

$$-(\widehat{k}_x A\widehat{k}_y+\widehat{k}_y A\widehat{k}_x) = \sin 2\phi\frac{\partial}{\partial\rho}\left(A\frac{\partial}{\partial\rho}\right)-\sin 2\phi\frac{A}{\rho^2}\frac{\partial^2}{\partial\phi^2}-\cos 2\phi\frac{A}{\rho^2}\frac{\partial}{\partial\phi}$$

$$+\cos 2\phi\frac{\partial}{\partial\rho}\left(\frac{A}{\rho}\frac{\partial}{\partial\phi}\right)+\cos 2\phi\frac{A}{\rho}\frac{\partial^2}{\partial\rho\partial\phi}-\sin 2\phi\frac{A}{\rho}\frac{\partial}{\partial\rho},$$

$$-(\widehat{k}_x A\widehat{k}_y-\widehat{k}_y A\widehat{k}_x) = \frac{A}{\rho^2}\frac{\partial}{\partial\phi}+\frac{\partial}{\partial\rho}\left(\frac{A}{\rho}\frac{\partial}{\partial\phi}\right)-\frac{A}{\rho}\frac{\partial^2}{\partial\rho\partial\phi}=\frac{1}{\rho}\frac{\partial A}{\partial\rho}\frac{\partial}{\partial\phi}$$

$$-(\widehat{k}_x A\widehat{k}_x-\widehat{k}_y A\widehat{k}_y) = \cos 2\phi\frac{\partial}{\partial\rho}\left(A\frac{\partial}{\partial\rho}\right)-\cos 2\phi\frac{A}{\rho^2}\frac{\partial^2}{\partial\phi^2}+\sin 2\phi\frac{A}{\rho^2}\frac{\partial}{\partial\phi}$$

$$-\sin 2\phi\frac{\partial}{\partial\rho}\left(\frac{A}{\rho}\frac{\partial}{\partial\phi}\right)-\sin 2\phi\frac{A}{\rho}\frac{\partial^2}{\partial\rho\partial\phi}-\cos 2\phi\frac{A}{\rho}\frac{\partial}{\partial\rho}.$$

Then

$$\widetilde{D}_{11} = -\frac{1}{2}\left\{\frac{\partial}{\partial\rho}\left((L_1+M_1)\frac{\partial}{\partial\rho}\right)+\frac{(L_1+M_1)}{\rho}\frac{\partial}{\partial\rho}-\frac{(F_z-J_z)^2}{\rho^2}(L_1+M_1)\right.$$

$$\left.+2\frac{\partial}{\partial z}M_2\frac{\partial}{\partial z}-\frac{(F_z-J_z)}{\rho}\frac{\partial(N_1-N_1')}{\partial\rho}\right\},$$

$$\widetilde{D}_{22} = -\frac{1}{2}\left\{\frac{\partial}{\partial\rho}\left((L_1+M_1)\frac{\partial}{\partial\rho}\right)+\frac{(L_1+M_1)}{\rho}\frac{\partial}{\partial\rho}-\frac{(F_z-J_z)^2}{\rho^2}(L_1+M_1)\right.$$

$$\left.+2\frac{\partial}{\partial z}M_2\frac{\partial}{\partial z}+\frac{(F_z-J_z)}{\rho}\frac{\partial(N_1-N_1')}{\partial\rho}\right\},$$

$$\widetilde{D}_{33} = -\left\{\frac{\partial}{\partial\rho}M_3\frac{\partial}{\partial\rho}+\frac{M_3}{\rho}\frac{\partial}{\partial\rho}-\frac{(F_z-J_z)^2}{\rho^2}M_3+\frac{\partial}{\partial z}L_2\frac{\partial}{\partial z}\right\},$$

and they remain unchanged in the SV representation:

$$\langle u_i'|\widetilde{D}_{ii}|u_i'\rangle = \langle u_i|\widetilde{D}_{ii}|u_i\rangle.$$

Note that, in the SV basis, we have

$$\langle u_i'|\widehat{k}_\pm A\widehat{k}_\pm|u_i'\rangle = -\left\{-\frac{A(\mathbf{r})}{\rho}\frac{\partial}{\partial\rho}\pm\frac{(F_z-J_z)}{\rho^2}A+\frac{\partial}{\partial\rho}\left(A\frac{\partial}{\partial\rho}\right)+\frac{(F_z-J_z)^2}{\rho^2}A\right.$$

$$\mp (F_z - J_z)\frac{\partial}{\partial \rho}\left(\frac{A}{\rho}\cdot\right) \mp \frac{A}{\rho}(F_z - J_z)\frac{\partial}{\partial \rho}\Bigg\}$$

$$= \left\{\frac{\partial}{\partial \rho}\left(A\frac{\partial}{\partial \rho}\right) \mp (F_z - J_z)\frac{\partial}{\partial \rho}\left(\frac{A}{\rho}\cdot\right) \mp \frac{A}{\rho}[(F_z - J_z) \pm 1]\frac{\partial}{\partial \rho}\right.$$

$$\left. + \frac{A}{\rho^2}(F_z - J_z)[(F_z - J_z) \pm 1]\right\},$$

$$\langle u_i' | (\widehat{k}_\pm A \widehat{k}_z + \widehat{k}_z A' \widehat{k}_\pm) | u_i' \rangle = -\left\{\frac{\partial}{\partial \rho}\left(A\frac{\partial}{\partial z}\right) + \frac{\partial}{\partial z}\left(A'\frac{\partial}{\partial \rho}\right)\right.$$

$$\left. \mp (F_z - J_3)\left[\frac{A}{\rho}\frac{\partial}{\partial z} + \frac{\partial}{\partial z}\left(\frac{A'}{\rho}\cdot\right)\right]\right\}.$$

Thus, for example,

$$\langle u_1' | \widetilde{D} | u_2' \rangle = \frac{1}{2}\left\{\frac{\partial}{\partial \rho}\left((L_1 - M_1)\frac{\partial}{\partial \rho}\right) + \left(F_z + \frac{1}{2}\right)\frac{\partial}{\partial \rho}\left(\frac{(L_1 - M_1)}{\rho}\right)\right.$$

$$\left. + \left(F_z - \frac{1}{2}\right)\frac{(L_1 - M_1)}{\rho}\frac{\partial}{\partial \rho} + \frac{(F_z + \frac{1}{2})(F_z - \frac{1}{2})}{\rho^2}(L_1 - M_1)\right\},$$

$$\langle u_1' | \widetilde{D} | u_3' \rangle = \frac{1}{\sqrt{2}}\left\{\frac{\partial}{\partial \rho}\left(N_2\frac{\partial}{\partial z}\right) + \frac{\partial}{\partial z}\left(N_2'\frac{\partial}{\partial \rho}\right) + (F_z - J_3)\left[\frac{N_2}{\rho}\frac{\partial}{\partial z} + \frac{\partial}{\partial z}\left(\frac{N_2'}{\rho}\right)\right]\right\}$$

$$= \frac{1}{\sqrt{2}}\left\{\frac{\partial}{\partial \rho}\left(N_2\frac{\partial}{\partial z}\right) + \frac{\partial}{\partial z}\left(N_2'\frac{\partial}{\partial \rho}\right) + (F_z - \frac{1}{2})\left[\frac{N_2}{\rho}\frac{\partial}{\partial z} + \frac{\partial}{\partial z}\left(\frac{N_2'}{\rho}\right)\right]\right\},$$

$$\langle u_2' | \widetilde{D} | u_3' \rangle = -\frac{1}{\sqrt{2}}\left\{\frac{\partial}{\partial \rho}\left(N_2\frac{\partial}{\partial z}\right) + \frac{\partial}{\partial z}\left(N_2'\frac{\partial}{\partial \rho}\right) - (F_z - J_z)\left[\frac{N_2}{\rho}\frac{\partial}{\partial z} + \frac{\partial}{\partial z}\left(\frac{N_2'}{\rho}\right)\right]\right\}$$

$$= -\frac{1}{\sqrt{2}}\left\{\frac{\partial}{\partial \rho}\left(N_2\frac{\partial}{\partial z}\right) + \frac{\partial}{\partial z}\left(N_2'\frac{\partial}{\partial \rho}\right) - (F_z - \frac{1}{2})\left[\frac{N_2}{\rho}\frac{\partial}{\partial z} + \frac{\partial}{\partial z}\left(\frac{N_2'}{\rho}\right)\right]\right\}.$$

Finally, the lower 3×3 subblock is related to the upper one except for changing the values of J_z and reordering. Thus,

$$S_{44} \sim S_{22},\, S_{55} \sim S_{11},\, S_{66} \sim S_{33},$$
$$S_{45} \sim S_{21},\, S_{46} \sim S_{23},$$
$$S_{54} \sim S_{12},\, S_{56} \sim S_{13},$$
$$S_{64} \sim S_{32},\, S_{65} \sim S_{31}.$$

Since strain is spin independent, there is no coupling between opposite spin states. It is also necessary to recast the strain Hamiltonian into the SV basis. For this, we write the strain tensor in cylindrical polar coordinates [262]:

$$\begin{aligned}
\varepsilon_{\rho\rho} &= \frac{\partial u_\rho}{\partial \rho}, \ \varepsilon_{\phi\phi} = \frac{1}{\rho}\frac{\partial u_\phi}{\partial \phi} + \frac{u_\rho}{\rho}, \ \varepsilon_{zz} = \frac{\partial u_z}{\partial z}, \\
2\varepsilon_{\rho z} &= \frac{\partial u_\rho}{\partial z} + \frac{\partial u_z}{\partial \rho}, \ 2\varepsilon_{\phi z} = \frac{1}{\rho}\frac{\partial u_z}{\partial \phi} + \frac{\partial u_\phi}{\partial z}, \\
2\varepsilon_{\rho\phi} &= \frac{\partial u_\phi}{\partial \rho} - \frac{u_\phi}{\rho} + \frac{1}{\rho}\frac{\partial u_\rho}{\partial \phi}.
\end{aligned} \tag{12.206}$$

We will also need

$$\frac{\partial}{\partial x} = \cos\phi \frac{\partial}{\partial \rho} - \frac{\sin\phi}{\rho}\frac{\partial}{\partial \phi}, \ \frac{\partial}{\partial y} = \sin\phi \frac{\partial}{\partial \rho} + \frac{\cos\phi}{\rho}\frac{\partial}{\partial \phi}.$$

Next, we convert cartesian components of vectors into the cylindrical components. Using

$$\begin{aligned}
\widehat{\rho} &= \cos\phi\widehat{\mathbf{x}} + \sin\phi\widehat{\mathbf{y}}, \\
\widehat{\phi} &= -\sin\phi\widehat{\mathbf{x}} + \cos\phi\widehat{\mathbf{y}},
\end{aligned}$$

then

$$\begin{aligned}
\mathbf{u} &= u_x\widehat{\mathbf{x}} + u_y\widehat{\mathbf{y}} + u_z\widehat{\mathbf{z}} = u_\rho\widehat{\rho} + u_\phi\widehat{\phi} + u_z\widehat{\mathbf{z}} \\
&= u_\rho(\cos\phi\widehat{\mathbf{x}} + \sin\phi\widehat{\mathbf{y}}) + u_\phi(-\sin\phi\widehat{\mathbf{x}} + \cos\phi\widehat{\mathbf{y}}) + u_z\widehat{\mathbf{z}},
\end{aligned}$$

giving

$$u_x = u_\rho\cos\phi - u_\phi\sin\phi, \ u_y = u_\rho\sin\phi + u_\phi\cos\phi.$$

One can now relate the strain tensor components in cartesian and cylindrical polar coordinates:

$$\begin{aligned}
\varepsilon_{xx} &= \frac{\partial u_x}{\partial x} = \left(\cos\phi\frac{\partial}{\partial\rho} - \frac{\sin\phi}{\rho}\frac{\partial}{\partial\phi}\right)(u_\rho\cos\phi - u_\phi\sin\phi) \\
&= \cos^2\phi\frac{\partial u_\rho}{\partial\rho} + \frac{\sin^2\phi}{\rho}\frac{\partial u_\phi}{\partial\phi} + \frac{\sin\phi\cos\phi}{\rho}u_\phi - \sin\phi\cos\phi\frac{\partial u_\phi}{\partial\rho} \\
&\quad + \frac{\sin^2\phi}{\rho}u_\rho - \frac{\sin\phi\cos\phi}{\rho}\frac{\partial u_\rho}{\partial\phi}, \\
\varepsilon_{yy} &= \frac{\partial u_y}{\partial y} = \left(\sin\phi\frac{\partial}{\partial\rho} + \frac{\cos\phi}{\rho}\frac{\partial}{\partial\phi}\right)(u_\rho\sin\phi + u_\phi\cos\phi) \\
&= \sin^2\phi\frac{\partial u_\rho}{\partial\rho} + \frac{\cos^2\phi}{\rho}\frac{\partial u_\phi}{\partial\phi} - \frac{\sin\phi\cos\phi}{\rho}u_\phi + \sin\phi\cos\phi\frac{\partial u_\phi}{\partial\rho}
\end{aligned}$$

$$
\begin{aligned}
&\quad + \frac{\cos^2\phi}{\rho} u_\rho + \frac{\sin\phi\cos\phi}{\rho}\frac{\partial u_\rho}{\partial \phi},\\
\varepsilon_{xx} + \varepsilon_{yy} &= \varepsilon_{\rho\rho} + \varepsilon_{\phi\phi},\\
2\varepsilon_{xy} &= \frac{\partial u_y}{\partial x} + \frac{\partial u_x}{\partial y}\\
&= 2\sin\phi\cos\phi\frac{\partial u_\rho}{\partial \rho} + \frac{1}{\rho}u_\phi - \frac{2\sin\phi\cos\phi}{\rho}\frac{\partial u_\phi}{\partial \phi} + \cos 2\phi\frac{\partial u_\phi}{\partial \rho}\\
&\quad - \frac{2\sin\phi\cos\phi}{\rho}u_\rho + \cos 2\phi\frac{\partial u_\rho}{\partial \phi},\\
\varepsilon_{xx} - \varepsilon_{yy} \pm 2\mathrm{i}\varepsilon_{xy} &= \mathrm{e}^{\pm 2\mathrm{i}\phi}\left[\frac{\partial u_\rho}{\partial \rho} - \frac{1}{\rho}\frac{\partial u_\phi}{\partial \phi} - \mathrm{i}\frac{u_\phi}{\rho} + \mathrm{i}\frac{\partial u_\phi}{\partial \rho} - \frac{u_\rho}{\rho} + \frac{\mathrm{i}}{\rho}\frac{\partial u_\rho}{\partial \phi}\right]\\
&= \mathrm{e}^{\pm 2\mathrm{i}\phi}\left[\varepsilon_{\rho\rho} - \varepsilon_{\phi\phi} - 2\mathrm{i}\varepsilon_{\rho\phi}\right],\\
\varepsilon_{zx} &= \frac{1}{2}\left(\frac{\partial u_x}{\partial z} + \frac{\partial u_z}{\partial x}\right)\\
&= \frac{1}{2}\left\{\cos\phi\left(\frac{\partial u_\rho}{\partial z} + \frac{\partial u_z}{\partial \rho}\right) - \sin\phi\left(\frac{\partial u_\phi}{\partial z} + \frac{1}{\rho}\frac{\partial u_z}{\partial \phi}\right)\right\},\\
\varepsilon_{zy} &= \frac{1}{2}\left(\frac{\partial u_y}{\partial z} + \frac{\partial u_z}{\partial y}\right)\\
&= \frac{1}{2}\left\{\sin\phi\left(\frac{\partial u_\rho}{\partial z} + \frac{\partial u_z}{\partial \rho}\right) + \cos\phi\left(\frac{\partial u_\phi}{\partial z} + \frac{1}{\rho}\frac{\partial u_z}{\partial \phi}\right)\right\},\\
\varepsilon_{zx} \pm \varepsilon_{zy} &= \frac{\mathrm{e}^{\pm\mathrm{i}\phi}}{2}\left\{\left(\frac{\partial u_\rho}{\partial z} + \frac{\partial u_z}{\partial \rho}\right) \pm \mathrm{i}\left(\frac{\partial u_\phi}{\partial z} + \frac{1}{\rho}\frac{\partial u_z}{\partial \phi}\right)\right\}\\
&= \mathrm{e}^{\pm\mathrm{i}\phi}\left\{\varepsilon_{\rho z} \pm \mathrm{i}\varepsilon_{\phi z}\right\}.
\end{aligned}
$$

It is now necessary to convert the strain Hamiltonian into the SV basis. For example,

$$
\begin{aligned}
S_{11}^{\varepsilon} &= \langle u_1(\mathbf{r})|\widehat{H}^{\varepsilon}|u_1(\mathbf{r})\rangle = \langle u_1'|\mathrm{e}^{-\mathrm{i}(F_z-J_1)\phi}\widehat{H}^{\varepsilon}\mathrm{e}^{\mathrm{i}(F_z-J_1)\phi}|u_1'\rangle\\
&= \frac{1}{2}\langle (X+\mathrm{i}Y)'|\mathrm{e}^{-\mathrm{i}(F_z-J_1)\phi}\widehat{H}^{\varepsilon}\mathrm{e}^{\mathrm{i}(F_z-J_1)\phi}|(X+\mathrm{i}Y)'\rangle\\
&= \frac{1}{2}(H_{11}^{\varepsilon} + H_{22}^{\varepsilon} + \mathrm{i}(H_{12}^{\varepsilon} - H_{21}^{\varepsilon})) = \frac{1}{2}\left\{(l_1+m_1)(\varepsilon_{xx}+\varepsilon_{yy}) + 2m_2\varepsilon_{zz}\right\}\\
&= \frac{1}{2}\left\{(l_1+m_1)(\varepsilon_{\rho\rho}+\varepsilon_{\phi\phi}) + 2m_2\varepsilon_{zz}\right\},
\end{aligned}
$$

$$
\begin{aligned}
S_{22}^{\varepsilon} &= \frac{1}{2}\langle (X-\mathrm{i}Y)'|\mathrm{e}^{-\mathrm{i}(F_z-J_2)\phi}\widehat{H}^{\varepsilon}\mathrm{e}^{\mathrm{i}(F_z-J_2)\phi}|(X-\mathrm{i}Y)'\rangle \\
&= \frac{1}{2}(H_{11}^{\varepsilon}+H_{22}^{\varepsilon}-\mathrm{i}(H_{12}^{\varepsilon}-H_{21}^{\varepsilon})) = S_{11}^{\varepsilon}, \\
S_{12}^{\varepsilon} &= -\frac{1}{2}\langle (X+\mathrm{i}Y)'|\mathrm{e}^{-\mathrm{i}(F_z-J_1)\phi}\widehat{H}^{\varepsilon}\mathrm{e}^{\mathrm{i}(F_z-J_2)\phi}|(X-\mathrm{i}Y)'\rangle \\
&= -\frac{1}{2}(H_{11}^{\varepsilon}-H_{22}^{\varepsilon}-\mathrm{i}(H_{12}^{\varepsilon}+H_{21}^{\varepsilon})) \\
&= -\frac{1}{2}\left\{(l_1-m_1)(\varepsilon_{xx}-\varepsilon_{yy})-2\mathrm{i}\varepsilon_{xy}n_1\right\} \\
&= -\frac{n_1}{2}\left\{\varepsilon_{xx}-\varepsilon_{yy}-2\mathrm{i}\varepsilon_{xy}\right\}, \\
S_{21}^{\varepsilon} &= S_{12}^{\varepsilon\,*}, \\
S_{31}^{\varepsilon} &= S_{13}^{\varepsilon\,*}, \\
S_{32}^{\varepsilon} &= S_{23}^{\varepsilon\,*}, \\
S_{45}^{\varepsilon} &= -\frac{1}{2}\langle (X+\mathrm{i}Y)'|\mathrm{e}^{-\mathrm{i}(F_z-J_4)\phi}\widehat{H}^{\varepsilon}\mathrm{e}^{\mathrm{i}(F_z-J_5)\phi}|Z'\rangle = S_{12}^{\varepsilon\,*}, \\
S_{46}^{\varepsilon} &= \frac{1}{\sqrt{2}}\langle (X-\mathrm{i}Y)'|\mathrm{e}^{-\mathrm{i}(F_z-J_4)\phi}\widehat{H}^{\varepsilon}\mathrm{e}^{\mathrm{i}(F_z-J_6)\phi}|Z'\rangle = \frac{1}{\sqrt{2}}(H_{13}^{\varepsilon}+\mathrm{i}H_{23}^{\varepsilon}) \\
&= \frac{n_2}{\sqrt{2}}(\varepsilon_{xz}+\mathrm{i}\varepsilon_{yz}) = S_{23}^{\varepsilon}, \\
S_{54}^{\varepsilon} &= S_{45}^{\varepsilon\,*}, \\
S_{56}^{\varepsilon} &= -\frac{1}{\sqrt{2}}\langle (X+\mathrm{i}Y)'|\mathrm{e}^{-\mathrm{i}(F_z-J_5)\phi}\widehat{H}^{\varepsilon}\mathrm{e}^{\mathrm{i}(F_z-J_6)\phi}|Z'\rangle = -\frac{1}{\sqrt{2}}(H_{13}^{\varepsilon}-\mathrm{i}H_{23}^{\varepsilon}) \\
&= -\frac{n_2}{\sqrt{2}}(\varepsilon_{xz}-\mathrm{i}\varepsilon_{yz}) = S_{13}^{\varepsilon}.
\end{aligned}
$$

The final Hamiltonian is given in Table 12.11, which is the sum of the previously-obtained MU Hamiltonian in SV basis together with the strain Hamiltonian.

The matrix elements are:

$$
\begin{aligned}
S_{11} = -\frac{\hbar^2}{2m_0}\frac{1}{2}\Bigg\{&\frac{\partial}{\partial\rho}\left((L_1+M_1)\frac{\partial}{\partial\rho}\right)+\frac{(L_1+M_1)}{\rho}\frac{\partial}{\partial\rho}+2\frac{\partial}{\partial z}M_2\frac{\partial}{\partial z} \\
&-\frac{(F_z-J_1)}{\rho}\frac{\partial(N_1-N_1')}{\partial\rho}-\frac{(F_z-J_1)^2}{\rho^2}(L_1+M_1)\Bigg\}, \\
S_{22} = -\frac{\hbar^2}{2m_0}\frac{1}{2}\Bigg\{&\frac{\partial}{\partial\rho}\left((L_1+M_1)\frac{\partial}{\partial\rho}\right)+\frac{(L_1+M_1)}{\rho}\frac{\partial}{\partial\rho}+2\frac{\partial}{\partial z}M_2\frac{\partial}{\partial z}
\end{aligned}
$$

Table 12.11 MU Hamiltonian in SV basis for strained wurtzite heterostructures

$$H = \begin{pmatrix} |u_1'\rangle & |u_2'\rangle & u_3'\rangle & |u_4'\rangle & |u_5'\rangle & |u_6'\rangle \\ S_{11} + \Delta_1 + \Delta_2 + S_{11}^{\varepsilon} & S_{12} + S_{12}^{\varepsilon} & S_{13} + S_{13}^{\varepsilon} & 0 & 0 & 0 \\ S_{21} + S_{12}^{\varepsilon *} & S_{22} + \Delta_1 - \Delta_2 + S_{22}^{\varepsilon} & S_{23} + S_{23}^{\varepsilon} & 0 & 0 & \Delta \\ S_{31} + S_{13}^{\varepsilon} & S_{32} + S_{23}^{\varepsilon} & S_{33} + S_{33}^{\varepsilon} & 0 & \Delta & 0 \\ 0 & 0 & 0 & S_{44} + \Delta_1 + \Delta_2 + S_{22}^{\varepsilon} & S_{45} + S_{12}^{\varepsilon *} & S_{46} + S_{23}^{\varepsilon} \\ 0 & 0 & \Delta & S_{54} + S_{12}^{\varepsilon} & S_{55} + \Delta_1 - \Delta_2 + S_{11}^{\varepsilon} & S_{56} + S_{13}^{\varepsilon} \\ 0 & \Delta & 0 & S_{64} + S_{23}^{\varepsilon *} & S_{65} + S_{13}^{\varepsilon *} & S_{66} + S_{33}^{\varepsilon} \end{pmatrix},$$

$$+\frac{(F_z - J_2)}{\rho}\frac{\partial(N_1 - N_1')}{\partial\rho} - \frac{(F_z - J_2)^2}{\rho^2}(L_1 + M_1)\Bigg\},$$

$$S_{33} = -\frac{\hbar^2}{2m_0}\left\{\frac{\partial}{\partial\rho}\left(M_3\frac{\partial}{\partial\rho}\right) + \frac{M_3}{\rho}\frac{\partial}{\partial\rho} + \frac{\partial}{\partial z}L_2\frac{\partial}{\partial z} - \frac{(F_z - J_3)^2}{\rho^2}M_3\right\},$$

$$S_{12} = \frac{\hbar^2}{2m_0}\frac{1}{2}\left\{\frac{\partial}{\partial\rho}\left((L_1 - M_1)\frac{\partial}{\partial\rho}\right) + \left(F_z + \frac{1}{2}\right)\frac{\partial}{\partial\rho}\left(\frac{(L_1 - M_1)}{\rho}\cdot\right)\right.$$

$$\left. + \left(F_z - \frac{1}{2}\right)\frac{(L_1 - M_1)}{\rho}\frac{\partial}{\partial\rho} + \frac{(F_z + \frac{1}{2})(F_z - \frac{1}{2})}{\rho^2}(L_1 - M_1)\right\},$$

$$S_{21} = \frac{\hbar^2}{2m_0}\frac{1}{2}\left\{\frac{\partial}{\partial\rho}\left((L_1 - M_1)\frac{\partial}{\partial\rho}\right) - \left(F_z - \frac{3}{2}\right)\frac{\partial}{\partial\rho}\left(\frac{(L_1 - M_1)}{\rho}\cdot\right)\right.$$

$$\left. - \left(F_z - \frac{1}{2}\right)\frac{(L_1 - M_1)}{\rho}\frac{\partial}{\partial\rho} + \frac{(F_z - \frac{3}{2})(F_z - \frac{1}{2})}{\rho^2}(L_1 - M_1)\right\},$$

$$S_{13} = \frac{\hbar^2}{2m_0}\frac{1}{\sqrt{2}}\left\{\frac{\partial}{\partial\rho}\left(N_2\frac{\partial}{\partial z}\right) + \frac{\partial}{\partial z}\left(N_2'\frac{\partial}{\partial\rho}\right) + \left(F_z - \frac{1}{2}\right)\left[\frac{\partial}{\partial z}\left(\frac{N_2'}{\rho}\cdot\right) + \frac{N_2}{\rho}\frac{\partial}{\partial z}\right]\right\},$$

$$S_{31} = \frac{\hbar^2}{2m_0}\frac{1}{\sqrt{2}}\left\{\frac{\partial}{\partial\rho}\left(N_2'\frac{\partial}{\partial z}\right) + \frac{\partial}{\partial z}\left(N_2\frac{\partial}{\partial\rho}\right) - \left(F_z - \frac{3}{2}\right)\left[\frac{\partial}{\partial z}\left(\frac{N_2}{\rho}\cdot\right) + \frac{N_2'}{\rho}\frac{\partial}{\partial z}\right]\right\},$$

$$S_{23} = -\frac{\hbar^2}{2m_0}\frac{1}{\sqrt{2}}\left\{\frac{\partial}{\partial\rho}\left(N_2\frac{\partial}{\partial z}\right) + \frac{\partial}{\partial z}\left(N_2'\frac{\partial}{\partial\rho}\right) - \left(F_z - \frac{1}{2}\right)\left[\frac{\partial}{\partial z}\left(\frac{N_2'}{\rho}\cdot\right) + \frac{N_2}{\rho}\frac{\partial}{\partial z}\right]\right\},$$

$$S_{32} = -\frac{\hbar^2}{2m_0}\frac{1}{\sqrt{2}}\left\{\frac{\partial}{\partial\rho}\left(N_2'\frac{\partial}{\partial z}\right) + \frac{\partial}{\partial z}\left(N_2\frac{\partial}{\partial\rho}\right) + \left(F_z + \frac{1}{2}\right)\left[\frac{\partial}{\partial z}\left(\frac{N_2}{\rho}\cdot\right) + \frac{N_2'}{\rho}\frac{\partial}{\partial z}\right]\right\},$$

$$S_{44} = -\frac{\hbar^2}{2m_0}\frac{1}{2}\left\{\frac{\partial}{\partial\rho}\left((L_1 + M_1)\frac{\partial}{\partial\rho}\right) + \frac{(L_1 + M_1)}{\rho}\frac{\partial}{\partial\rho} + 2\frac{\partial}{\partial z}M_2\frac{\partial}{\partial z}\right.$$

$$\left. + \frac{(F_z - J_4)}{\rho}\frac{\partial(N_1 - N_1')}{\partial\rho} - \frac{(F_z - J_4)^2}{\rho^2}(L_1 + M_1)\right\},$$

$$S_{55} = -\frac{\hbar^2}{2m_0}\frac{1}{2}\left\{\frac{\partial}{\partial\rho}\left((L_1 + M_1)\frac{\partial}{\partial\rho}\right) + \frac{(L_1 + M_1)}{\rho}\frac{\partial}{\partial\rho} + 2\frac{\partial}{\partial z}M_2\frac{\partial}{\partial z}\right.$$

$$\left. - \frac{(F_z - J_5)}{\rho}\frac{\partial(N_1 - N_1')}{\partial\rho} - \frac{(F_z - J_5)^2}{\rho^2}(L_1 + M_1)\right\},$$

$$S_{66} = -\frac{\hbar^2}{2m_0}\left\{\frac{\partial}{\partial\rho}\left(M_3\frac{\partial}{\partial\rho}\right) + \frac{M_3}{\rho}\frac{\partial}{\partial\rho} + \frac{\partial}{\partial z}L_2\frac{\partial}{\partial z} - \frac{(F_z - J_6)^2}{\rho^2}M_3\right\},$$

$$S_{45} = \frac{\hbar^2}{2m_0}\frac{1}{2}\left\{\frac{\partial}{\partial\rho}\left((L_1 - M_1)\frac{\partial}{\partial\rho}\right) - \left(F_z - \frac{1}{2}\right)\frac{\partial}{\partial\rho}\left(\frac{(L_1 - M_1)}{\rho}\cdot\right)\right.$$

$$-\left(F_z+\frac{1}{2}\right)\frac{(L_1-M_1)}{\rho}\frac{\partial}{\partial\rho}+\frac{(F_z-\frac{1}{2})(F_z-\frac{3}{2})}{\rho^2}(L_1-M_1)\Big\},$$

$$S_{46}=-\frac{\hbar^2}{2m_0}\frac{1}{\sqrt{2}}\left\{\frac{\partial}{\partial\rho}\left(N_2\frac{\partial}{\partial z}\right)+\frac{\partial}{\partial z}\left(N_2'\frac{\partial}{\partial\rho}\right)-(F_z+\frac{1}{2})\left[\frac{\partial}{\partial z}\left(\frac{N_2'\cdot}{\rho}\right)+\frac{N_2}{\rho}\frac{\partial}{\partial z}\right]\right\},$$

$$S_{54}=\frac{\hbar^2}{2m_0}\frac{1}{2}\left\{\frac{\partial}{\partial\rho}\left((L_1-M_1)\frac{\partial}{\partial\rho}\right)+\left(F_z+\frac{3}{2}\right)\frac{\partial}{\partial\rho}\left(\frac{(L_1-M_1)}{\rho}\cdot\right)\right.$$
$$\left.+\left(F_z+\frac{1}{2}\right)\frac{(L_1-M_1)}{\rho}\frac{\partial}{\partial\rho}+\frac{(F_z+\frac{3}{2})(F_z+\frac{5}{2})}{\rho^2}(L_1-M_1)\right\},$$

$$S_{56}=\frac{\hbar^2}{2m_0}\frac{1}{\sqrt{2}}\left\{\frac{\partial}{\partial\rho}\left(N_2\frac{\partial}{\partial z}\right)+\frac{\partial}{\partial z}\left(N_2'\frac{\partial}{\partial\rho}\right)+\left(F_z+\frac{1}{2}\right)\left[\frac{\partial}{\partial z}\left(\frac{N_2'\cdot}{\rho}\right)+\frac{N_2}{\rho}\frac{\partial}{\partial z}\right]\right\},$$

$$S_{64}=-\frac{\hbar^2}{2m_0}\frac{1}{\sqrt{2}}\left\{\frac{\partial}{\partial\rho}\left(N_2'\frac{\partial}{\partial z}\right)+\frac{\partial}{\partial z}\left(N_2\frac{\partial}{\partial\rho}\right)+\left(F_z+\frac{3}{2}\right)\left[\frac{\partial}{\partial z}\left(\frac{N_2\cdot}{\rho}\right)+\frac{N_2'}{\rho}\frac{\partial}{\partial z}\right]\right\},$$

$$S_{65}=\frac{\hbar^2}{2m_0}\frac{1}{\sqrt{2}}\left\{\frac{\partial}{\partial\rho}\left(N_2'\frac{\partial}{\partial z}\right)+\frac{\partial}{\partial z}\left(N_2\frac{\partial}{\partial\rho}\right)-\left(F_z-\frac{1}{2}\right)\left[\frac{\partial}{\partial z}\left(\frac{N_2\cdot}{\rho}\right)+\frac{N_2'}{\rho}\frac{\partial}{\partial z}\right]\right\},$$

$$S_{11}^{\varepsilon}=\frac{1}{2}\left\{(l_1+m_1)(\varepsilon_{xx}+\varepsilon_{yy})+2m_2\varepsilon_{zz}\right\}=\frac{1}{2}\left\{(l_1+m_1)(\varepsilon_{\rho\rho}+\varepsilon_{\phi\phi})+2m_2\varepsilon_{zz}\right\},$$
$$S_{22}^{\varepsilon}=S_{11}^{\varepsilon},$$
$$S_{33}^{\varepsilon}=m_3(\varepsilon_{xx}+\varepsilon_{yy})+l_2\varepsilon_{zz}=m_3(\varepsilon_{\rho\rho}+\varepsilon_{\phi\phi})+l_2\varepsilon_{zz},$$

Table 12.12 Parameters for wurtzite [57]

$L_1=A_2+A_4+A_5,$
$L_2=A_1,$
$M_1=A_2+A_4-A_5,$
$M_2=A_1+A_3,$
$M_3=A_2,$
$N_1=3A_5-(A_2+A_4)+1,$
$N_1'=-A_5+A_2+A_4-1,$
$N_2=1-(A_1+A_3)+\sqrt{2}A_6,$
$N_2'=A_1+A_3-1,$
$l_1=D_2+D_4+D_5,$
$l_2=D_1,$
$m_1=D_2+D_4-D_5,$
$m_2=D_1+D_3,$
$m_3=D_2,$
$n_1=2D_5,$
$n_2=\sqrt{2}D_6.$

$$S_{12}^{\varepsilon} = -\frac{n_1}{2}(\varepsilon_{xx} - \varepsilon_{yy} - 2\mathrm{i}\varepsilon_{xy}) = -\frac{n_1}{2}\left(\varepsilon_{\rho\rho} - \varepsilon_{\phi\phi} - 2\mathrm{i}\varepsilon_{\rho\phi}\right),$$
$$S_{13}^{\varepsilon} = -\frac{n_2}{\sqrt{2}}(\varepsilon_{zx} - \mathrm{i}\varepsilon_{zy}) = -\frac{n_2}{\sqrt{2}}(\varepsilon_{\rho z} - \mathrm{i}\varepsilon_{\phi z}),$$
$$S_{23}^{\varepsilon} = -S_{13}^{\varepsilon *}.$$

A comparison of the new wurtzite parameters and the RSP ones is given in Table 12.12.

12.6 Arbitrary Nanostructure Orientation

In applications of the $k \cdot p$ theory to QW's, it became useful to rotate coordinates such that the growth direction of the structure is denoted as z', when its direction in terms of the conventional cubic directions is $[hkl]$. It is thus necessary to relate the unprimed to the primed coordinates.

12.6.1 Overview

Heterostructures grown along directions other than [100] were probably first considered by Mailhiot and Smith in 1987 when they studied [111] superlattices [263]. Batty et al. [264] studied laser gain in [111] GaAs/AlGaAs QW's. Others were Vurgaftman et al. [265]. The rotation of the four-band Hamiltonian was carried out in 1991 by Xia for $(11l)$ $(l = 0, 1, 2, 3)$ directions [266], for [100], [110], and [111] by Ikonic et al. [267], for [113] by Goldoni and Peeters [268], for arbitrary $(11l)$ directions by Henderson and Towe [269], and for (hhk) by Fishman [270]. It was extended to a general (hkl) direction by Henderson and Towe [271]. Kajikawa [272] and, more recently, Seo and Donegan[273] considered the six-band Hamiltonian along an arbitrary $(11l)$ direction. Los et al. did the eight-band problem for arbitrary high-index planes with strain [274]. Quantum wires along different directions have also been studied [239, 265, 275].

12.6.2 Rotation Matrix

We first establish the coordinate rotation matrix (Fig. 12.4). This was given by, e.g., Henderson and Towe [269]:

$$U(\theta, \phi) = \begin{pmatrix} \cos\phi\cos\theta & \sin\phi\cos\theta & -\sin\theta \\ -\sin\phi & \cos\phi & 0 \\ \cos\phi\sin\theta & \sin\phi\sin\theta & \cos\theta \end{pmatrix}, \tag{12.207}$$

where the new and old vectors are related by

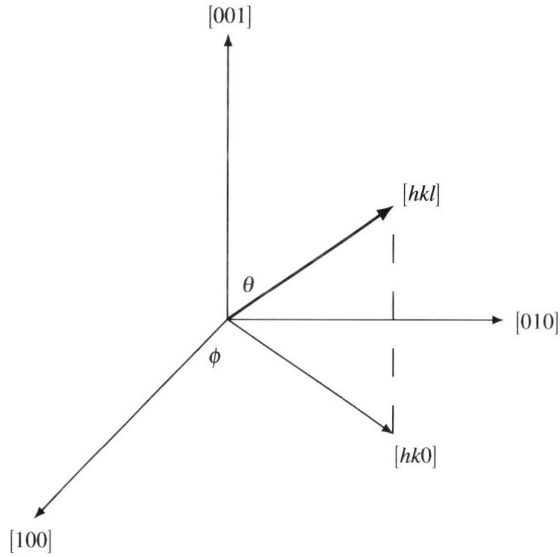

Fig. 12.4 Rotation of axes

$$x'_j = U_{ji} x_i. \tag{12.208}$$

For example, if we want to rotate the coordinate axes such that the new z axis, z', is along the $[1\bar{1}0]$ direction, we have

$$\cos\theta = \frac{l}{\sqrt{h^2+k^2+l^2}} = 0 \Rightarrow \quad \theta = 90^0,$$
$$\cos\phi = \frac{h}{\sqrt{h^2+k^2}} = \frac{1}{\sqrt{2}} \Rightarrow \quad \phi = -45^0, \text{ (for } k = -1).$$

Thus,

$$U[1\bar{1}0] = \begin{pmatrix} 0 & 0 & -1 \\ \frac{1}{\sqrt{2}} & \frac{1}{\sqrt{2}} & 0 \\ \frac{1}{\sqrt{2}} & -\frac{1}{\sqrt{2}} & 0 \end{pmatrix}. \tag{12.209}$$

The new axes, x', y', z' in terms of the old coordinate system are

$$x' = \begin{pmatrix} 0 \\ 0 \\ -1 \end{pmatrix}, y' = \begin{pmatrix} 1 \\ 1 \\ 0 \end{pmatrix}, z' = \begin{pmatrix} 1 \\ -1 \\ 0 \end{pmatrix}.$$

Common rotated axes are given in Table 12.13.

It will also be necessary to write the old wave-vector components in terms of the new ones:

Table 12.13 Rotated axes for common growth directions

$[hkl]$	x'	y'	z'
$[1\bar{1}0]$	$[00\bar{1}]$	$[110]$	$[1\bar{1}0]$
$[111]$	$\frac{1}{\sqrt{6}}[11\bar{2}]$	$\frac{1}{\sqrt{2}}[\bar{1}10]$	$\frac{1}{\sqrt{3}}[111]$

$$\mathbf{k} = U^T[1\bar{1}0]\mathbf{k}' = \begin{pmatrix} 0 & \frac{1}{\sqrt{2}} & \frac{1}{\sqrt{2}} \\ 0 & \frac{1}{\sqrt{2}} & -\frac{1}{\sqrt{2}} \\ -1 & 0 & 0 \end{pmatrix} \begin{pmatrix} k'_x \\ k'_y \\ k'_z \end{pmatrix} = \begin{pmatrix} \frac{1}{\sqrt{2}}k'_y + \frac{1}{\sqrt{2}}k'_z \\ \frac{1}{\sqrt{2}}k'_y - \frac{1}{\sqrt{2}}k'_z \\ -k'_x \end{pmatrix}. \tag{12.210}$$

12.6.3 General Theory

We break down the process of obtaining the Hamiltonian in the new rotated coordinate system into three steps. The first involves rotating the basis functions of the three-band Hamiltonian. The second involves replacing the unprimed wave vector by the primed one. The third involves re-expressing the matrix in terms of the $|J\,M_J\rangle$ states. Note that other procedures exist; see, e.g., [266] for an alternative approach. One can give a succinct presentation of the procedure.

Given

$$x'_j = U_{ji}x_i, \; k'_j = U_{ji}k_i,$$

and the original unrotated three-band Hamiltonian matrix

$$D_{mn}(\mathbf{k}) = \sum_{ij} D^{ij}_{mn}k_ik_j.$$

The matrix in terms of the rotated basis (i.e., X', Y', Z') is given by

$$\begin{aligned} D'_{m'n'} &= \sum_{mn} \left(\langle m|U^T_{mm'}\right) DU_{n'n}|n\rangle \sum_{mn} U^T_{mm'}D_{mn}U_{n'n} = \sum_{mn} U_{m'm}D_{mn}U^T_{nn'} \\ &= [U\,DU^T]_{m'n'}. \end{aligned} \tag{12.211}$$

We write, in component form,

$$D'_{m'n'}(\mathbf{k}) = \sum_{mn}\sum_{ij} U_{m'm}U_{n'n}D^{ij}_{mn}k_ik_j. \tag{12.212}$$

Next,

$$D'_{m'n'}(\mathbf{k}') = \sum_{mn}\sum_{ij}\sum_{i'j'} U_{m'm}U_{n'n}U_{i'i}U_{j'j}D^{ij}_{mn}k_{i'}k_{j'}. \tag{12.213}$$

The final step of converting to $|J\,M_J\rangle$ states will not be written out explicitly.

12.6.4 [1$\bar{1}$0] *Quantum Wires*

We consider here in detail a not-so-trivial example that relates to so-called V-grooved quantum wires [276]. The latter have been grown with the wire axis along [1$\bar{1}$0].

12.6.4.1 Rotation of *D* Matrix

We will take as starting matrix the *D* matrix of Stravinou and van Dalen [232]. This was reproduced in Table 12.3. Using Eq. (12.211), we need first the product DU^T which is given in Table 12.14. Finally, this results in the matrix given in Table 12.15.

12.6.4.2 Wave-Vector Replacement

We now replace the unprimed wave vector by the primed wave vector. This results in the matrix given in Table 12.15.

12.6.4.3 JM_J Basis

Since all the quantities are now in the rotated system, one would now express the $|JM_J\rangle$ states in the rotated system in terms of the rotated LS basis. This is, of course, the same as the corresponding linear combination in the unprimed system; in other words, the Clebsch-Gordan coefficients are the same. Therefore, the structure of the matrix in JM_J basis is the same as in the unrotated system. Nevertheless, the individual matrix elements are different and are now given. Since these are the final expressions, we have also made the operator form of the wave vector explicit. Indeed, we have done so for all components so that the result is valid for three-dimensional quantization.

$$
\begin{aligned}
P' &= \frac{1}{2}(H_{11} + H_{22}) + \frac{\mathrm{i}}{2}(H_{12} - H_{21}) \\
&= \frac{1}{4}\left\{\widehat{k}'_y(A + 3B + C_1 - C_2)\widehat{k}'_y + \widehat{k}'_z(A + 3B - C_1 + C_2)\widehat{k}'_z + 2\widehat{k}'_x(A + B)\widehat{k}'_x\right\} \\
&\quad + \frac{\mathrm{i}}{2}\left\{\widehat{k}'_x(C_1 + C_2)\widehat{k}'_y - \widehat{k}'_y(C_1 + C_2)\widehat{k}'_x\right\},
\end{aligned} \tag{12.214}
$$

$$
\begin{aligned}
P'' &= \frac{1}{6}(H_{11} + H_{22}) + \frac{2}{3}H_{33} + \frac{\mathrm{i}}{6}(H_{12} - H_{21}) \\
&= \frac{1}{12}\left\{\widehat{k}'_y(5A + 7B - 3C_1 + 3C_2)\widehat{k}'_y + \widehat{k}'_z(5A + 7B + 3C_1 - 3C_2)\widehat{k}'_z\right. \\
&\quad \left. + 2\widehat{k}'_x(A + 5B)\widehat{k}'_x\right\} + \frac{\mathrm{i}}{6}\left\{\widehat{k}'_x(C_1 + C_2)\widehat{k}'_y - \widehat{k}'_y(C_1 + C_2)\widehat{k}'_x\right\},
\end{aligned} \tag{12.215}
$$

Table 12.14 DU^T

$$DU^T = \frac{\hbar^2}{2m_0}\begin{pmatrix} k_xAk_x + k_yBk_y + k_zBk_z & k_xC_1k_y - k_yC_2k_x & k_xC_1k_z - k_zC_2k_x \\ k_yC_1k_x - k_xC_2k_y & k_yAk_y + k_xBk_x + k_zBk_z & k_yC_1k_z - k_zC_2k_y \\ k_zC_1k_x - k_xC_2k_z & k_zC_1k_y - k_yC_2k_z & k_zAk_z + k_xBk_x + k_yBk_y \end{pmatrix}\begin{pmatrix} 0 & \frac{1}{\sqrt{2}} & \frac{1}{\sqrt{2}} \\ 0 & \frac{1}{\sqrt{2}} & -\frac{1}{\sqrt{2}} \\ -1 & 0 & 0 \end{pmatrix}$$

$$= \frac{\hbar^2}{2m_0}\begin{pmatrix} -k_xC_1k_z + k_zC_2k_x & \frac{1}{\sqrt{2}}[k_xAk_x + k_yBk_y + k_zBk_z + k_xC_1k_y - k_yC_2k_x] & \frac{1}{\sqrt{2}}[k_xAk_x + k_yBk_y + k_zBk_z - k_xC_1k_y + k_yC_2k_x] \\ -k_yC_1k_z + k_zC_2k_y & \frac{1}{\sqrt{2}}[k_yAk_y + k_xBk_x + k_zBk_z + k_yC_1k_x - k_xC_2k_y] & \frac{1}{\sqrt{2}}[-k_yAk_y - k_zBk_z - k_xBk_x + k_yC_1k_x - k_xC_2k_y] \\ -k_zAk_z - k_xBk_x - k_yBk_y & \frac{1}{\sqrt{2}}[k_zC_1k_x - k_xC_2k_z + k_zC_1k_y - k_yC_2k_z] & \frac{1}{\sqrt{2}}[k_zC_1k_x - k_xC_2k_z - k_zC_1k_y + k_yC_2k_z] \end{pmatrix}$$

Table 12.15 Stravinou and van Dalen Hamiltonian in $[1\bar{1}0]$ rotated LS basis

$$D'(\mathbf{k}) = \frac{\hbar^2}{2m_0}\begin{pmatrix} |X'\rangle & |Y'\rangle & |Z'\rangle \\ k_zAk_z + k_xBk_x + k_yBk_y & \frac{1}{\sqrt{2}}[-k_zC_1k_x + k_xC_2k_z - k_zC_1k_y + k_yC_2k_z] & \frac{1}{\sqrt{2}}[-k_zC_1k_x + k_xC_2k_z + k_zC_1k_y - k_yC_2k_z] \\ \frac{1}{\sqrt{2}}[-k_xC_1k_z + k_zC_2k_x - k_yC_1k_z + k_zC_2k_y] & \frac{1}{2}[k_x(A+B)k_x + k_y(A+B)k_y + 2k_zBk_z & \frac{1}{2}[k_x(A-B)k_x - k_y(A-B)k_y \\ & +k_x(C_1-C_2)k_y + k_y(C_1-C_2)k_x] & -k_x(C_1+C_2)k_y + k_y(C_1+C_2)k_x] \\ \frac{1}{\sqrt{2}}[-k_xC_1k_z + k_zC_2k_x + k_yC_1k_z - k_zC_2k_y] & \frac{1}{2}[k_x(A-B)k_x - k_y(A-B)k_y & \frac{1}{2}[k_x(A+B)k_x + k_y(A+B)k_y + 2k_zBk_z \\ & +k_x(C_1+C_2)k_y - k_y(C_1+C_2)k_x] & -k_x(C_1-C_2)k_y - k_y(C_1-C_2)k_x] \end{pmatrix},$$

$$D'(\mathbf{k}') = \frac{\hbar^2}{2m_0}\begin{pmatrix} |X'\rangle & |Y'\rangle & |Z'\rangle \\ k'_xAk'_x + k'_yBk'_y + k'_zBk'_z & k'_xC_1k'_y - k'_yC_2k'_x & k'_xC_1k'_z - k'_zC_2k'_x \\ k'_yC_1k'_x - k'_xC_2k'_y & \frac{1}{2}[k'_y(A+B+C_1-C_2)k'_y + k'_z(A+B-C_1+C_2)k'_z] & \frac{1}{2}[k'_y(A-B+C_1+C_2)k'_z + k'_z(A-B-C_1-C_2)k'_y] \\ & +k'_xBk'_x & \\ k'_zC_1k'_x - k'_xC_2k'_z & \frac{1}{2}[k'_z(A-B+C_1+C_2)k'_y + k'_y(A-B-C_1-C_2)k'_z] & \frac{1}{2}[k'_y(A+B-C_1+C_2)k'_y + k'_z(A+B+C_1-C_2)k'_z] \\ & & +k'_xBk'_x \end{pmatrix}.$$

$$
\begin{aligned}
P''' &= \frac{1}{3}(H_{11} + H_{22} + H_{33}) + \frac{\mathrm{i}}{3}(H_{12} - H_{21}) \\
&= \frac{1}{3}\left\{\widehat{k}'_x(A+2B)\widehat{k}'_x + \widehat{k}'_y(A+2B)\widehat{k}'_y + \widehat{k}'_z(A+2B)\widehat{k}'_z\right\} \\
&\quad + \frac{\mathrm{i}}{3}\left\{\widehat{k}'_x(C_1+C_2)\widehat{k}'_y - \widehat{k}'_y(C_1+C_2)\widehat{k}'_x\right\},
\end{aligned} \tag{12.216}
$$

$$
\begin{aligned}
S_- &= -\frac{1}{\sqrt{3}}(H_{13} - \mathrm{i}H_{23}) \\
&= -\frac{1}{\sqrt{3}}[\widehat{k}'_x C_1 \widehat{k}'_z - \widehat{k}'_z C_2 \widehat{k}'_x] + \frac{\mathrm{i}}{2\sqrt{3}}[\widehat{k}'_y(A - B + C_1 + C_2)\widehat{k}'_z \\
&\quad + \widehat{k}'_z(A - B - C_1 - C_2)\widehat{k}'_y],
\end{aligned} \tag{12.217}
$$

$$
\begin{aligned}
R &= \frac{1}{2\sqrt{3}}(H_{11} - H_{22} - \mathrm{i}H_{12} - \mathrm{i}H_{21}) \\
&= \frac{1}{2\sqrt{3}}\left\{\widehat{k}'_x(A-B)\widehat{k}'_x - \frac{1}{2}\widehat{k}'_y(A - B + C_1 - C_2)\widehat{k}'_y - \frac{1}{2}\widehat{k}'_z(A - B - C_1 + C_2)\widehat{k}'_z\right\} \\
&\quad - \frac{\mathrm{i}}{2\sqrt{3}}\left\{\widehat{k}'_x(C_1 - C_2)\widehat{k}'_y + \widehat{k}'_y(C_1 - C_2)\widehat{k}'_x\right\},
\end{aligned} \tag{12.218}
$$

$$
\begin{aligned}
C &= -\frac{1}{3}(-H_{13} + H_{31} + \mathrm{i}H_{23} - \mathrm{i}H_{32}) \\
&= \frac{1}{3}[\widehat{k}'_x(C_1+C_2)\widehat{k}'_z - \widehat{k}'_z(C_1+C_2)\widehat{k}'_x] - \frac{\mathrm{i}}{3}[\widehat{k}'_y(C_1+C_2)\widehat{k}'_z - \widehat{k}'_z(C_1+C_2)\widehat{k}'_y],
\end{aligned} \tag{12.219}
$$

$$
\begin{aligned}
Q &= -\frac{1}{6}(H_{11} + H_{22} + \mathrm{i}H_{12} - \mathrm{i}H_{21}) + \frac{1}{3}H_{33} \\
&= \frac{1}{12}\left\{-2\widehat{k}'_x(A-B)\widehat{k}'_x + \widehat{k}'_y(A - B - 3C_1 + 3C_2)\widehat{k}'_y + \widehat{k}'_z(A - B + 3C_1 - 3C_2)\widehat{k}'_z\right\} \\
&\quad - \frac{\mathrm{i}}{6}\left\{\widehat{k}'_x(C_1+C_2)\widehat{k}'_y - \widehat{k}'_y(C_1+C_2)\widehat{k}'_x\right\},
\end{aligned} \tag{12.220}
$$

$$
\begin{aligned}
\Sigma_- &= -\frac{1}{3\sqrt{3}}(H_{13} + 2H_{31} - \mathrm{i}H_{23} - 2\mathrm{i}H_{32}) \\
&= -\frac{1}{3\sqrt{3}}[\widehat{k}'_x(C_1 - 2C_2)\widehat{k}'_z + \widehat{k}'_z(2C_1 - C_2)\widehat{k}'_x] \\
&\quad + \frac{\mathrm{i}}{6\sqrt{3}}[\widehat{k}'_y(3A - 3B - C_1 - C_2)\widehat{k}'_z + \widehat{k}'_z(3A - 3B + C_1 + C_2)\widehat{k}'_y].
\end{aligned} \tag{12.221}
$$

The combinations of Stravinou-van Dalen parameters can be reduced to Luttinger parameters; this is given in Table 12.16.

12.6.4.4 Symmetrized Hamiltonian

We now give the corresponding Hamiltonian within the symmetrized Luttinger–Kohn theory. The procedure is to first treat the wave vector as c-numbers, combine the identical terms, and then symmetrize the resulting expression. By inspection,

Table 12.16 Combinations of Stravinou-van Dalen parameters in terms of Luttinger parameters

$A+B$	$-2(\gamma_1+\gamma_2)$,
$A+2B$	$-3\gamma_1$,
$A+3B$	$-2(2\gamma_1-\gamma_2)$,
$A+5B$	$-6(\gamma_1-\gamma_2)$,
$5A+7B$	$-6(2\gamma_1+\gamma_2)$,
$A-B$	$-6\gamma_2$,
C_1+C_2	$2(1+\gamma_1-2\gamma_2-3\gamma_3)$,
C_1-C_2	$-6\gamma_3$,
C_1-2C_2	$-(1+\gamma_1-2\gamma_2+6\gamma_3)$.

one finds that the following matrix elements are unchanged:

$$R,$$

and the following change:

$$\begin{aligned}
P' &= \frac{1}{4}\left\{\widehat{k}'_y(A+3B+C_1-C_2)\widehat{k}'_y+\widehat{k}'_z(A+3B-C_1+C_2)\widehat{k}'_z+2\widehat{k}'_x(A+B)\widehat{k}'_x\right\},\\
P'' &= \frac{1}{12}\left\{\widehat{k}'_y(5A+7B-3C_1+3C_2)\widehat{k}'_y+\widehat{k}'_z(5A+7B+3C_1-3C_2)\widehat{k}'_z+2\widehat{k}'_x(A+5B)\widehat{k}'_x\right\},\\
P''' &= \frac{1}{3}\left\{\widehat{k}'_x(A+2B)\widehat{k}'_x+\widehat{k}'_y(A+2B)\widehat{k}'_y+\widehat{k}'_z(A+2B)\widehat{k}'_z\right\},\\
S_- &= -\frac{1}{\sqrt{3}}[\widehat{k}'_x C_1\widehat{k}'_z-\widehat{k}'_z C_2\widehat{k}'_x]+\frac{i}{2\sqrt{3}}[\widehat{k}'_y(A-B+C_1+C_2)\widehat{k}'_z+\widehat{k}'_z(A-B-C_1-C_2)\widehat{k}'_y]\\
&\to -\frac{1}{\sqrt{3}}\widehat{k}'_x\widehat{k}'_z(C_1-C_2)+\frac{i}{2\sqrt{3}}\widehat{k}'_y\widehat{k}'_z 2(A-B)\\
&\to -\frac{1}{2\sqrt{3}}[\widehat{k}'_x(C_1-C_2)\widehat{k}'_z+\widehat{k}'_z(C_1-C_2)\widehat{k}'_x]+\frac{i}{2\sqrt{3}}[\widehat{k}'_y(A-B)\widehat{k}'_z+\widehat{k}'_z(A-B)\widehat{k}'_y],\\
C &= 0,\\
Q &= \frac{1}{12}[-2\widehat{k}'_x(A-B)\widehat{k}'_x+\widehat{k}'_y(A-B-3C_1+3C_2)\widehat{k}'_y+\widehat{k}'_z(A-B+3C_1-3C_2)\widehat{k}'_z]\\
\Sigma_- &= -\frac{1}{3\sqrt{3}}[\widehat{k}'_x(C_1-2C_2)\widehat{k}'_z+\widehat{k}'_z(2C_1-C_2)\widehat{k}'_x]\\
&\quad+\frac{i}{6\sqrt{3}}[\widehat{k}'_y(3A-3B-C_1-C_2)\widehat{k}'_z+\widehat{k}'_z(3A-3B+C_1+C_2)\widehat{k}'_y]\\
&\to -\frac{1}{\sqrt{3}}\widehat{k}'_x\widehat{k}'_z(C_1-C_2)+\frac{i}{\sqrt{3}}\widehat{k}'_y\widehat{k}'_z(A-B)\\
&\to -\frac{1}{2\sqrt{3}}[\widehat{k}'_x(C_1-C_2)\widehat{k}'_z+\widehat{k}'_z(C_1-C_2)\widehat{k}'_x]+\frac{i}{2\sqrt{3}}[\widehat{k}'_y(A-B)\widehat{k}'_z+\widehat{k}'_z(A-B)\widehat{k}'_y].
\end{aligned}$$

We note that the S term leads to hh–lh and hh–sh coupling, the C term leads to lh–lh and sh–sh coupling, the Q term leads to lh–sh coupling, and the Σ term leads to lh–sh coupling.

12.6.4.5 Analysis

In general, the band structure in the plane perpendicular to the Z' axis is not isotropic. One quick check is to note that two in-plane directions are [110] and [001]. It is already known in the bulk that the valence band displays warping. One can also demonstrate the warping explicitly by looking at the matrix elements. For example, the coefficient of $k_x'^2$ in P' is

$$2(A+B) = -12(\gamma_1 - \gamma_2),$$

while that of $k_y'^2$ is

$$A + 3B + C_1 - C_2 = -2(2\gamma_1 - \gamma_2 + 3\gamma_3).$$

But they are equal if one makes the axial approximation: $\gamma_2 = \gamma_3$. Similarly, one can show that, under this approximation, all the matrix elements of the Hamiltonian only depend upon the magnitude of the in-plane wave vector $k_{||}$ (in addition to k_z of course) and not on the orientation in the plane.

At the zone center of the quantum wire dispersion relation, the following matrix elements are zero: S_-, C, Σ_-. Nevertheless, there is still a difference between the BF and LK Hamiltonians, contrary to the case at $k_{||} = 0$ for a $[1\bar{1}0]$ QW. The nonzero matrix elements are:

$$P = \frac{1}{4}\left\{\widehat{k}_y'(A+3B+C_1-C_2)\widehat{k}_y' + 2\widehat{k}_x'(A+B)\widehat{k}_x'\right\}$$
$$+\frac{i}{2}\left\{\widehat{k}_x'(C_1+C_2)\widehat{k}_y' - \widehat{k}_y'(C_1+C_2)\widehat{k}_x'\right\}, \qquad (12.222)$$

$$P'' = \frac{1}{12}\left\{\widehat{k}_y'(5A+7B-3C_1+3C_2)\widehat{k}_y' + 2\widehat{k}_x'(A+5B)\widehat{k}_x'\right\}$$
$$+\frac{i}{6}\left\{\widehat{k}_x'(C_1+C_2)\widehat{k}_y' - \widehat{k}_y'(C_1+C_2)\widehat{k}_x'\right\}, \qquad (12.223)$$

$$P''' = \frac{1}{3}\left\{\widehat{k}_x'(A+2B)\widehat{k}_x' + \widehat{k}_y'(A+2B)\widehat{k}_y'\right\}$$
$$+\frac{i}{3}\left\{\widehat{k}_x'(C_1+C_2)\widehat{k}_y' - \widehat{k}_y'(C_1+C_2)\widehat{k}_x'\right\}, \qquad (12.224)$$

$$R = \frac{1}{2\sqrt{3}}\left\{\widehat{k}_x'(A-B)\widehat{k}_x' - \frac{1}{2}\widehat{k}_y'(A-B+C_1-C_2)\widehat{k}_y'\right\}$$
$$-\frac{i}{2\sqrt{3}}\left\{\widehat{k}_x'(C_1-C_2)\widehat{k}_y' + \widehat{k}_y'(C_1-C_2)\widehat{k}_x'\right\}, \qquad (12.225)$$

$$Q = \frac{1}{12}\left\{-2\widehat{k}'_x(A-B)\widehat{k}'_x + \widehat{k}'_y(A-B-3C_1+3C_2)\widehat{k}'_y\right\}$$
$$-\frac{\mathrm{i}}{6}\left\{\widehat{k}'_x(C_1+C_2)\widehat{k}'_y - \widehat{k}'_y(C_1+C_2)\widehat{k}'_x\right\}.$$

We note that the Hamiltonian also block diagonalizes:

$$H_{\mathrm{BF1}} = \begin{pmatrix} \left|\frac{3}{2}\,\frac{3}{2}\right\rangle & \left|\frac{3}{2}\,-\frac{1}{2}\right\rangle & \left|\frac{1}{2}\,-\frac{1}{2}\right\rangle & \left|\frac{3}{2}\,\frac{1}{2}\right\rangle & \left|\frac{3}{2}\,-\frac{3}{2}\right\rangle & \left|\frac{1}{2}\,\frac{1}{2}\right\rangle \\ P' & -R & -\sqrt{2}R & & & \\ & P''^* & -\sqrt{2}Q^* & & & \\ & & P'''^* - \Delta_0 & & & \\ & & & P'' & R & -\sqrt{2}Q \\ & & & & P'^* & \sqrt{2}R^\dagger \\ & & & & & P''' - \Delta_0 \end{pmatrix}. \quad (12.226)$$

12.6.4.6 Boundary Conditions

The standard technique is to integrate the differential equations across an interface. In the process, contributions arise from, for example, $\widehat{k}'_i\mathcal{A}(\mathbf{r}')f(\mathbf{r}')$ terms, where $\mathcal{A}(\mathbf{r}')$ are position-dependent material parameters (within LK theory, they are related to the Luttinger parameters) and $f(\mathbf{r}')$ is the envelope function. For QW's, this corresponds to the term $\widehat{k}'_z\mathcal{A}(z')f(z')$; when applied to the interaction between heavy-hole and light-hole states of the same spin, it only leads to π and δ contributions using the BF Hamiltonian [232], but it also includes the σ contribution if the LK Hamiltonian is used. In the quantum-wire case, the same matrix element contains the term $\widehat{k}'_x C_1 f(z')$ which includes the σ contribution; this is the s-like coupling term absent from previous implementations of the BF theory, but which is present in the LK theory [34]. One can similarly analyze the other matrix elements. Overall, one expects that, at finite k_z the difference between BF and LK might be reduced.

12.6.4.7 Calculations

An example of a valence-band structure for a 2.5 nm V-groove GaAs/$Al_{0.3}Ga_{0.7}As$ $[1\bar{1}0]$ wire using four-band BF (solid lines) and LK (dashed lines) Hamiltonians is shown in Fig. 12.5. For this particular geometry and structure, the difference between the two Hamiltonians is less than 1 meV. The sum squared of the envelope-function components for the lowest electron and hole states are given in Fig. 12.6, showing the confinement of these states inside the quantum wire. The electron state was obtained using a one-band model.

The above calculations can be repeated for a lattice-matched $In_{0.53}Ga_{0.47}As$/InP quantum wire (Fig. 12.7); V-groove quantum wires based upon these materials have

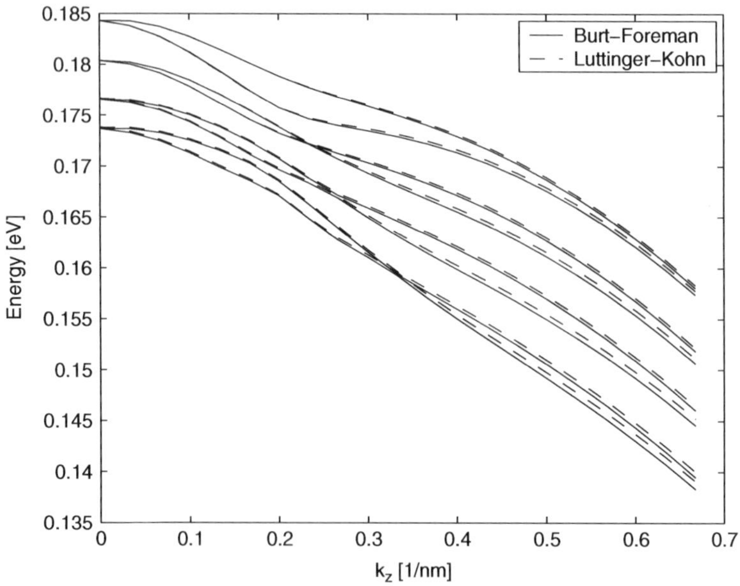

Fig. 12.5 Valence-band structure for 2.5 nm V-groove GaAs/$Al_{0.3}Ga_{0.7}As$ $[1\bar{1}0]$ wire using four-band BF (*solid*) and LK (*dashed*) Hamiltonians. Using parameters of Vouilloz et al. [276]

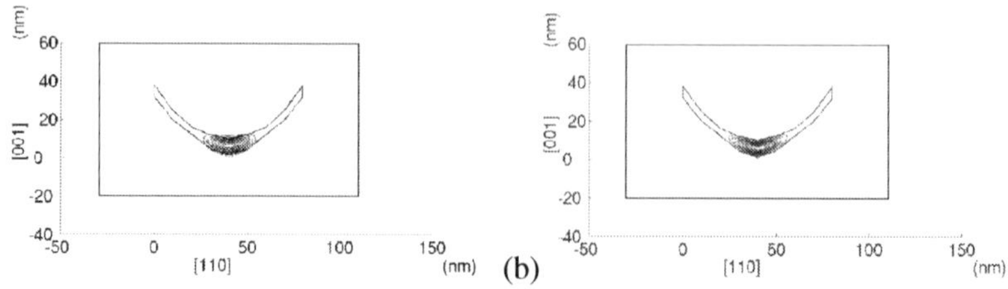

Fig. 12.6 Geometry of embedded wire and first **(a)** electron, **(b)** hole probability density for the GaAs/AlGaAs quantum wire

also been made [277], though the experimental structures are probably all strained. It is also interesting because there is a larger difference in the Luttinger parameters. Indeed, this translates into a correspondingly larger difference in LK and BF results (Fig. 12.7).

12.7 Spurious Solutions

One of the difficulties faced by the $k \cdot p$ theory for heterostructures is the generation of so-called spurious solutions. It was first discussed by White and Sham in 1981 [213]. Spurious solutions arising from small Hamiltonian matrix elements of

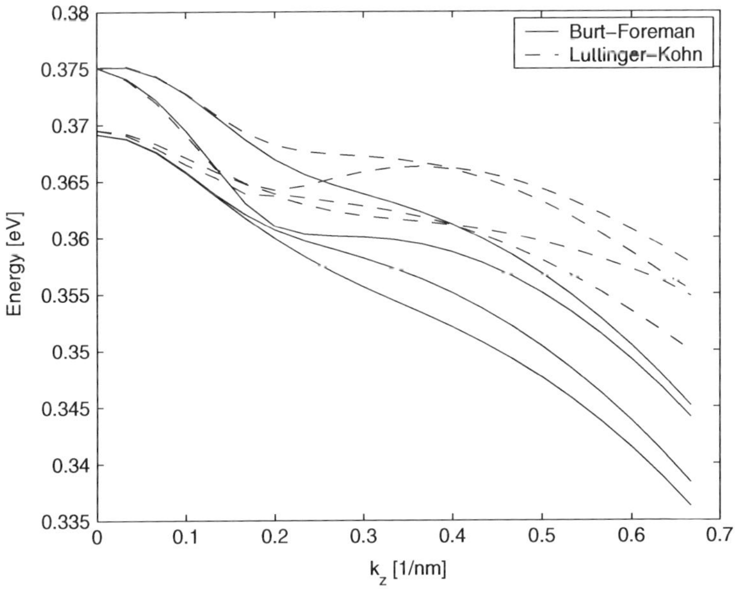

Fig. 12.7 Valence-band structure for 2.5 nm V-groove $In_{0.53}Ga_{0.47}As$ $[1\bar{1}0]$ wire using four-band BF (*solid lines*) and LK (*dashed lines*) Hamiltonians

order k^2 are well known [228, 233, 275, 278–286]. Some of the methods proposed to solve the spurious solution problem include: integral representation in momentum space [287], secular matrix [288–290], and adding k^4 term [291]. Foreman [292] has recently presented a method for elimination of spurious states in envelope-function theory by using a unitary transformation of basis states to modify the conduction-valence band $k \cdot p$ interaction by a small amount. While the transformation applied is equivalent to a gauge transformation (which, in exact theories, is known to leave physical variables invariant), the fact that envelope-function methods are only approximate leads to small changes in physical variables subject to such a unitary transformation. In effect, the conduction-valence band $k \cdot p$ interaction can be either strengthened or weakened by applying this unitary transformation. The precise value is determined by setting the partially renormalized conduction-band mass to zero [292]. The suggested change of basis states applied to first-principles envelope-function theory led to good agreement with density-functional theory calculations on InGaAs/InP superlattice structures.

12.8 Summary

Three main approaches to the electronic structure of heterostructures were discussed: one-band models, the Bastard model, and the Burt–Foreman model. In addition, a number of other formulations of the $k \cdot p$ theory have emerged. Early on Mailhiot and Smith proposed a model that is based on a single set of zone-center basis functions [78, 293, 294]. However, it is fairly complex in that is requires a starting point of an empirical pseudopotential band structure for a fictitious average material, a full-zone $k \cdot p$ representation of the constituent materials derived from the former, a complex band structure from the latter, and finally wave function matching for heterostructures. More recent models include the pseudopotential-based full-zone $k \cdot p$ model of Wang and Zunger [295], the atomic-scale model of Yi and Razeghi (similar to Wang and Zunger) [296], the theory of Takhtamirov and Volkov [243], and the complete Fourier representation theory of Stubner et al. [297]. The work of Wang and Zunger, in particular, showed that the standard LK eight-band model works surprisingly well for the lowest few hole states of $(GaAs)_p(AlAs)_p$ superlattices (within 200 meV of top of valence band) down to a few nanometers but not so in the conduction band if intervalley coupling effects are important. They also showed that the six-band model, applied to InP and CdSe quantum dots, " ... overestimates significantly ..." the hole and electron confinement energies and can even produce a reverse order of s- and p-like valence states [298]. However, this view is being disputed by others [299, 300].

Chapter 13
Heterostructures: Further Topics

13.1 Overview

There are a number of properties that are important for heterostructures. For example, there are new contributions to spin splitting and the dimensionality of excitons is effectively reduced in a nanostructure. Hence, even though the basic theories for bulk band structures have already been presented, it is useful to show the modifications for heterostructures. Such research is clearly ongoing and particularly in terms of applications; the literature on the application of $k \cdot p$ theory to heterostructures numbers in the thousands. No attempt will be made to cite them all.

In the following, we will start each section by providing an overview of the topic and then discuss a couple of examples in more detail.

13.2 Spin Splitting

In a heterostructure, there are now two contributions to spin splitting: the Dresselhaus contribution that was already present in the bulk (hence, also labeled as the bulk inversion asymmetry or BIA contribution) and the Rashba contribution which is a new one resulting from the macroscopic asymmetry introduced by the interfaces (hence also known as the structural inversion asymmetry or SIA contribution). Sometimes, one identifies a third (structural) contribution due to the atomic character of the interface. The $k \cdot p$ theory of spin splitting has been extensively studied for heterostructures [21, 301]. This is still widely studied in quantum wells [302–305] and quantum dots [306]. A recent review article is due to Zawadzki and Pfeffer [307].

13.2.1 Zincblende Superlattices

In zincblende superlattice structures, it is possible to have linear-in-k spin-splitting components for the Γ_{6c} conduction band due to the superlattice confinement even in the absence of the Rashba (macroscopic) structural-inversion-asymmetry effect.

L.C. Lew Yan Voon, M. Willatzen, *The* k · p *Method*,
DOI 10.1007/978-3-540-92872-0_13, © Springer-Verlag Berlin Heidelberg 2009

In this section, we shall consider spin splittings in a single quantum-well and in superlattice structures [308, 309].

For the Γ_{6c} spin splittings in a superlattice zincblende structure, the Hamiltonian in Eq. (6.4) is used with k_z replaced by $-\mathrm{i}\frac{\partial}{\partial z}$ where z is the superlattice growth direction. We note that the off-diagonal terms in Eq. (6.4) have opposite signs as compared to [308, 309] due to a sign change in the choice of the spin-down S basis state. This will clearly not affect the discussion of spin-splitting results. Furthermore, γ_{6c} in this work equals $\gamma/2$ in [308, 309].

Following the usual symmetrization principle in $k \cdot p$ heterostructure theory accounting for a position-dependent spin-splitting coefficient γ_{6c}:

$$\gamma_{6c}\frac{\partial}{\partial z} \to \frac{1}{2}\left(\gamma_{6c}\frac{\partial}{\partial z} + \frac{\partial}{\partial z}\gamma_{6c}\right),$$
$$\gamma_{6c}\frac{\partial^2}{\partial z^2} \to \frac{\partial}{\partial z}\gamma_{6c}\frac{\partial}{\partial z}, \tag{13.1}$$

Equation (6.4) is rewritten as the matrix equation:

$$H_{ij}^{6c6c}(\mathbf{k}_\parallel) = \frac{\partial}{\partial z}A_{ij}(\mathbf{k}_\parallel)\frac{\partial}{\partial z} + \frac{1}{2}\left(B_{ij}(\mathbf{k}_\parallel)\frac{\partial}{\partial z} + \frac{\partial}{\partial z}B_{ij}(\mathbf{k}_\parallel)\right) + C_{ij}(\mathbf{k}_\parallel), \tag{13.2}$$
$$H^{6c6c}(\mathbf{k}_\parallel)F(\mathbf{k}_\parallel) = E(\mathbf{k}_\parallel)F(\mathbf{k}_\parallel), \tag{13.3}$$

where $A_{ij}(\mathbf{k}_\parallel)$, $B_{ij}(\mathbf{k}_\parallel)$, $C_{ij}(\mathbf{k}_\parallel)$ are position-dependent coefficients, $F(\mathbf{k}_\parallel)$ is the z-dependent two-component superlattice envelope-function part, and $E(\mathbf{k}_\parallel)$ is the associated two-component energy with the upper (lower) component being the spin-up (spin-down) components. Hence, the following continuity conditions must apply everywhere in the heterostructure and in particular at the material interfaces:

$$F(\mathbf{k}_\parallel)\Big|_L = F(\mathbf{k}_\parallel)\Big|_R, \tag{13.4}$$
$$\left(A_{ij}(\mathbf{k}_\parallel)\frac{\partial}{\partial z} + \frac{1}{2}B_{ij}(\mathbf{k}_\parallel)\right)\Big|_L F(\mathbf{k}_\parallel)\Big|_L = \left(A_{ij}(\mathbf{k}_\parallel)\frac{\partial}{\partial z} + \frac{1}{2}B_{ij}(\mathbf{k}_\parallel)\right)\Big|_R F(\mathbf{k}_\parallel)\Big|_R,$$

with L (R) denoting to the immediate left (right) of the material interface.

Using degenerate perturbation theory to the first order, the spin splittings $\Delta E(\mathbf{k}_\parallel) = E_1(\mathbf{k}_\parallel) - E_2(\mathbf{k}_\parallel)$ are given by

$$\Delta E(\mathbf{k}_\parallel) = 2\sqrt{V_{11}^2 + |V_{12}|^2}, \tag{13.5}$$
$$V_{ij} = \langle F_i(\mathbf{k}_\parallel)|H^{6c6c}(\mathbf{k}_\parallel)|F_j(\mathbf{k}_\parallel)\rangle. \tag{13.6}$$

For completeness, we note that in the bulk case (with k_z replacing $-\mathrm{i}\frac{\partial}{\partial z}$),

$$\Delta E(\mathbf{k}) = 2\gamma_{6c}\sqrt{k^2\left(k_x^2k_y^2 + k_y^2k_z^2 + k_z^2k_x^2\right) - 9k_x^2k_y^2k_z^2}. \tag{13.7}$$

For superlattices grown along the [001] direction, the above formalism leads to the spin-splitting result:

$$\Delta E(\mathbf{k}_\parallel) = |2\overline{\gamma_{6c}k_z^2}(k_x + \mathrm{i}k_y) - 2\overline{\gamma}_{6c}\mathrm{i}k_x k_y(k_x - \mathrm{i}k_y)|. \tag{13.8}$$

Similarly, for superlattices grown along the [110] direction, the result is

$$\Delta E(\mathbf{k}_\parallel) = | - \overline{\gamma_{6c}k_z^2}\frac{(1-\mathrm{i})}{\sqrt{2}}k_y + \overline{\gamma}_{6c}\frac{(1-\mathrm{i})}{\sqrt{2}}k_y(k_y^2 - 2k_x^2)|, \tag{13.9}$$

where k_x and k_y are along the $[00\overline{1}]$ and $[\overline{1}10]$ directions, respectively. Note that reflection along the superlattice direction is a symmetry for zincblende superlattices grown along the [001] and [110] directions. Thus, envelope functions are parity eigenfunctions and matrix elements V_{11} and V_{22} vanish but V_{12} and V_{21} contribute.

For superlattices grown along the [111] direction, the above formalism leads to the spin-splitting result:

$$\Delta E(\mathbf{k}_\parallel) = |\frac{1}{\sqrt{3}}\Big[16(\overline{\gamma_{6c}k_z^2})^2(k_x^2 + k_y^2) - 8\overline{\gamma}_{6c}\overline{\gamma_{6c}k_z^2}(k_x^2 + k_y^2)^2 + \overline{\gamma}_{6c}^2\left(k_x^6 + 21k_x^4k_y^2 - 9k_x^2k_y^4 + 3k_y^6\right)\Big]^{1/2}|, \tag{13.10}$$

where k_x and k_y are along the $[11\overline{2}]$ and $[\overline{1}10]$ directions, respectively.

In the above expressions, we have introduced the averaged γ_{6c} coefficients:

$$\overline{\gamma}_{6c} = \frac{1}{L}\int_{-L/2}^{L/2} \mathrm{d}z\, F_1^*(\mathbf{k}_\parallel)\gamma_{6c}F_1(\mathbf{k}_\parallel),$$

$$\overline{\gamma_{6c}k_z^2} = \frac{1}{L}\int_{-L/2}^{L/2} \mathrm{d}z\, F_1^*(\mathbf{k}_\parallel)\widehat{k}_z\gamma_{6c}\widehat{k}_z F_1(\mathbf{k}_\parallel), \tag{13.11}$$

using the normalization

$$\frac{1}{L}\int_{-L/2}^{L/2} \mathrm{d}z\, F_1^*(\mathbf{k}_\parallel)F_1(\mathbf{k}_\parallel) = 1, \tag{13.12}$$

where $L = \infty$ in the quantum-well case and the superlattice unit cell extends from $-L/2$ to $L/2$ in the superlattice case (with L finite). Note that the unperturbed envelope-function components are the same: $F_1(\mathbf{k}_\parallel) = F_2(\mathbf{k}_\parallel)$ thus $F(\mathbf{k}_\parallel) = \big(F_1(\mathbf{k}_\parallel), F_2(\mathbf{k}_\parallel)\big) = \big(F_1(\mathbf{k}_\parallel), F_1(\mathbf{k}_\parallel)\big)$.

In the general case with the superlattice growth direction along $[n11]$ the Hamiltonian $H_{[n11]}^{6c6c}$ can be found by transforming the Hamiltonian H^{6c6c} in Eq. (6.4) according to [309]

$$H_{[n11]}^{6c6c}(\mathbf{k}_\parallel) = H_{[001]}^{6c6c}(\mathrm{R}\mathbf{k}_\parallel), \tag{13.13}$$

where the rotation matrix reads

$$\mathsf{R} = n_q \begin{pmatrix} -\sqrt{2} & 0 & n \\ \frac{n}{\sqrt{2}} & -\frac{1}{\sqrt{2}n_q} & 1 \\ \frac{n}{\sqrt{2}} & \frac{1}{\sqrt{2}n_q} & 1 \end{pmatrix}, n_q = \frac{1}{\sqrt{n^2+2}}. \tag{13.14}$$

By applying the above procedure, we obtain the following expressions:

$$\begin{aligned} V_{11} = &-2\overline{\gamma}_{6c}\left[-\frac{k_y^3}{4\sqrt{2}} + \frac{k_x k_y^2 n n_q}{4\sqrt{2}} + \left(\frac{k_x^2 k_y}{\sqrt{2}} + \frac{k_x^2 k_y n^2}{4\sqrt{2}}\right) n_q^2 + \left(\frac{k_x^3 n}{\sqrt{2}} - \frac{k_x^3 n^3}{4\sqrt{2}}\right) n_q^3\right] \\ &-2\overline{\gamma_{6c} k_z^2}\left[\left(\frac{k_y}{2\sqrt{2}} + \frac{k_y n^2}{2\sqrt{2}}\right) n_q^2 + \left(\frac{-7k_x n}{2\sqrt{2}} + \frac{k_x n^3}{2\sqrt{2}}\right) n_q^3\right], \\ V_{12} = &\ 2\overline{\gamma}_{6c}\left[-\mathrm{i}\frac{k_y^3}{4\sqrt{2}} - \mathrm{i}\frac{k_x k_y^2 n n_q}{4\sqrt{2}} + \left(\mathrm{i}\frac{k_x^2 k_y}{\sqrt{2}} - \sqrt{2}k_x^2 k_y n + \mathrm{i}\frac{k_x^2 k_y n^2}{4\sqrt{2}}\right) n_q^2\right. \\ &\left.+ \left(-\mathrm{i}\frac{k_x^3 n}{\sqrt{2}} + \mathrm{i}\frac{k_x^3 n^3}{4\sqrt{2}}\right) n_q^3\right] + 2\overline{\gamma_{6c} k_z^2}\left[\left(\mathrm{i}\frac{k_y}{2\sqrt{2}} + \sqrt{2}k_y n + \mathrm{i}\frac{k_y n^2}{2\sqrt{2}}\right) n_q^2\right. \\ &\left.+ \left(\frac{7\mathrm{i}k_x n}{2\sqrt{2}} - \mathrm{i}\frac{k_x n^3}{2\sqrt{2}}\right) n_q^3\right]. \end{aligned} \tag{13.15}$$

In Fig. 13.1, spin-splitting results are plotted in contour diagrams for the quantum-well growth directions: [001], [111], [011], [211], and [311] directions using a γ_{6c} value equal to 8.5 eV Å^3 [310]. Along the two figure axes, superlattice in-plane **k** components (perpendicular to one another) are varied in the range $[-0.06; 0.06]$ in units of $2\pi/a_0$ where a_0 is the lattice constant. Note that the spin splittings depend strongly on the in-plane direction even for smaller in-plane **k** components in the cases where the superlattice is grown along the [011], [211], and [311] directions. This anisotropic tendency becomes pronounced for the [001] and [111] superlattice directions only at higher in-plane **k** component values. This result is important for the interpretation of spin-flip Raman scattering results [309].

Linear spin-splitting terms in the dispersion of the Γ_{6c} band may also occur in bulk zincblende structures due to strain effects. This result can be understood as stemming from the replacement of two wave-vector components in the general third-order spin-splitting result by a symmetry-equivalent strain tensor component, i.e., replacing $k_i k_j$ by ε_{ij} [104]. Finally, it should be mentioned that the present $k \cdot p$ results are in good agreement with more detailed ab initio LMTO and tight-binding calculations results in bulk and superlattice structures as long as the superlattice material layers are not too small [104, 309].

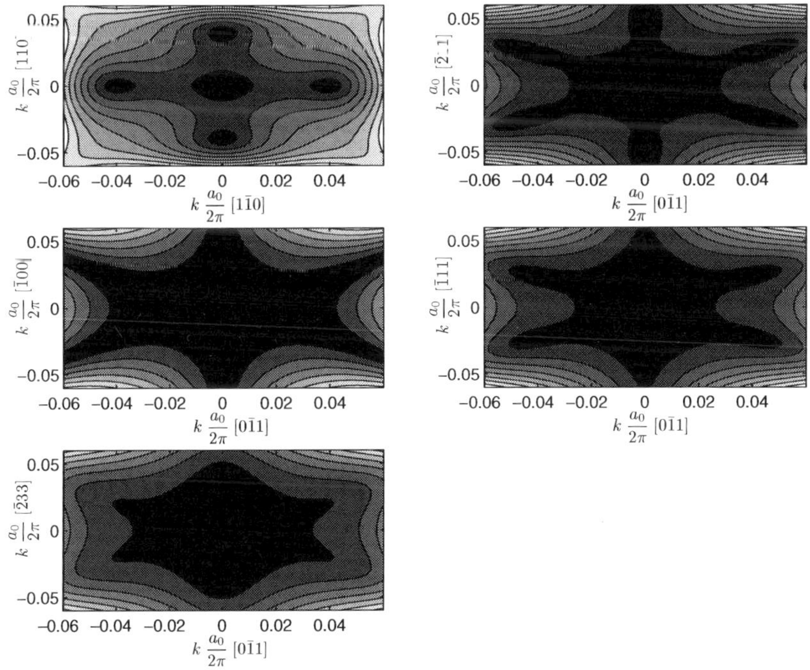

Fig. 13.1 Spin-splitting results for different superlattice growth directions. Adapted from [309]. ©1995 by the American Physical Society

13.3 Strain in Heterostructures

Strain is introduced in semiconductor heterostructures due to either an externally-applied stress or to the misfit in bulk lattice constants of the materials forming the heterostructure.

13.3.1 External Stress

The interest in applying an external stress to a nanostructure is the same as for a bulk crystal, i.e., to attempt to extract more information about the electronic structure through distortion of the electronic structure and/or symmetry reduction. One difference compared to the bulk case is the fact that a uniaxial stress does not lower the symmetry of a zincblende-based superlattice if applied along the growth direction.

The problem of external strain is modeled within $k \cdot p$ theory in the same fashion as for the unstrained problem except that the starting point is one of the strain Hamiltonians from Chap. 7. For instance, within the four-band LK model, a uniaxial stress applied along the growth direction of an [001] QW gives rise to an additional strain Hamiltonian given by [199]

$$H_\varepsilon = \begin{pmatrix} -\xi & & & \\ & \xi & & \\ & & \xi & \\ & & & -\xi \end{pmatrix}, \tag{13.16}$$

where

$$\xi = b(S_{11} - S_{12})X,$$

and the S's are compliance constants and X is the applied stress.

As an example, Fig. 13.2 shows how a compressive stress affects the electronic structure of a GaAs-$Al_{0.3}Ga_{0.7}As$ [001] quantum well.

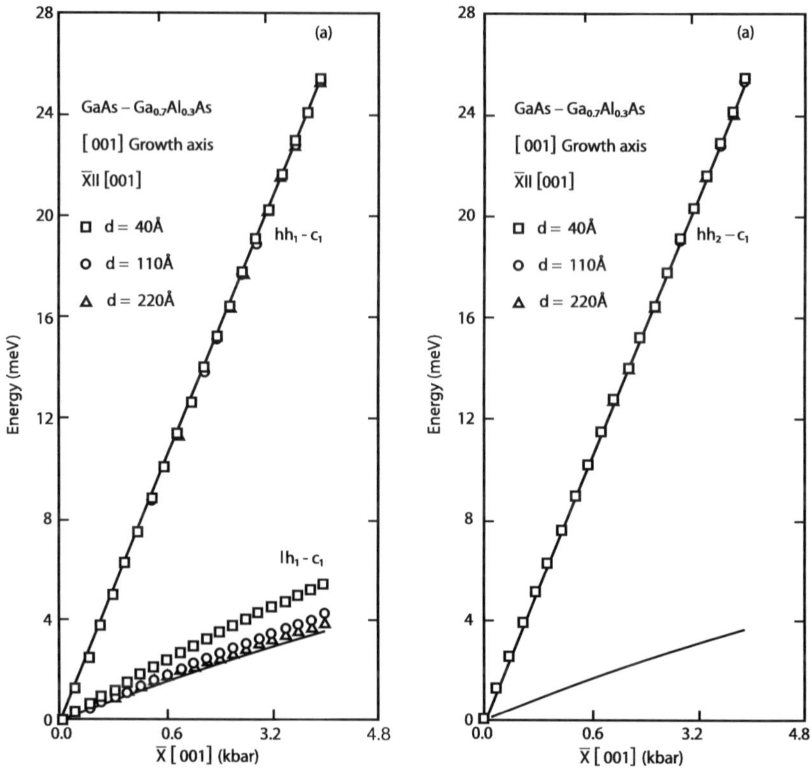

Fig. 13.2 Transition energies as a function of a compressive stress along [001] for a GaAs-$Al_{0.3}Ga_{0.7}As$ [001] quantum well. Reprinted with permission from [311]. ©1987 by the American Physical Society

13.3.2 Strained Heterostructures

Strained-layer epitaxy is based on the fact that a certain amount of elastic strain in heterostructures is possible without introducing dislocations or defects [312]. The energy so as to accomodate a stable strained epitaxial layer depends on both the epitaxial-layer thickness and the size of the lattice mismatch of the constituents. The strained heterostructure, if stable, does not transform to another structure configuration with dislocations etc. due to the energy barrier in making the transformation [313]. It is well known [314, 315] that growth of an overlayer bulk substrate may occur in one of three modes: Frank-van der Merwe growth (FM), Volmer-Weber growth (VW), or Stranski-Krastanov growth (SK). In the first case (FM), one monolayer is grown at a time in a two-dimensional manner where no atoms occupy layer i before layer $i-1$ is completed. On the other hand, VW growth occurs by the nucleation of solid clusters that condense as three-dimensional islands. Over time these islands form a continuous film. The latter category of growth (SK) is a mixture between the FM and VW growth mechanisms: First, a few monolayers are grown as following the two-dimensional layer-by-layer stacking principle after which nucleation of three-dimensional islands occurs. The SK growth method is the mechanism for growing quantum-dot wetting-layer structures of importance for present-day optoelectronic applications [316]. Coherent strain in multilayer structures implies that the in-plane lattice constants of the materials forming the multilayer are the same and an intermediate between the bulk lattice constants of the constituents. Matthews and Blakeslee [317] first obtained an expression for the in-plane lattice constant $a_\perp$ by minimizing the total strain energy of a single pair of layers. If the bulk lattice constants of material A and B are a_A and a_B, respectively, then

$$a_\perp = a_A \left[1 + \frac{f_0}{1 + \frac{G_A h_A}{G_B h_B}} \right], \tag{13.17}$$

where G_i, h_i are the shear moduli and the thickness of layer i, respectively, and $f_0 = (a_B - a_A)/a_A$ is the misfit. Note that this result implies that the average strain in the bilayer is

$$\langle \varepsilon_{||} \rangle = \frac{\sum \varepsilon_{||,i} h_i}{h_i}. \tag{13.18}$$

The average strain is zero under equilibrium conditions where the bilayer is free-standing or grown on a substrate with the bulk lattice constant a_s. The expression in Eq. (13.18) is carried over to the case with more than two layers in the multilayer.

An important type of strain for quantum wells is the biaxial strain resulting from the lattice mismatch between the structure and the substrate. The strain tensor is

$$\varepsilon_{||} = \varepsilon_{xx} = \varepsilon_{yy} = \frac{a_s - a_e}{a_e},$$

$$\varepsilon_{\perp} = \varepsilon_{zz} = -2\frac{C_{12}}{C_{11}}\varepsilon_{\|}, \tag{13.19}$$
$$\varepsilon_{i \neq j} = 0.$$

If the in-plane strain is compressive (i.e., $a_s < a_e$), the hydrostatic component increases the band gap (i.e., lowers the valence band) by

$$2a\left(1 - \frac{C_{12}}{C_{11}}\right)\varepsilon_{\|}, \tag{13.20}$$

and, to first order in ζ/Δ_0, the shear component splits the hh and lh bands by 2ζ, where

$$\zeta = -b\left(1 + 2\frac{C_{12}}{C_{11}}\right)\varepsilon_{\|}, \tag{13.21}$$

and $b < 0$ for compressive strain. A representation of the band-edge shifts is shown in Fig. 13.3. A study of the impact of the splitting and mixing of hh and lh states was initially carried out by O'Reilly and Winslow [318].

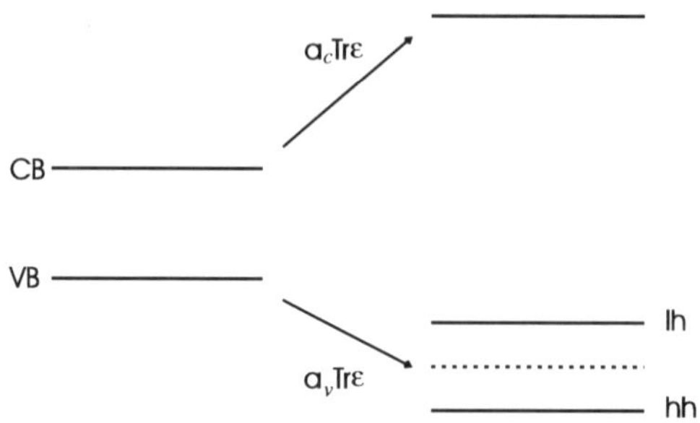

Fig. 13.3 Band edges under a compressive biaxial strain (CB: conduction band, VB: valence band)

A general strain calculation [57, 319–327] is based on solving the elastic equations (Navier's equations) using the definition of the strain tensor, constitutive relations describing the connection between stress and strain accounting for possible electromechanical effects such as piezoelectricity. The heterostructure geometry affects the boundary conditions to be imposed.

The rich variety of material combinations possible for strained epitaxial heterostructures allows for much freedom in controlling and adjusting electronic and optical properties. Effects induced by the strain include changes in energy gaps, splittings due to the lowering of symmetry, and variations in effective masses [1, 120, 125, 321]. Deformation potentials for the relevant conduction and valence bands together with the position-dependent strain tensor give all the information necessary to compute the effect of strain on heterostructure band structures and consequences for device applications. Heterostructure band-structure calculations accounting for strain in zincblende and wurtzite can be found in, e.g., [90, 311, 323, 325, 328–330]. In general, once the strain tensor is obtained, the calculation of the electronic properties follow the formalism provided for incorporating strain into the appropriate Hamiltonian.

13.4 Impurity States

The formulation of the problem is very similar to the bulk case except that the Coulomb potential has a relative permittivity that is position dependent and is also now a function of the location of the impurity site.

It is a well-established fact that the use of the envelope-function approximation to the problem of hydrogen-like bound states in semiconductor heterostructures gives good agreement with experimental results [7]. The specific requirements to the envelope functions at material interfaces are continuity of the N-component envelope function and the component associated with the normal-velocity operator acting on the N-component envelope function where N is the number of electron (or hole) states solved for [20].

13.4.1 Donor States

For a quantum well, one can write the effective-mass Hamiltonian as

$$H = \frac{p_x^2 + p_y^2}{2m_e} + \frac{p_z^2}{2m_e} + V(z) - \frac{e^2}{4\pi\varepsilon_r\varepsilon_0\sqrt{\rho^2 + (z - z_0)^2}}, \tag{13.22}$$

where $V(z)$ is the confinement potential, z_0 is the impurity location, and $\rho^2 = (x^2 + y^2)$. It has been common to solve Eq. (13.22) variationally.

Bastard [331] calculated the dependence of the binding energy of donor-bound electron on quantum-well thickness for an electron situated at the center of the quantum well or at the interface using an infinite-barrier model. Mailhiot et al. computed for the donor problem the dependence of the first few eigenstates with varying quantum-well thickness taking into account the influence of a position-dependent dielectric constant [332]. The donor problem in the presence of a magnetic field and strain was addressed by Greene and Bajaj [333].

The hydrogenic-donor binding energies and infrared transition energies were calculated in [334] for cylindrical quantum wires using a variational approach. Kim et al. [335] investigated electron transport in a quasi-one-dimensional constriction with an attractive, finite-size impurity in the ballistic limit theoretically using the envelope-function approximation. Novel coherent effects were discovered related to Fano resonances and the appearance of discrete levels in the continuum. Impurity and polaron effects in quantum dots are discussed in [336]. Amado et al. recently studied donor-bound electron states magnetic-field effects in two-dimensional quantum rings of finite width in the framework of the envelope-function approximation. It is demonstrated that electron-confinement properties change rapidly with the external magnetic field [337]. With the technological advances in the growth of quantum-dot structures including charged-particle control through the addition of charged donors and acceptors, the study of electronic eigenstates in quantum dots with impurities will continue to attract attention.

13.4.2 Acceptor States

The acceptor problem intrinsically involves a multiband formulation. Probably the first attempt to solve this problem was by Masselink et al. [338] for a carbon impurity at the center of a GaAs-AlGaAs quantum well. They used the variational method. Their binding-energy results were in excellent agreement with experimental data (Fig. 13.4). Masselink et al. also addressed the hole-acceptor problem and magnetic fields including band-mixing effects with special emphasis to the symmetry characteristics and splitting effects of the upper *hh*/*lh* states [339].

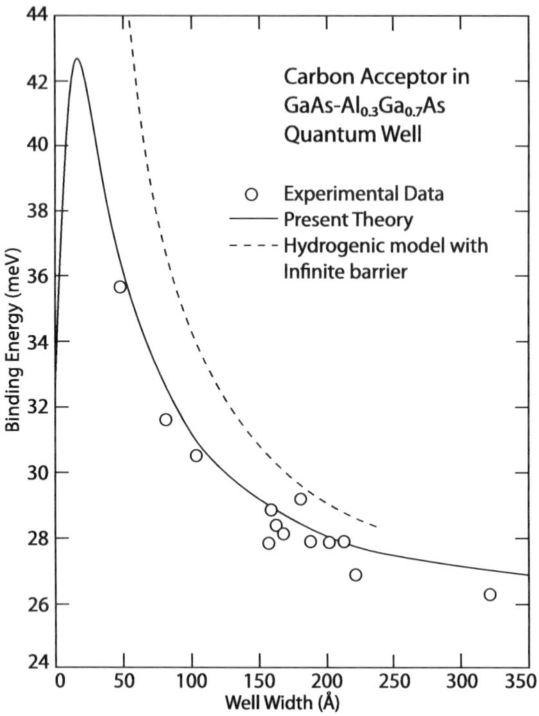

Fig. 13.4 Binding energies for an acceptor in a quantum well using a four-band model. Reprinted with permission from [338]. ©1983 by the American Physical Society

13.5 Excitons

$k \cdot p$ exciton studies of semiconductor heterostructures [7, 14, 340] are based on adding an electron-hole Coulomb interaction term characterized by an effective dielectric constant to the one-particle electron and hole Hamiltonians. The latter are for diamond and zincblende structures described by, e.g., a one-band model in the s conduction states and a six-band model in the p valence states including electron and hole confinement potentials, respectively. Typical solution methods are variational analysis or perturbation theory [57, 341]. We note that perturbation theory in the Coulomb interaction is possible and particularly convenient when addressing (smaller) heterostructures with strong particle confinement since the Coulomb energy varies slowly with distance (increases as $1/r$) as opposed to particle confinement energies (increases as $1/r^2$). A two-particle Schrödinger equation describing excitons can be derived from the many-electron state in the Hartree-Fock approximation. The polarization in many-particle physics obeys a Bethe-Salpeter equation, which in the so-called ladder approximation [342, 343], leads to the classic Elliott expression for absorption [344]. Alternatively, the use of semiconductor Bloch equations [345] in the density matrix allows for studies of excitons, biexcitons, etc. in the Hartree-Fock approximation or including many-particle interactions beyond the Hartree-Fock approximation.

Exciton studies are important for understanding, e.g., optical spectra of quantum-confined structures especially in the low-excitation regime [345] and leads to significant changes in absorption below and above band gap. Time-resolved photoluminescence of heterostructures allows for a detailed study of radiative exciton decay mechanisms [346, 347]. In [57], exciton effects including absorption spectra and radiative lifetimes were computed using a multiband single-particle hole Hamiltonian for zincblende and wurtzite pyramidal quantum-dot structures taking into account strain and piezoelectric effects. Sidor et al. [348] computed, based on a one-band model, magneto-exciton effects in V-shaped GaAs/AlGaAs and InAs/InP rectangular quantum wires (these shapes were addressed experimentally in [349–351]). Elliott's [344] expression for the absorption coefficient in bulk structures has been generalized to the case of an ideal two-dimensional semiconductors by Shinada and Sugano [352]. Polaron-exciton effects are discussed by Elliott and Haken [353, 354]. Stahl and Balslev [355] and Zimmermann [356] report on exciton effects in low-dimensional semiconductors subject to high-excitation conditions relevant for lasers. There have been numerous calculations for quantum wells [357–359], quantum wires [360–362], and quantum dots [169, 258, 363–374]. Excitons in other geometries have been studied using the effective-mass approximation, such as for uniaxial crystals [375] and on a spherical surface [376].

13.5.1 One-Band Model

Consider now a quantum-well heterostructure consisting of a material A sandwiched between two semi-infinite layers of a material B. For simplicity, we shall neglect differences in electron and hole effective masses in the two materials. The idealized exciton Hamiltonian problem reads

$$\left[E_0 + \frac{p_e^2}{2m_e} + \frac{p_h^2}{2m_h} - \frac{e^2}{4\pi\varepsilon|\mathbf{r}_e - \mathbf{r}_h|} + V_e\Theta\left(|z_e| - \frac{L}{2}\right) + V_h\Theta\left(|z_h| - \frac{L}{2}\right)\right]$$
$$\times\psi(\mathbf{r}_e, \mathbf{r}_h) = E\psi(\mathbf{r}_e, \mathbf{r}_h), \tag{13.23}$$

where E_0, V_e (V_h), and Θ are the energy gap in material A, the conduction (valence) band-edge potential difference between materials B and A, and Heaviside's function, respectively. Two-particle eigenstates in the absence of Coulomb interaction reads:

$$\psi_0(\mathbf{r}_e, \mathbf{r}_h) = \mathrm{e}^{\mathrm{i}(\mathbf{k}_{\|,e}\cdot\mathbf{r}_{\|,e} + \mathbf{k}_{\|,h}\cdot\mathbf{r}_{\|,h})}\chi_{n,e}(z_e)\chi_{m,h}(z_h), \tag{13.24}$$

where $\mathbf{r}_\| = (x, y)$, $\mathbf{k}_\| = (k_x, k_y)$, $\chi_{n,e}$, and $\chi_{m,h}$ are the z-part (quantum-well growth axis) electron and hole wave-function components corresponding to level n and m, respectively. The associated non-interacting two-particle energy is

$$E = E_0 + E_{e,n} + E_{h,m} + \frac{h^2 k_{\|,e}^2}{2m_e} + \frac{h^2 k_{\|,h}^2}{2m_h}. \tag{13.25}$$

where $E_{e,n}$ and $E_{h,m}$ are the electron and hole z-axis quantization energies corresponding to level n and m, respectively.

Introducing relative and center-of-mass coordinates for the in-plane position coordinates

$$\mathbf{r}_\| = \mathbf{r}_{\|,e} - \mathbf{r}_{\|,h}, \tag{13.26}$$

$$\mathbf{R}_\| = \frac{m_e\mathbf{r}_{\|,e} + m_h\mathbf{r}_{\|,h}}{m_e + m_h}, \tag{13.27}$$

allows Eq. (13.24) to be written as:

$$\left[E_0 + \frac{P_\|^2}{2(m_e + m_h)} + \frac{p_\|^2}{2\mu} + \frac{p_{z,e}^2}{2m_e} + \frac{p_{z,h}^2}{2m_h} - \frac{e^2}{4\pi\varepsilon\sqrt{r_\|^2 + (z_e - z_h)^2}}\right.$$
$$\left. + V_e\Theta\left(|z_e| - \frac{L}{2}\right) + V_h\Theta\left(|z_h| - \frac{L}{2}\right)\right]\psi(\mathbf{r}_e, \mathbf{r}_h) = E\psi(\mathbf{r}_e, \mathbf{r}_h), \tag{13.28}$$

for which solutions can be found of the form

$$\psi(\mathbf{r}_e, \mathbf{r}_h) = \mathrm{e}^{\mathrm{i}\mathbf{K}_\|\cdot\mathbf{R}_\|}\phi(z_e, z_h, \mathbf{r}_\|), \tag{13.29}$$

with $\mathbf{P}_\|$, $\mathbf{p}_\|$ being the in-plane center-of-mass and relative momenta, respectively, and $\hbar\mathbf{K}_\|$ is the eigenvalue of $\mathbf{P}_\|$.

Following [7, 377], ground-state solutions to this problem are sought variationally by using the following ansätze:

$$(I)\quad \phi(z_e, z_h, \mathbf{r}_\parallel) = \chi_{1,e}(z_e)\chi_{1,h}(z_h)\mathrm{e}^{-r_\parallel/\lambda}, \tag{13.30}$$

$$(II)\quad \phi(z_e, z_h, \mathbf{r}_\parallel) = \chi_{1,e}(z_e)\chi_{1,h}(z_h)\mathrm{e}^{-\sqrt{r_\parallel^2+(z_e-z_h)^2}/\lambda}, \tag{13.31}$$

where λ is the variational parameter. The subscripts 1 on the z-part envelope functions refer to the lowest subband indices $n = m - 1$.

In this simple idealized picture of exciton states, one should note that various forms can be formed such as *e-hh* and *e-lh* excitons. The various exciton forms are decoupled. Thus, it is meaningful to assign an effective Bohr radius and an effective Rydberg energy for each exciton form. For the *e-hh* and *e-lh* excitons, we have:

$$a_{hh}^{\mathrm{ex}} = \frac{4\pi\varepsilon\hbar^2}{\mu_{hh}e^2},\ R_{hh}^{\mathrm{ex}} = \frac{\mu_{hh}e^4}{2(4\pi\varepsilon)^2\hbar^2}, \tag{13.32}$$

$$a_{lh}^{\mathrm{ex}} = \frac{4\pi\varepsilon\hbar^2}{\mu_{lh}e^2},\ R_{lh}^{\mathrm{ex}} = \frac{\mu_{lh}e^4}{2(4\pi\varepsilon)^2\hbar^2}, \tag{13.33}$$

and

$$\mu_{hh} = \frac{m_e m_{hh}}{m_e + m_{hh}},\ \mu_{lh} = \frac{m_e m_{lh}}{m_e + m_{lh}}. \tag{13.34}$$

Some trivial conclusions are immediately apparent. Since the *lh* mass is smaller than the *hh* mass, the reduced *lh* mass is smaller than the reduced *hh* mass, and the *e-lh* Bohr radius is larger than the *e-hh* Bohr radius. Similarly, the *e-lh* Rydberg energy is smaller than the *e-hh* Rydberg energy. The intuitive explanation is that the smaller hole mass the weaker the exciton is bound. We point out that the *e-lh* reduced mass need not be smaller than the *e-hh* mass if multiband-coupling effects are taken into account as will be shown.

In Fig. 13.5, we plot the exciton binding energy as a function of the quantum-well width L, i.e., the thickness of the B layer in the case with infinite electron and hole barrier potentials. Evidently, the binding energy increases from the bulk value R_∞ corresponding to $L \to \infty$) to $4\,R_\infty$ as $L \to 0$, a result that is qualitatively similar to the hydrogenic impurity binding energy vs. electron-impurity distance. The similarity with the impurity problem is maintained also in the case with finite-barrier potentials: The binding energy increases as L decreases from infinity, reaches a maximum, and then drops to the Rydberg energy of the barrier material.

Excitons in quantum dots have been reviewed in a number of publications [188, 340, 378, 379].

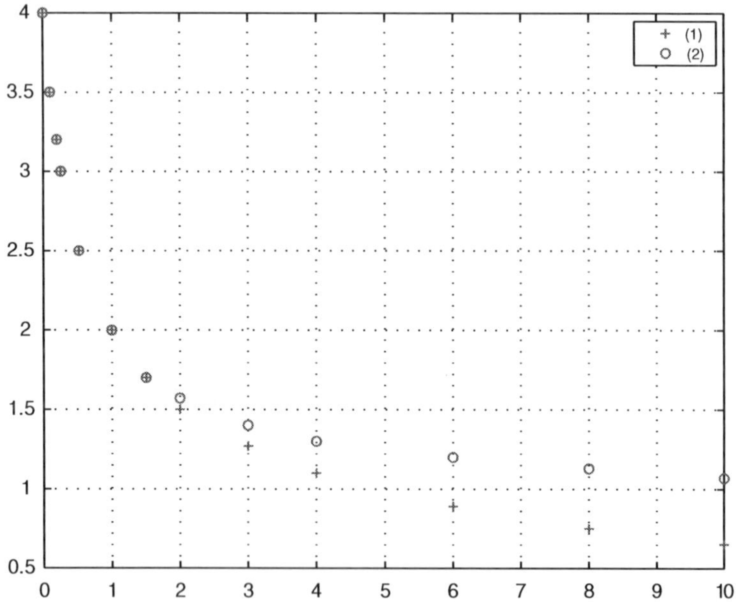

Fig. 13.5 Exciton binding energy (in units of R_∞) calculated as a function of the quantum-well width L (in units of the bulk Bohr radius a_∞). The curves labelled (1) and (2) correspond to the variational expressions: Eqs. (13.30) and (13.31), respectively. Adapted with permission from [357]. ©1982 by the American Physical Society

13.5.2 Type-II Excitons

A type-II quantum-well exciton is characterized by having the electron and hole forming the exciton spatially separated. For instance, in a GaSb-InAs-GaSb quantum well and in the absence of Coulomb interaction, holes are free to move in the semi-infinite GaSb layers where the valence-band maximum is located. They will, unless the hole kinetic energy is large, be confined to one of the two semi-infinite regions assuming tunneling through the InAs barrier is weak. The electron, however, is confined to the InAs quantum-well region where the conduction-band minimum is located. The material combination GaSb-InAs-GaSb is interesting because the valence-band maximum in GaSb is higher than the conduction-band minimum in InAs.

For type-II excitons, the idealized Hamiltonian reads

$$\left[E_A + \frac{p_e^2}{2m_e} + \frac{p_h^2}{2m_h} - \frac{e^2}{4\pi\varepsilon|\mathbf{r}_e - \mathbf{r}_h|} + V_e\Theta\left(|z_e| - \frac{L}{2}\right) - V_h\Theta\left(|z_h| - \frac{L}{2}\right)\right] \times\psi(\mathbf{r}_e, \mathbf{r}_h) = E\psi(\mathbf{r}_e, \mathbf{r}_h), \quad (13.35)$$

where the sign of the hole-confinement term: $V_h\Theta(|z_h| - L/2)$ is now negative.

Following [377], a variational solution, $\psi(\mathbf{r}_e, \mathbf{r}_h)$ [$= \exp(\mathrm{i}\mathbf{K}_\parallel \cdot \mathbf{R}_\parallel)\phi(z_e, z_h, \mathbf{r}_\parallel)$], for the ground state is sought as

$$\phi(z_e, z_h, \mathbf{r}_\parallel) = \chi_{1,e}(z_e)\left[\left(z_h - \frac{L}{2}\right)\mathrm{e}^{-\frac{b}{2}(z_h-\frac{L}{2})}\Theta\left(z_h - \frac{L}{2}\right)\right.$$
$$\left. - \left(z_h + \frac{L}{2}\right)\mathrm{e}^{-\frac{b}{2}(z_h+\frac{L}{2})}\Theta\left(-z_h - \frac{L}{2}\right)\right]\mathrm{e}^{-r_\parallel/\lambda}, \quad (13.36)$$

where b and λ are variational parameters and the A-B material interfaces are located at $z = -L/2$ and $z = L/2$. The electron z-part wave function is assumed to be the lowest quantum-well state corresponding to complete confinement:

$$\chi_{1,e}(z_e) = \cos\left(\frac{\pi z_e}{L}\right). \quad (13.37)$$

This choice reflects that the exciton ground state is expected to be an even function in the z_e, z_h coordinates while the hole probability is highest near the A–B interfaces due to the Coulomb attraction with the quantum-well confined electron.

In Fig. 13.6, the exciton binding energy (in units of the bulk InAs Rydberg exciton binding energy R_∞) is plotted as a function of the quantum-well width L for the GaSb-InAs-GaSb type-II heterostructure in units of the bulk InAs exciton Bohr radius a_∞. The dashed part of the line indicates that the electron energy E_1 lies below the valence-band maximum. Note by comparison with Fig. 13.5 that the binding energy (for a given quantum-well width L) in the type-II case is considerably smaller than for a type-I system due to the spatial separation (and hence reduced Coulomb interaction) of the electron and hole in type-II systems. In actual fact, exciton binding energies are smaller in the type-II heterostructure case as compared to the bulk.

13.5.3 Multiband Theory of Excitons

In [380], exciton states in CdS quantum dots were calculated by diagonalizing the full Hamiltonian (in a finite set of basis states found in the case without Coulomb interaction) consisting of a one-band electron Hamiltonian in the two S spin states, a 6×6 Hamiltonian in the hole P states, and the Coulomb interaction between the electron and hole. The model accounts for position-dependent permittivity and effective-mass parameters and particle self-energy effects. Exciton ground-state results for spherical and tetrahedal quantum-dot structures are qualitatively alike with binding energies in the sphere case increasing from approximately 100 meV at a radius of 4.5 nm to approximately 350 meV at a radius of 3.0 nm. A nice review of multiband calculations of excitonic effects in quantum dots up to 2000 was provided by Efros and Rosen [188].

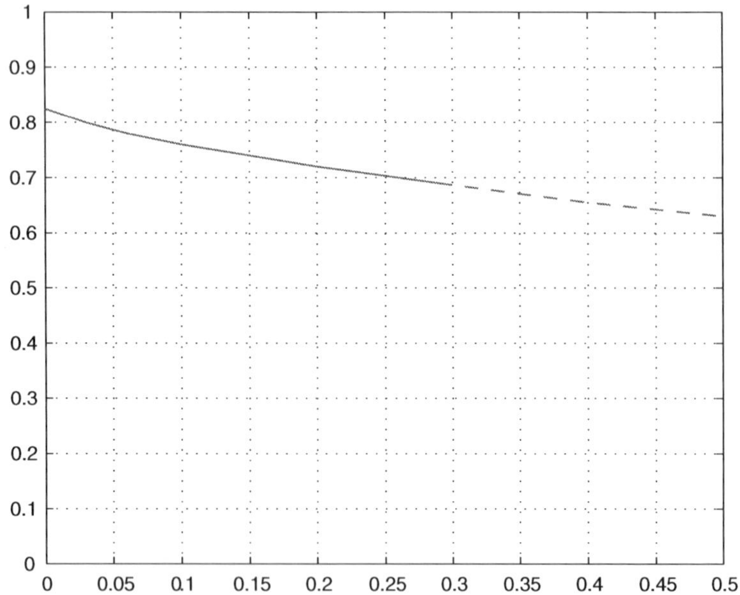

Fig. 13.6 Exciton binding energy (in units of R_∞) as a function of quantum-well width L (in units of a_∞) for a GaSb/InAs/GaSb double heterostructure. The dashed line indicates that the ground-state electron subband energy E_1 lies below the top of the GaSb valence band. Adapted with permission from [357]. ©1982 by the American Physical Society

13.6 Magnetic Problem

We recall that the simple, parabolic-band model for bulk can be solved exactly in the presence of a static magnetic field but not the multiband problem. Of course, for a nanostructure, exact solutions are, in general, not available even in the one-band case.

The study of magnetic field within a one-band approach has been carried out for graded semiconductors [172], quantum wells [91, 333, 381–383], quantum wires [384–387], quantum rings [388, 389], and quantum dots [372]. The study of magnetic field within a multiband approach has been carried out for quantum wells [195, 390–398], quantum wires, quantum rings [255, 257], and quantum dots [256, 399–402]. Altarelli and Platero [394] did the first multiband calculation with parallel magnetic field; this was done using the axial approximation and with $p_z = 0$. A Burt–Foreman formulation has been derived [403].

A few examples of the above works will now be discussed in order to illustrate the types of equations and solutions to the magnetic problem. A useful but dated review was provided by Maan [404].

13.6.1 One-Band Model

Consider first a single, isotropic and parabolic band in the bulk, subject to a confinement potential $V(z)$ and in a static and uniform magnetic field $\mathbf{B} = (0, B_{\parallel}, B_{\perp}) = B(0, \sin\theta, \cos\theta)$ where θ is measured from the z axis and $B_{\parallel}$ ($B_{\perp}$) means the component parallel (perpendicular) to the QW plane. Note that there is an analytic solution for a parabolic quantum well with an arbitrary orientation of the magnetic field [381].

13.6.1.1 Equation

The equation for an electron in a magnetic field is obtained by first writing down the effective Hamiltonian in the absence of a field (Chap. 12) then adding the magnetic field via the minimal-coupling procedure (Chap. 9), and, if necessary, adding a Zeeman term. The Hamiltonian for a quantum well can then be written as (assuming m_e to be independent of z):

$$H = H_{\parallel} + H_{\perp} + W, \tag{13.38}$$

where

$$H_{\parallel}(z) = \frac{p_z^2}{2m_e} + V(z) + \frac{e^2 B_{\parallel}^2 z^2}{2m_e} \pm \frac{1}{2} g^* \mu_B B, \tag{13.39}$$

$$H_{\perp}(x) = \frac{p_x^2}{2m_e} + \frac{1}{2m_e}\left(\hbar k_y + eB_{\perp}x\right)^2, \tag{13.40}$$

$$W(x, z) = p_x z \frac{eB_{\parallel}}{m_e}, \tag{13.41}$$

which is a separable problem in the absence of the interaction term W. Indeed, analytic solutions have been proposed for the two separate cases of parallel and perpendicular (also known as in-plane) magnetic field. For completeness, we note that if the spatial dependence of the mass is taken into consideration (or if one goes beyond the parabolic-band approximation), then the problem is no longer separable even in the absence of a magnetic field [405].

13.6.1.2 Perpendicular Magnetic Field

Analytic solutions have been proposed in a number of articles for when the magnetic field is in the growth direction [406]. Here, $W = 0$, $H_{\parallel}$ is the QW problem, and $H_{\perp}$ is the simple harmonic oscillator problem. Let

$$H_{\parallel}\chi_m(z) = \xi_m \chi_m(z), \tag{13.42}$$

$$H_{\perp}\phi_n(x) = \epsilon_n \phi_n(x). \tag{13.43}$$

The eigenvalues of the ϕ_n equation (being independent of the value of k_y) are [refer to Eq. (9.27)]:

$$\varepsilon_n = \left(n + \frac{1}{2}\right) \hbar\omega_{c\perp}, \tag{13.44}$$

where $\omega_{c\perp} = eB_\perp/m_e$.

Each zero-field subband splits into a Landau fan chart (Fig. 13.7). Labeling the Landau levels by $|m, n\rangle$ (leaving out the degeneracy with respect to k_y and the Zeeman splitting for simplicity), one notes that there can be a crossing of levels from different subbands if

$$\xi_m' - \xi_m = (n - n')\hbar\omega_{c\parallel}.$$

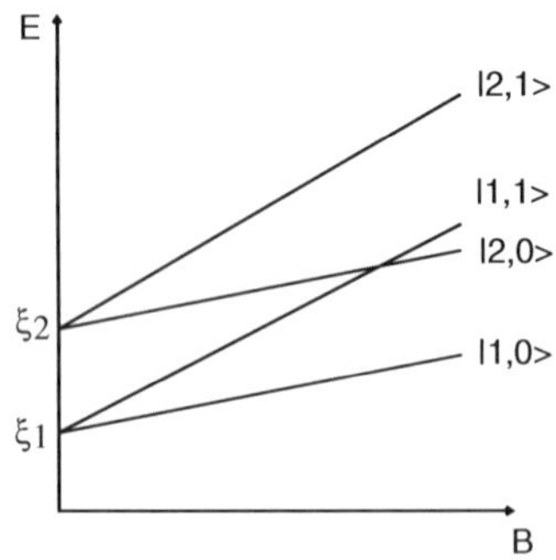

Fig. 13.7 Landau levels for a quantum well in a perpendicular magnetic field within the one-band model

13.6.1.3 In-Plane Magnetic Field

Here the magnetic field is in the QW plane and $B_\perp = 0$. We can rewrite Eq. (13.38) as (leaving out the Zeeman term for simplicity)

$$H = \frac{1}{2m_e}(p_y^2 + p_z^2) + \frac{1}{2m_e}(p_x + eB_\parallel z)^2 + V(z). \tag{13.45}$$

Let the solution

$$f(\mathbf{r}) = \chi(z)\mathrm{e}^{\mathrm{i}(k_x x + k_y y)}, \tag{13.46}$$

giving the following differential equation:

$$\frac{\mathrm{d}^2\chi}{\mathrm{d}z^2} + \frac{2m_e}{\hbar^2}\left[E - \frac{\hbar^2 k_y^2}{2m_e} - \frac{m_e\omega_c^2}{2}(z + z_0)^2 - V(z)\right]\chi = 0, \tag{13.47}$$

with $z_0 = a_B^2 k_y$ (orbit center) and $a_B^2 = \hbar/(eB_\parallel)$ (a_B is known as the magnetic length). The solution has been shown to reduce to the Weber functions [407].

Two interesting effects for a superlattice structure are when the cyclotron radius,

$$R_c = \frac{1}{\sqrt{s}},\ s = \frac{eB}{\hbar},$$

is comparable to the layer thicknesses and when the cyclotron energy is comparable to the miniband width. For example, a magnetic field of 10 T gives a cyclotron radius of approximately 10 nm. It has been observed experimentally, for a 1,12–3.92 nm $Al_{0.4}Ga_{0.6}As$–GaAs superlattice in a magnetic field of 15–20 T, that there are peaks in the luminescence at low energies due to interband Landau-level transitions but that they disappear at higher energies [382]. This observation has been correlated to the fact that cyclotron motion is allowed when the energies fall within the zero-field minibands (giving rise to flat Landau levels, i.e., the levels are not dependent on the orbit center) but that when the energies fall within the zero-field minigaps, the carriers are reflected by the barriers and the Landau levels become dispersive and making these transitions less observable.

13.6.1.4 Arbitrary Magnetic Field

This has also been refered to as the tilted magnetic field case [408]. Note that the problem is separable if the potential is parabolic [409]. Otherwise, a number of numerical methods have been proposed such as variational [410], finite and reduced bases [408], and perturbation [411].

We will present in detail the perturbative approach with W as the perturbation. In actual fact, if $W = 0$ eigenfunctions f can be written as

$$f_{mn}(x,z) = \chi_m(z)\phi_n(x)\,, \tag{13.48}$$

with χ and ϕ as defined in Eqs. (13.42) and (13.43).

Since

$$\langle \phi_n | p_x | \phi_n \rangle = 0\,, \tag{13.49}$$

W must be treated using second-order perturbation theory. According to second-order perturbation theory, the energy shifts δE_W stemming from W, can be obtained as

$$\delta E_W = \sum_{(m',n')\neq(m,n)} \frac{\left(\frac{eB_\parallel}{m_e}\right)^2 |\langle \chi_m | z | \chi_{m'} \rangle|^2 |\langle \phi_n | p_x | \phi_{n'} \rangle|^2}{\xi_m - \xi_{m'} + (n-n')\hbar\omega_{c\perp}}\,, \tag{13.50}$$

assuming the energy denominator does not become too small corresponding to near-degeneracy of unperturbed states. In the case of near-degeneracy, W must be treated using degenerate perturbation theory. For a square quantum well, parity dictates that there is no coupling (i.e., $\delta E_W = 0$) unless m and m' refer to states of opposite parity. In addition, the matrix element $\langle \phi_n | p_x | \phi_{n'} \rangle$ requires n and n' to differ by one. When these two conditions are satisfied, such crossings as in Fig. 13.7 become anti-crossings.

Alternatively, the term $e^2 B_{\parallel}^2 z^2/(2m_e)$ in $H_{\parallel}$ can be solved for using first-order perturbation theory. The energy shifts $\delta E_{\parallel}$ become

$$\delta E_{\parallel} = \frac{e^2 B_{\parallel}^2}{2m_e} \langle \chi'_m | z^2 | \chi'_m \rangle, \tag{13.51}$$

where now χ'_m are eigenstates of the Hamiltonian

$$H'_{\parallel}(z) = \frac{p_z^2}{2m_e} + V(z) \pm \frac{1}{2} g^* \mu_B B. \tag{13.52}$$

It is sometimes convenient to treat $e^2 B_{\parallel}^2 z^2/(2m_e)$ as a first-order perturbation since the states χ'_m can be easily found and henceforth used as an expansion set for the full Hamiltonian problem [Eq. (13.38)]. For a 100 Å GaAs quantum well in a magnetic field of 10 T, Bastard [7] has estimated $\delta E_{\parallel} < 1$ meV.

13.6.2 Multiband Model

There are two basic approaches to the multiband problem.

13.6.2.1 Luttinger–Kohn Approach

The procedure for incorporating the magnetic field is similar to the one-band case [89, 390, 393], whereby the c-number $\mathbf{k}$ in a bulk multiband Hamiltonian becomes an operator, the magnetic field is introduced via the minimal coupling and a confinement potential $V(\mathbf{r})$ is added to the Hamiltonian [95]. For concreteness, we will focus on a quantum well grown in the z direction and with the magnetic field in the same direction. Then, the vector potential is in the x–y plane and one can introduce harmonic-oscillator operators as was first done by LK:

$$a^\dagger = \frac{1}{\sqrt{2s}} k_+, \, a = \frac{1}{\sqrt{2s}} k_-, \tag{13.53}$$

where $k_\pm = k_x \pm i k_y$. As an example, the Hamiltonian used by Yang, Broido and Sham [393] is reproduced here:

$$H_{YBS} = \begin{bmatrix} -P - Q - \frac{3}{2}\kappa B & -R & S & 0 \\ -R^{\dagger} & -P + Q + \frac{1}{2}\kappa B & 0 & S \\ S^{\dagger} & 0 & -P + Q - \frac{1}{2}\kappa B & -R \\ 0 & S^{\dagger} & -R^{\dagger} & -P - Q + \frac{3}{2}\kappa B \end{bmatrix}, \tag{13.54}$$

with

$$\begin{aligned} P &= \frac{\gamma_1}{2}(k_z^2 + 2\omega_c a^{\dagger} a + \omega_c), \\ Q &= \frac{\gamma_2}{2}(-2k_z^2 + 2\omega_c a^{\dagger} a + \omega_c), \\ R &= -\sqrt{3}\overline{\gamma}\omega_c a^2 - \sqrt{3}\mu\omega_c a^{\dagger 2}, \\ S &= \sqrt{6}\gamma_3 k_z \omega_c a, \\ \overline{\gamma} &= \frac{1}{2}(\gamma_1 + \gamma_2), \; \mu = \frac{1}{2}(\gamma_3 - \gamma_2). \end{aligned} \tag{13.55}$$

This is obtained by starting with the LK Hamiltonian and keeping k_z as an (differential) operator. As we have seen in the discussion without a magnetic field, this "canonical" quantization procedure has an operator ordering problem when the $k \cdot p$ parameters are position dependent. There are two possible resolutions. One is assume the $k \cdot p$ parameters constant (as in the above equations). The other is to impose the symmetrization principle as in the zero-field case. Yang, Broido and Sham [393] provides a detailed discussion of the solution to H_{YBS}. The problem with a magnetic field perpendicular to the growth direction can be similarly treated [394, 412].

13.6.2.2 Burt–Foreman Model

A Burt–Foreman formulation of the multiband problem has been derived [403]. It was found that there can be significant disagreement with the conventional LK formulation at large magnetic field and that the BF model agrees more with a single-band calculation (Fig. 13.8).

Note, however, that the recent first-principles theory of Foreman [413] predict additional interface terms not present in the theory of Mlinar et al. This is to be expected since, as already discussed in the previous chapter, the BF Hamiltonian neglects a number of interfacial terms in the exact envelope-function theory.

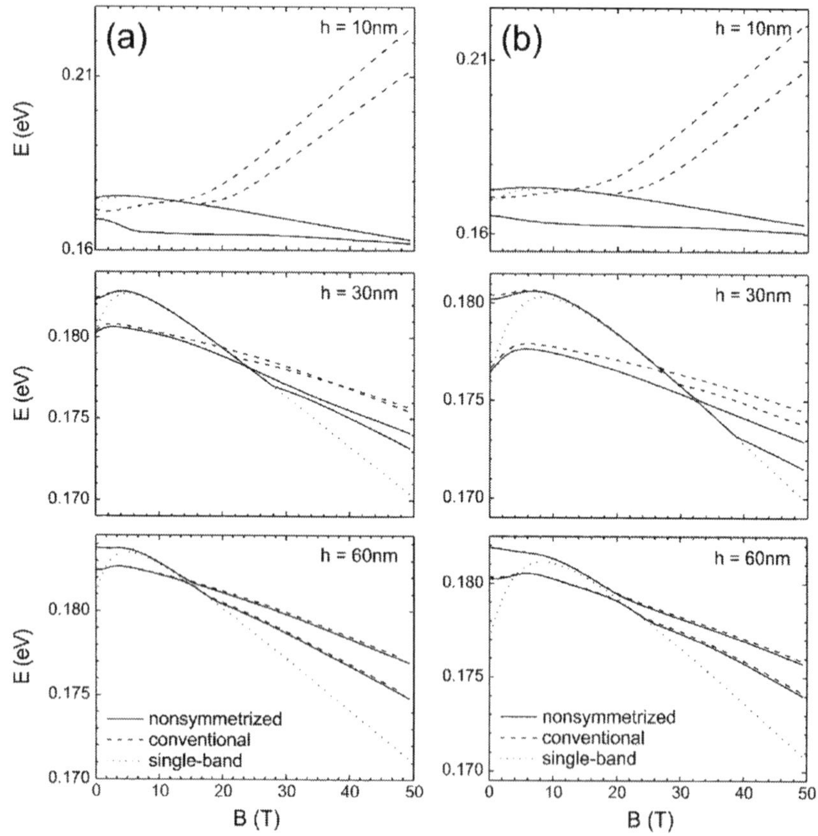

Fig. 13.8 Comparison of LK and BF magnetic Hamiltonians for quantum dots. Reprinted with permission from [403]. ©2005 by the American Physical Society

13.7 Static Electric Field

In a quantum well, there are two kinds of DC Stark effects, either with the electric field along the growth direction (longitudinal Stark effect) or perpendicular to it (transverse Stark effect); these configurations are depicted in Fig. 13.9.

13.7.1 Transverse Stark Effect

This problem is mathematically almost identical to the bulk problem (Sect. 10.2). Within a one-band model, the problem is separable. Consider a quantum well grown along the z direction and an external DC electric field $\mathcal{E}$ applied along the x direction. The effective-mass Hamiltonian of an electron reads

$$H = \frac{p_x^2}{2m^*} + e\mathcal{E}x + \frac{p_y^2}{2m^*} + \frac{p_z^2}{2m^*} + V(z), \tag{13.56}$$

where parameters take their usual meaning. Evidently, in the classical case, the energy E_x associated with motion along the x direction requires $x < E_x/e\mathcal{E}$. The wave function can be sought as

$$\psi(\mathbf{r}) = \phi(x)\chi_n(z)\mathrm{e}^{\mathrm{i}k_y y}, \tag{13.57}$$

where χ_n is the n'th quantum-well bound state with energy E_n. Writing for the total energy

$$E = E_x + \frac{\hbar^2 k_y^2}{2m^*} + E_n, \tag{13.58}$$

and ϕ is the solution to the differential equation

$$\left(\frac{p_x^2}{2m^*} + e\mathcal{E}x\right)\phi(x) = E_x\phi(x), \tag{13.59}$$

with the boundary condition

$$\lim_{x\to\pm\infty} |\phi(x)| < \infty. \tag{13.60}$$

This differential equation is solved by transforming x into η defined as

$$x = \frac{E_x}{e\mathcal{E}} - \left(\frac{\hbar^2}{2m^* e\mathcal{E}}\right)^{1/3} \eta, \tag{13.61}$$

and Eq. (13.59) becomes the Airy equation:

$$\frac{d^2\phi}{d\eta^2} + \eta\phi = 0, \tag{13.62}$$

where

$$\lim_{\eta\to\pm\infty} |\phi(\eta)| < \infty. \tag{13.63}$$

The solutions are given in standard textbooks:

$$\phi(\eta) = \sqrt{\frac{|\eta|}{3\pi}} K_{1/3}\left[\frac{2}{3}|\eta|^{3/2}\right], \quad \eta \le 0,$$

$$\phi(\eta) = \frac{1}{3}\sqrt{\pi\eta}\left[J_{-1/3}\left(\frac{2}{3}\eta^{3/2}\right) + J_{1/3}\left(\frac{2}{3}\eta^{3/2}\right)\right], \quad \eta \ge 0, \tag{13.64}$$

where J_ν and K_ν are the Bessel J and K functions of order ν, respectively. The asymptotic properties of this solution are

$$\phi(\eta) \approx |\eta|^{-1/4} \exp\left[-\frac{2}{3}|\eta|^{3/2}\right], \quad \eta \to -\infty,$$

$$\phi(\eta) \approx |\eta|^{-1/4} \sin\left[\frac{2}{3}\eta^{3/2} - \frac{\pi}{4}\right], \quad \eta \to \infty. \tag{13.65}$$

Since the potential term $e\mathcal{E}x$ is arbitrarily negative as $x \to -\infty$, the spectrum is continuous. In particular, bounded solutions with any $E_x < 0$ are allowed whenever $\mathcal{E} \neq 0$. Similarly, for holes, a continuous spectrum exists above the valence-band edge whenever $\mathcal{E} \neq 0$. In turn, this implies that band-to-band optical absorption at a photon energy $\hbar\omega$ can occur below the $\mathcal{E} = 0$ limit: $E_0 + E_1 + H_1$, where E_0, E_1, H_1 are the bulk energy gap, the first electron and first hole quantization energy measured above the conduction and below the valence-band edge, respectively. The absorption coefficient oscillates as a function of ω if $\mathcal{E} \neq 0$ an effect known as the Franz-Keldysh effect. Further, the conclusions obtained here for the quantum-well case can be carried over to the bulk case by replacing E_n and H_m with $\hbar^2 k_z^2/(2m_e^*)$ and $\hbar^2 k_z^2/(2m_h^*)$, respectively.

13.7.2 Longitudinal Stark Effect

Consider next the case where the electric field is parallel to the quantum-well growth direction. A longitudinal field (Fig. 13.9) leads to a spatial separation of photoexcited electrons and holes and a consequent reduction in both the transition energy and absorption coefficient. The Hamiltonian reads

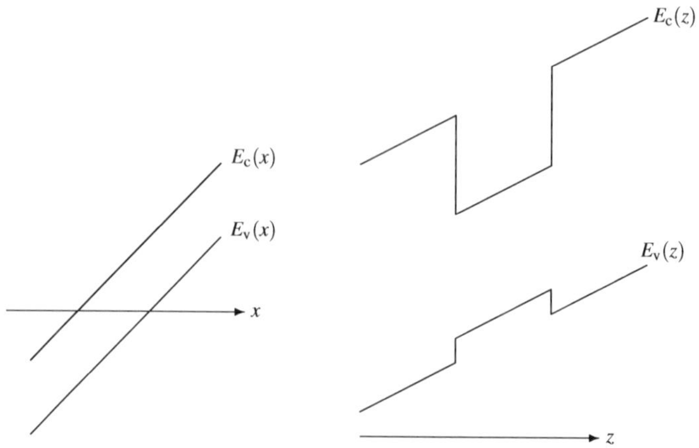

Fig. 13.9 Static electric fields in a quantum well: **(a)** transverse, **(b)** longitudinal

$$H = \frac{p_x^2 + p_y^2}{2m^*} + \frac{p_z^2}{2m^*} + e\mathcal{E}z + V(z). \tag{13.66}$$

The in-plane motion is now free and the general solution can be written as

$$\psi(\mathbf{r}) = e^{i(k_x x + k_y y)} \chi_n(z). \tag{13.67}$$

If the electric field is small enough, it is possible to treat the Stark term perturbatively. In first-order perturbation theory for symmetric confinement potentials $V(z)$, the Stark effect vanishes since the Stark term has odd parity and eigenstates are all parity eigenstates. For an effect we must resort to second-order perturbation theory, i.e., the energy shift becomes

$$\Delta E_n = e^2\mathcal{E}^2 \sum_{m \neq n} \frac{|z_{nm}|^2}{E_n^0 - E_m^0}, \tag{13.68}$$

where

$$z_{nm} = \int_{-\infty}^{\infty} \mathrm{d}z\, \chi_m^0(z) z \chi_n^0(z), \tag{13.69}$$

and χ_i^0, E_i^0 are the unperturbed i'th eigenstate and energy eigenvalue, respectively. The change in groundstate energy satisfies

$$\Delta E_1 \sim m^* L^4 \mathcal{E}^2, \tag{13.70}$$

being quadratic in the electric field and growing to the fourth power in the quantum-well width. Thus, as the electric field (or the quantum-well width) increases, Eq. (13.70) applies only as long as $\Delta E_1 << E_2^0 - E_1^0$. In the general case, where perturbation theory does not necessarily apply, one can assume a variational solution for the groundstate, e.g., in the form

$$\chi_1(z) = \chi_1^0(z) e^{-\beta z}. \tag{13.71}$$

For electrons and holes variational solutions are sought as $\beta = \beta_e$ and $\beta = -\beta_h$ ($\beta_e, \beta_h > 0$), respectively, due to the opposite electric-field force directions on the electron and the hole. If one introduces the carrier characteristic penetration length κ^{-1} of the ground state into the barrier region and the quantum-well barrier height V_B for the electron, different qualitative regimes exist for the electron ground state depending on

$$(a)\ \mathcal{E} = 0,$$

$$(b)\ e^2\mathcal{E}^2 \sum_{m \neq n} \frac{|z_{nm}|^2}{E_n^0 - E_m^0} << E_2^0 - E_1^0,$$

$$(c)\ e\mathcal{E}\kappa^{-1} << V_B - E_1^0,$$

$$(d)\ e\mathcal{E}\kappa^{-1} \sim> V_B - E_1^0. \tag{13.72}$$

In case (a), for a symmetric quantum-well confinement potential, the ground state is even and symmetrically positioned with respect to the center of the quantum well. As the electric field increases, the electron ground state is linearly shifted with $\mathcal{E}$ towards one end of the quantum well [case (b)]. As the electric field increases further [case (c)], electrons accumulate near one end of the quantum well still confined within the quantum well, i.e., the ground state decays exponentially outside the quantum-well region. In case (d), the electric field is sufficiently strong that electrons escape the quantum well region and solutions oscillate in the barrier regions. The same qualitative behavior is found for hole solutions except that they are shifted by the electric field towards the other end of the quantum well. Finally, it should be mentioned that the longitudinal Stark effect problem can be solved exactly by writing solutions in the various regions: to the left of the quantum well, in the quantum well, and to the right of the quantum well as a linear combination of Airy functions and matching boundary conditions at the quantum-well interfaces.

13.7.3 Multiband Theory

There have been a few treatments of electric fields within a multiband $k \cdot p$ theory in nanostructures [414].

A multiband analysis of Stark effects in heterostructures has been carried out by Dupertuis et al. [415] revealing that two-dimensional confinement in GaAs/AlGaAs V-groove structures can lead to wave function splitting and cascading. In [415], a one-band (four-band) model is used for the conduction electrons (*hh* and *lh* states) and consequences of the changes in wave function character on Stark shifts, oscillation strengths, and electroabsorption spectra examined. In particular, it is demonstrated that electroabsorption spectra show duplicate peaks as a consequence of the wave function changes.

The quantum-confined Stark effects has also been explored in Ge quantum wells grown on virtual SiGe substrates using an eight-band $k \cdot p$ analysis [416]. As is well known, due to the indirect band gap of Si, electro-optical effects are of much less significance in Si than in the AlGaAs and InGaAsP material systems. It is shown [416] that eight-band modeling of compressively-strained Ge quantum-well structures subject to varying degrees of externally-applied electric fields leads to excellent agreement with experiments and indicates direct band-gap absorption at low electric fields.

Spin splittings due to interface inversion asymmetry have been demonstrated in envelope-function approximation calculations in the late 1990s [417, 418]. These results were further supported by group-theoretical analyses [245] and measurements [419]. Structural inversion asymmetry in two-dimensional structures has been intensively studied for electrons and holes, particularly, the determination of the effective electron Rashba coefficient [420–422],

$$\langle \beta \mathcal{E}_z \rangle, \tag{13.73}$$

where $\langle \cdot \rangle$ denotes averaging over the well and barrier material regions. The Rashba coefficient appears in the electron dispersion,

$$E_{\pm} = \alpha k_{\perp}^2 \pm \langle \beta \mathcal{E}_z \rangle k_{\parallel}, \tag{13.74}$$

stemming from the Rashba Hamiltonian,

$$H = \beta \mathcal{E}_z \mathrm{i} \left(k_- \sigma_+ - k_+ \sigma_- \right), \tag{13.75}$$

where $2\alpha/\hbar^2$ and ε_z denote the reciprocal effective electrin mass and the electric field, respectively. Due to the fourfold degeneracy of the Γ_8^v valence bands, structural inversion asymmetry hole spin splittings show different and more complicated characteristics as compared to the inversion asymmetry electron spin splitting [89, 423, 424].

Recently, the effect of an electric field on spin splitting in Si/SiGe quantum wells has been examined [425] using a combination of the $sp^3d^5s^*$ tight-binding method and a generalized envelope-function approximation extended to asymmetrical quantum wells. It is demonstrated [425] that the electric field modifies the interface-induced spin splittings and gives rise to a Rashba-type contribution to the electron Hamiltonian. The valley splittings and spin splittings are shown to oscillate with the quantum-well width due to intervalley reflection of the electron wave at the interfaces. It is found that spin-splitting oscillations are suppressed in rather low electric fields.

Chapter 14
Conclusion

In this book, we have tried to show (at times in great detail) how to apply the $k \cdot p$ theory to semiconductors with cubic and hexagonal structures. We have then shown how the theory gets modified by the inclusion of a variety of external fields and deformations.

The attention devoted to the old but classic works of Dresselhaus, Kip, Kittel, Luttinger, Kohn, Kane, Cardona, Pikus, Bir and many others was intentional because, all too often now, these pioneering work get relegated to a footnote in modern texts or even altogether forgotten. Surprisingly, a lot of this early work is still valid today and those Hamiltonians are still being used in the research literature. For example, as discussed in Chap. 10, recent work on the treatment of electric field in periodic solids has pointed to the amazingly versatile nature of the Luttinger–Kohn basis. Still, the goal in writing this book was also to report on some more recent work that has received little attention beyond the original literature. One such example is the Burt–Foreman theory, where the only other extensive discussion appears to be a section in the book by Ridley [17].

We also hope this book is only the starting point of a journey into the intricacies and beauty of the $k \cdot p$ method. We are convinced the reader will find places in the book where they believe the theory is inadequate and attempt to correct that. One such example that quickly comes to mind is in the heterostructure theory.

There are many topics, and papers, that simply had to be left out to make this project manageable. We would like to mention a few of those. First is the fact that we have presented one form of the $k \cdot p$ theory. Thus, we did not address the issue of how to formulate the theory for a many-particle problem in general (a brief discussion was given within the context of the exciton problem); this was addressed by Kohn [426]. Also, we did not consider the theory for other types of one-electron Hamiltonians, such as the Dirac equation [427]. Next, a major restriction was to the three main types of crystal structures. Certainly, applications to other crystal structures have also been carried out [428–433]. Examples here include general studies [430–432], the surface problem (particularly via a complex-band-structure formulation [434]), the important problem of cubic rocksalt structures [428, 429, 435, 436] and the more recent graphene and nanotube structures [437, 438]. In addition, there is also work for a $k \cdot p$ expansion starting from

L.C. Lew Yan Voon, M. Willatzen, *The* k · p *Method*,
DOI 10.1007/978-3-540-92872-0_14, © Springer-Verlag Berlin Heidelberg 2009

a point other than the zone center of the Brillouin zone [439, 440]. Finally, another major omission was of course in terms of applications. We have already listed a number of books in the Introduction that covers various applications from optical to transport. In most cases, the band-structure theory is only a passive component of the calculation (such as providing the energy levels and momentum matrix elements for an optical-coefficient calculation). Other times, the electronic structure is modified by the problem at hand, such as in a self-consistent charge calculation [173] or in the presence of electromechanical phenomena [49, 57, 323, 441, 442].

Appendix A
Quantum Mechanics and Group Theory

A.1 Löwdin Perturbation Theory

P.O. Löwdin introduced a perturbation scheme for treating the mutual influence of two classes of unperturbed states [443]. The usual nondegenerate theory applies when one of the classes has only one state.

A.1.1 Variational Principle

Given a set of unperturbed eigensolutions, one can write the exact solution as

$$\psi \approx \sum_{n=1}^{N} \psi_n^{(0)} C_n. \tag{A.1}$$

This is an equality only if $\{\psi_n^{(0)}\}$ is a complete set, in which case the C_n's are fixed by the choice of the basis set. If the set is incomplete, then the relation is only approximate, and the C_n's can be chosen in order to minimize the error. This is the variational approach.

Defining the Hamiltonian matrix element

$$H_{mn} \equiv \int \mathbf{dr}\, \psi_m^{(0)*} H \psi_n^{(0)}, \tag{A.2}$$

then the energy is given by the expectation value

$$E = \frac{\int \mathbf{dr}\, \psi^* H \psi}{\int \mathbf{dr}\, \psi^* \psi} = \frac{\sum_{mn} C_m^* H_{mn} C_n}{\sum_n C_n^* C_n}. \tag{A.3}$$

The variational principle says that the energy is an extremum when the coefficients C_m are modified, i.e., $\delta E / \delta C_m = 0$, giving

$$\sum_{n=1}^{N}\Big[H_{mn} - E\delta_{mn}\Big]C_n = 0,\ m = 1, N. \tag{A.4}$$

This is a set of N linear equations.

A.1.2 Perturbation Formula

One starts by dividing the basis $\psi_n^{(0)}$ into two classes:

$$\psi = \sum_{m\in A}\psi_m^{(0)} + \sum_{n\in B}\psi_n^{(0)}. \tag{A.5}$$

Set A is chosen to be the normal set and one treats the influence of states in set B as a perturbation on the states in A. For example, the usual approach leads to a secular determinant with dimension dim(A)+dim(B). But, Löwdin perturbation has a determinant of dimension dim(A). This is appropriate when we only need the lowest set of dim(A) eigenvalues. In order to get the same set of lower eigenvalues, however, one needs a new set of states $\widetilde{A}$ perturbed from A by set B. For this, one obviously has to work with the secular equation rather than the secular determinant. We rewrite Eq. (A.4) as

$$(E - H_{mm})C_m = \sum_{n\in A} H'_{mn}C_n + \sum_{n\in B} H'_{mn}C_n, \tag{A.6}$$

with

$$H'_{mn} \equiv H_{mn}(1 - \delta_{mn}). \tag{A.7}$$

The coefficients can, therefore, be rewritten as

$$C_m = \left(\sum_{n\in A} + \sum_{n\in B}\right)\frac{H'_{mn}}{E - H_{mm}}C_n \equiv \left(\sum_{n\in A} + \sum_{n\in B}\right)h'_{mn}C_n. \tag{A.8}$$

Since we want to set up a secular determinant solely in terms of A-like states ($\widetilde{A}$), we need to iterate away all states B:

$$\begin{aligned} C_m &= \sum_{n\in A} h'_{mn}C_n + \sum_{n\in B} h'_{mn}C_n \\ &= \sum_{n\in A} h'_{mn}C_n + \sum_{n\in B} h'_{mn}\left[\sum_{\alpha\in A} h'_{n\alpha}C_\alpha + \sum_{\alpha\in B} h'_{n\alpha}C_\alpha\right] \\ &= \sum_{n\in A} h'_{mn}C_n + \sum_{n\in A}\sum_{\alpha\in B} h'_{\alpha n}h'_{m\alpha}C_n + \sum_{n\in A}\sum_{\alpha,\beta\in B} h'_{\beta n}h'_{m\alpha}h'_{\alpha\beta}C_n + \cdots \end{aligned}$$

$$= \sum_{n\in A}\left[h'_{mn} + \sum_{\alpha\in B} h'_{m\alpha}h'_{\alpha n} + \sum_{\alpha,\beta\in B} h'_{m\alpha}h'_{\alpha\beta}h'_{\beta n} + \cdots\right] C_n \tag{A.9}$$

$$= \frac{1}{(E-H_{mm})}\sum_{n\in A}\left[H'_{mn} + \sum_{\alpha\in B}\frac{H'_{m\alpha}H'_{\alpha n}}{(E-H_{\alpha\alpha})} + \sum_{\alpha,\beta\in B}\frac{H'_{m\alpha}H'_{\alpha\beta}H'_{\beta n}}{(E-H_{\alpha\alpha})(E-H_{\beta\beta})} + \cdots\right] C_n$$

$$\equiv \frac{1}{(E-H_{mm})}\sum_{n\in A}\left[U^A_{mn} - H_{mn}\delta_{mn}\right] C_n, \tag{A.10}$$

where

$$U^A_{mn} \equiv H_{mn} + \sum_{\alpha\in B}\frac{H'_{m\alpha}H'_{\alpha n}}{(E-H_{\alpha\alpha})} + \sum_{\alpha,\beta\in B}\frac{H'_{m\alpha}H'_{\alpha\beta}H'_{\beta n}}{(E-H_{\alpha\alpha})(E-H_{\beta\beta})}. \tag{A.11}$$

Now, for $m \in A$,

$$(E-H_{mm})C_m = \sum_{n\in A}\left[U^A_{mn} - H_{mn}\delta_{mn}\right] C_n,$$

$$\Longrightarrow \quad \sum_{m,n\in A}\left[U^A_{mn} - E\delta_{mn}\right] C_n = 0. \tag{A.12}$$

Hence, we get a set of secular equations with the original Hamiltonian replaced by the renormalized Hamiltonian U^A. Since $m, n \in A$ in Eq. (A.12) and $\alpha \in B$ in Eq. (A.11), the primes on the Hamiltonian matrix elements in the latter can be dropped, giving

$$U^A_{mn} \equiv H_{mn} + \sum_{\alpha\in B}\frac{H_{m\alpha}H_{\alpha n}}{(E-H_{\alpha\alpha})} + \sum_{\alpha,\beta\in B}\frac{H_{m\alpha}H_{\alpha\beta}H_{\beta n}}{(E-H_{\alpha\alpha})(E-H_{\beta\beta})}. \tag{A.13}$$

This is the form of the renormalized Hamiltonian that will be used in applications. Note that it is a function of the desired eigenvalues E; hence, in practice, a self-consistent solution is needed.

For $m = \alpha \in B$, we have in Eq. (A.10),

$$\delta_{\alpha n} = 0\ \forall\, n \in A,$$

$$\Longrightarrow \quad C_\alpha = \frac{1}{(E-H_{\alpha\alpha})}\sum_{n\in A} U^A_{\alpha n}C_n. \tag{A.14}$$

This procedure is valid (i.e., U can be truncated) if

$$|h'_{mn}| << 1, \;\; m, n \in B,$$

i.e., when the spread in B-state energies is much less than the average separation of A and B energies. We now give an alternative operator form of Eq. (A.9):

$$\mathbf{C} = \left[\mathbf{h}' + \mathbf{h}' \cdot \mathbf{h}' + \mathbf{h}' \cdot \mathbf{h}' \cdot \mathbf{h}' + \cdots\right] \cdot \mathbf{C}.$$

We, finally, show how the Löwdin perturbation relates to other special cases.

A.1.2.1 A = Single, Nondegenerate State k

There is, thus, no summation in Eq. (A.12) and $m, n = k$. Hence,

$$U_{kk}^A - E = 0,$$

and

$$E = U_{kk}^A = H_{kk} + \sum_{n \in B} \frac{H'_{kn} H'_{nk}}{(E - H_{nn})} + \sum_{m,n \in B} \frac{H'_{km} H'_{mn} H'_{nk}}{(E - H_{mm})(E - H_{nn})} + \cdots \quad \text{(A.15)}$$

But

$$H_{mn} = H_{mn}\delta_{mn} + H'_{mn} = H_{mn}^0 \delta_{mn} + V_{mn},$$

and so we have

$$E = E_0 + V_{kk} + \sum_{n \in B} \frac{V_{kn} V_{nk}}{(E - E_n^0)} + \sum_{m,n \in B} \frac{V_{km} V_{mn} V_{nk}}{(E - E_m^0)(E - E_n^0)} + \cdots \quad \text{(A.16)}$$

which is the usual nondegenerate perturbation theory result.

A.1.2.2 A = Class of Degenerate States

Let us first recall the conventional procedure: one diagonalizes the degenerate states to remove the degeneracy and then solve by nondegenerate perturbation theory. According to the Löwdin procedure, we first incorporate B states into the A set by writing down U_{mn}^A. Then, one diagonalizes by using Eq. (A.12) to give $C_m (m \in A)$. Finally, Eq. (A.14) gives $C_m (m \in B)$. For example, to second order,

$$U_{mn}^A \approx H_{mn} + \sum_{\alpha \in B} \frac{H'_{m\alpha} H'_{\alpha n}}{(E - H_{\alpha\alpha})}.$$

A.1.2.3 A = Class of Arbitrary States

For a nondegenerate $k \in A$, one first eliminates system B by using Eq. (A.14), then Eq. (A.12) says that U_{mn}^A is the effective Hamiltonian in the A basis. Now, since k is nondegenerate, the new basis is equivalent to the one treated in the nondegenerate case with k being the sole state in A and the rest of A being equivalent to the B states. Thus, Eq. (A.15) applies for the energy of state k if we write

$$H_{mn} = U^A_{mn},$$

giving

$$E = U^A_{kk} + \sum_{n \in A}' \frac{U^A_{kn} U^A_{nk}}{(E - U^A_{nn})} + \sum_{m,n \in A}' \frac{U^A_{km} U^A_{mn} U^A_{nk}}{(E - U^A_{mm})(E - U^A_{nn})} + \cdots \tag{A.17}$$

A.2 Group Representation Theory

A summary of some useful properties of group representations and their characters is given in this section [444]. Let the group be denoted by $\mathcal{G}$, its order (i.e., the number of elements) by $l_g = \mathcal{O}(\mathcal{G})$, a group element by g, and the matrix representation of the element by $\Gamma^{(\mu)}(g)$.

A.2.1 Great Orthogonality Theorem

Lemma A.1 (Schur's Second Lemma). *A matrix which commutes with all the matrices of an irreducible representation is a multiple of the unit matrix.*

Lemma A.2 (Schur's First Lemma). *Given two irreducible representations $\Gamma^{(1)}$ and $\Gamma^{(2)}$ of dimensions l_1 and l_2, respectively, a rectangular matrix M that satisfies*

$$\Gamma^{(1)}(g)M = M\Gamma^{(2)}(g),$$

for any $g \in \mathcal{G}$ must be either (1) the zero matrix, or (2) a square matrix ($l_1 = l_2$) and $|M| \neq 0$.

Theorem A.1 (Great Orthogonality Theorem). *Given two irreducible representations of a group,*

$$\sum_g \Gamma^{(\mu)}(g^{-1})_{mn} \Gamma^{(\nu)}(g)_{m'n'} = \frac{l_g}{d_\mu} \delta_{\mu\nu} \delta_{m'm} \delta_{nn'}, \tag{A.18}$$

where $l_g = \mathcal{O}(\mathcal{G})$, $d_\mu = \mathrm{Tr}[\Gamma^{(\mu)}(e)]$, $g \in \mathcal{G}$.

For a multiplicative group, $\Gamma^{(\mu)}(g^{-1})_{mn}$ is replaced by $\left[\Gamma^{(\mu)}(g)_{mn}\right]^{-1}$. For unitary representations, Eq. (A.18) becomes

$$\sum_g \Gamma^{(\mu)*}(g)_{mn} \Gamma^{(\nu)}(g)_{m'n'} = \frac{l_g}{d_\mu} \delta_{\mu\nu} \delta_{m'm} \delta_{nn'}.$$

A.2.2 Characters

A.2.2.1 Definitions

An invariant of a matrix representation of an operator is the character:

$$\chi(g) = \sum_m \Gamma_{mm}(g). \tag{A.19}$$

Note that the characters of elements of a class are all the same. In addition, we have

$$\chi(e) = d_\mu, \tag{A.20}$$

where d_μ is the dimension of the matrix representation.

Characters for direct-product groups are given as follows:

$$\begin{aligned} \chi^{\alpha\times\beta}(g) &= \chi^\alpha(g)\chi^\beta(g), \\ \chi^{[\alpha\times\alpha]}(g) &= \frac{1}{2}[\chi^\alpha(g)^2 + \chi^\alpha(g^2)], \\ \chi^{\{\alpha\times\alpha\}}(g) &= \frac{1}{2}[\chi^\alpha(g)^2 - \chi^\alpha(g^2)], \end{aligned} \tag{A.21}$$

where the last two equations are for symmetrized and anti-symmetrized direct products.

A.2.2.2 Theorems for Characters

1. A representation is irreducible iff

$$\sum_g |\chi(g)|^2 = l_g. \tag{A.22}$$

2. The number of times n_μ that a given irreducible representation $\Gamma^{(\mu)}$ appears in the reduction of a reducible representation Γ is

$$n_\mu = \frac{1}{l_g}\sum_g \chi^{\mu *}(g)\,\chi(g). \tag{A.23}$$

3. The number of inequivalent irreducible representation of a group equals the number of classes.
4. The dimensions d_μ of the irreducible representations satisfy the following sum rule:

$$\sum_\mu d_\mu^2 = l_g. \tag{A.24}$$

5. Irreducible representations satisfy the following orthogonality equations:

$$\sum_g \chi^{\mu *}(g)\chi^{\nu}(g) = l_g\, \delta_{\mu\nu}, \tag{A.25}$$

$$\sum_\mu \chi^{\mu *}(f)\chi^{\mu}(g) = l_g\, \delta_{fg}. \tag{A.26}$$

A.3 Angular-Momentum Theory

Basic properties of angular-momentum theory of use in this book are included here for completeness.

A.3.1 Angular Momenta

The angular-momentum operator **L** has the following properties:

$$\begin{aligned}
&\mathbf{L} \times \mathbf{L} = \mathrm{i}\hbar\mathbf{L},\\
&L_z|lm\rangle = m\hbar|lm\rangle,\\
&L^2|lm\rangle = l(l+1)\hbar^2|lm\rangle,\\
&L_\pm|lm\rangle = (L_x \pm \mathrm{i}L_y)|lm\rangle = \sqrt{l(l+1) - m(m\pm 1)}\hbar|lm \pm 1\rangle.
\end{aligned} \tag{A.27}$$

A.3.2 Spherical Tensors

Cartesian tensors are related to spherical tensors of rank γ via the following unitary transformation [81, 444]:

$$\begin{aligned}
X_1 &= -\frac{1}{\sqrt{2}}(X_1^{(1)} - X_{-1}^{(1)}),\\
X_2 &= \frac{\mathrm{i}}{\sqrt{2}}(X_1^{(1)} + X_{-1}^{(1)}),\\
X_3 &= X_0^{(1)},\\
X_{11} &= \frac{1}{2}(X_2^{(2)} + X_{-2}^{(2)}) - \frac{1}{\sqrt{6}}X_0^{(2)},\\
X_{22} &= -\frac{1}{2}(X_2^{(2)} + X_{-2}^{(2)}) - \frac{1}{\sqrt{6}}X_0^{(2)},\\
X_{33} &= \sqrt{\frac{2}{3}}X_0^{(2)},
\end{aligned} \tag{A.28}$$

$$X_{23} = \frac{\mathrm{i}}{2}(X_1^{(2)} + X_{-1}^{(2)}),$$
$$X_{31} = -\frac{1}{2}(X_1^{(2)} - X_{-1}^{(2)}),$$
$$X_{12} = -\frac{\mathrm{i}}{2}(X_2^{(2)} - X_{-2}^{(2)}).$$

A.3.3 Wigner-Eckart Theorem

The essense of the Wigner-Eckart theorem is that the matrix element of an irreducible tensor (i.e., one that transforms according to the irreducible representation of the group) with states that also transform according to the irreducible representations of the group can be written as a product of a "reduced matrix element" that is independent of the rows of the various irreducible representations and of a Clebsch-Gordan coefficient:

$$\langle njm|T_\kappa^{(k)}|n'j'm'\rangle = \frac{\langle nj||T^{(k)}||n'j'\rangle}{\sqrt{2j+1}}\langle j'k;m'\kappa|j'k;jm\rangle. \tag{A.29}$$

Following Trebin et al. [81], we will also use the Wigner-Eckart theorem in the following form

$$(X_q^{(\gamma)})_{mm'} = (-)^{j-m}\langle\alpha j||X^{(\gamma)}||\beta j'\rangle\begin{pmatrix} j & \gamma & j' \\ -m & q & m' \end{pmatrix}, \tag{A.30}$$

which is in terms of the Wigner $3j$ symbols.

A.3.4 Wigner 3j Symbols

The Wigner $3j$ symbols are given by [444]

$$\begin{pmatrix} j_1 & j_2 & j_3 \\ m_1 & m_2 & m_3 \end{pmatrix} = \frac{(-1)^{j_1-j_2-m_3}}{\sqrt{2j_3+1}}a_{j_3m_1m_2}^{(j_1j_2)}\delta_{m_1+m_2+m_3,0}, \tag{A.31}$$

where

$$\begin{aligned} &a_{j_3m_1m_2}^{(j_1j_2)} \\ &= \frac{[(j_3+j_1-j_2)!(j_3-j_1+j_2)!(j_3+m_1+m_2)!(j_3-m_1-m_2)!(j_1+j_2-J)!]^{\frac{1}{2}}}{[(j_3+j_1+j_2+1)!(j_1-m_1)!(j_1+m_1)!(j_2-m_2)!(j_2+m_2)!]^{\frac{1}{2}}} \\ &\times \sum_{j_3+m_1+m_2\geq\kappa\geq 0}\frac{(-1)^{\kappa+j_2+m_2}\sqrt{(2j_3+1)}(j_3+j_2+m_1-\kappa)!(j_1-m_1+\kappa)!}{\kappa!(j_3-j_1+j_2-\kappa)!(j_3+m_1+m_2-\kappa)!(\kappa+j_1-j_2-m_1-m_2)!}. \end{aligned} \tag{A.32}$$

Appendix B
Symmetry Properties

B.1 Introduction

We are interested in: crystal symmetry, group table, action of symmetry operators on real-space coordinates, first Brillouin zone (FBZ), group of wave vector $\mathcal{G}_{\mathbf{k}}$, irreducible representations of $\mathcal{G}_{\mathbf{k}}$, character tables of $\mathcal{G}_{\mathbf{k}}$, compatibility relations, and double groups. Most of the following information can be obtained from standard sources (see, e.g., Bassani [445] and Birman [446]).

B.2 Zincblende

- Description
  Two interpenetrating face-centered cubic (FCC) sublattices (Fig. B.1).
- Crystal lattice
  FCC

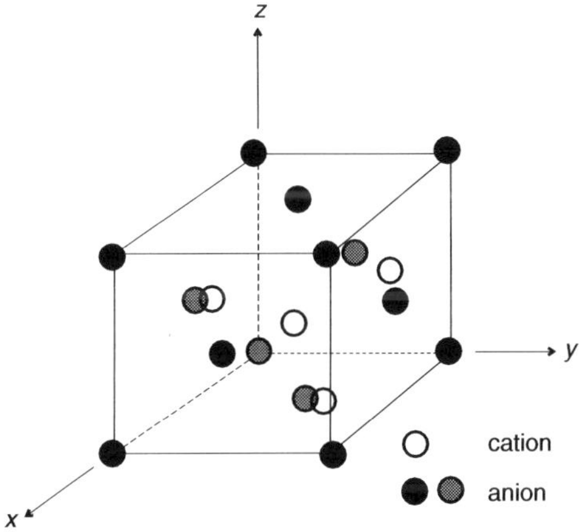

Fig. B.1 Zincblende crystal structure

- Space group
  T_d^2 — $F\bar{4}3m$ (symmorphic)
- Point group
  T_d — $\bar{4}3m$ (inversion asymmetric)
- Cubic vector set

$$\mathbf{a}_1 = a_0[100], \mathbf{a}_2 = a_0[010], \mathbf{a}_3 = a_0[001]. \tag{B.1}$$

- Primitive vector set

$$\mathbf{t}_1 = \frac{a_0}{2}[110], \mathbf{t}_2 = \frac{a_0}{2}[101], \mathbf{t}_3 = \frac{a_0}{2}[011]. \tag{B.2}$$

- Fourier vector set

$$\mathbf{B}_1 = \frac{2\pi}{a_0}[11\bar{1}], \mathbf{B}_2 = \frac{2\pi}{a_0}[1\bar{1}1], \mathbf{B}_3 = \frac{2\pi}{a_0}[\bar{1}11]. \tag{B.3}$$

The corresponding Brillouin zone is displayed in Fig. B.2. The labeling of symmetry points are given in Table B.1.

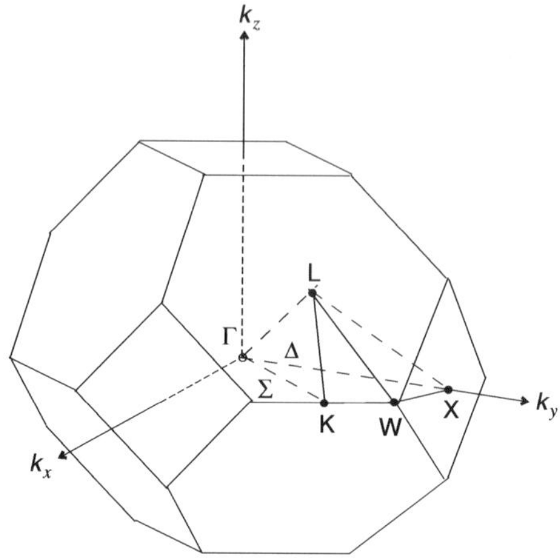

Fig. B.2 First Brillouin zone of the face-centered cubic lattice

B.2.1 Point Group

The symmetry operations of the T_d point group are given in Table B.2. The definition of the operators follows that of Bassani [445] (but is opposite to that of Birman [446]).

Table B.1 Labeling of symmetry points in the first Brillouin zone of zincblende

	$\mathbf{k}$ $(2\pi/a_0)$
Γ	0,0,0
X	1,0,0
L	$\frac{1}{2},\frac{1}{2},\frac{1}{2}$
W	$1,\frac{1}{2},0$
K	$0,\frac{3}{4},\frac{3}{4}$
K'	$1,-\frac{1}{4},-\frac{1}{4}$
$K'' = U$	$1,\frac{1}{4},\frac{1}{4}$

Table B.2 Symmetry operations of the point group T_d

Class	Operation R	$R^{-1}\mathbf{r}$
E	$\{E\|0\}$	$x\ y\ z$
C_4^2	$\{C_{2x}\|0\}$	$x\ \overline{y}\ \overline{z}$
	$\{C_{2y}\|0\}$	$\overline{x}\ y\ \overline{z}$
	$\{C_{2z}\|0\}$	$\overline{x}\ \overline{y}\ z$
IC_4	$\{IC_{4x}\|0\}$	$\overline{x}\ \overline{z}\ y$
	$\{IC_{4x}^{-1}\|0\}$	$\overline{x}\ z\ \overline{y}$
	$\{IC_{4y}\|0\}$	$z\ \overline{y}\ \overline{x}$
	$\{IC_{4y}^{-1}\|0\}$	$\overline{z}\ \overline{y}\ x$
	$\{IC_{4z}\|0\}$	$\overline{y}\ x\ \overline{z}$
	$\{IC_{4z}^{-1}\|0\}$	$y\ \overline{x}\ \overline{z}$
IC'_2	$\{IC_{2xy}\|0\}$	$\overline{y}\ \overline{x}\ z$
	$\{IC_{2x\overline{y}}\|0\}$	$y\ x\ z$
	$\{IC_{2yz}\|0\}$	$x\ \overline{z}\ \overline{y}$
	$\{IC_{2y\overline{z}}\|0\}$	$x\ z\ y$
	$\{IC_{2xz}\|0\}$	$\overline{z}\ y\ \overline{x}$
	$\{IC_{2x\overline{z}}\|0\}$	$z\ y\ x$
C_3	$\{C_{3xyz}\|0\}$	$y\ z\ x$
	$\{C_{3xyz}^{-1}\|0\}$	$z\ x\ y$
	$\{C_{3x\overline{y}z}\|0\}$	$\overline{y}\ \overline{z}\ x$
	$\{C_{3x\overline{y}z}^{-1}\|0\}$	$z\ \overline{x}\ \overline{y}$
	$\{C_{3xy\overline{z}}\|0\}$	$y\ \overline{z}\ \overline{x}$
	$\{C_{3xy\overline{z}}^{-1}\|0\}$	$\overline{z}\ x\ \overline{y}$
	$\{C_{3x\overline{y}\overline{z}}\|0\}$	$\overline{y}\ z\ \overline{x}$
	$\{C_{3x\overline{y}\overline{z}}^{-1}\|0\}$	$\overline{z}\ \overline{x}\ y$

B.2.2 Irreducible Representations

The character tables of the irreducible representations of the double group at various k points are given in Table B.3.

For example, at $\mathbf{k} = \mathbf{0}$, the group is T_d. Various notations and basis functions are given in Table B.4.

Table B.3 Character tables of the double group of wave vector for zincblende

T'_d	E	$\overline{E}$	$3C_4^2, 3\overline{C}_4^2$	$6IC_4$	$6\overline{IC}_4$	$6IC_2, 6\overline{IC}_2$	$8C_3$	$8\overline{C}_3$
Γ_1	1	1	1	1	1	1	1	1
Γ_2	1	1	1	−1	−1	−1	−1	1
Γ_{12}	2	2	2	0	0	0	−1	−1
Γ_{15}	3	3	−1	−1	−1	1	0	0
Γ_{25}	3	3	−1	1	1	−1	0	0
Γ_6	2	−2	0	$\sqrt{2}$	$-\sqrt{2}$	0	1	−1
Γ_7	2	−2	0	$-\sqrt{2}$	$\sqrt{2}$	0	1	−1
Γ_8	4	−4	0	0	0	0	−1	1

$$\Gamma_1 \sim s \to \Gamma_6$$

$$\Gamma_{15} \sim (x, y, z),\ (xy, yz, zx) \to \Gamma_7(\sim Y_{\frac{1}{2}m}) + \Gamma_8(\sim Y_{\frac{3}{2}m})$$

$$\Gamma_{12} \sim (x^2 - y^2, 3z^2 - r^2) \to \Gamma_8$$

C'_{2v}	E	$\overline{E}$	$C_{2x}, \overline{C}_{2x}$	$IC_{2yz}, \overline{IC}_{2yz}$	$IC_{2y\bar{z}}, \overline{IC}_{2y\bar{z}}$
Δ_1	1	1	1	1	1
Δ_2	1	1	1	−1	−1
Δ_3	1	1	−1	1	−1
Δ_4	1	1	−1	−1	1
Δ_5	2	−2	0	0	0

D'_{2d}	E	$\overline{E}$	$C_{2x}, \overline{C}_{2x}$	$C_{2y}, C_{2z}, \overline{C}_{2y}, \overline{C}_{2z}$	IC_{4x}, IC_{4x}^{-1}	$\overline{IC}_{4x}, \overline{IC}_{4x}^{-1}$	$IC_{2yz}, IC_{2y\bar{z}}, \overline{IC}_{2yz}, \overline{IC}_{2y\bar{z}}$
X_1	1	1	1	1	1	1	1
X_2	1	1	1	1	−1	−1	−1
X_3	1	1	1	−1	−1	−1	1
X_4	1	1	1	−1	1	1	−1
X_5	2	2	−2	0	0	0	0
X_6	2	−2	0	0	$\sqrt{2}$	$-\sqrt{2}$	0
X_7	2	−2	0	0	$-\sqrt{2}$	$\sqrt{2}$	0

C'_{3v}	E	$\overline{E}$	$IC_{2x\bar{y}}, IC_{2x\bar{z}}, IC_{2y\bar{z}}$	$\overline{IC}_{2x\bar{y}}, \overline{IC}_{2x\bar{z}}, \overline{IC}_{2y\bar{z}}$	C_{3xyz}, C_{3xyz}^{-1}	$\overline{C}_{3xyz}, \overline{C}_{3xyz}^{-1}$
Λ_1, L_1	1	1	1	1	1	1
Λ_2, L_2	1	1	−1	−1	1	1
Λ_3, L_3	2	2	0	0	−1	−1
Λ_4, L_4	1	−1	i	−i	−1	1
Λ_5, L_5	1	−1	−i	i	−1	1
Λ_6, L_6	2	−2	0	0	1	−1

Table B.4 Zincblende: Γ point – T_d. Representations and basis functions

KDWS [33] W03 [21]	D55 [32]	BSW [447], P55 [448]	T64 [444]	L56 [4]	IP95 [20]		
Γ_1	Γ_1	Γ_1	A_1	A_1	A_1	s	
Γ_2	Γ_2	Γ_2	A_2	A_2	A_2		$[J_x J_y J_z]_s$
Γ_3	Γ_3	Γ_{12}	E	E	E	$x^2 - y^2, 3z^2 - r^2$	
Γ_4	Γ_5	Γ_{25}	T_1	T_2	F_1		$J_x, J_y, J_z; J_x^3, J_y^3, J_z^3$
Γ_5	Γ_4	Γ_{15}	T_2	T_1	F_2	$x, y, z; yz, zx, xy$	$[J_y J_z]_s, [J_z J_x]_s, [J_x J_y]_s$
Γ_6	Γ_7						
Γ_7	Γ_6						
Γ_8	Γ_8						

B.3 Diamond

- Description
  Two interpenetrating FCC sublattices. The Brillouin zone is the same as for zincblende.
- Crystal lattice
  FCC
- Space group
  O_h^7 — $Fd3m$ (nonsymmorphic)
- Point group
  O_h — $m3m$ (inversion symmetric)

B.3.1 Symmetry Operators

The symmetry operators are given in Table B.5. The remaining description of the real-space lattice is identical to the zincblende case.

Table B.5 Rotational symmetry operations for diamond (O_h symmetry); $\boldsymbol{\tau} = (1, 1, 1)a_0/4$

Type $\{\mathbf{R}\vert\mathbf{0}\}$	$\{\mathbf{R}\vert\mathbf{0}\}^{-1}\mathbf{r}$	Type $\{\mathbf{R}\vert\boldsymbol{\tau}\}$	$\{\mathbf{R}\vert\boldsymbol{\tau}\}^{-1}\mathbf{r}$
E	$x\ y\ z$	I	$\overline{x}\ \overline{y}\ \overline{z}$
C_{2x}	$x\ \overline{y}\ \overline{z}$	IC_{2x}	$\overline{x}\ y\ z$
C_{2y}	$\overline{x}\ y\ \overline{z}$	IC_{2y}	$x\ \overline{y}\ z$
C_{2z}	$\overline{x}\ \overline{y}\ z$	IC_{2z}	$x\ y\ \overline{z}$
IC_{4x}	$\overline{x}\ \overline{z}\ y$	C_{4x}	$x\ z\ \overline{y}$
IC_{4x}^{-1}	$\overline{x}\ z\ \overline{y}$	C_{4x}^{-1}	$x\ \overline{z}\ y$
IC_{4y}	$z\ \overline{y}\ \overline{x}$	C_{4y}	$\overline{z}\ y\ x$
IC_{4y}^{-1}	$\overline{z}\ \overline{y}\ x$	C_{4y}^{-1}	$z\ y\ \overline{x}$
IC_{4z}	$\overline{y}\ x\ \overline{z}$	C_{4z}	$y\ \overline{x}\ z$
IC_{4z}^{-1}	$y\ \overline{x}\ \overline{z}$	C_{4z}^{-1}	$\overline{y}\ x\ z$
IC_{2xy}	$\overline{y}\ \overline{x}\ z$	C_{2xy}	$y\ x\ \overline{z}$
$IC_{2x\overline{y}}$	$y\ x\ z$	$C_{2x\overline{y}}$	$\overline{y}\ \overline{x}\ \overline{z}$
IC_{2yz}	$x\ \overline{z}\ \overline{y}$	C_{2yz}	$\overline{x}\ z\ y$
$IC_{2y\overline{z}}$	$x\ z\ y$	$C_{2y\overline{z}}$	$\overline{x}\ \overline{z}\ \overline{y}$
IC_{2xz}	$\overline{z}\ y\ \overline{x}$	C_{2xz}	$z\ \overline{y}\ x$
$IC_{2x\overline{z}}$	$z\ y\ x$	$C_{2x\overline{z}}$	$\overline{z}\ \overline{y}\ \overline{x}$
C_{3xyz}	$y\ z\ x$	IC_{3xyz}	$\overline{y}\ \overline{z}\ \overline{x}$
C_{3xyz}^{-1}	$z\ x\ y$	IC_{3xyz}^{-1}	$\overline{z}\ \overline{x}\ \overline{y}$
$C_{3x\overline{y}z}$	$\overline{y}\ \overline{z}\ x$	$IC_{3x\overline{y}z}$	$y\ z\ \overline{x}$
$C_{3x\overline{y}z}^{-1}$	$z\ \overline{x}\ \overline{y}$	$IC_{3x\overline{y}z}^{-1}$	$\overline{z}\ x\ y$
$C_{3xy\overline{z}}$	$y\ \overline{z}\ \overline{x}$	$IC_{3xy\overline{z}}$	$\overline{y}\ z\ x$
$C_{3xy\overline{z}}^{-1}$	$\overline{z}\ x\ \overline{y}$	$IC_{3xy\overline{z}}^{-1}$	$z\ \overline{x}\ y$
$C_{3x\overline{y}\overline{z}}$	$\overline{y}\ z\ \overline{x}$	$IC_{3x\overline{y}\overline{z}}$	$y\ \overline{z}\ x$
$C_{3x\overline{y}\overline{z}}^{-1}$	$\overline{z}\ \overline{x}\ y$	$IC_{3x\overline{y}\overline{z}}^{-1}$	$z\ x\ \overline{y}$

B.3.2 Irreducible Representations

The character table of the irreducible representations of the O_h' double group at the Γ point is given in Table B.6. Various notations and basis functions are given in Table B.7.

B.4 Wurtzite

- Description
  The primitive cell is a hexagonal prism with four atoms (Fig. B.3). It is characterized by three lattice parameters: the length of the c axis, the c/a ratio, and the bond-length parameter u.
- Crystal lattice
  Hexagonal close packed (HCP)

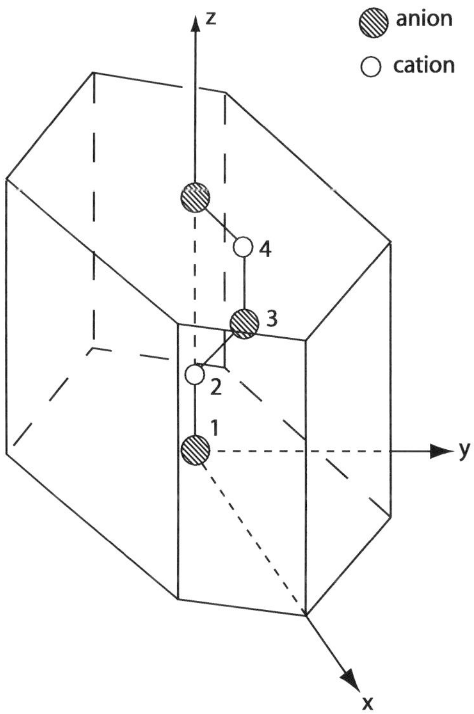

Fig. B.3 Primitive cell for wurtzite. Atomic coordinates: $\mathbf{r}_1 = (0, 0, 0)$, $\mathbf{r}_2 = (0, 0, uc)$, $\mathbf{r}_3 = (0, \frac{a}{\sqrt{3}}, \frac{c}{2})$, $\mathbf{r}_4 = (0, \frac{a}{\sqrt{3}}, \frac{c}{2} + uc)$. Reprinted with permission from [48]. ©1996 by the American Physical Society

Table B.6 Character table of the double group O'_h

O'_h	E	$\overline{E}$	$8C_3$	$8\overline{C}_3$	$3C_2, 3\overline{C}_2$	$6C_4$	$6\overline{C}_4$	$6C'_2, 6\overline{C}'_2$	I	$\overline{I}$	$8S_6$	$8\overline{S}_6$	$3\sigma_h, 3\overline{\sigma}_h$	$6S_4$	$6\overline{S}_4$	$6\sigma_d, 6\overline{\sigma}_d$
Γ_1	1	1	1	1	1	1	1	1	1	1	1	1	1	1	1	1
Γ_2	1	1	1	1	1	−1	−1	−1	1	1	1	1	1	−1	−1	−1
Γ_{12}	2	2	−1	−1	2	0	0	0	2	2	−1	−1	2	0	0	0
$\Gamma_{15'}$	3	3	0	0	−1	1	1	−1	3	3	0	0	−1	1	1	−1
$\Gamma_{25'}$	3	3	0	0	−1	−1	−1	1	3	3	0	0	−1	−1	−1	1
$\Gamma_{1'}$	1	1	1	1	1	1	1	1	−1	−1	−1	−1	−1	−1	−1	−1
$\Gamma_{2'}$	1	1	1	1	1	−1	−1	−1	−1	−1	−1	−1	−1	1	1	1
$\Gamma_{12'}$	2	2	−1	−1	2	0	0	0	−2	−2	1	1	−2	0	0	0
Γ_{15}	3	3	0	0	−1	1	1	−1	−3	−3	0	0	1	−1	−1	1
Γ_{25}	3	3	0	0	−1	−1	−1	1	−3	−3	0	0	1	1	1	−1
Γ_6^+	2	−2	1	−1	0	$\sqrt{2}$	$-\sqrt{2}$	0	2	−2	1	−1	0	$\sqrt{2}$	$-\sqrt{2}$	0
Γ_7^+	2	−2	1	−1	0	$-\sqrt{2}$	$\sqrt{2}$	0	2	−2	1	−1	0	$-\sqrt{2}$	$\sqrt{2}$	0
Γ_8^+	4	−4	−1	1	0	0	0	0	4	−4	−1	1	0	0	0	0
Γ_6^-	2	−2	1	−1	0	$\sqrt{2}$	$-\sqrt{2}$	0	−2	2	1	1	0	$-\sqrt{2}$	$\sqrt{2}$	0
Γ_7^-	2	−2	1	−1	0	$-\sqrt{2}$	$\sqrt{2}$	0	−2	2	−1	1	0	$\sqrt{2}$	$-\sqrt{2}$	0
Γ_8^-	4	−4	−1	1	0	0	0	0	−4	4	1	−1	0	0	0	0

$$\Gamma_1 \sim s \rightarrow \Gamma_6^+$$

$$\Gamma_{15} \sim (x, y, z) \rightarrow \Gamma_6^- + \Gamma_8^-$$

$$\Gamma_{25'} \sim (xy, yz, zx) \rightarrow \Gamma_7^+ + \Gamma_8^+$$

$$\Gamma_{12} \sim (x^2 - y^2, 3z^2 - r^2) \rightarrow \Gamma_8^+$$

Table B.7 Diamond: Γ point – O_h. l designation in last column refers to lowest l value for that representation as given by von der Lage and Bethe [27]

KDWS [33]	DKK [2]	BSW [447] E54 [35]	VLB [27]	Heine [449]	L56 [4]	BP [1]	
Γ_1^+	Γ_1^+	Γ_1	α	A_{1g}	A_1	A_1^+	s
Γ_2^+	Γ_2^+	Γ_2	β'	A_{2g}	A_2	A_2^+	$(l=6)$
Γ_3^+	Γ_{12}^+	Γ_{12}	γ	E_g	E	E^+	$x^2-y^2, 3z^2-r^2$
Γ_4^+	Γ_{15}^+	Γ_{15}'	δ'	T_{1g}	T_2	F_1^+	R_x, R_y, R_z $(l-4)$
Γ_5^+	Γ_{25}^+	Γ_{25}'	ϵ	T_{2g}	T_1	F_2^+	xy, yz, zx
Γ_1^-	Γ_1^-	Γ_1'	α'	A_{1u}	A_1	A_1^-	$(l=9)$
Γ_2^-	Γ_2^-	Γ_2'	β	A_{2u}	A_2	A_2^-	xyz
Γ_3^-	Γ_{12}^-	Γ_{12}'	γ'	E_u	E	E^-	$(l=5)$
Γ_4^-	Γ_{15}^-	Γ_{15}	δ	T_{1u}	T_2	F_2^-	x, y, z
Γ_5^-	Γ_{25}^-	Γ_{25}	ε'	T_{2u}	T_1	F_1^-	$(l=3)$

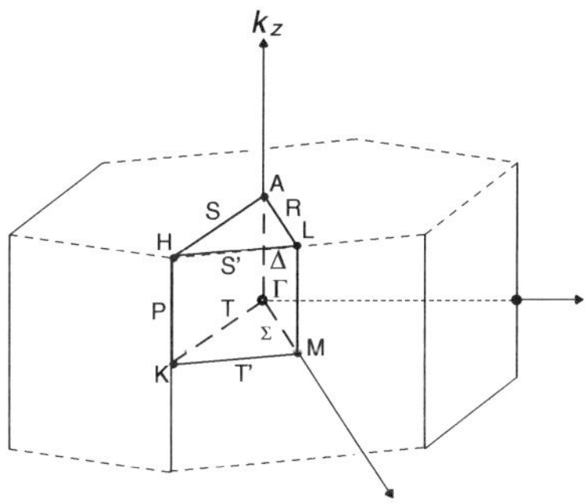

Fig. B.4 First Brillouin zone for wurtzite

Table B.8 Labeling of symmetry points in the first Brillouin zone of wurtzite [33]

	k (2π)
Γ	0,0,0
A	$0, 0, \frac{1}{2c}$
L	$0, \frac{1}{\sqrt{3}a}, \frac{1}{2c}$
M	$0, \frac{1}{\sqrt{3}a}, 0$
H	$-\frac{1}{3a}, \frac{1}{\sqrt{3}a}, \frac{1}{2c}$
K	$-\frac{1}{3a}, \frac{1}{\sqrt{3}a}, 0$

- Space group
  $C_{6v}^4 - P6_3mc$ (non-symmorphic)
- Point group
  $C_{6v} - 6mm$

- Hexagonal vector set

$$\mathbf{a}_1 = a[100], \mathbf{a}_2 = a[010], \mathbf{a}_3 = c[001]. \tag{B.4}$$

- Primitive vector set

$$\mathbf{t}_1 = \frac{a}{2}[\sqrt{3}\bar{1}0], \mathbf{t}_2 = a[010], \mathbf{t}_3 = c[001]. \tag{B.5}$$

- Fourier vector set

$$\mathbf{B}_1 = \frac{2\pi}{a}[\frac{2}{\sqrt{3}}00], \mathbf{B}_2 = \frac{2\pi}{a}[\frac{1}{\sqrt{3}}10], \mathbf{B}_3 = \frac{2\pi}{c}[001]. \tag{B.6}$$

The corresponding Brillouin zone is displayed in Fig. B.4. The labeling of symmetry points is given in Table B.8.

B.4.1 Irreducible Representations

The character table of the irreducible representations of the C_{6v} double group at the Γ point is given in Table B.9. Note that the ordering differs slightly from, for example, Rashba [450]. Various notations are given in Table B.10.

Table B.9 Character table for the single and double group of **k**, C_{6v} ($6mm$), at the zone center (Γ point) of the Brillouin zone for the wurtzite crystal structure

C_{6v}	E	C_2	$2C_3$	$2C_6$	$3\sigma_v$	$3\sigma_d$			
Γ_1	1	1	1	1	1	1			
Γ_2	1	1	1	1	−1	−1			
Γ_3	1	−1	1	−1	−1	1			
Γ_4	1	−1	1	−1	1	−1			
Γ_5	2	−2	−1	1	0	0			
Γ_6	2	2	−1	−1	0	0			
	E	$\overline{E}$	$C_2, \overline{C}_2$	$2C_3$	$2\overline{C}_3$	$2C_6$	$2\overline{C}_6$	$3\sigma_v, 3\overline{\sigma}_v$	$3\sigma_d, 3\overline{\sigma}_d$
Γ_7	2	−2	0	1	−1	$\sqrt{3}$	$-\sqrt{3}$	0	0
Γ_8	2	−2	0	1	−1	$-\sqrt{3}$	$\sqrt{3}$	0	0
Γ_9	2	−2	0	−2	2	0	0	0	0

Table B.10 Wurtzite: Γ point – C_{6v} ($6mm$). Notation for representations and corresponding basis functions [20, 33]

KDWS[33]	IP95[20]	Basis
Γ_1	A_1	$s, z, 3z^2 - r^2$
Γ_2	A_3	R_z
Γ_3	A_4	$x^3 - 3xy^2$
Γ_4	A_2	$y^3 - 3yx^2$
Γ_5	E_2	$(R_x, R_y), (x, y), (zx, yz)$
Γ_6	E_1	$(x^2 - y^2, xy)$
Γ_7	E'_1	α, β
Γ_8	E'_2	$(x + \mathrm{i}y)^3\alpha, (x - \mathrm{i}y)^3\beta$
Γ_9	E'_3	$(x + \mathrm{i}y)\alpha, (x - \mathrm{i}y)\beta$

Appendix C
Hamiltonians

C.1 Basis Matrices

In the method of invariants, the matrix representation of the effective $k \cdot p$ Hamiltonian is achieved with the use of basis matrices. Three sets are used in this book and will be given using Eq. (A.27).

C.1.1 $s = \frac{1}{2}$

For a basis of spinors quantized along the z direction, the Pauli spin matrices are:

$$\sigma_x = \begin{pmatrix} 0 & 1 \\ 1 & 0 \end{pmatrix}, \quad \sigma_y = \begin{pmatrix} 0 & -\mathrm{i} \\ \mathrm{i} & 0 \end{pmatrix}, \quad \sigma_z = \begin{pmatrix} 1 & 0 \\ 0 & -1 \end{pmatrix}. \tag{C.1}$$

C.1.2 $l = 1$

In the basis $|l\ m\rangle$ ($|1, 1\rangle$, $|1, 0\rangle$, $|1, -1\rangle$) and writing $I_i = L_i/\hbar$,

$$I_x = \frac{1}{\sqrt{2}} \begin{pmatrix} 0 & 1 & 0 \\ 1 & 0 & 1 \\ 0 & 1 & 0 \end{pmatrix}, \quad I_y = \frac{\mathrm{i}}{\sqrt{2}} \begin{pmatrix} 0 & -1 & 0 \\ 1 & 0 & -1 \\ 0 & 1 & 0 \end{pmatrix}, \quad I_z = \begin{pmatrix} 1 & 0 & 0 \\ 0 & 0 & 0 \\ 0 & 0 & -1 \end{pmatrix}, \tag{C.2}$$

$$I_+ = \begin{pmatrix} 0 & 1 & 0 \\ 0 & 0 & 1 \\ 0 & 0 & 0 \end{pmatrix}, \quad I_- = \begin{pmatrix} 0 & 0 & 0 \\ 1 & 0 & 0 \\ 0 & 1 & 0 \end{pmatrix}, \tag{C.3}$$

C.1.3 $J = \frac{3}{2}$

In the $|J\ M_J\rangle$ basis with $M_J = \frac{3}{2}, \frac{1}{2}, -\frac{1}{2}, -\frac{3}{2}$,

$$J_x = \begin{pmatrix} 0 & \frac{\sqrt{3}}{2} & 0 & 0 \\ \frac{\sqrt{3}}{2} & 0 & 1 & 0 \\ 0 & 1 & 0 & \frac{\sqrt{3}}{2} \\ 0 & 0 & \frac{\sqrt{3}}{2} & 0 \end{pmatrix}, \; J_y = \begin{pmatrix} 0 & -\mathrm{i}\frac{\sqrt{3}}{2} & 0 & 0 \\ \mathrm{i}\frac{\sqrt{3}}{2} & 0 & -\mathrm{i} & 0 \\ 0 & \mathrm{i} & 0 & -\mathrm{i}\frac{\sqrt{3}}{2} \\ 0 & 0 & \mathrm{i}\frac{\sqrt{3}}{2} & 0 \end{pmatrix}, \; J_z = \begin{pmatrix} \frac{3}{2} & 0 & 0 & 0 \\ 0 & \frac{1}{2} & 0 & 0 \\ 0 & 0 & -\frac{1}{2} & 0 \\ 0 & 0 & 0 & -\frac{3}{2} \end{pmatrix},$$

$$J_+ = \begin{pmatrix} 0 & \sqrt{3} & 0 & 0 \\ 0 & 0 & 2 & 0 \\ 0 & 0 & 0 & \sqrt{3} \\ 0 & 0 & 0 & 0 \end{pmatrix}, \; J_- = \begin{pmatrix} 0 & 0 & 0 & 0 \\ \sqrt{3} & 0 & 0 & 0 \\ 0 & 2 & 0 & 0 \\ 0 & 0 & \sqrt{3} & 0 \end{pmatrix}. \tag{C.4}$$

C.2 $|JM_J\rangle$ States

For the cubic semiconductors, the $\left|J = \frac{3}{2}\, M_J\right\rangle$ states are useful for describing the states at $\mathbf{k} = \mathbf{0}$. One choice is given in Table C.1. A comparison of different choices is given in Table C.2.

Table C.1 $|J, M_J\rangle$ states

$$\begin{aligned}
\left|\frac{3}{2}\;\frac{3}{2}\right\rangle &= \left|\frac{1}{\sqrt{2}}\right\rangle |(x+\mathrm{iy})\uparrow\rangle\,, \\
\left|\frac{3}{2}\;\frac{1}{2}\right\rangle &= \frac{1}{\sqrt{6}}|(x+\mathrm{iy})\downarrow\rangle - \sqrt{\frac{2}{3}}|z\uparrow\rangle, \\
\left|\frac{3}{2}\;-\frac{1}{2}\right\rangle &= -\frac{1}{\sqrt{6}}|(x-\mathrm{iy})\uparrow\rangle - \sqrt{\frac{2}{3}}|z\downarrow\rangle, \\
\left|\frac{3}{2}\;-\frac{3}{2}\right\rangle &= \frac{1}{\sqrt{2}}|(x-\mathrm{iy})\downarrow\rangle, \\
\left|\frac{1}{2}\;\frac{1}{2}\right\rangle &= \frac{1}{\sqrt{3}}|(x+\mathrm{iy})\downarrow\rangle + \frac{1}{\sqrt{3}}|z\uparrow\rangle, \\
\left|\frac{1}{2}\;-\frac{1}{2}\right\rangle &= -\frac{1}{\sqrt{3}}|(x-\mathrm{iy})\uparrow\rangle + \frac{1}{\sqrt{3}}|z\downarrow\rangle, \\
&\mathrm{i}|s\uparrow\rangle, \\
&\mathrm{i}|s\downarrow\rangle.
\end{aligned}$$

C.3 Hamiltonians

A representative set of Hamiltonians described in this book is collected together and repeated in this appendix for ease of reference. Some of the larger ones (e.g., the 14-band extended Kane Hamiltonian) are not repeated here; they are clearly displayed within the book.

Table C.2 Clebsch–Gordan transformation. The phase factors given represent the phase difference compared to our choice given in Table C.1. Coefficients in parentheses are not present for all the works listed

	E54 [35]	LK [6] PB66 [86] BP75 [1]	DKK [2]	RLZ [25]	K56 [3]	K66 [31] P90 [125]	K82 [427]	KDWS [33] W78 [98] F88 [24]	BS85 [89] W03 [21]	Cho [83] C95 [15]	SV90 [36]	S87 [94]
$\lvert\frac{3}{2}\ \frac{3}{2}\rangle$	+1	+1	+1	+1	−1	+1	+1	−i	−1	−1	−1	$e^{-2i\phi}$
$\lvert\frac{3}{2}\ \frac{1}{2}\rangle$	+1	+i	+1	−1	−1	+1	+1	−i	−1	−1	−1	$ie^{-i\phi}$
$\lvert\frac{3}{2}\ -\frac{1}{2}\rangle$	−1	−1	+1	−1	−1	+1	+1	−i	−1	−1	−1	+1
$\lvert\frac{3}{2}\ -\frac{3}{2}\rangle$	+1	+i	−1	+1	−1	−1	+1	+i	+1	+1	+1	$ie^{i\phi}$
$\lvert\frac{1}{2}\ \frac{1}{2}\rangle$	+1	+1	−1	+1	−1	+1	+1	−i	(−1)	+1	−1	$ie^{-i\phi}$
$\lvert\frac{1}{2}\ -\frac{1}{2}\rangle$	−1	+i	+1	−1	−1	−1	+1	+i	(+1)	−1	+1	+1
$i\lvert s\uparrow\rangle$		(+1)				−i	−i	(−i)	(+1)	(+1)	(−i)	
$i\lvert s\downarrow\rangle$		(+1)				−i	−i	(−i)	(+1)	(+1)	(−i)	

C.3.1 Notations

$$H\psi_{n\mathbf{k}}(\mathbf{r}) = E_n(\mathbf{k})\psi_{n\mathbf{k}}(\mathbf{r})\ , \tag{C.5}$$

$$H = \frac{p^2}{2m_0} + V(\mathbf{r}) + \frac{\hbar}{4m_0^2c^2}(\sigma \times \nabla V)\cdot \mathbf{p} = H_0 + H_{\mathrm{so}}\ , \tag{C.6}$$

$$H_0 = \frac{p^2}{2m_0} + V(\mathbf{r})\ , \tag{C.7}$$

$$H_{\mathrm{so}} = \frac{\hbar}{4m_0c^2}(\sigma \times \nabla V)\cdot \mathbf{p} = H_{\mathrm{s},i}\sigma_i\ , \tag{C.8}$$

$$H(\mathbf{k})u_{n\mathbf{k}} = \mathcal{E}_n(\mathbf{k})u_{n\mathbf{k}}\ , \tag{C.9}$$

$$H(\mathbf{k}) = H_0 + H_{\mathrm{so}}(\mathbf{k}) + H_{k\cdot p}\ , \tag{C.10}$$

$$H_{\mathrm{so}}(\mathbf{k}) = H_{\mathrm{so}} + H_{\mathrm{so},k}\ , \tag{C.11}$$

$$H_{\mathrm{so},\mathbf{k}} = \frac{\hbar}{4m_0^2c^2}(\sigma \times \nabla V)\cdot \mathbf{k}. \tag{C.12}$$

C.3.2 Diamond

The valence-band Hamiltonians of Dresselhaus, Kip and Kittel [2], of Chuang [15], and of Cardona and Pollak [5] are given in Tables C.3, C.4, C.5, and C.6.

C.3.3 Zincblende

The Hamiltonians within the Kane model for zincblende are given in Tables C.7, and C.8.

C.3.4 Wurtzite

The commonly-used basis functions for bulk wurtzite semiconductors are given in Table C.9.

Three valence-band Hamiltonians for wurtzite are given in Tables C.10, C.11, and C.12. The four-band Hamiltonian by Andreev and O'Reilly is given in Table C.13.

C.3.5 Heterostructures

We now give the Hamiltonians for various nanostructures (Tables C.13, C.14, C.15, C.16, C.17, and C.18).

C.4 Summary of $k \cdot p$ Parameters

A few of the valence-band parameters are compared in Table C.19. Example band parameters for a few semiconductors are given in Tables C.20 and C.21.

Table C.3 Dresselhaus–Kip–Kittel Hamiltonian without spin for diamond

$$H_{\mathrm{DKK}}(\mathbf{k}) = \begin{pmatrix} |yz\rangle & |zx\rangle & |xy\rangle \\ Lk_x^2 + M\left(k_y^2 + k_z^2\right) & Nk_xk_y & Nk_xk_z \\ Nk_xk_y & Lk_y^2 + M\left(k_z^2 + k_x^2\right) & Nk_yk_z \\ Nk_xk_z & Nk_yk_z & Lk_z^2 + M\left(k_x^2 + k_y^2\right) \end{pmatrix},$$

with

$$\begin{aligned} L &= \frac{\hbar^2}{m_0^2} \sum_{\alpha \in \Gamma_2^-, \Gamma_{12}^-}{}' \frac{|\langle xy|p_z|l\alpha\nu\rangle|^2}{E_{\Gamma_{25}^+} - E_{l\alpha}} \\ &= F + 2G, \\ M &= \frac{\hbar^2}{m_0^2} \sum_{\alpha \in \Gamma_{15}^-, \Gamma_{25}^-}{}' \frac{|\langle xy|p_y|l\alpha\nu\rangle|^2}{E_{\Gamma_{25}^+} - E_{l\alpha}} \\ &= H_1 + H_2, \\ N &= \frac{\hbar^2}{m_0^2} \sum_{\alpha \in \Gamma_2^-, \Gamma_{12}^-, \Gamma_{15}^-, \Gamma_{25}^-}{}' \frac{\langle xy|p_z|l\alpha\nu\rangle\langle l\alpha\nu|p_x|yz\rangle + \langle xy|p_x|l\alpha\nu\rangle\langle l\alpha\nu|p_z|yz\rangle}{E_{\Gamma_{25}^+} - E_{l\alpha}} \\ &= F - G + H_1 - H_2, \end{aligned}$$

$$\begin{aligned} F &= \frac{\hbar^2}{m_0^2} \sum_{l \in \Gamma_2^-}{}' \frac{|\langle yz|p_x|\beta_l^-\rangle|^2}{E_{\Gamma_{25}^+} - E_l}, \\ G &= \frac{\hbar^2}{m_0^2} \sum_{l \in \Gamma_{12}^-}{}' \frac{|\langle yz|p_x|\gamma_{\nu l}^-\rangle|^2}{E_{\Gamma_{25}^+} - E_l}, \\ H_1 &= \frac{\hbar^2}{m_0^2} \sum_{l \in \Gamma_{15}^-}{}' \frac{|\langle xy|p_y|\delta_{3l}^-\rangle|^2}{E_{\Gamma_{25}^+} - E_l}, \\ H_2 &= \frac{\hbar^2}{m_0^2} \sum_{l \in \Gamma_{25}^-}{}' \frac{|\langle xy|p_y|\epsilon_{3l}^-\rangle|^2}{E_{\Gamma_{25}^+} - E_l}. \end{aligned}$$

Table C.4 Dresselhaus–Kip–Kittel Hamiltonian in $|JM_J\rangle$ basis for diamond

$$H(\mathbf{k}) = \begin{pmatrix} |\frac{3}{2}\,\frac{3}{2}\rangle & |\frac{3}{2}\,\frac{1}{2}\rangle & |\frac{3}{2}\,-\frac{1}{2}\rangle & |\frac{3}{2}\,-\frac{3}{2}\rangle & |\frac{1}{2}\,\frac{1}{2}\rangle & |\frac{1}{2}\,-\frac{1}{2}\rangle \\ \frac{1}{2}P & R & S & 0 & -\frac{1}{\sqrt{2}}R & \sqrt{2}S \\ R^* & \frac{1}{6}P+\frac{2}{3}Q & 0 & -S & \frac{1}{3\sqrt{2}}(P-2Q) & -\sqrt{\frac{3}{2}}R \\ S^* & 0 & \frac{1}{6}P+\frac{2}{3}Q & R & \sqrt{\frac{3}{2}}R^* & \frac{1}{3\sqrt{2}}(P-2Q) \\ 0 & -S^* & R^* & \frac{1}{2}P & -\sqrt{2}S^* & -\frac{1}{\sqrt{2}}R^* \\ -\frac{1}{\sqrt{2}}R^* & \frac{1}{3\sqrt{2}}(P-2Q) & \sqrt{\frac{3}{2}}R & -\sqrt{2}S & \frac{1}{3}(P+Q)-\Delta_0 & 0 \\ \sqrt{2}S^* & -\sqrt{\frac{3}{2}}R^* & \frac{1}{3\sqrt{2}}(P-2Q) & -\frac{1}{\sqrt{2}}R & 0 & \frac{1}{3}(P+Q)-\Delta_0 \end{pmatrix}$$

with

$$\begin{aligned} P &= (L+M)(k_x^2+k_y^2)+2Mk_z^2+\frac{\hbar^2k^2}{m_0}, \\ Q &= M(k_x^2+k_y^2)+Lk_z^2+\frac{\hbar^2k^2}{2m_0}, \\ R &= -\frac{N}{\sqrt{3}}(k_x-\mathrm{i}k_y)k_z, \\ S &= -\frac{1}{2\sqrt{3}}\left[(L-M)(k_x^2-k_y^2)-2\mathrm{i}Nk_xk_y\right]. \end{aligned}$$

Table C.5 Six-band Hamiltonian in $|JM_J\rangle$ basis using Chuang's notation for diamond

$$H(\mathbf{k}) = -\begin{pmatrix} |\frac{3}{2}\,\frac{3}{2}\rangle & |\frac{3}{2}\,\frac{1}{2}\rangle & |\frac{3}{2}\,-\frac{1}{2}\rangle & |\frac{3}{2}\,-\frac{3}{2}\rangle & |\frac{1}{2}\,\frac{1}{2}\rangle & |\frac{1}{2}\,-\frac{1}{2}\rangle \\ P+Q & -S & R & 0 & \frac{1}{\sqrt{2}}S & \sqrt{2}R \\ -S^* & P-Q & 0 & -R & \sqrt{2}Q & \sqrt{\frac{3}{2}}S \\ R^* & 0 & P-Q & -S & -\sqrt{\frac{3}{2}}S^* & \sqrt{2}Q \\ 0 & -R^* & -S^* & P+Q & -\sqrt{2}R^* & \frac{1}{\sqrt{2}}S^* \\ \frac{1}{\sqrt{2}}S^* & \sqrt{2}Q & -\sqrt{\frac{3}{2}}S & -\sqrt{2}R & P+\Delta_0 & 0 \\ \sqrt{2}R^* & \sqrt{\frac{3}{2}}S^* & \sqrt{2}Q & \frac{1}{\sqrt{2}}S & 0 & P+\Delta_0 \end{pmatrix}$$

with

$$\begin{aligned} P &= \frac{\hbar^2\gamma_1}{2m_0}(k_x^2+k_y^2+k_z^2), \\ Q &= \frac{\hbar^2\gamma_2}{2m_0}(k_x^2+k_y^2-2k_z^2), \\ S &= \frac{\hbar^2\gamma_2}{2m_0}2\sqrt{3}\gamma_3(k_x-\mathrm{i}k_y)k_z, \\ R &= \frac{\hbar^2}{2m_0}[-\sqrt{3}\gamma_2(k_x^2-k_y^2)+2\mathrm{i}\sqrt{3}\gamma_3k_xk_y]. \end{aligned}$$

Parameters:

$$\gamma_1,\gamma_2,\gamma_3,\Delta_0.$$

Table C.6 15-band Cardona–Pollak Hamiltonian

$$
\left[\begin{array}{ccccccccccccccc}
 & \Gamma^l_{25'} & & & \mathrm{i}\Gamma_{15} & & & \Gamma^u_{25'} & & \Gamma^u_1 & \Gamma^l_1 & \mathrm{i}\Gamma^l_{2'} & \mathrm{i}\Gamma_{12'} & & \mathrm{i}\Gamma^u_{2'} \\
\mathrm{yz} & \mathrm{zx} & xy & \mathrm{i}x & \mathrm{i}y & \mathrm{i}z & yz & zx & xy & s^+ & s^+ & \mathrm{i}s^- & d_1^- & \mathrm{i}d_2^- & \mathrm{i}s^- \\
0 & 0 & 0 & 0 & \frac{Q}{2}k_z & \frac{Q}{2}k_y & 0 & 0 & 0 & 0 & 0 & \frac{P}{2}k_x & \frac{R}{\sqrt{2}}k_x & 0 & \frac{P''}{2}k_x \\
0 & 0 & 0 & \frac{Q}{2}k_z & 0 & \frac{Q}{2}k_x & 0 & 0 & 0 & 0 & 0 & \frac{P}{2}k_y & -\frac{R}{2\sqrt{2}}k_y & -\frac{\sqrt{3}}{2\sqrt{2}}Rk_y & \frac{P''}{2}k_y \\
0 & 0 & 0 & \frac{Q}{2}k_y & \frac{Q}{2}k_x & 0 & 0 & 0 & 0 & 0 & 0 & \frac{P}{2}k_z & -\frac{R}{2\sqrt{2}}k_z & \frac{\sqrt{3}}{2\sqrt{2}}Rk_z & \frac{P''}{2}k_z \\
 & & & E(\Gamma_{15}) & 0 & 0 & 0 & \frac{Q'}{2}k_z & \frac{Q'}{2}k_y & \frac{T}{2}k_x & \frac{T'}{2}k_x & 0 & 0 & 0 & 0 \\
 & & & 0 & E(\Gamma_{15}) & 0 & \frac{Q'}{2}k_z & 0 & \frac{Q'}{2}k_x & \frac{T}{2}k_y & \frac{T'}{2}k_y & 0 & 0 & 0 & 0 \\
 & & & 0 & 0 & E(\Gamma_{15}) & \frac{Q'}{2}k_y & \frac{Q'}{2}k_x & 0 & \frac{T}{2}k_z & \frac{T'}{2}k_z & 0 & 0 & 0 & 0 \\
 & & & & & & E(\Gamma^u_{25'}) & 0 & 0 & 0 & 0 & \frac{P'}{2}k_x & \frac{R'}{\sqrt{2}}k_x & 0 & \frac{P'''}{2}k_x \\
 & & & & & & 0 & E(\Gamma^u_{25'}) & 0 & 0 & 0 & \frac{P'}{2}k_y & -\frac{R'}{2\sqrt{2}}k_y & -\frac{\sqrt{3}}{2\sqrt{2}}R'k_y & \frac{P'''}{2}k_y \\
 & & & & & & 0 & 0 & E(\Gamma^u_{25'}) & 0 & 0 & \frac{P'}{2}k_z & -\frac{1}{2\sqrt{2}}R'k_z & \frac{\sqrt{3}}{2\sqrt{2}}R'k_z & \frac{P'''}{2}k_z \\
 & & & & & & & & & E(\Gamma^u_1) & 0 & 0 & 0 & 0 & 0 \\
 & & & & & & & & & & E(\Gamma^l_1) & 0 & 0 & 0 & 0 \\
 & & & & & & & & & & & E(\Gamma^l_{2'}) & 0 & 0 & 0 \\
 & & & & & & & & & & & & E(\Gamma_{12'}) & 0 & 0 \\
 & & & & & & & & & & & & & E(\Gamma_{12'}) & 0 \\
 & & & & & & & & & & & & & & E(\Gamma^u_{2'})
\end{array}\right] + \frac{\hbar^2 k^2}{2m_0},
$$

where

$$P = 2\mathrm{i}\frac{\hbar}{m_0}\langle \Gamma^l_{25'}|p|\Gamma^l_{2'}\rangle,\ P' = 2\mathrm{i}\frac{\hbar}{m_0}\langle \Gamma^l_{25'}|p|\Gamma^l_{2'}\rangle,$$

$$P'' = 2\mathrm{i}\frac{\hbar}{m_0}\langle \Gamma^l_{25'}|p|\Gamma^u_{2'}\rangle,\ P''' = 2\mathrm{i}\frac{\hbar}{m_0}\langle \Gamma^u_{25'}|p|\Gamma^u_{2'}\rangle,$$

$$Q = 2\mathrm{i}\frac{\hbar}{m_0}\langle \Gamma^l_{25'}|p|\Gamma_{15}\rangle,\ Q' = 2\mathrm{i}\frac{\hbar}{m_0}\langle \Gamma^u_{25'}|p|\Gamma_{15}\rangle,$$

$$R = 2\frac{\hbar}{m_0}\langle (yz)^l|p_x|\gamma_1^-\rangle,\ R' = 2\frac{\hbar}{m_0}\langle (yz)^u|p_x|\gamma_1^-\rangle,$$

$$T = 2\mathrm{i}\frac{\hbar}{m_0}\langle \Gamma^u_1|p|\Gamma_{15}\rangle,\ T' = 2\mathrm{i}\frac{\hbar}{m_0}\langle \Gamma^l_1|p|\Gamma_{15}\rangle.$$

Table C.7 First-order eight-band Kane Hamiltonian in $|J M_J\rangle$ basis for zincblende

$$\begin{pmatrix}
\mathrm{i}|S\uparrow\rangle & |\frac{3}{2}\,\frac{1}{2}\rangle & |\frac{3}{2}\,\frac{3}{2}\rangle & |\frac{1}{2}\,\frac{1}{2}\rangle & \mathrm{i}|S\downarrow\rangle & |\frac{3}{2}\,-\frac{1}{2}\rangle & |\frac{3}{2}\,-\frac{3}{2}\rangle & |\frac{1}{2}\,-\frac{1}{2}\rangle \\
\epsilon(\mathbf{k})+E_0 & -\sqrt{\frac{2}{3}}Pk_z & \frac{P}{\sqrt{2}}k_+ & \sqrt{\frac{1}{3}}Pk_z & 0 & -\sqrt{\frac{1}{6}}Pk_- & 0 & -\sqrt{\frac{1}{3}}Pk_- \\
-\sqrt{\frac{2}{3}}Pk_z & \epsilon(\mathbf{k}) & 0 & 0 & \sqrt{\frac{1}{6}}Pk_- & 0 & 0 & 0 \\
\frac{P}{\sqrt{2}}k_- & 0 & \epsilon(\mathbf{k}) & 0 & 0 & 0 & 0 & 0 \\
\sqrt{\frac{1}{3}}Pk_z & 0 & 0 & \epsilon(\mathbf{k})-\Delta_0 & \sqrt{\frac{1}{3}}Pk_- & 0 & 0 & 0 \\
0 & \sqrt{\frac{1}{6}}Pk_+ & 0 & \sqrt{\frac{1}{3}}Pk_+ & \epsilon(\mathbf{k})+E_0 & -\sqrt{\frac{2}{3}}Pk_z & \frac{P}{\sqrt{2}}k_- & \sqrt{\frac{1}{3}}Pk_z \\
-\sqrt{\frac{1}{6}}Pk_+ & 0 & 0 & 0 & -\sqrt{\frac{2}{3}}Pk_z & \epsilon(\mathbf{k}) & 0 & 0 \\
0 & 0 & 0 & 0 & \frac{P}{\sqrt{2}}k_+ & 0 & \epsilon(\mathbf{k}) & 0 \\
-\sqrt{\frac{1}{3}}Pk_+ & 0 & 0 & 0 & \sqrt{\frac{1}{3}}Pk_z & 0 & 0 & \epsilon(\mathbf{k})-\Delta_0
\end{pmatrix},$$

where

$$\begin{aligned}
\epsilon(\mathbf{k}) &= \frac{\hbar^2 k^2}{2m_0}, \\
k_\pm &\equiv k_x \pm \mathrm{i}k_y, \\
P &= -\mathrm{i}\frac{\hbar}{m_0}\langle Sc|p_x|Xv\rangle.
\end{aligned}$$

Table C.8 Second-order four-band Kane Hamiltonian for zincblende

$$H(\mathbf{k}) = \epsilon(\mathbf{k})1 + \begin{pmatrix} S & X & Y & Z \\ A'k^2 + E_0 & Bk_yk_z + \mathrm{i}k_xP & Bk_xk_z + \mathrm{i}k_yP & Bk_xk_y + \mathrm{i}k_zP \\ & L'k_x^2 + M(k_y^2 + k_z^2 & N'k_xk_y & N'k_xk_z \\ & \dagger & L'k_y^2 + M(k_x^2 + k_z^2) & N'k_yk_z \\ & & & L'k_z^2 + M(k_x^2 + k_y^2) \end{pmatrix},$$

where

$$\epsilon(\mathbf{k}) = \frac{\hbar^2 k^2}{2m_0}, \quad P = -\mathrm{i}\frac{\hbar}{m_0}\langle S|p_x|X\rangle,$$

$$L' = F' + 2G,$$

$$M = H_1 + H_2,$$

$$N' = F' - G + H_1 - H_2,$$

$$F' = \frac{\hbar^2}{m_0^2} \sum_l^{\Gamma_1;\Gamma_2^-}{}' \frac{|\langle X|p_x|u_l\rangle|^2}{E_{\Gamma_v} - E_l},$$

$$G \equiv \frac{\hbar^2}{2m_0^2} \sum_l^{\Gamma_{12};\Gamma_{12}^-}{}' \frac{|\langle X|p_x|u_l\rangle|^2}{E_{\Gamma_v} - E_l},$$

$$H_1 = \frac{\hbar^2}{m_0^2} \sum_l^{\Gamma_{15};\Gamma_{15}^-}{}' \frac{|\langle X|p_y|u_l\rangle|^2}{E_v - E_l},$$

$$H_2 = \frac{\hbar^2}{m_0^2} \sum_l^{\Gamma_{25};\Gamma_{25}^-}{}' \frac{|\langle X|p_y|u_l\rangle|^2}{E_v - E_l},$$

$$A' = \frac{\hbar^2}{m_0^2} \sum_l^{\Gamma_{15};\Gamma_{25}^+}{}' \frac{|\langle S|p_x|u_l\rangle|^2}{E_c - E_l},$$

$$B = \frac{2\hbar^2}{m_0^2} \sum_{l\in\Gamma_{15}}{}' \frac{\langle S|p_x|u_l\rangle\langle u_l|p_y|Z\rangle}{\frac{E_0}{2} - E_l},$$

Table C.9 Basis sets for wurtzite

CC [46]				AO [49]		GJ [42]			LWCC [48]
Γ_5	$\vert u_1\rangle$	$-\frac{1}{\sqrt{2}}\vert (X+\mathrm{i}Y)\uparrow\rangle$	$Y_{1\ 1}\uparrow$	$\vert\widetilde{u}_1\rangle$	$\vert S\uparrow\rangle$	Γ_5	$\vert u_5^*\uparrow\rangle$	$\vert (x-\mathrm{i}y)\uparrow\rangle$	$\vert e\uparrow\rangle = q_s\vert s\uparrow\rangle + q_z\vert z\uparrow\rangle$
Γ_5	$\vert u_2\rangle$	$\frac{1}{\sqrt{2}}\vert (X-\mathrm{i}Y)\uparrow\rangle$	$Y_{1\,-1}\uparrow$	$\vert\widetilde{u}_2\rangle$	$\frac{1}{\sqrt{2}}\vert (X+\mathrm{i}Y)\uparrow\rangle$	Γ_1	$\vert u_1\downarrow\rangle$	$\vert z\downarrow\rangle$	$\vert Z\uparrow\rangle = -q_z\vert s\uparrow\rangle + q_s\vert z\uparrow\rangle$
Γ_1	$\vert u_3\rangle$	$\vert Z\uparrow\rangle$	$Y_{1\ 0}\uparrow$	$\vert\widetilde{u}_3\rangle$	$\frac{1}{\sqrt{2}}\vert (X-\mathrm{i}Y)\uparrow\rangle$	Γ_6	$\vert u_6\uparrow\rangle$	$\vert (x+\mathrm{i}y)^2\uparrow\rangle$	$\vert e\downarrow\rangle = q_s\vert s\downarrow\rangle + q_z\vert z\downarrow\rangle$
Γ_5	$\vert u_4\rangle$	$\frac{1}{\sqrt{2}}\vert (X-\mathrm{i}Y)\downarrow\rangle$	$Y_{1\,-1}\downarrow$	$\vert\widetilde{u}_4\rangle$	$\vert Z\uparrow\rangle$	Γ_3	$\vert u_3\downarrow\rangle$	$\vert x(3y^2-x^2)\downarrow\rangle$	$\vert Z\downarrow\rangle = -q_z\vert s\downarrow\rangle + q_s\vert z\downarrow\rangle$
Γ_5	$\vert u_5\rangle$	$-\frac{1}{\sqrt{2}}\vert (X+\mathrm{i}Y)\downarrow\rangle$	$Y_{1\ 1}\downarrow$			Γ_5	$\vert u_5\uparrow\rangle$	$\vert (x+\mathrm{i}y)\uparrow\rangle$	$\vert 7\uparrow\rangle = \sqrt{1-q_7^2}\vert u_5\downarrow\rangle - q_7\vert u_1\uparrow\rangle$
Γ_1	$\vert u_6\rangle$	$\vert Z\downarrow\rangle$	$Y_{1\ 0}\downarrow$			Γ_6	$\vert u_6\downarrow\rangle$	$\vert (x+\mathrm{i}y)^2\downarrow\rangle$	$\vert 9\uparrow\rangle = \sqrt{1-q_9^2}\vert u_5\uparrow\rangle + q_9\vert u_6\downarrow\rangle$
						Γ_5	$\vert u_5\downarrow\rangle$	$\vert (x+\mathrm{i}y)\downarrow\rangle$	$\vert 7\downarrow\rangle = -\sqrt{1-q_7^2}\vert u_5^*\uparrow\rangle - q_7\vert u_1\downarrow\rangle$
						Γ_1	$\vert -u_1\uparrow\rangle$	$\vert -z\uparrow\rangle$	$\vert 9\downarrow\rangle = \sqrt{1-q_9^2}\vert u_5^*\downarrow\rangle - q_9\vert u_6^*\uparrow\rangle$
						Γ_6	$\vert u_6^*\downarrow\rangle$	$\vert (x-\mathrm{i}y)^2\downarrow\rangle$	$\vert 7'\uparrow\rangle = q_7\vert u_5\downarrow\rangle + \sqrt{1-q_7^2}\vert u_1\uparrow\rangle$
						Γ_3	$\vert -u_3\uparrow\rangle$	$\vert x(x^2-3y^2)\uparrow\rangle$	$\vert 9'\uparrow\rangle = -q_9\vert u_5\uparrow\rangle + \sqrt{1-q_9^2}\vert u_6\downarrow\rangle$
						Γ_5	$\vert u_5^*\downarrow\rangle$	$\vert (x-\mathrm{i}y)\downarrow\rangle$	$\vert 7'\downarrow\rangle = -q_7\vert u_5^*\uparrow\rangle + \sqrt{1-q_7^2}\vert u_1\downarrow\rangle$
						Γ_6	$\vert -u_6^*\uparrow\rangle$	$\vert -(x-\mathrm{i}y)^2\uparrow\rangle$	$\vert 9'\downarrow\rangle = -q_9\vert u_5^*\downarrow\rangle - \sqrt{1-q_9^2}\vert u_6^*\uparrow\rangle$

SJKLS [43]

$$|\tfrac{3}{2}\,\tfrac{3}{2}\rangle = |1\,1\rangle|\uparrow\rangle,$$
$$|\tfrac{3}{2}\,\tfrac{1}{2}\rangle = \tfrac{1}{\sqrt{3}}|1\,1\rangle|\downarrow\rangle + \sqrt{\tfrac{2}{3}}|1\,0\rangle|\uparrow\rangle,$$
$$|\tfrac{3}{2}\,-\tfrac{1}{2}\rangle = \sqrt{\tfrac{2}{3}}|1\,0\rangle|\downarrow\rangle + \tfrac{1}{\sqrt{3}}|1\,-1\rangle|\uparrow\rangle,$$
$$|\tfrac{3}{2}\,-\tfrac{3}{2}\rangle = |1\,-1\rangle|\downarrow\rangle,$$
$$|\tfrac{1}{2}\,\tfrac{1}{2}\rangle = \sqrt{\tfrac{2}{3}}|1\,1\rangle|\downarrow\rangle - \tfrac{1}{\sqrt{3}}|1\,0\rangle|\uparrow\rangle,$$
$$|\tfrac{1}{2}\,-\tfrac{1}{2}\rangle = \tfrac{1}{\sqrt{3}}|1\,0\rangle|\downarrow\rangle - \sqrt{\tfrac{2}{3}}|1\,-1\rangle|\uparrow\rangle.$$

Table C.10 Six-band Rashba–Sheka–Pikus Hamiltonian in LS basis for wurtzite

$$H_{\mathrm{RSP}}(\mathbf{k}) = \begin{pmatrix} |Y_{1\,1}\uparrow\rangle & |Y_{1\,1}\downarrow\rangle & |Y_{1\,0}\uparrow\rangle & |Y_{1\,0}\downarrow\rangle & |Y_{1\,-1}\uparrow\rangle & |Y_{1\,-1}\downarrow\rangle \\ F & 0 & -H^* & 0 & k^* & 0 \\ 0 & G & \Delta & -H^* & 0 & k^* \\ -H & \Delta & \lambda & 0 & I^* & 0 \\ 0 & -H & 0 & \lambda & \Delta & I^* \\ k & 0 & I & \Delta & G & 0 \\ 0 & k & 0 & I & 0 & F \end{pmatrix},$$

where

$$\begin{aligned} \Delta &= \sqrt{2}\Delta_3, \\ F &= \Delta_1 + \Delta_2 + \lambda + \theta, \\ G &= \Delta_1 - \Delta_2 + \lambda + \theta, \\ k &= A_5 k_+^2 + D_5 \varepsilon_+, \\ H &= \mathrm{i}\,(A_6 k_+ k_z + D_6 \varepsilon_{z+} + A_7 k_+), \\ I &= \mathrm{i}\,(A_6 k_+ k_z + D_6 \varepsilon_{z+} - A_7 k_+), \\ \lambda &= A_1 k_z^2 + A_2 k_\perp^2 + D_1 \varepsilon_{zz} + D_2 \varepsilon_\perp, \\ \theta &= A_3 k_z^2 + A_4 k_\perp^2 + D_3 \varepsilon_{zz} + D_4 \varepsilon_\perp. \end{aligned}$$

Parameters:

$$\Delta_1, \Delta_2, \Delta_3, A_1, A_2, A_3, A_4, A_5, A_6, A_7,$$
$$D_1, D_2, D_3, D_4, D_5, D_6.$$

Table C.11 Six-band Hamiltonian in LS basis for wurtzite according to Sirenko and coworkers [43]

$$H_{SJKLS}(\mathbf{k}) = -\begin{pmatrix} |Y_{1\,1}\uparrow\rangle & |Y_{1\,1}\downarrow\rangle & |Y_{1\,0}\uparrow\rangle & |Y_{1\,0}\downarrow\rangle & |Y_{1\,-1}\uparrow\rangle & |Y_{1\,-1}\downarrow\rangle \\ P+Q-\Delta_2 & 0 & T+S & 0 & R & 0 \\ 0 & P+Q+\Delta_2 & -\sqrt{2}\Delta_3 & T+S & 0 & R \\ T^*+S^* & -\sqrt{2}\Delta_3 & P & 0 & T-S & 0 \\ 0 & T^*+S^* & 0 & P & -\sqrt{2}\Delta_3 & T-S \\ R^* & 0 & T^*-S^* & -\sqrt{2}\Delta_3 & P+Q+\Delta_2 & 0 \\ 0 & R^* & 0 & T^*-S^* & 0 & P+Q-\Delta_2 \end{pmatrix},$$

where

$$\begin{aligned} P &= \Delta_1 + \Delta_2 + B_1 k_z^2 + B_3 k_\perp^2 + C_1 \varepsilon_{zz} + C_3 \varepsilon_\perp, \\ Q &= -\Delta_1 + B_2 k_z^2 + B_4 k_\perp^2 + C_2 \varepsilon_{zz} + C_4 \varepsilon_\perp, \\ R &= B_5 k_-^2 + C_5 \varepsilon_-, \\ S &= B_6 k_z k_- + C_6 \varepsilon_{-z}, \\ T &= \mathrm{i}\frac{\hbar^2 \mathcal{K}}{2m_0} k_-. \end{aligned}$$

Parameters:

$$\Delta_1, \Delta_2, \Delta_3, B_1, B_2, B_3, B_4, B_5, B_6, \mathcal{K},$$
$$C_1, C_2, C_3, C_4, C_5, C_6.$$

Table C.12 Six-band Chuang–Chang Hamiltonian in u basis for wurtzite

$$H_{\mathrm{CC}}(\mathbf{k}) = \begin{pmatrix} |u_1\rangle & |u_2\rangle & |u_3\rangle & |u_4\rangle & |u_5\rangle & |u_6\rangle \\ F & -K^* & -H^* & 0 & 0 & 0 \\ -K & G & H & 0 & 0 & \Delta \\ -H & H^* & \lambda & 0 & \Delta & 0 \\ 0 & 0 & 0 & F & -K & H \\ 0 & 0 & \Delta & -K^* & G & -H^* \\ 0 & \Delta & 0 & H^* & -H & \lambda \end{pmatrix},$$

where

$$F = \Delta_1 + \Delta_2 + \lambda + \theta,$$

$$G = \Delta_1 - \Delta_2 + \lambda + \theta,$$

$$\lambda = L_2 k_z^2 + M_3(k_x^2 + k_y^2),$$

$$\theta = (M_2 - L_2)k_z^2 + \left(\frac{L_1 + M_1}{2} - M_3\right)\left(k_x^2 + k_y^2\right),$$

$$K = \frac{N_1}{2}\left(k_x + \mathrm{i}k_y\right)^2,$$

$$H = \frac{N_2}{\sqrt{2}}(k_x + \mathrm{i}k_y)k_z,$$

$$\Delta = \sqrt{2}\Delta_3.$$

Parameters:

$$\Delta_1, \Delta_2, \Delta_3,$$
$$L_1, L_2, M_1, M_2, N_1, N_2.$$

Table C.13 Four-band Andreev–O'Reilly Hamiltonian for wurtzite

$$H_{\mathrm{AO}}(\mathbf{k}) = \begin{pmatrix} |\widetilde{u}_1\rangle & |\widetilde{u}_2\rangle & |\widetilde{u}_3\rangle & |\widetilde{u}_4\rangle \\ E_0 & \frac{1}{\sqrt{2}}P_\perp k_+ & \frac{1}{\sqrt{2}}P_\perp k_- & P_{||}k_z \\ \frac{1}{\sqrt{2}}P_\perp k_- & F & K^* & -H^* \\ \frac{1}{\sqrt{2}}P_\perp k_+ & K & F & H \\ P_{||}k_z & -H & H^* & \lambda \end{pmatrix},$$

where

$$\begin{aligned} F &= \Delta_1 + \lambda + \theta, \\ K &= \widetilde{A}_5 \left(k_x + \mathrm{i}k_y\right)^2, \\ H &= \mathrm{i}\widetilde{A}_6 \left(k_x + \mathrm{i}k_y\right) k_z, \\ \theta &= \widetilde{A}_3 k_z^2 + \widetilde{A}_4 \left(k_x^2 + k_y^2\right), \\ \lambda &= -\Delta_1 + \widetilde{A}_1 k_z^2 + \widetilde{A}_2 \left(k_x^2 + k_y^2\right), \end{aligned}$$

and

$$\begin{aligned} \widetilde{A}_1 &= A_1 + \frac{2m_0}{\hbar^2}\frac{P_\perp^2}{E_0}, \\ \widetilde{A}_2 &= A_2, \\ \widetilde{A}_3 &= A_3 - \frac{2m_0}{\hbar^2}\frac{P_\perp^2}{E_0}, \\ \widetilde{A}_4 &= A_4 + \frac{m_0}{\hbar^2}\frac{P_{||}^2}{E_0}, \\ \widetilde{A}_5 &= A_5 + \frac{m_0}{\hbar^2}\frac{P_{||}^2}{E_0}, \\ \widetilde{A}_6 &= A_6 + \frac{\sqrt{2}m_0}{\hbar^2}\frac{P_{||}P_\perp}{E_0}, \\ P_{||}^2 &= \frac{\hbar^2}{2m_0}\left(\frac{m_0}{m_c^{||}} - 1\right)\frac{(E_0+\Delta_1+\Delta_2)(E_0+2\Delta_2) - 2\Delta_3^2}{E_0 + 2\Delta_2}, \\ P_\perp^2 &= \frac{\hbar^2}{2m_0}\left(\frac{m_0}{m_c^{\perp}} - 1\right)\frac{\left[(E_0+\Delta_1+\Delta_2)(E_0+2\Delta_2) - 2\Delta_3^2\right]E_0}{(E_0+\Delta_2)(E_0+\Delta_1+\Delta_2) - \Delta_3^2}. \end{aligned}$$

Parameters:

$$\begin{aligned} &\Delta_1, \Delta_2, \Delta_3, E_0, \\ &A_1, A_2, A_3, A_4, A_5, A_6, A_7, \\ &m_c^{||}, m_c^{\perp}. \end{aligned}$$

Table C.14 Six-band Burt–Foreman Hamiltonian in $|JM_J\rangle$ basis for arbitrary quantization for zincblende

$$H_{\mathrm{BF1}} = \begin{pmatrix} |\frac{3}{2}\,\frac{3}{2}\rangle & |\frac{3}{2}\,\frac{1}{2}\rangle & |\frac{3}{2}\,-\frac{1}{2}\rangle & |\frac{3}{2}\,-\frac{3}{2}\rangle & |\frac{1}{2}\,\frac{1}{2}\rangle & |\frac{1}{2}\,-\frac{1}{2}\rangle \\ P' & S_- & -R & 0 & -\frac{1}{\sqrt{2}}S_- & -\sqrt{2}R \\ S_-^\dagger & P'' & -C & R & -\sqrt{2}Q & -\sqrt{\frac{3}{2}}\Sigma_- \\ -R^\dagger & -C^\dagger & P''^* & S_+^\dagger & \sqrt{\frac{3}{2}}\Sigma_+ & -\sqrt{2}Q^* \\ 0 & R^\dagger & S_+ & P'^* & \sqrt{2}R^\dagger & -\frac{1}{\sqrt{2}}S_+ \\ -\frac{1}{\sqrt{2}}S_-^\dagger & -\sqrt{2}Q^\dagger & \sqrt{\frac{3}{2}}\Sigma_+^\dagger & -\sqrt{2}R & P''' - \Delta_0 & C \\ -\sqrt{2}R^\dagger & -\sqrt{\frac{3}{2}}\Sigma_-^\dagger & -\sqrt{2}Q & -\frac{1}{\sqrt{2}}S_+^\dagger & C^\dagger & P'''^* - \Delta_0 \end{pmatrix}$$

with

$$P' = \frac{1}{2}\left\{\widehat{k}_x(L+M)\widehat{k}_x + \widehat{k}_y(L+M)\widehat{k}_y + \widehat{k}_z 2M\widehat{k}_z\right\}$$

$$+\frac{\mathrm{i}}{2}\left\{\widehat{k}_x(F-G-H_1+H_2)\widehat{k}_y - \widehat{k}_y(F-G-H_1+H_2)\widehat{k}_x\right\},$$

$$P'' = \frac{1}{6}\left\{\widehat{k}_x(L+5M)\widehat{k}_x + \widehat{k}_y(L+5M)\widehat{k}_y + 2\widehat{k}_z(2L+M)\widehat{k}_z\right\}$$

$$+\frac{\mathrm{i}}{6}\left\{\widehat{k}_x(F-G-H_1+H_2)\widehat{k}_y - \widehat{k}_y(F-G-H_1+H_2)\widehat{k}_x\right\},$$

$$P''' = \frac{1}{3}\left\{\widehat{k}_x(L+2M)\widehat{k}_x + \widehat{k}_y(L+2M)\widehat{k}_y + \widehat{k}_z(L+2M)\widehat{k}_z\right\}$$

$$+\frac{\mathrm{i}}{3}\left\{\widehat{k}_x(F-G-H_1+H_2)\widehat{k}_y - \widehat{k}_y(F-G-H_1+H_2)\widehat{k}_x\right\},$$

$$Q = -\frac{1}{6}\left\{\widehat{k}_x(L-M)\widehat{k}_x + \widehat{k}_y(L-M)\widehat{k}_y + 2\widehat{k}_z(L-M)\widehat{k}_z\right.$$

$$\left. + \mathrm{i}[\widehat{k}_x(F-G-H_1+H_2)\widehat{k}_y - \widehat{k}_y(F-G-H_1+H_2)\widehat{k}_x]\right\},$$

$$R = \frac{1}{2\sqrt{3}}\left\{\widehat{k}_x(L-M)\widehat{k}_x - \widehat{k}_y(L-M)\widehat{k}_y - \mathrm{i}[\widehat{k}_x N\widehat{k}_y + \widehat{k}_y N\widehat{k}_x]\right\},$$

$$S_\pm = -\frac{1}{\sqrt{3}}\left\{\widehat{k}_\pm(F-G)\widehat{k}_z + \widehat{k}_z(H_1-H_2)\widehat{k}_\pm\right\},$$

$$C = -\frac{1}{3}\left\{\widehat{k}_z(F-G-H_1+H_2)\widehat{k}_- - \widehat{k}_-(F-G-H_1+H_2)\widehat{k}_z\right\},$$

$$\Sigma_- = -\frac{1}{3\sqrt{3}}\left\{\widehat{k}_-(F-G+2H_1-2H_2)\widehat{k}_z + \widehat{k}_z(2F-2G+H_1-H_2)\widehat{k}_-\right\}.$$

Table C.15 Six-band Burt–Foreman Hamiltonian in $|JM_J\rangle$ basis for [001] quantization for zincblende

$$
H_{\mathrm{BF2}} = -\begin{pmatrix}
|\frac{3}{2}\,\frac{3}{2}\rangle & |\frac{3}{2}\,\frac{1}{2}\rangle & |\frac{3}{2}\,-\frac{1}{2}\rangle & |\frac{3}{2}\,-\frac{3}{2}\rangle & |\frac{1}{2}\,\frac{1}{2}\rangle & |\frac{1}{2}\,-\frac{1}{2}\rangle \\
P+Q & -S_- & R & 0 & \frac{1}{\sqrt{2}}S_- & \sqrt{2}R \\
-S_-^\dagger & P-Q & C & -R & \sqrt{2}Q & \sqrt{\frac{3}{2}}\Sigma_- \\
R^\dagger & C^\dagger & P-Q & -S_+^\dagger & -\sqrt{\frac{3}{2}}\Sigma_+ & \sqrt{2}Q^* \\
0 & -R^\dagger & -S_+ & P+Q & -\sqrt{2}R^\dagger & \frac{1}{\sqrt{2}}S_+ \\
\frac{1}{\sqrt{2}}S_-^\dagger & \sqrt{2}Q & -\sqrt{\frac{3}{2}}\Sigma_+^\dagger & -\sqrt{2}R & P+\Delta_0 & -C \\
\sqrt{2}R^\dagger & \sqrt{\frac{3}{2}}\Sigma_-^\dagger & \sqrt{2}Q & \frac{1}{\sqrt{2}}S_+^\dagger & -C^\dagger & P+\Delta_0
\end{pmatrix}
$$

with

$$
\begin{aligned}
P &= \frac{\hbar^2}{2m_0}\left(\gamma_1 k_{\parallel}^2 + \widehat{k}_z \gamma_1 \widehat{k}_z\right), \\
Q &= \frac{\hbar^2}{2m_0}\left(\gamma_2 k_{\parallel}^2 - 2\widehat{k}_z \gamma_2 \widehat{k}_z\right), \\
R &= -\frac{\hbar^2}{2m_0}\sqrt{3}\left(\overline{\gamma} k_-^2 - \mu k_+^2\right), \\
S_\pm &= \frac{\hbar^2}{m_0}\sqrt{3}k_\pm\left[(\sigma-\delta)\widehat{k}_z + \widehat{k}_z \pi\right], \\
C &= \frac{\hbar^2}{m_0}k_-\left[\widehat{k}_z(\sigma-\delta-\pi) - (\sigma-\delta-\pi)\widehat{k}_z\right], \\
\Sigma_\pm &= \frac{\hbar^2}{m_0}\sqrt{3}k_\pm\left\{[\frac{1}{3}(\sigma-\delta) + \frac{2}{3}\pi]\widehat{k}_z + \widehat{k}_z[(\frac{2}{3}(\sigma-\delta) + \frac{1}{3}\pi]\right\}.
\end{aligned}
$$

Table C.16 Four-band Sercel–Vahala Hamiltonian within axial approximation for zincblende

$$
H_{SV}(F_z) = -\begin{pmatrix}
|\frac{3}{2}\,\frac{3}{2}\rangle|F_z-\frac{3}{2}\rangle & |\frac{3}{2}\,\frac{1}{2}\rangle|F_z-\frac{1}{2}\rangle & |\frac{3}{2}\,-\frac{1}{2}\rangle|F_z+\frac{1}{2}\rangle & |\frac{3}{2}\,-\frac{3}{2}\rangle|F_z+\frac{3}{2}\rangle \\
P' & -S & R & 0 \\
-S^* & P'' & 0 & -R \\
R^* & 0 & P'' & -S \\
0 & -R^* & -S^* & P'
\end{pmatrix}
$$

with

$$
\begin{aligned}
P' &= \frac{\hbar^2}{2m_0}\left\{(\gamma_1+\gamma_2)k_{\parallel}^2 + (\gamma_1 - 2\gamma_2)k_z^2\right\}, \\
P'' &= \frac{\hbar^2}{2m_0}\left\{(\gamma_1-\gamma_2)k_{\parallel}^2 + (\gamma_1 + 2\gamma_2)k_z^2\right\}, \\
S &= \frac{\hbar^2}{2m_0}2\sqrt{3}\widetilde{\gamma}k_{\parallel}k_z, \\
R &= -\frac{\hbar^2}{2m_0}\sqrt{3}\widetilde{\gamma}k_{\parallel}^2.
\end{aligned}
$$

Table C.17 Mireles–Ulloa Hamiltonian in L basis for wurtzite heterostructures

$$H = \begin{pmatrix} |X\rangle & |Y\rangle & |Z\rangle \\ \widehat{k}_x L_1 \widehat{k}_x + \widehat{k}_y M_1 \widehat{k}_y + \widehat{k}_z M_2 \widehat{k}_z & \widehat{k}_x N_1 \widehat{k}_y + \widehat{k}_y N_1' \widehat{k}_x & \widehat{k}_x N_2 \widehat{k}_z + \widehat{k}_z N_2' \widehat{k}_x \\ \widehat{k}_y N_1 \widehat{k}_x + \widehat{k}_x N_1' \widehat{k}_y & \widehat{k}_x M_1 \widehat{k}_x + \widehat{k}_y L_1 \widehat{k}_y + \widehat{k}_z M_2 \widehat{k}_z & \widehat{k}_y N_2 \widehat{k}_z + \widehat{k}_z N_2' \widehat{k}_y \\ \widehat{k}_z N_2 \widehat{k}_x + \widehat{k}_x N_2' \widehat{k}_z & \widehat{k}_z N_2 \widehat{k}_y + \widehat{k}_y N_2' \widehat{k}_z & \widehat{k}_x M_3 \widehat{k}_x + \widehat{k}_y M_3 \widehat{k}_y + \widehat{k}_z L_2 \widehat{k}_z \end{pmatrix}.$$

Table C.18 Mireles–Ulloa Hamiltonian in LS basis for wurtzite heterostructures

$$H = \begin{pmatrix} |u_1\rangle & |u_2\rangle & |u_3\rangle & |u_4\rangle & |u_5\rangle & |u_6\rangle \\ \widetilde{D}_{11} + \Delta_1 + \Delta_2 & \widetilde{D}_{12} & \widetilde{D}_{13} & 0 & 0 & 0 \\ \widetilde{D}_{12}^{\dagger} & \widetilde{D}_{22} + \Delta_1 - \Delta_2 & \widetilde{D}_{23} & 0 & 0 & \Delta \\ \widetilde{D}_{13}^{\dagger} & \widetilde{D}_{23}^{\dagger} & \widetilde{D}_{33} & 0 & \Delta & 0 \\ 0 & 0 & 0 & \widetilde{D}_{22} + \Delta_1 + \Delta_2 & \widetilde{D}_{12}^{\dagger} & \widetilde{D}_{23} \\ 0 & 0 & \Delta & \widetilde{D}_{12} & \widetilde{D}_{11} + \Delta_1 - \Delta_2 & \widetilde{D}_{13} \\ 0 & \Delta & 0 & \widetilde{D}_{23}^{\dagger} & \widetilde{D}_{13}^{\dagger} & \widetilde{D}_{33} \end{pmatrix},$$

where

$$\widetilde{D}_{11} = \frac{1}{2}\left\{\widehat{k}_x(L_1 + M_1)\widehat{k}_x + \widehat{k}_y(L_1 + M_1)\widehat{k}_y + 2\widehat{k}_z M_2 \widehat{k}_z + \mathrm{i}\left[\widehat{k}_x(N_1 - N_1')\widehat{k}_y - \widehat{k}_y(N_1 - N_1')\widehat{k}_x\right]\right\},$$

$$\widetilde{D}_{22} = \frac{1}{2}\left\{\widehat{k}_x(L_1 + M_1)\widehat{k}_x + \widehat{k}_y(L_1 + M_1)\widehat{k}_y + 2\widehat{k}_z M_2 \widehat{k}_z - \mathrm{i}\left[\widehat{k}_x(N_1 - N_1')\widehat{k}_y - \widehat{k}_y(N_1 - N_1')\widehat{k}_x\right]\right\},$$

$$\widetilde{D}_{33} = D_{33} = \widehat{k}_x M_3 \widehat{k}_x + \widehat{k}_y M_3 \widehat{k}_y + \widehat{k}_z L_2 \widehat{k}_z,$$

$$\widetilde{D}_{12} = -\frac{1}{2}\left\{\widehat{k}_x(L_1 - M_1)\widehat{k}_x - \widehat{k}_y(L_1 - M_1)\widehat{k}_y - \mathrm{i}\left[\widehat{k}_x(N_1 + N_1')\widehat{k}_y + \widehat{k}_y(N_1 + N_1')\widehat{k}_x\right]\right\},$$

$$\widetilde{D}_{13} = -\frac{1}{\sqrt{2}}\left\{\widehat{k}_- N_2 \widehat{k}_z + \widehat{k}_z N_2' \widehat{k}_-\right\},$$

$$\widetilde{D}_{23} = \frac{1}{\sqrt{2}}\left\{\widehat{k}_+ N_2 \widehat{k}_z + \widehat{k}_z N_2' \widehat{k}_+\right\},$$

$$\widetilde{D}_{21} = -\frac{1}{2}\widehat{k}_+(L_1 - M_1)\widehat{k}_+,$$

$$\widetilde{D}_{31} = -\frac{1}{\sqrt{2}}\left\{\widehat{k}_z N_2 \widehat{k}_+ + \widehat{k}_+ N_2' \widehat{k}_z\right\},$$

$$\widetilde{D}_{32} = \frac{1}{\sqrt{2}}\left\{\widehat{k}_z N_2 \widehat{k}_- + \widehat{k}_- N_2' \widehat{k}_z\right\}.$$

Table C.19 Comparison of valence $k \cdot p$ parameters

		DKK	Luttinger	Foreman
DKK	L	$F + 2G$		
	M	$H_1 + H_2$		
	N	$F - G + H_1 - H_2$		
Luttinger	γ_1	$-\frac{2m_0}{3\hbar^2}(L + 2M) - 1$		
	γ_2	$-\frac{2m_0}{6\hbar^2}(L - M)$		
	γ_3	$-\frac{2m_0}{6\hbar^2}N$		
Foreman	σ	$-\frac{2m_0}{6\hbar^2}F$	$-\frac{1}{18}(\gamma_1 - 8\gamma_2 - 12\gamma_3 + 1)$	
	π	$-\frac{2m_0}{6\hbar^2}H_1$	$\frac{1}{6}(\gamma_1 - 2\gamma_2 + 1)$	
	δ	$-\frac{2m_0}{6\hbar^2}G$	$\frac{1}{9}(\gamma_1 + \gamma_2 - 3\gamma_3 + 1)$	
Stravinou -van Dalen	A	$\frac{2m_0}{\hbar^2}(F + 2G) + 1$	$-(\gamma_1 + 4\gamma_2)$	$1 - 6\sigma - 12\delta$
	B	$\frac{2m_0}{\hbar^2}(H_1 + H_2) + 1$	$-(\gamma_1 - 2\gamma_2)$	$1 - 6\pi$
	C_1	$\frac{2m_0}{\hbar^2}(F - G)$	$\gamma_1 - 2\gamma_2 - 6\gamma_3 + 1$	$6\delta - 6\sigma$
	C_2	$-\frac{2m_0}{\hbar^2}(H_1 - H_2)$	$\gamma_1 - 2\gamma_2 + 1$	6π

Table C.20 Selection of band parameters for cubic semiconductors [85]

	γ_1	γ_2	γ_3	κ	q	E_p	m_c	g_c
Si	4.22	0.39	1.44	−0.26	0.01	21.6	0.23	1.96
Ge	13.35	4.25	5.69	3.41	0.07	26.3	0.038	−2.86
AlAs	4.04	0.78	1.57	0.12	0.03	21.1	0.22	1.52
GaAs	7.65	2.41	3.28	1.72	0.04	25.7	0.067	−0.06
CdTe	5.29	1.89	2.46	1.27	0.05	20.7	0.096	−1.12
HgTe	−18.68	−10.19	−9.56	−10.85	0.06	18.0	−0.031	59.0

Table C.21 Parameters for wurtzite

	InN	GaN		AlN	ZnO		MgO
	[51]	[51]	[57]	[57]	[53]	[130]	[130]
a (Å)			3.189	3.112	3.250	3.2505	3.199
c (Å)			5.185	4.982	5.210		
E_0 (eV)	0.7		3.475	6.23	3.34	3.2	5.289
Δ_{cr} (eV)			0.019	−0.164	0.0391	0.0305	
Δ_{so} (eV)			0.014	0.019	−0.0035	0.0126	
m_c						0.24	
$m_c^{\perp}$	0.071	0.212					
$m_c^{\parallel}$	0.067	0.190					
A_1	−15.230	−5.798	−6.56	−3.95	−6.68036	−3.78	
A_2	−0.520	−0.545	−0.91	−0.27	−0.45388	−0.44	
A_3	14.673	5.259	5.65	3.68	6.12750		
A_4	−7.012	−2.473	−2.83	−1.84	−2.70374		
A_5	−6.948	−2.491	−3.13	−1.95	−4.62566	−3.13	
A_6	−3.143	−9.794	−4.86	−2.91	−4.62566		
A_7 (eV Å)	0.174	0.049					
$E_p^{\perp}$ (eV)	8.89	16.22					
$E_p^{\parallel}$ (eV)	8.97	17.39					

References

1. G.L. Bir, G.E. Pikus, *Symmetry and Strain-Induced Effects in Semiconductors* (Wiley, New York, 1975)
2. G. Dresselhaus, A.F. Kip, C. Kittel, Phys. Rev. **98**(2), 368 (1955)
3. E.O. Kane, J. Phys. Chem. Sol. **1**, 82 (1956)
4. J.M. Luttinger, Phys. Rev. **102**(4), 1030 (1955)
5. M. Cardona, F.H. Pollak, Phys. Rev. **142**(2), 530 (1966)
6. J.M. Luttinger, W. Kohn, Phys. Rev. **97**(4), 869 (1955)
7. G. Bastard, *Wave Mechanics Applied to Semiconductor Heterostructures* (Les Editions de Physique, Les Ulis, 1988)
8. S. Groves, W. Paul, Phys. Rev. Lett. **11**(5), 194 (1963). DOI 10.1103/PhysRevLett.11.194.
9. S.H. Groves, R.N. Brown, C.R. Pidgeon, Phys. Rev. **161**(3), 779 (1967). DOI 10.1103/PhysRev.161.779.
10. C. Kittel, *Quantum Theory of Solids* (Wiley, New York, 1963)
11. J. Callaway, *Energy Band Theory, Pure and Applied Physics*, vol. 16 (Academic Press, New York, 1964)
12. H.J. Zeiger, G.W. Pratt, *Magnetic Interactions in Solids* (Clarendon Press, Oxford, 1973)
13. I.M. Tsidilkovski, *Band Structure of Semiconductors, International Series in the Science of the Solid State*, vol. 19 (Pergamon, Oxford, 1982)
14. C. Weisbuch, B. Vinter, *Quantum Semiconductor Structures* (Academic Press, San Diego, 1991)
15. S.L. Chuang, *Physics of Optoelectronic Devices* (Wiley, New York, 1995)
16. P. Yu, M. Cardona, *Fundamentals of Semiconductors* (Springer, Berlin, 1995)
17. B.K. Ridley, *Electrons and Phonons in Semiconductor Multilayers* (Cambridge University Press, Cambridge, 1997)
18. V.V. Mitin, V.A. Kochelap, M.A. Stroscio, *Quantum Heterostructures* (Cambridge University Press, Cambridge, 1999)
19. J. Singh, *Electronic and Optoelectronic Properties of Semiconductor Structures* (Cambridge University Press, Cambridge, 2003)
20. E.L. Ivchenko, G.E. Pikus, *Superlattices and Other Heterostructures* (Springer, Berlin, 1995)
21. R. Winkler, *Spin-Orbit Coupling in Two-Dimensional Electron and Hole Systems* (Springer, Berlin, 2003)
22. F.T. Vasko, A.V. Kuznetsov, *Electronic States and Optical Transitions in Semiconductor Heterostructures* (Springer, New York, 1999)
23. U. Rössler, Solid State Commun. **49**(10), 943 (1984)
24. G. Fishman, *Énergie et Fonction D'Onde des Semiconducteurs* (Les Éditions de Physique, Paris, 1988)
25. L.M. Roth, B. Lax, S. Zwerdling, Phys. Rev. **114**(1), 90 (1959). DOI 10.1103/PhysRev.114.90.

26. D. Helmholz, L.C. Lew Yan Voon, Phys. Rev. B **65**, 3204 (2002)
27. F.C. von der Lage, H.A. Bethe, Phys. Rev. **71**(9), 612 (1947)
28. F. Herman, Phys. Rev. **93**(6), 1214 (1954)
29. M. Willatzen, M. Cardona, N.E. Christensen, Phys. Rev. B **50**(24), 18054 (1994)
30. A. Baldereschi, N. Lipari, Phys. Rev. B **8**(6), 2697 (1973)
31. E.O. Kane, in *Physics of III-V Compounds*, *Semiconductors and Semimetals*, vol. 1, ed. by R.K. Willardson, A.C. Beer (Academic, New York, 1966), Chap. 3, pp. 75–100
32. G. Dresselhaus, Phys. Rev. **100**(2), 580 (1955)
33. G.F. Koster, J.O. Dimmock, R.G. Wheeler, H. Statz, *Properties of the Thirty-Two Point Groups* (MIT, Cambridge, 1963)
34. B.A. Foreman, Phys. Rev. B **48**(7), 4964 (1993)
35. R.J. Elliott, Phys. Rev. **96**(2), 266 (1954)
36. P.C. Sercel, K.J. Vahala, Phys. Rev. B **42**(6), 3690 (1990)
37. G. Dresselhaus, Phys. Rev. **105**(1), 135 (1957)
38. J.L. Birman, Phys. Rev. Lett. **2**(4), 157 (1959)
39. R.C. Casella, Phys. Rev. **114**(6), 1514 (1959)
40. R.C. Casella, Phys. Rev. Lett. **5**(8), 371 (1960)
41. G.E. Pikus, Sov. Phys. JETP **14**(4), 898 (1962)
42. E. Gutsche, F. Jahne, Phys. Stat. Sol. **19**, 823 (1967)
43. Y.M. Sirenko, J. Jeon, K.W. Kim, M.A. Littlejohn, Phys. Rev. B **53**(4), 1997 (1996)
44. Y.M. Sirenko, J. Jeon, K.W. Kim, M.A. Littlejohn, M.A. Stroscio, Appl. Phys. Lett. **69**(17), 2504 (1996)
45. J. Jeon, Y.M. Sirenko, K.W. Kim, M.A. Littlejohn, M.A. Stroscio, Solid State Commun. **99**(6), 423 (1996)
46. S.L. Chuang, C.S. Chang, Phys. Rev. B **54**(4), 2491 (1996)
47. M. Kumagai, S.L. Chuang, H. Ando, Phys. Rev. B **57**(24), 15303 (1998)
48. L.C. Lew Yan Voon, M. Willatzen, M. Cardona, N.E. Christensen, Phys. Rev. B **52**(3), 10 703 (1996)
49. A.D. Andreev, E.P. O'Reilly, Phys. Rev. B **62**(23), 15851 (2000)
50. M. Winkelnkemper, A. Schliwa, D. Bimberg, Phys. Rev. B **74**(15), 155322 (2006)
51. P. Rinke, M. Scheffler, A. Qteish, M. Winkelnkemper, D. Bimberg, J. Neugebauer, Appl. Phys. Lett. **89**(16), 161919 (2006)
52. D.J. Dugdale, S. Brand, R.A. Abram, Phys. Rev. B **61**(19), 12933 (2000)
53. W.J. Fan, J.B. Xia, P.A. Agus, S.T. Tan, S.F. Yu, X.W. Sun, J. Appl. Phys.**99**(1), 013702 (2006)
54. M. Suzuki, T. Uenoyama, A. Yanase, Phys. Rev. B **52**(11), 8132 (1995)
55. F. Mireles, S.E. Ulloa, Phys. Rev. B **60**(19), 13659 (1999)
56. F. Mireles, S.E. Ulloa, Phys. Rev. B **62**(4), 2562 (2000)
57. V.A. Fonoberov, A.A. Balandin, J. Appl. Phys. **94**(11), 7178 (2003)
58. L.C. Lew Yan Voon, C. Galeriu, B. Lassen, M. Willatzen, R. Melnik, Appl. Phys. Lett. **87**, 041906 (2005)
59. J. Wu, W. Walukiewicz, K.M. Yu, J.W.A. III, E.E. Haller, H. Lu, W.J. Schaff, Y. Saito, Y. Nanishi, Appl. Phys. Lett. **80**(21), 3967 (2002). DOI 10.1063/1.1482786. URL http://link.aip.org/link/?APL/80/3967/1
60. P. Rinke, M. Winkelnkemper, A. Qteish, D. Bimberg, J. Neugebauer, M. Scheffler, Phys. Rev. B: Condens. Matter Mater. Phys. **77**(7), 075202 (2008). DOI 10.1103/PhysRevB.77.075202. http://link.aps.org/abstract/PRB/v77/e075202
61. G.E. Pikus, Sov. Phys. JETP **14**(5), 1075 (1962)
62. K. Kim, W.R.L. Lambrecht, B. Segall, M. van Schilfgaarde, Phys. Rev. B **56**(12), 7363 (1997). DOI 10.1103/PhysRevB.56.7363.
63. J.P. Loehr, Phys. Rev. B **52**(4), 2374 (1995)
64. K. Boujdaria, S. Ridene, G. Fishman, Phys. Rev. B **63**(23), 235302 (2001)
65. F.H. Pollak, C.W. Higginbotham, M. Cardona, J. Phys. Soc. Jpn. **21**(Suppl), 20 (1966)
66. W. Brinkman, B. Goodman, Phys. Rev. **149**(2), 597 (1966). DOI 10.1103/PhysRev.149.597.

67. F.H. Pollak, M. Cardona, C.W. Higginbotham, F. Herman, J.P. Van Dyke, Phys. Rev. B **2**(2), 352 (1970). DOI 10.1103/PhysRevB.2.352.
68. J. Stanley, N. Goldsman, Phys. Rev. B **51**(8), 4931 (1995). DOI 10.1103/PhysRevB.51.4931.
69. N. Cavassilas, F. Aniel, K. Boujdaria, G. Fishman, Phys. Rev. B **64**(11), 115207 (2001)
70. S.B. Radhia, S. Ridene, K. Boujdaria, H. Bouchriha, G. Fishman, J. Appl. Phys. **92**(8), 4422 (2002)
71. S.B. Radhia, K. Boujdaria, S. Ridene, H. Bouchriha, G. Fishman, J. Appl. Phys. **94**(9), 5726 (2003)
72. S. Richard, F. Aniel, G. Fishman, Phys. Rev. B **70**(23), 235204 (2004)
73. K. Boujdaria, O. Zitouni, Solid State Commun. **129**(3), 205 (2004)
74. O. Zitouni, K. Boujdaria, H. Bouchriha, Semicond. Sci. Technol. **20**(9), 908 (2005)
75. D. Rideau, M. Feraille, L. Ciampolini, M. Minondo, C. Tavernier, H. Jaouen, A. Ghetti, Phys. Rev. B **74**(19), 195208 (2006)
76. S.B. Radhia, N. Fraj, I. Saidi, K. Boujdaria, Semicond. Sci. Technol. **22**(4), 427 (2007)
77. N. Fraj, I. Saïdi, S.B. Radhia, K. Boujdaria, J. Appl. Phys. **102**, 053703 (2007)
78. C. Mailhiot, T.C. McGill, D.L. Smith, J. Vac. Sci. Technol. B **2**(3), 371 (1984)
79. S. Richard, F. Aniel, G. Fishman, Phys. Rev. B **72**(24), 245316 (2005)
80. K. Suzuki, J.C. Hensel, Phys. Rev. B **9**(10), 4184 (1974)
81. H.R. Trebin, U. Rössler, R. Ranvaud, Phys. Rev. B **20**(2), 686 (1979)
82. K. Cho, Phys. Rev. B **14**(10), 4463 (1976). DOI 10.1103/PhysRevB.14.4463.
83. K. Cho, in *Excitons*, *Topics in Current Physics*, vol. 14, ed. by K. Cho (Springer, Berlin, 1979), Chap. 2, pp. 15–53
84. J.P. Elliot, P.G. Dawber, *Symmetry in Physics*, vol. 1 (Oxford University Press, New York, 1979)
85. P. Lawaetz, Phys. Rev. B **4**(10), 3460 (1971)
86. C.R. Pidgeon, R.N. Brown, Phys. Rev. **146**(2), 575 (1966)
87. J.C. Hensel, K. Suzuki, Phys. Rev. Lett. **22**(16), 838 (1969). DOI 10.1103/PhysRevLett.22.838.
88. E.O. Kane, Phys. Rev. B **11**(10), 3850 (1975)
89. D. Broido, L.J. Sham, Phys. Rev. B **31**(2), 888 (1985)
90. J. Lee, M.O. Vassell, Phys. Rev. B **37**(15), 8855 (1988)
91. U. Ekenberg, M. Altarelli, Phys. Rev. B **32**(6), 3712 (1985)
92. M. Altarelli, N.O. Lipari, Phys. Rev. Lett. **36**(11), 619 (1976)
93. M. Altarelli, U. Ekenberg, A. Fasolino, Phys. Rev. B **32**(8), 5138 (1985)
94. C.M. de Sterke, Phys. Rev. B **36**(12), 6574 (1987)
95. L.J. Sham, in *Physics of Low-Dimensional Semiconductor Structures*, ed. by P. Butcher, N.H. March, M.P. Tosi, *Physics of Solids and Liquids* (Plenum Press, New York, 1993), Chap. 1, pp. 1–56
96. K.H. Hellwege (ed.), *Landolt-Börnstein, Numerical Data and Functional Relationships in Science and Technology, New Series, Group III*, vol. 22a (Springer-Verlag, Berlin, 1982)
97. C.R. Pidgeon, S.H. Groves, Phys. Rev. **186**(3), 824 (1969). DOI 10.1103/PhysRev.186.824.
98. M.H. Weiler, R.L. Aggrawal, B. Lax, Phys. Rev. B **17**(8), 3269 (1978)
99. M.H. Weiler, *Magnetooptical Studies of Small-Gap Semiconductors: $Hg_{1-x}Cd_xTe$ and InSb*, Phd (MIT, Cambridge, 1977)
100. M.H. Weiler, in *Defects, (HgCd)Se, (HgCd)Te, Semiconductors and Semimetals*, vol. 16, ed. by R.K. Willardson, A.C. Beer (Academic Press, New York, 1981), Chap. 3, pp. 119–191
101. H. Mayer, U. Rössler, Phys. Rev. B **44**(16), 9048 (1991). DOI 10.1103/PhysRevB.44.9048.
102. P. Pfeffer, W. Zawadzki, Phys. Rev. B **41**(3), 1561 (1990)
103. P. Pfeffer, W. Zawadzki, Phys. Rev. B **53**(19), 12813 (1996)
104. M. Cardona, N.E. Christensen, G. Fasol, Phys. Rev. B **38**(3), 1806(1988)
105. Y.M. Sirenko, J. Jeon, B.C. Lee, K.W. Kim, M.A. Littlejohn, M.A. Stroscio, G.J. Iafrate, Phys. Rev. B **55**(7), 4360 (1997)

106. D.W. Langer, R.N. Euwema, K. Era, T. Koda, Phys. Rev. B **2**(10), 4005 (1970). DOI 10.1103/PhysRevB.2.4005.
107. M. Braun, Rössler, J. Phys. C **18**(17), 3365 (1985)
108. M. Willatzen, M. Cardona, N.E. Christensen, Phys. Rev. B **51**(19), 13150 (1995). DOI 10.1103/PhysRevB.51.13150.
109. A.N. Chantis, M. Cardona, N.E. Christensen, D.L. Smith, M. van Schilfgaarde, T. Kotani, A. Svane, R.C. Albers, Phys. Rev. B **78**(7), 075208 (2008)
110. O.K. Andersen, Phys. Rev. B **12**(8), 3060 (1975). DOI 10.1103/PhysRevB.12.3060.
111. M. Methfessel, M. van Schilfgaarde, R.A. Casali, *Electronic Structure and Physical Properties of Solids: The Uses of the LMTO Method*, *Lecture Notes in Physics*, vol. 535 (Springer-Verlag, Berlin, 2000)
112. J.J. Hopfield, D.G. Thomas, Phys. Rev. **132**, 563 (1963)
113. G. Mahan, J.J. Hopfield, Phys. Rev. **135**, A428 (1964)
114. J.J. Hopfield, J. Appl. Phys. **32**, 2277 (1961)
115. H. Brooks, in *Advances in Electronics and Electron Physics*, vol. 7, ed. by L. Marton (Academic Press, New York, 1955)
116. C. Herring, E. Vogt, Phys. Rev. **101**(3), 944 (1956). DOI 10.1103/PhysRev.101.944.
117. W.H. Kleiner, L.M. Roth, Phys. Rev. Lett. **2**(8), 334 (1959). DOI 10.1103/PhysRevLett.2.334.
118. H. Hasegawa, Phys. Rev. **129**, 1029 (1963)
119. E.O. Kane, Phys. Rev. **178**(3), 1368 (1969). DOI 10.1103/PhysRev.178.1368.
120. I. Balslev, in *Optical Properties of III-V Compounds*, *Semiconductors and Semimetals*, vol. 9, ed. by R.K. Willardson, A.C. Beer (Academic Press, New York, 1972), Chap. 5, pp. 403–456
121. F.H. Pollak, Surf. Sci. **37**, 863 (1973)
122. T.B. Bahder, Phys. Rev. B **41**(17), 11992 (1990)
123. D.E. Aspnes, M. Cardona, Phys. Rev. B **17**(2), 726 (1978)
124. J.M. Hinckley, J. Singh, Phys. Rev. B **41**(5), 2912 (1990). DOI 10.1103/PhysRevB.41.2912.
125. F.H. Pollak, in *Strained-layer Superlattices: Physics*, *Semiconductors and Semimetals*, vol. 32, ed. by T.P. Pearsall (Academic, New York, 1990), Chap. 2, pp. 17–53
126. M. Chandrasekhar, F.H. Pollak, Phys. Rev. B **15**(4), 2127 (1977). DOI 10.1103/PhysRevB.15.2127.
127. C. van de Walle, Phys. Rev. B **39**(3), 1871 (1989)
128. J.Y. Marzin, J.M. Gérard, P. Voisin, J.A. Brum, in *Semiconductors and Semimetals*, vol. 32, ed. by R.K. Willardson, A.C. Beer (Academic, New York, 1990), Chap. 3, pp. 56–118
129. A. Blacha, H. Presting, M. Cardona, Phys. Status Solidi b **126**(1), 11 (1984)
130. S.H. Park, K.J. Kim, S.N. Yi, D. Ahn, S.J. Lee, J. Korean Phys. Soc. **47**(3), 448 (2005)
131. G.H. Wannier, Phys. Rev. **52**(3), 191 (1937)
132. E.N. Adams II, J. Chem. Phys. **21**(11), 2013 (1953)
133. C. Kittel, A.H. Mitchell, Phys. Rev. **96**(4), 1488 (1954)
134. E.I. Blount, in *Solid State Physics*, *Solid State Physics*, vol. 13, ed. by F. Seitz, D. Turnbull (Academic, New York, 1962), Chap. 2, pp. 305–373
135. B.A. Foreman, Phys. Rev. Lett. **84**(11), 2505 (2000)
136. B.A. Foreman, Phys. Rev. Lett. **84**(19), 4513 (2000)
137. J.M. Luttinger, Phys. Rev. **84**(4), 814 (1951)
138. W. Kohn, D. Schechter, Phys. Rev. **99**(6), 1903 (1955). DOI 10.1103/PhysRev.99.1903.
139. J. Serrano, M. Cardona, T. Ruf, Solid State Commun. **113**(7), 411 (2000)
140. A. Baldereschi, N. Lipari, Phys. Rev. B **9**(4), 1525 (1974)
141. R. Ranvaud, H.R. Trebin, U. Rössler, F.H. Pollak, Phys. Rev. B **20**(2), 701 (1979)
142. B. Lax, J.G. Mavroides, in *Optical Properties of III-V Compounds*, *Semiconductors and Semimetals*, vol. 3, ed. by R.K. Willardson, A.C. Beer (Academic, New York, 1967), Chap. 8, pp. 321–401
143. R.L. Aggarwal, in *Optical Properties of III-V Compounds*, *Semiconductors and Semimetals*, vol. 9, ed. by R.K. Willardson, A.C. Beer (Academic Press, New York, 1972), Chap. 2, pp. 151–256

144. W. Zawadzki, S. Klahn, U. Merkt, Phys. Rev. Lett. **55**(9), 983 (1985). DOI 10.1103/PhysRevLett.55.983.
145. J.C. Hensel, K. Suzuki, Phys. Rev. B **9**(10), 4219 (1974)
146. R.L. Bell, K.T. Rogers, Phys. Rev. **152**(2), 746 (1966). DOI 10.1103/PhysRev.152.746.
147. C. Hermann, C. Weisbuch, Phys. Rev. B **15**(2), 823 (1977)
148. G.H. Wannier, Phys. Rev. **117**(2), 432 (1960)
149. I. Souza, J. Iniguez, D. Vanderbilt, Phys. Rev. Lett. **89**(11), 117602 (2002)
150. H. Fu, L. Bellaiche, Phys. Rev. Lett. **91**(5), 057601 (2003)
151. B.A. Foreman, J. Phys.: Condens. Matter **12**, R435 (2000)
152. J. Zak, Phys. Rev. Lett. **54**, 1075 (1985)
153. M.G. Burt, Semicond. Sci. Technol. **3**, 739 (1988)
154. M. Glück, A.R. Kolovsky, H.J. Korsch, Phys. Rep. **366**(3), 103 (2002)
155. D. Ahn, S.L. Chuang, Phys. Rev. B **34**(12), 9034 (1986)
156. K.I. Hino, Solid State Commun. **128**(1), 9 (2003)
157. G.G. Macfarlane, T.P. McLean, J.E. Quarrington, V. Roberts, Phys. Rev. **108**(6), 1377 (1957)
158. G. Dresselhaus, J. Phys. Chem. Sol. **1**, 14 (1956)
159. A. Baldereschi, N. Lipari, Phys. Rev. B **3**, 439 (1971)
160. L.D. Laude, F.H. Pollak, M. Cardona, Phys. Rev. B **3**(8), 2623 (1971)
161. M. Altarelli, N.O. Lipari, Phys. Rev. B **7**(8), 3798 (1973). DOI 10.1103/PhysRevB.7.3798.
162. M. Altarelli, N.O. Lipari, Phys. Rev. B **8**(8), 4046 (1973). DOI 10.1103/PhysRevB.8.4046.2.
163. K. Cho, S. Suga, W. Dreybrodt, F. Willmann, Phys. Rev. B **11**(4), 1512 (1975). DOI 10.1103/PhysRevB.11.1512.
164. M. Altarelli, N.O. Lipari, Phys. Rev. B **15**(10), 4898 (1977)
165. H. Haken, *Quantum Field Theory of Solids* (North-Holland, Amsterdam, 1988)
166. J.O. Dimmock, in *Optical Properties of III-V Compounds*, *Semiconductors and Semimetals*, vol. 3, ed. by R.K. Willardson, A.C. Beer (Academic, New York, 1967), Chap. 7, pp. 259–319
167. L.D. Landau, E.M. Lifshitz, *Quantum Mechanics*, *Course of Theoretical Physics*, vol. 3, 3rd edn. (Butterworth-Heinemann, Oxford, 2002)
168. G. Blattner, G. Kurtze, G. Schmieder, C. Klingshirn, Phys. Rev. B **25**(12), 7413 (1982)
169. R. Bardoux, T. Guillet, B. Gil, P. Lefebvre, T. Bretagnon, T. Taliercio, S. Rousset, F. Semond, Phys. Rev. B **77**(23), 235315 (2008)
170. R. Dingle, W. Wiegmann, C.H. Henry, Phys. Rev. Lett. **33**(14), 827 (1974). DOI 10.1103/PhysRevLett.33.827.
171. L.L. Chang, L. Esaki, R. Tsu, Appl. Phys. Lett. **24**(12), 593 (1974). DOI 10.1063/1.1655067. http://link.aip.org/link/?APL/24/593/1
172. G. Bastard, J.K. Furdyna, J. Mycielski, Phys. Rev. B **24**(10), 4356(1975)
173. T. Ando, A.B. Fowler, F. Stern, Rev. Mod. Phys. **54**(2), 437 (1982). DOI 10.1103/RevModPhys.54.437.
174. W. Pötz, W. Porod, D.K. Ferry, Phys. Rev. B **32**(6), 3868 (1985). DOI 10.1103/PhysRevB.32.3868.
175. G. Bastard, Phys. Rev. B **24**, 5693 (1981)
176. M. Altarelli, Phys. Rev. B **28**(2), 842 (1983)
177. A.L. Éfros, A.L. Éfros, Sov. Phys. Semicond. **16**(7), 772 (1982)
178. S.S. Nedorezov, Sov. Phys. Solid State **12**(8), 1814 (1971)
179. G. Bastard, Phys. Rev. B **25**(12), 7584 (1982). DOI 10.1103/PhysRevB.25.7584.
180. M.G. Burt, Semicond. Sci. Technol. **2**, 460 (1987)
181. M.G. Burt, Semicond. Sci. Technol. **2**, 701 (1987)
182. G. Bastard, J.A. Brum, IEEE J. Quantum Electron **22**, 1625 (1986)
183. G. Bastard, J.A. Brum, R. Ferreira, in *Semiconductor Heterostructures and Nanostructures*, *Solid State Physics*, vol. 44, ed. by H. Ehrenreich, D. Turnbull, *Solid State Physics*, (Academic Press, Boston, 1991), pp. 229–415
184. D.L. Smith, C. Mailhiot, Rev. Mod. Phys. **62**(1), 173 (1990). DOI 10.1103/RevModPhys.62.173.
185. M.G. Burt, J. Phys.: Condensed Matter **4**, 6651 (1992)

186. M.G. Burt, J. Phys.: Condensed Matter **11**, R53 (1999)
187. L.C. Lew Yan Voon, Y. Zhang, B. Lassen, M. Willatzen, Q. Xiong, P. Eklund, J. Nanosci. Nanotechnol. **8**, 1 (2008)
188. A.I. Efros, M. Rosen, Annu. Rev. Mater. Sci. **30**, 475 (2000)
189. T. Ando, H. Akera, Phys. Rev. B **40**(17), 11619 (1989)
190. L.E. Bremme, P.C. Klipstein, Phys. Rev. B **66**(23), 235316 (2002). DOI 10.1103/PhysRevB.66.235316.
191. W.A. Harrison, J. Vac. Sci. Technol. **14**(4), 1016 (1977). DOI 10.1116/1.569312. http://link.aip.org/link/?JVS/14/1016/1
192. M. Cardona, N.E. Christensen, Phys. Rev. B **35**(12), 6182 (1987)
193. D. Gershoni, C.H. Henry, G. Baraff, IEEE J. Quantum Electron **29**(9), 2433 (1993)
194. C. Galeriu, L.C. Lew Yan Voon, R.N. Melnik, M. Willatzen, Comp. Phys. Commun. **157**, 147 (2004)
195. A. Fasolino, M. Altarelli, in *Two-Dimensional Systems, Heterostructures, and Superlattices, Solid-State Sciences*, vol. 53, ed. by G. Bauer, F. Kuchar, H. Heinrich, *Solid-State Sciences*, (Springer-Verlag, Berlin, 1984), pp. 176–182
196. T. Ando, S. Wakahara, H. Akera, Phys. Rev. B **40**(17), 11609 (1989)
197. T.L. Li, K.J. Kuhn, Phys. Rev. B **49**(4), 2608 (1994). DOI 10.1103/PhysRevB.49.2608.
198. Y.C. Chang, J.N. Schulman, Phys. Rev. B **31**(4), 2069 (1985)
199. L.C. Andreani, A. Pasquarello, F. Bassani, Phys. Rev. B **36**(11), 5887 (1987). DOI 10.1103/PhysRevB.36.5887.
200. R.Q. Yang, J.M. Xu, M. Sweeny, Phys. Rev. B **50**(11), 7474 (1994). DOI 10.1103/PhysRevB.50.7474.
201. S. Jorda, U. Rössler, Superlattices Microstruct. **8**(4), 481 (1990)
202. O. von Roos, Phys. Rev. B **27**(12), 7547 (1983)
203. C.M. Van Vliet, A.H. Marshak, Phys. Rev. B **29**(10), 5960 (1984). DOI 10.1103/PhysRevB.29.5960.
204. K. Kawamura, R.A. Brown, Phys. Rev. B **37**(8), 3932 (1988). DOI 10.1103/PhysRevB.37.3932.
205. L. Chetouani, L. Dekar, T.F. Hammann, Phys. Rev. A **52**(1), 82 (1995). DOI 10.1103/PhysRevA.52.82.
206. J.M. Lévy-Leblond, Phys. Rev. A **52**(3), 1845 (1995). DOI 10.1103/PhysRevA.52.1845.
207. T. Gora, F. Williams, Phys. Rev. **177**(3), 1179 (1969). DOI 10.1103/PhysRev.177.1179.
208. T.L. Li, K.J. Kuhn, Phys. Rev. B **47**(19), 12760 (1993)
209. F.S.A. Cavalcante, R.N. Costa Filho, J.R. Filho, C.A.S. de Almeida, V.N. Freire, Phys. Rev. B **55**(3), 1326 (1997). DOI 10.1103/PhysRevB.55.1326.
210. T. Ando, S. Mori, Surf. Sci. **113**(1–3), 124 (1982)
211. Q.G. Zhu, H. Kroemer, Phys. Rev. B **27**(6), 3519 (1983). DOI 10.1103/PhysRevB.27.3519.
212. D.J. BenDaniel, C.B. Duke, Phys. Rev. **152**(2), 683 (1966). DOI 10.1103/PhysRev.152.683.
213. S.R. White, L.J. Sham, Phys. Rev. Lett. **47**(12), 879 (1981)
214. I. Galbraith, G. Duggan, Phys. Rev. B **38**(14), 10057 (1988). DOI 10.1103/PhysRevB.38.10057.
215. G.T. Einevoll, P.C. Hemmer, J. Thomsen, Phys. Rev. B **42**(6), 3485 (1990). DOI 10.1103/PhysRevB.42.3485.
216. R.A. Morrow, K.R. Brownstein, Phys. Rev. B **30**(2), 678 (1984)
217. R. Balian, D. Bessis, G.A. Mezincescu, Phys. Rev. B **51**(24), 17624 (1995)
218. M.E. Pistol, Phys. Rev. B **60**(20), 14269 (1999)
219. L.C. Lew Yan Voon, Superlattices Microstruct. **31**, 269 (2002)
220. M.G. Burt, Semicond. Sci. Technol. **3**, 1224 (1988)
221. M.G. Burt, Semicond. Sci. Technol. **8**(7), 1393 (1993)
222. M.G. Burt, Phys. Rev. B **50**(11), 7518 (1994)
223. M.G. Burt, Semicond. Sci. Technol. **10**(4), 412 (1995)
224. M.G. Burt, Superlattices Microstruct. **23**(2), 531 (1998)
225. B.A. Foreman, Phys. Rev. B **49**(3), 1757 (1994)

226. B.A. Foreman, Phys. Rev. B **52**(16), 12241 (1995)
227. B.A. Foreman, Phys. Rev. B **54**(3), 1909 (1996)
228. B.A. Foreman, Phys. Rev. B **56**(20), R12748 (1997)
229. B.A. Foreman, Phys. Rev. Lett. **81**(2), 425 (1998)
230. B.A. Foreman, Phys. Rev. Lett. **80**(17), 3823 (1998)
231. B.A. Foreman, Phys. Rev. Lett. **86**(12), 2641 (2001)
232. P.N. Stravinou, R. van Dalen, Phys. Rev. B **55**(23), 15456 (1997)
233. A.T. Meney, B. Gonul, E.P. O'Reilly, Phys. Rev. B **50**(15), 10893 (1994)
234. K. Boujdaria, S. Ridene, S.B. Radhia, H. Bouchriha, G. Fishman, Solid State Commun. **129**(3), 221 (2004)
235. R. van Dalen, P.N. Stravinou, Semicond. Sci. Technol. **13**, 11 (1998)
236. E.P. Pokalitov, V.A. Fonoberov, V.M. Fomin, J.T. Devreese, Phys. Rev. B **64**(24), 245328 (2001)
237. E.P. Pokalitov, V.A. Fonoberov, V.M. Fomin, J.T. Devreese, Phys. Rev. B **64**(24), 245329 (2001)
238. C. Galeriu, L.C. Lew Yan Voon, M. Willatzen, in *Proc. Workshop on Intersubband Transitions in Quantum Wells*, ed. by H. Sigg (Evolene, Switzerland, 2003)
239. B. Lassen, R. Melnik, M. Willatzen, L.C. Lew Yan Voon, in *Proc. Workshop on Intersubband Transitions in Quantum Wells*, ed. by H. Sigg (Evolene, Switzerland, 2003)
240. L.C. Lew Yan Voon, B. Lassen, R. Melnik, M. Willatzen, Nano Lett. **4**(2), 289 (2004)
241. E.E. Takhtamirov, V.A. Volkov, Semicond. Sci. Technol. **12**(1), 77 (1997). http://stacks.iop.org/0268-1242/12/77
242. V.A. Volkov, E.E. Takhtamirov, Phys. Usp. **40**(10), 1071 (1997)
243. E.E. Takhtamirov, V.A. Volkov, J. Exp. Theor. Phys. **89**(5), 1000 (1999)
244. E.E. Takhtamirov, V.A. Volkov, J. Exp. Theor. Phys. **90**(6), 1063 (2000)
245. E.L. Ivchenko, A.Y. Kaminski, U. Rössler, Phys. Rev. B **54**, 5852 (1996)
246. O. Krebs, P. Voisin, Phys. Rev. B **61**(11), 7265 (2000). DOI 10.1103/PhysRevB.61.7265.
247. H.J. Haugan, F. Szmulowicz, G.J. Brown, K. Mahalingam, J. Appl. Phys. **96**(5), 2580 (2004). DOI 10.1063/1.1776321. http://link.aip.org/link/?JAP/96/2580/1
248. I. Semenikhin, A. Zakharova, K.A. Chao, Phys. Rev. B **77**(11), 113307 (2008)
249. B.A. Foreman, Phys. Rev. B **72**(16), 165345 (2005)
250. B.A. Foreman, Phys. Rev. B **76**, 045327 (2007)
251. K.J. Vahala, P.C. Sercel, Phys. Rev. Lett. **65**(2), 239 (1990)
252. P.C. Sercel, K.J. Vahala, Appl. Phys. Lett. **57**(6), 545 (1990)
253. P.C. Sercel, K.J. Vahala, Phys. Rev. B **44**(11), 5681 (1991)
254. D. Katz, T. Wizansky, O. Millo, E. Rothenberg, T. Mokari, U. Banin, Phys. Rev. Lett. **89**(8), 086801 (2002)
255. J. Planelles, W. Jaskólski, J.I. Aliaga, Phys. Rev. B **65**(3), 033306 (2002)
256. J. Planelles, J.G. Díaz, J. Climente, W. Jaskólski, Phys. Rev. B **65**(24), 245302 (2002)
257. J.I. Climente, J. Planelles, W. Jaskólski, Phys. Rev. B **68**(7), 075307 (2003)
258. F.B. Pedersen, Y.C. Chang, Phys. Rev. B **53**(3), 1507 (1996)
259. C. Galeriu, *$k \cdot p$ Theory of Semiconductor Nanostructures*, Phd (WPI, Worcester 2005)
260. L.C. Lew Yan Voon, B. Lassen, R. Melnik, M. Willatzen, J. Appl. Phys. **96**, 4660 (2004)
261. L.C. Lew Yan Voon, M. Willatzen, J. Appl. Phys. **93**, 9997 (2003)
262. L.D. Landau, E.M. Lifshitz, *Theory of Elasticity*, *Course of Theoretical Physics*, vol. 7, 3rd edn. (Butterworth–Heinemann, Oxford, 1999)
263. C. Mailhiot, D.L. Smith, Phys. Rev. B **35**(3), 1242 (1987)
264. W. Batty, U. Ekenberg, A. Ghit, E.P. O'Reilly, Semicond. Sci. Technol. **4**(11), 904 (1989)
265. I. Vurgaftman, J. Hinckley, J. Singh, IEEE J. Quantum Electron **30**(1), 75 (1994)
266. J.B. Xia, Phys. Rev. B **43**(12), 9856 (1991)
267. Z. Ikonic, V. Milanovic, D. Tjapkin, Phys. Rev. B **46**(7), 4285 (1992)
268. G. Goldoni, F.M. Peeters, Phys. Rev. B **51**(24), 17806 (1995)
269. R.H. Henderson, E. Towe, J. Appl. Phys. **78**(4), 2447 (1995)
270. G. Fishman, Phys. Rev. B **52**(15), 11132 (1995)

271. R.H. Henderson, E. Towe, J. Appl. Phys. **79**(4), 2029 (1996)
272. Y. Kajikawa, J. Appl. Phys. **86**(10), 5663 (1999)
273. W.H. Seo, J.F. Donegan, Phys. Rev. B **68**(7), 075318 (2003)
274. J. Los, A. Fasolino, A. Castellani, Phys. Rev. B **53**(8), 4630 (1996)
275. B. Lassen, M. Willatzen, R. Melnik, L.C. Lew Yan Voon, J. Mat. Res. **21**(11), 2927 (2006)
276. F. Vouilloz, D.Y. Oberli, M.A. Dupertius, A. Gustafsson, F. Reinhardt, E. Kapon, Phys. Rev. Lett. **78**(8), 1580 (1997)
277. M. Kappelt, M. Grundmann, A. Krost, V. Türck, D. Bimberg, Appl. Phys. Lett. **68**(25), 3596 (1996)
278. M.F.H. Schuurmans, G.W. 't Hooft, Phys. Rev. B **31**(12), 8041 (1985). DOI 10.1103/PhysRevB.31.8041.
279. F. Szmulowicz, G.J. Brown, Phys. Rev. B **51**(19), 13203 (1995)
280. M.J. Godfrey, A.M. Malik, Phys. Rev. B **53**(24), 16504 (1996). DOI 10.1103/PhysRevB.53.16504.
281. L.W. Wang, Phys. Rev. B **61**(11), 7241 (2000). DOI 10.1103/PhysRevB.61.7241.
282. A.V. Rodina, A.Y. Alekseev, A.L. Efros, M. Rosen, B.K. Meyer, Phys. Rev. B **65**(12), 125302 (2002). DOI 10.1103/PhysRevB.65.125302.
283. X. Cartoixa, D.Z.Y. Ting, T.C. McGill, J. Appl. Phys. **93**(7), 3974 (2003). DOI 10.1063/1.1555833. http://link.aip.org/link/?JAP/93/3974/1
284. F. Szmulowicz, Europhys. Lett. **69**(2), 249 (2005)
285. R.G. Veprek, S. Steiger, B. Witzigmann, Phys. Rev. B **76**(16), 165320 (2007)
286. B. Lassen, M. Willatzen, R. Melnik, L.C.L.Y. Voon, AIP Conf. Proc. **893**, 393 (2007)
287. R. Winkler, U. Rössler, Phys. Rev. B **48**(12), 8918 (1993)
288. F. Szmulowicz, Phys. Rev. B **51**(3), 1613 (1995)
289. F. Szmulowicz, Phys. Rev. B **54**(16), 11539 (1996)
290. F. Szmulowicz, Superlattices Microstruct. **22**(3), 295 (1997)
291. K.I. Kolokolov, J. Li, C.Z. Ning, Phys. Rev. B **68**(16), 161308 (2003)
292. B.A. Foreman, Phys. Rev. B **75**, 235331 (2007)
293. D.L. Smith, C. Mailhiot, Phys. Rev. B **33**(12), 8345 (1986). DOI 10.1103/PhysRevB.33.8345.
294. C. Mailhiot, D.L. Smith, Phys. Rev. B **33**(12), 8360 (1986). DOI 10.1103/PhysRevB.33.8360.
295. L.W. Wang, A. Zunger, Phys. Rev. B **54**(16), 11417 (1996)
296. H. Yi, M. Razeghi, Phys. Rev. B **56**(7), 3933 (1997)
297. R. Stubner, R. Winkler, O. Pankratov, Phys. Rev. B **62**(3), 1843 (2000)
298. H. Fu, L.W. Wang, A. Zunger, Appl. Phys. Lett. **71**(23), 3433 (1997)
299. A.L. Efros, M. Rosen, Appl. Phys. Lett. **73**(8), 1155 (1998)
300. H. Fu, L.W. Wang, A. Zunger, Appl. Phys. Lett. **73**(8), 1157 (1998)
301. L. Wissinger, U. Rössler, R. Winkler, B. Jusserand, D. Richards, Phys. Rev. B **58**(23), 15375 (1998)
302. R. Winkler, Phys. Rev. B **62**(7), 4245 (2000). DOI 10.1103/PhysRevB.62.4245.
303. J.P. Stanley, N. Pattinson, C.J. Lambert, J.H. Jefferson, Physica E **20**(3/4), 433 (2003)
304. W. Zawadzki, P. Pfeffer, Physica E **20**(3/4), 392 (2003)
305. X. Cartoixa, D.Z.Y. Ting, T.C. McGill, Nanotechnology **14**(2), 308 (2003). http://stacks.iop.org/0957-4484/14/308
306. F.M. Alves, C. Trallero-Giner, V. Lopez-Richard, G.E. Marques, Phys. Rev. B **77**(3), 035434 (2008)
307. W. Zawadzki, P. Pfeffer, Semicond. Sci. Technol. **19**, R1 (2004)
308. R. Eppenga, M. Schuurmans, Phys. Rev. B **37**, 10923 (1988)
309. P. Santos, M. Willatzen, M. Cardona, A. Cantarero, Phys. Rev. B **51**, 5121 (1995)
310. N. Christensen, M. Cardona, Solid State Commun. **51**, 491 (1984)
311. C. Mailhiot, D.L. Smith, Phys. Rev. B **36**(5), 2942 (1987). DOI 10.1103/PhysRevB.36.2942.
312. G.C. Osbourn, P.L. Gourley, I.J. Fritz, R.M. Biefeld, L.R. Dawson, T.E. Zipperian, in *Applications of Multiquantum Wells, Selective Doping, and Superlattices*, *Semiconductors and*

Semimetals, vol. 24, ed. by R. Dingle (Academic Press, New York, 1987), Chap. 8, pp. 459–505
313. F.C. Frank, J.H. van der Merwe, Proc. Roy. Soc. Lond. **A198**, 216 (1949)
314. R.W. Vook, Int. Met. Rev. **27**, 209 (1982)
315. R. People, Phys. Rev. B **32**, 1405 (1985)
316. Y. Androussi, A. Lefebvre, B. Courboule's, N. Grandjean, J. Massies, T. Bouhacina, J.P. Aimé, Appl. Phys. Lett. **65**, 1162 (1994)
317. J.W. Matthews, A.E. Blakeslee, J. Cryst. Growth **27**, 118 (1974)
318. E.P. O'Reilly, G.P. Witchlow, Phys. Rev. B **34**(8), 6030 (1986). DOI 10.1103/PhysRevB.34.6030.
319. J.D. Eshelby, Proc. Roy. Soc. Lond. **A241**, 376 (1957)
320. E. O'Reilly, Sem. Sci. Tech. **4**, 121 (1989)
321. E. O'Reilly, A.R. Adams, IEEE J. Quantum. Electron **30**, 366 (1994)
322. T.J. Gosling, J.R. Willis, J. Appl. Phys. **77**, 5061 (1995)
323. M. Grundmann, O. Stier, D. Bimberg, Phys. Rev. B **52**, 11969 (1995)
324. J.R. Downes, D.A. Faux, E.P. O'Reilly, J. Appl. Phys. **81**, 6700 (1997)
325. A.D. Andreev, J.R. Downes, D.A. Faux, E.P. O'Reilly, Phys. Rev. B **86**, 297 (1999)
326. B. Jogai, J. Albrecht, E. Pan, J. Appl. Phys. **94**, 6566 (2003)
327. M. Willatzen, B. Lassen, L.C. Lew Yan Voon, R. Melnik, J. Appl. Phys. **100**(2), 024302 (2006)
328. G.D. Sanders, Y.C. Chang, Phys. Rev. B **32**, 4282 (1985)
329. J. Lee, M.O. Vassell, Phys. Rev. B **37**(15), 8861 (1988)
330. B. Lassen, M. Willatzen, R. Melnik, L.C. Lew Yan Voon, J. Math. Phys. **46**(11), 112102 (2005)
331. G. Bastard, Phys. Rev. B **24**(8), 4714 (1981). DOI 10.1103/PhysRevB.24.4714.
332. C. Mailhiot, Y.C. Chang, T.C. McGill, Phys. Rev. B **26**(8), 4449 (1982). DOI 10.1103/PhysRevB.26.4449.
333. R.L. Greene, K.K. Bajaj, Phys. Rev. B **31**(2), 913 (1985)
334. A. Latgé, N. Porras-Montenegro, L.E. Oliveira, Phys. Rev. B **45**, 9420 (1992)
335. C.S. Kim, A.M. Satanin, Y.S. Joe, R.M. Cosby, Phys. Rev. B **60**, 10962 (1999)
336. C.M. Lee, C.C. Lam, S.W. Gu, Phys. Rev. B **61**, 10376 (2000)
337. M. Amado, R.P.A. Lima, C. González-Santander, F. Domínguez-Adame, Phys. Rev. B **76**, 073312 (2007)
338. W.T. Masselink, Y.C. Chang, H. Morkoç, Phys. Rev. B **28**(12), 7373 (1983). DOI 10.1103/PhysRevB.28.7373.
339. W.T. Masselink, Y. Chang, H. Morkoc, Phys. Rev. B **32**(8), 5190 (1985)
340. S. Glutsch, *Excitons in Low-Dimensional Semiconductors, Solid State Physics* (Springer, Berlin, 2004)
341. D.K. Kim, D.S. Citrin, Phys. Rev. B **76**, 125305 (2007)
342. H. Stolz, Phys. Stat. Sol. **32**, 631 (1969)
343. H. Haug, S. Schmitt-Rink, Prog. Quantum. Electron **9**, 3 (1984)
344. R.J. Elliott, Phys. Rev. B **108**, 1384 (1957)
345. H. Haug, S.W. Koch, *Quantum Theory of the Optical and Electronic Properties of Semiconducors* (World Scientific, Singapore, 1994)
346. L. Andreani, F. Tassone, F. Bassani, Solid State Commun. **77**, 641 (1991)
347. D.S. Citrin, Phys. Rev. B **47**, 3832 (1993)
348. Y. Sidor, B. Partoens, F.M. Peeters, Phys. Rev. B **71**, 165323 (2005)
349. Y. Nagamune, Y. Arakawa, S. Tsukamoto, M. Nishioka, S. Sasaki, N. Miura, Phys. Rev. Lett. **69**, 2963 (1992)
350. B. Alén, J. Martínez-Pastor, A. García-Cristobal, L. González, J.M. García, Appl. Phys. Lett. **78**, 4025 (2001)
351. J. Martínez-Pastor, L. González, J.M. García, S.I. Molina, A. Ponce, R. García, Phys. Rev. B **65**, 241301 (2002)
352. M. Shinada, S. Sugano, J. Phys. Soc. Jpn. **21**, 1936 (1966)

353. R.J. Elliott, in *Polarons and Excitons*, ed. by C.G. Kuper, G.D. Whitfield (Plenum Press, New York, 1963), pp. 269–294
354. H. Haken, in *Polarons and Excitons*, ed. by C.G. Kuper, G.D. Whitfield (Plenum Press, New York, 1963), pp. 295–322
355. A. Stahl, I. Balslev, *Electrodynamics of the Semiconductor Band Edge* (Springer, Berlin, 1987)
356. R. Zimmermann, *Many-Particle Theory of Highly Excited Semiconductors* (Teubner, Leipzig, 1988)
357. G. Bastard, E.E. Mendez, L.L. Chang, L. Esaki, Phys. Rev. B **26**(4), 1974 (1982). DOI 10.1103/PhysRevB.26.1974.
358. R. Atanasov, F. Bassani, A. D'Andrea, N. Tomassini, Phys. Rev. B **51**(19), 14381 (1994)
359. E.P. Pokatilov, D.L. Nika, V.M. Fomin, J.T. Devreese, Phys. Rev. B: Condens. Matter Mater. Phys. **77**(12), 125328 (2008). DOI 10.1103/PhysRevB.77.125328. http://link.aps.org/abstract/PRB/v77/e125328
360. T. Ogawa, T. Takagahara, Phys. Rev. B **44**(15), 8138 (1991)
361. K. Chang, J.B. Xia, Phys. Rev. B **58**(4), 2031 (1998)
362. E.A. Muljarov, E.A. Zhukov, V.S. Dneprovskii, Y. Masumoto, Phys. Rev. B **62**(11), 7420 (2000)
363. G.W. Bryant, Phys. Rev. B **37**(15), 8763 (1988)
364. S.L. Goff, B. Stébé, Phys. Rev. B **47**(3), 1383 (1993)
365. T. Takagahara, Phys. Rev. B **47**(8), 4569 (1993)
366. S. Nomura, Y. Segawa, T. Kobayashi, Phys. Rev. B **49**(19), 13571 (1994). DOI 10.1103/PhysRevB.49.13571.
367. G. Lamouche, Y. Lépine, Phys. Rev. B **54**(7), 48114819 (1996)
368. A.L. Efros, M. Rosen, Phys. Rev. B **54**(7), 4843 (1996)
369. J.M. Ferreyra, C.R. Proetto, Phys. Rev. B **60**(15), 10672 (1999)
370. J. Li, J.B. Xia, Phys. Rev. B **62**(19), 12613 (2000)
371. J. Li, J.B. Xia, Phys. Rev. B **61**(23), 15880 (2000)
372. K.L. Janssens, F.M. Peeters, V.A. Schweigert, Phys. Rev. B **63**(20), 205311 (2001)
373. V.A. Fonoberov, E.P. Pokatilov, A.A. Balandin, Phys. Rev. B **66**(08), 085310 (2002)
374. J.I. Climente, M. Korkusinski, G. Goldoni, P. Hawrylak, Phys. Rev. B **78**(11), 115323 (2008)
375. Y. Zhang, A. Mascarenhas, E.D. Jones, J. Appl. Phys. **83**(1), 448 (1998). DOI 10.1063/1.366659. http://link.aip.org/link/?JAP/83/448/1
376. D. Viri, R.D. Sole, Solid State Commun. **97**(11), 985 (1996)
377. F. Fang, W. Howard, Phys. Rev. Lett. **16**, 797 (1966)
378. S.V. Gaponenko, *Optical Properties of Semiconductor Nanocrystals* (Cambridge University Press, Cambridge, 1998)
379. L. Jacak, P. Hawrylak, A. Wójs, *Quantum Dots* (Springer, Berlin, 1998)
380. V. Fonoberov, E.P. Pokatilov, A.A. Balandin, Phys. Rev. B **66**, 085310 (2002)
381. J.C. Maan, in *Two-Dimensional Systems, Heterostructures, and Superlattices*, *Solid-State Sciences*, vol. 53, ed. by G. Bauer, F. Kuchar, H. Heinrich, *Solid-State Sciences* (Springer-Verlag, Berlin, 1984), pp. 183–191
382. G. Belle, J.C. Maan, G. Weimann, Surf. Sci. **170**, 611 (1986)
383. J.C. Maan, Superlattices Microstruct. **2**(6), 557 (1986)
384. M. Masale, N.C. Constantinou, D.R. Tilley, Phys. Rev. B **46**(23), 15432 (1992)
385. E.X. Ping, V. Dalal, J. Appl. Phys. **76**(4), 2547 (1994). DOI 10.1063/1.357570. http://link.aip.org/link/?JAP/76/2547/1
386. N. Villamil, N. Porras-Montenegro, J.C. Granada, Phys. Rev. B **59**(3), 1605 (1999)
387. D. Csontos, U. Zülicke, Phys. Rev. B **76**(7), 073313 (2007)
388. J.B. Xia, S.S. Li, Phys. Rev. B **66**(03), 035311 (2002)
389. O. Voskoboynikov, Y. Li, H.M. Lu, C.F. Shih, C.P. Lee, Phys. Rev. B **66**(15), 155306 (2002)
390. A. Fasolino, M. Altarelli, Surf. Sci. **142**, 322 (1984)

391. M. Altarelli, J. Luminescence **30**, 472 (1985)
392. M. Altarelli, Festkörperprobleme **XXV**, 381 (1985)
393. S.R.E. Yang, D.A. Broido, L.J. Sham, Phys. Rev. B **56**(7), 6630 (1985)
394. M. Altarelli, G. Platero, Surf. Sci. **196**, 540 (1988)
395. F. Ancilotto, A. Fasolino, J.C. Maan, Phys. Rev. B **38**(3), 1788 (1988)
396. S. Lamari, L.J. Sham, Phys. Rev. B **38**(14), 9810 (1988)
397. G.Y. Wu, T.C. McGill, C. Mailhiot, D.L. Smith, Phys. Rev. B **39**(9), 6060 (1989)
398. B.E. Cole, J.M. Chamberlain, M. Henini, T. Cheng, W. Batty, A. Wittlin, J.A.A.J. Perenboom, A. Ardavan, A. Polisski, J. Singleton, Phys. Rev. B **55**(4), 2503 (1997)
399. M. Pacheco, Z. Barticevic, J. Phys.: Condens. Matter **11**(4), 1079 (1999)
400. J. Planelles, W. Jaskólski, J. Phys.: Condens. Matter **15**(2), L67 (2002)
401. S.J. Prado, C. Trallero-Giner, V. López-Richard, A.M. Alcaldea, G.E. Marques, Physica E **20**(3/4), 286 (2003)
402. M. Tadić, F.M. Peeters, Phys. Rev. B **71**(12), 125342 (2005)
403. V. Mlinar, M. Tadi, B. Partoens, F.M. Peeters, Phys. Rev. B **71**(20), 205305 (2005)
404. J.C. Maan, in *Physics of Low-Dimensional Semiconductor Structures*, ed. by P. Butcher, N.H. March, M.P. Tosi, *Physics of Solids and Liquids* (Plenum Press, New York, 1993), Chap. 9, pp. 333–374
405. X.H. Wang, B.Y. Gu, G.Z. Yang, Phys. Rev. B **56**(15), 9224 (1997). DOI 10.1103/PhysRevB.56.9224.
406. S. Živanović, V. Milanović, Z. Ikonić, Phys. Rev. B **52**(11), 8305 (1995). DOI 10.1103/PhysRevB.52.8305.
407. H.R. Lee, H.G. Oh, T.F. George, C.I. Um, J. Appl. Phys. **66**(6), 2442 (1989). DOI 10.1063/1.344254. http://link.aip.org/link/?JAP/66/2442/1
408. D.M. Mitrinović, V. Milanović, Z. Ikonić, Phys. Rev. B **54**(11), 7666 (1996). DOI 10.1103/PhysRevB.54.7666.
409. M. Shayegan, T. Sajoto, J. Jo, M. Santos, H.D. Drew, Phys. Rev. B **40**(5), 3476 (1989). DOI 10.1103/PhysRevB.40.3476.
410. T. Chakraborty, P. Pietiläinen, Phys. Rev. B **39**(11), 7971 (1989). DOI 10.1103/PhysRevB.39.7971.
411. M.A. Brummell, M.A. Hopkins, R.J. Nicholas, J.C. Portal, K.Y. Cheng, A.Y. Cho, Journal of Physics C: Solid State Phys. **19**(5), L107 (1986). http://stacks.iop.org/0022-3719/19/L107
412. G.Y. Wu, K.M. Hung, C.J. Chen, Phys. Rev. B **46**(3), 1521 (1992)
413. B.A. Foreman, J. Phys.: Condens. Matter **18**, 1335 (2006)
414. B.F. Zhu, Y.C. Chang, Phys. Rev. B **50**(16), 11932 (1994). DOI 10.1103/PhysRevB.50.11932.
415. M.A. Dupertuis, E. Martinet, H. Weman, E. Kapon, Europhys. Lett. **44**, 759 (1998)
416. D.J. Paul, Phys. Rev. B: Condens. Matter Mater. Phys. **77**(15), 155323 (2008). DOI 10.1103/PhysRevB.77.155323. http://link.aps.org/abstract/PRB/v77/e155323
417. L. Vervoort, R. Ferreira, P. Voisin, Phys. Rev. B **56**, R12744 (1997)
418. L. Vervoort, R. Ferreira, P. Voisin, Semicond. Sci. Technol. **14**, 227 (1999)
419. T. Guettler, A.L.C. Triques, L. Vervoort, R. Ferreira, P. Roussignol, P. Voisin, D. Rondi, J.C. Harmand, Phys. Rev. B **58**, R10179 (1998)
420. R. Lassnig, Phys. Rev. B **31**, 8076 (1985)
421. L.G. Gerchikov, A.V. Subashiev, Sov. Phys. Semicond. **26**, 73 (1992)
422. P. Pfeffer, W. Zawadzki, Phys. Rev. B **52**(19), 14332 (1995)
423. R. Winkler, M. Merkler, T. Darnhofer, U. Rössler, Phys. Rev. B **53**, 10858 (1996)
424. O. Mauritz, U. Ekenberg, Phys. Rev. B **55**, 10729 (1997)
425. M.O. Nestoklon, E.L. Ivchenko, J.M. Jancu, P. Voisin, Phys. Rev. B **77**, 155328 (2008)
426. W. Kohn, Phys. Rev. **105**(2), 509 (1957). DOI 10.1103/PhysRev.105.509.
427. E.O. Kane, in *Handbook of Semiconductors*, vol. 1, ed. by W. Paul (North-Holland, Amsterdam, 1982), pp. 193–217
428. J.O. Dimmock, G.B. Wright, Phys. Rev. **135**(3A), A821 (1964). DOI 10.1103/PhysRev.135.A821.

429. D.L. Mitchell, R.F. Wallis, Phys. Rev. **151**(2), 581 (1966). DOI 10.1103/PhysRev.151.581.
430. J.L. Birman, T.K. Lee, R. Berenson, Phys. Rev. B **14**(2), 318 (1976)
431. T.K. Lee, J.L. Birman, S.J. Williamson, Phys. Rev. Lett. **39**(13), 839 (1977)
432. T.K. Lee, J.L. Birman, Phys. Rev. B **17**(12), 4931 (1978)
433. J.C.Y. Teo, L. Fu, C.L. Kane, Phys. Rev. B **78**(4), 045426 (2008)
434. C.M. Chaves, N. Majlis, M. Cardona, Solid State Commun. **4**(6), 271 (1966)
435. G.E. Tudury, M.V. Marquezini, L.G. Ferreira, L.C. Barbosa, C.L. Cesar, Phys. Rev. B **62**(11), 7357 (2000). DOI 10.1103/PhysRevB.62.7357.
436. A.C. Bartnik, F.W. Wise, A. Kigel, E. Lifshitz, Phys. Rev. B **75**(24), 245424 (2007)
437. H. Ajiki, T. Ando, J. Phys. Soc. Jpn. **62**(7), 2470 (1993)
438. R. Tamura, M. Tsukada, J. Phys. Soc. Jpn. **68**(3), 910 (1999)
439. A.K. Bhattacharjee, Phys. Rev. B **41**(9), 5696 (1990). DOI 10.1103/PhysRevB.41.5696.
440. K. Ando, H. Saito, M.C. Debnath, V. Zayets, A.K. Bhattacharjee, Phys. Rev. B **77**(12), 125123 (2008)
441. B. Jogai, J. Appl. Phys. **91**(6), 3721 (2002)
442. B. Lassen, D. Barettin, M. Willatzen, L.C. Lew Yan Voon, Microelectron. J. **31**, 1226 (2008)
443. P.O. Löwdin, J. Chem. Phys. **19**(11), 1396 (1951)
444. M. Tinkham, *Group Theory and Quantum Mechanics* (McGraw-Hill, New York, 1964)
445. F. Bassani, in *Physics of III-V Compounds, Semiconductors and Semimetals*, vol. 1, ed. by R.K. Willardson, A.C. Beer (Academic, New York, 1966), Chap. 2, pp. 21–74
446. J.L. Birman, *Theory of Crystal Space Groups and Lattice Dynamics* (Springer-Verlag, Berlin, 1984)
447. L.P. Bouckaert, R. Smoluchowski, E. Wigner, Phys. Rev. **50**(1), 58 (1936)
448. R.H. Parmenter, Phys. Rev. **100**(2), 573 (1955). DOI 10.1103/PhysRev.100.573.
449. V. Heine, *Group Theory in Quantum Mechanics* (Pergamon Press, London, 1960)
450. E.I. Rashba, Sov. Phys. Solid State **1**, 368 (1959)

Index

26758135R00264

Printed in Poland
by Amazon Fulfillment
Poland Sp. z o.o., Wrocław